T0280264

Bryophyte Ecology and Climate Change

Bryophytes, especially mosses, represent a largely untapped resource for monitoring and indicating effects of climate change on the living environment. They are tied very closely to the external environment and have been likened to "canaries in the coal mine." *Bryophyte Ecology and Climate Change* is the first book to bring together a diverse array of research in bryophyte ecology, including physiology, desiccation tolerance, photosynthesis, and temperature and UV responses, under the umbrella of climate change. It covers a great variety of ecosystems in which bryophytes are important, including aquatic, desert, tropical, boreal, alpine, Antarctic, and *Sphagnum*-dominated wetlands, and considers the effects of climate change on the distribution of common and rare species as well as the computer modeling of future changes. This book should be of particular value to individuals, libraries, and research institutions interested in global climate change.

ZOLTÁN TUBA (1951–2009) was an internationally known ecophysiologist based at Szent Istvan University, Gödöllö, Hungary. He established the first experimental Hungarian research station and field laboratory at Gödöllö for research on global climate change. His research covered a broad range of topics and he was one of the first to work on desiccation tolerance of bryophytes under elevated CO_2.

NANCY SLACK teaches Bryophyte Ecology at the Humboldt Field Research Institute (ME) and is Professor of Biology emerita at the Sage Colleges, Troy, NY. She has conducted research in bryology and plant ecology in the USA, Canada, and Sweden, especially on peatland and alpine ecosystems. She was recently President of the American Bryological and Lichenological Society (ABLS).

LLOYD STARK is a plant reproductive ecologist interested in explanations of unbalanced sex ratios in bryophytes, how mosses respond to abiotic stress and climate change, and the factors limiting sexual reproduction in mosses. Lloyd is currently an Associate Professor in the School of Life Sciences at the University of Nevada, Las Vegas, where he has recently been honored as the College of Sciences Teacher of the Year for his courses in ecology and general biology.

Bryophyte Ecology and Climate Change

ZOLTÁN TUBA,
NANCY G. SLACK,
AND
LLOYD R. STARK

Shaftesbury Road, Cambridge CB2 8EA, United Kingdom

One Liberty Plaza, 20th Floor, New York, NY 10006, USA

477 Williamstown Road, Port Melbourne, VIC 3207, Australia

314–321, 3rd Floor, Plot 3, Splendor Forum, Jasola District Centre, New Delhi – 110025, India

103 Penang Road, #05–06/07, Visioncrest Commercial, Singapore 238467

Cambridge University Press is part of Cambridge University Press & Assessment,
a department of the University of Cambridge.

We share the University's mission to contribute to society through the pursuit of education, learning and research at the highest international levels of excellence.

www.cambridge.org
Information on this title: www.cambridge.org/9780521757775

First published 2011

A catalogue record for this publication is available from the British Library

Library of Congress Cataloging-in-Publication data
Bryophyte ecology and climate change / [edited by] Zoltán Tuba,
Nancy G. Slack, Lloyd R. Stark.
p. cm.
Includes bibliographical references and index.
ISBN 978-0-521-76763-7 – ISBN 978-0-521-75777-5 (pbk.)
1. Bryophytes – Climatic factors. 2. Bryophytes – Ecology. I. Tuba, Zoltán.
II. Slack, Nancy G. III. Stark, Lloyd R. IV. Title.
QK533.B717 2011
588′.1722–dc22
2010021884

ISBN 978-0-521-76763-7 Hardback
ISBN 978-0-521-75777-5 Paperback

This book is dedicated to Zoltán Tuba (1951–2009)

Contents

Contributors

Anderson, Barbara J.
UKPopNet, Biology Department University of York, P.O. Box 373, York YO10 5YW, UK

Bates, Jeffrey W.
Department of Life Sciences, Silwood Park Campus, Imperial College London, Ascot, Berkshire SL5 7PY, UK

Björk, Robert G.
Department of Plant and Environmental Sciences, Göteborg University, P.O. Box 461, SE 405 30 Gothenburg, Sweden

Black, Kelly
Department of Mathematics and Computer Science, Clarkson University, Potsdam, NY 13699, USA

Brault, Erin
Department of Biology, Villanova University, Villanova, PA 19085, USA

Brinda, John C.
School of Life Sciences, University of Nevada, 4505 Maryland Parkway, Las Vegas, NV 89154–4004, USA

Callaghan, Terry
Abisko Scientific Research Station, 981 07 Abisko, Sweden

Fernando, Catherine
School of Life Sciences, University of Nevada, 4505 Maryland Parkway, Las Vegas, NV 89154–4004, USA

Figueira, Rui
Instituto de Investigação Científica Tropical, Jardim Botânico Tropical, Trav. Conde da Ribeira, 9, 1300–142 Lisboa, Portugal

Gignac, L. Dennis
Campus Saint-Jean, University of Alberta, Edmonton, Alberta, Canada T6C4G9

Glime, Janice M.
Michigan Technological University, 219 Hubbell Street, Houghton, MI 49931, USA

Gottfried, Michael
University of Vienna, Department of Conservation Biology, Vegetation Ecology & Landscape Biology, Rennweg 14, 1030 Vienna, Austria

Grabherr, Georg
University of Vienna, Department of Conservation Biology, Vegetation Ecology & Landscape Biology, Rennweg 14, 1030 Vienna, Austria

Gradstein, S. Robbert
Institute of Plant Sciences, University of Göttingen, Untere Karspüle 2, 37073 Goettingen, Germany

Hohenwallner, Daniela
Federal Environment Agency Austria, Spittelauer Lände 5, 1090 Wien, Austria

Huttunen, Satu
Department of Biology, University of Oulu, P.O. Box 3000, FIN-90014 University of Oulu, Finland

Hyyryläinen, Anna
Department of Biology, University of Oulu, P.O. Box 3000, FIN-90014 University of Oulu, Finland

Jácome, Jorge
Departamento de Biología, Pontificia Universidad Javeriana, Carrera 7 No 43–82, Bogotá, Colombia

Jägerbrand, Annika K.
Swedish National Road and Transport Research Institute, SE-581 95 Linköping, Sweden

Jakab, Gusztáv
Institute of Environmental Sciences, Water and Environmental Management Faculty, Szent István University, 5540-Szarvas, Szabadság Str. 1–3, Hungary

Jócsák, Ildikó
Institute of Botany and Ecophysiology, (Biological PhD School, Ecological Programme), Szent István University, Páter K. u. 1., H-2103 Gödöllő, Hungary

Kessler, Michael
Institute of Systematic Botany, University of Zürich, Zollikerstraße 107, 8008 Zürich, Switzerland

Kosykh, N. P.
Institute of Soil Science and Agrochemistry SB RAS, Novosibirsk, Russia

Lappalainen, Niina M.
Department of Biology, University of Oulu, P.O. Box 3000, FIN-90014 University of Oulu, Finland

Mango, Jesse
Department of Biological Sciences, Union College, Schenectady, NY 12308, USA

Martínez-Abaigar, Javier
Universidad de La Rioja, Complejo Científico-Tecnológico, Avda. Madre de Dios 51, 26006 Logroño (La Rioja), Spain

McLetchie, D. Nicholas
Department of Biology, 101 Morgan Bld., University of Kentucky, Lexington, KY 40506–0225, USA

Menezes, Rui
CERENA – Centro de Recursos Naturais e Ambiente, Instituto Superior Técnico, Av. Rovisco Pais, 1049–001 Lisboa, Portugal

Moser, Dietmar
Federal Environment Agency Austria, Spittelauer Lände 5, 1090 Wien, Austria

Nagy, János
Institute of Botany and Plant Ecophysiology, Szent István University, Gödöllő, 2103, Hungary

Naumov, Aleksei V.
Institute of Soil Science and Agrochemistry SB RAS, Novosibirsk, Russia

Neal, Nathali
Department of Biological Sciences, Union College, Schenectady, NY 12308, USA

Núñez-Olivera, Encarnación
Universidad de La Rioja, Complejo Científico-Tecnológico, Avda. Madre de Dios 51, 26006 Logroño (La Rioja), Spain

Ohlemüller, Ralf
Institute of Hazard and Risk Research (IHRR) & School of Biological and Biomedical Sciences, Durham University, South Road, Durham DH1 3LE, UK

Oliver, Melvin J.
Plant Genetics Research Group, USDA-ARS-MWA, University of Missouri, 204 Curtis Hall, Columbia, MO 65211, USA

Ötvös, Edit
Institute of Botany and Ecophysiology, Szent István University, Páter K. u. 1., H-2103 Gödöllő, Hungary

Pauli, Harald
University of Vienna, Department of Conservation Biology, Vegetation Ecology & Landscape Biology, Rennweg 14, 1030 Vienna, Austria

Pócs, Tamás
Department of Botany, Eszterházy College, Eger, Pf. 43, H-3301, Hungary

Preston, Christopher D.
Biological Records Centre, CEH Maclean Building, Benson Lane, Crowmarsh Gifford, Wallingford, Oxfordshire OX10 8BB, UK

Proctor, Michael C. F.
School of Biosciences, University of Exeter, The Geoffrey Pope Building, Stocker Road, Exeter EX4 4QD, UK

Reiter, Karl
University of Vienna, Department of Conservation Biology, Vegetation Ecology & Landscape Biology, Rennweg 14, 1030 Vienna, Austria

Rice, Steven
Department of Biological Sciences, Union College, Schenectady, NY 12308, USA

Scott, Kimberli D.
Department of Biology, Villanova University, Villanova, PA 19085, USA

Seppelt, Rodney D.
Australian Antarctic Division, Channel Highway, Kingston 7050, Tasmania, Australia

Sérgio, Cecília
Universidade de Lisboa, Museu Nacional de História Natural, Jardim Botânico. CBA-Centro de Biologia Ambiental, Rua da Escola Politécnica 58, 1269–170 Lisboa, Portugal

Slack, Nancy G.
Department of Biology, The Sage Colleges, 45 Ferry Street, Troy, NY 12180, USA

Smith, Stanley D.
School of Life Sciences, University of Nevada, 4505 Maryland Parkway, Las Vegas, NV 89154–4004, USA

Stark, Lloyd R.
School of Life Sciences, University of Nevada, 4505 Maryland Parkway, Las Vegas, NV 89154–4004, USA

Sümegi, Pál
Department of Geology and Palaeontology, University of Szeged, 6722-Szeged, Egyetem Str. 2–6, Hungary

Tuba, Zoltán
Institute of Botany and Ecophysiology, Szent István University, Páter K. u. 1., H-2103 Gödöllő, Hungary

Vile, Melanie A.
Department of Biology, Villanova University, Villanova, PA 19085, USA

Vitt, Dale H.
Department of Plant Biology, Southern Illinois University, Carbondale, IL 62901, USA

Wieder, R. Kelman
Department of Biology, Villanova University, Villanova, PA 19085, USA

Zechmeister, Harald Gustav
University of Vienna, Department of Conservation Biology, Vegetation Ecology & Landscape Biology, Rennweg 14, 1030 Vienna, Austria

Preface

This book is dedicated to Zoltán Tuba. Its origin was a symposium entitled *Ecological Responses of Bryophytes to Changing Climate*. It was presented at the American Bryological and Lichenological Society (ABLS) meeting with the Botanical Society of America (BSA) in Chico, California, in 2006. Nancy Slack, then president of ABLS, and Zoltán Tuba of Gödöllő University, Hungary, organized the symposium, which included speakers from many different countries. An editor at Cambridge University Press (England) saw the program on the Internet and asked the organizers to write a book on this subject. All the symposium speakers agreed to contribute chapters; subsequently, others doing important work in this field were asked to join them. Zoltán Tuba worked on the book with Nancy Slack from 2006 until shortly before his untimely death at 58 in July 2009. In the fall of 2009 Lloyd R. Stark, an active researcher in this field and co-author of two of the chapters, agreed to work with Nancy Slack to finish the book. Zoltán was a major contributor to research in ecophysiology of bryophytes in relation to climate change, as well as in other fields. He will be greatly missed as a scientist as well as a friend and co-worker.

A number of people have written to the present editors about Zoltán. In addition, part of an obituary by Zoltán's mentor, Professor Gábor Fekete: In Memoriam Zoltán Tuba (1951–2009), in *Acta Botanica Hungarica* vol. 52/1–2 (2010), is quoted here:

> On July 4, 2009, Professor Zoltán Tuba, a leading expert in plant ecophysiology, left us forever. Even though his tolerance and desire for life had been as strong as the intensity with which he had lived his life, the horrifying illness won over him after one-and-a-half years of heroic battle.

Zoltán was born in 1951 in Sátoraljaújhely in northern Hungary, and received a degree from József Attila University in Szeged. In 1976, wrote Professor Fekete,

> A young college student showed up in my office at the Department of Botany of the Hungarian Natural History Museum. He was Zoltán Tuba. His eyes reflected intelligence and his words were full of ambition.

Later, in 1978 they became colleagues at the Botanical Research Institute of the Hungarian Academy of Sciences, where Fekete was the leader of a research team. Professor Fekete wrote:

> Zoltán Tuba started his work at an extraordinary pace. He wanted to become a great researcher. It did not matter whether there was twilight, rain, summer or winter, he always completed his scheduled fieldwork. In 1985 he was invited to the University of Agriculture at Gödöllő where he was a full professor from 1992, and a department chair.

Zoltán Tuba received a Doctor of Science degree in 1998 and directed doctoral studies in biology at Szent István University. Fekete continued,

> His dreams came true one after another. The establishment of a great scientific school without any local tradition… is unprecedented. Foreign scientists were a daily sight in the internationally renowned plant physiological laboratory… He developed the only postgraduate doctoral program in ecophysiology in Hungary. His exceptional achievements were acknowledged by the Hungarian Academy of Sciences with the establishment of a research group in plant ecology within his department.

Tuba himself was a visiting professor at the University of Karlsruhe and at Edinburgh and Exeter. His research had an enormous range, including much work on the desiccation tolerance strategies of plants. In terms of this book, one of his most important accomplishments was, in Fekete's words:

> He was among the first in Central and Eastern Europe to launch an experimental research program to study the ecological effects of climate change (rising atmospheric CO_2 levels and temperature). He established in 1993 an experimental research station and a field laboratory with CO_2 fumigation equipment at Gödöllő, which was regarded as one of the core projects of the Global Change and Terrestrial Ecosystems international research program.

Zoltán was one of the first to work on poikilohydric and desiccation-tolerant cryptogams [mainly bryophytes] under elevated CO_2 concentrations, the subject of one of his chapters in this book.

Fekete continued:

> Zoltán was not an easygoing man. He did want to accomplish his far-reaching goals and ideas ... He always focused on scientific goals and not on personal success. He was aware of the importance of teamwork ... He generously supported his young colleagues, helped to build their careers, and introduced them into the international community.

Michael C. F. Proctor of Exeter University, who was Zoltán's friend and colleague and the author of one of the chapters in this book, wrote:

> I first met Zoltán Tuba at the meeting of the International Association of Bryologists in Budapest in 1985. He was an outgoing, energetic and enthusiastic young bryological physiologist who worked tirelessly to make sure we all had a good meeting and enjoyed Budapest. I met him again... when he was visiting the Department of Ecology and Natural Resources in Edinburgh. During the 1990s he and his colleagues spent time in my laboratory in Exeter and I made a number of visits to his department in Gödöllő. His department was small but enthusiastic and productive. Its interest ranged widely across ecophysiology, including bioindication of heavy metal and sulphur pollution, plant responses to elevated CO_2, aquatic vegetation, effects of drought in Hungarian calcareous and steppe grasslands, and desiccation tolerance. The department had an open-top chamber and later a free-air carbon-enrichment facility. The lab was like a family, with stresses and strains like all families, but close-knit nevertheless. Zoltán Tuba was the Professor, tireless in working for his department and its place in the world, ambitious for it, sometimes exasperating, quick to spot and try new opportunities, techniques or ideas, a driving force, an enabler, a catalyst. He got things done. He had the good fortune (or good judgment) to be supported by a strong, diverse and loyal team, whose aptitudes were complementary to one another and to his own.
>
> Zoltán was a larger-than life person, sometimes impulsive, but kind, warm and very human. He was very much a family man. I look back with pleasure at time spent with Zoltán, Ildiko and their two boys... and on expeditions into the Hungarian countryside, and no less on their

> visits to England. One of my abiding memories is of the four of them standing in the evening sunshine above the cliffs at Land's End, looking out over the Atlantic. Zoltán will be sorely missed.

Zoltán's close friend from Budapest and author of another chapter in this book, Tamás Pócs, wrote:

> I knew Zoltán for more than 30 years, as we worked together at the Vacratot Ecological and Botanical Research Institute and even after he left to head the Gödöllő department, we kept close contact and were good friends until his death. He was a very straightforward man, who did not keep his opinion in secret... He supported all good ideas and involved in his research many talented young students and colleagues. It is typical for him, when he received the highest Hungarian award, the Szechenyi Prize for his scientific results, he did not use it for himself nor for his own research but started a foundation to support the higher education of talented secondary school students at his home school... He was a many-sided man, fond of geography, ethnography, history, literature, especially in his homeland from which he edited a monograph of 1800 pages, entitled "Bodrogkoz." He was a man with a good sense of humour, always keeping up the mood of his company. He liked to live, liked good food and was a connoisseur of good wines ... Before his death he was proposed to be a member of the Hungarian Academy of Sciences. To our greatest sorrow he could not live to see this honour.

Zoltán Tuba was fond of fieldwork, and continued to participate in field research until almost the end of his life. In recent years he traveled to India, Madagascar, and Brazil. In 1997, after the International Congress of Bryology near Beijing, many field trips were provided by the organizers. Zoltán Tuba and I (Nancy Slack) together with two other westerners and a larger number of Chinese scientists, chose to go to the far reaches of Sichuan Province. We traveled by mini-bus from Chengdu over the mountains to Jiuzhaigou, a national reserve with magnificent forests full of bryophytes. On our way, we came to an area of loess hills (loess is a deposit formed from wind-blown silt). Zoltán, a large man, hurried out of the mini-bus and scrambled up a loess hill, followed by myself. He was very excited because the vegetation was so similar to the loess vegetation he had studied in Hungary.

Even though much of his professional life took place in the laboratory and as an administrator, he was very much in touch with plant ecology and phytosociology in the field, in Hungary and elsewhere.

Zoltán Tuba was a man of great energy and enthusiasm. For Nancy Slack it was a pleasure to work on this book with him, and so sad that his life was cut short. As you can read in his introductory chapter, many of the aspects of bryophyte ecology with which he was concerned in his laboratory are presented by the authors of this book.

Michael Proctor,
Tamás Pócs, and
Nancy Slack

I INTRODUCTORY CHAPTERS

1

The Ecological Value of Bryophytes as Indicators of Climate Change

NANCY G. SLACK

Bryophytes are the most successful group of plants other than angiosperms in terms of their numbers of species, geographical distribution on all continents, and their habitat diversification. There are at least 10,000 species of mosses and over 6000 liverworts. All three groups of bryophytes, also including the hornworts, were the earliest green plants to move to the land; each group has had a very long evolutionary history, probably more than 400 million years. All three groups, derived from a green algal ancestor, evolved separately from one another and from vascular plants through this long period. Although the great diversity of tropical bryophytes is often cited, Rydin (2009) pointed out their important contribution to biodiversity in northern ecosystems: 7.5% of the world's bryophyte species are found in Sweden, whereas only 0.8% of vascular plant species are found there.

Bryophytes are unique among land plants in that their dominant stage is the haploid green gametophyte rather than the much shorter-lived diploid sporophyte. They differ from vascular plants in other ways as well, in aspects that make them excellent environmental monitors. They inhabit a very wide range of ecosystems, habitats, and specific microhabitats, including substrates on which vascular plants cannot live. Many species are able to live in nutrient-poor conditions, and are adapted to respond rapidly physiologically to intermittent periods favorable for photosynthesis.

Morphology and physiology

Bryophytes lack the roots, xylem, and phloem of vascular plants. The great majority are ectohydric, that is, without internal conducting tissues. They absorb water and nutrients over the whole surface of the gametophyte. Water

Bryophyte Ecology and Climate Change, eds. Zoltán Tuba, Nancy G. Slack and Lloyd R. Stark. Published by Cambridge University Press.

moves through externally, in capillary spaces around leaf bases, in hairs (tomentum) on stems, and in paraphyllia on the stems of some bryophytes. A few groups, particularly the Polytrichaceae and the larger Mniaceae, do have an internal water-conducting system composed of a central strand of hydroids (Hébant 1977). Some complex thallose liverworts in the Marchantiales conduct water internally around and within cell walls. Mosses of genera like *Polytrichum* and *Dawsonia* are thus able to grow tall, but they have not developed lignin; there are no bryophyte trees. The great majority are indeed very small, although both aquatic bryophytes such as species of *Fontinalis* and many pendant epiphytic bryophytes may grow very long. Bryophytes are very specific to particular microhabitats; they have diversified greatly ecologically in the course of their evolution. Many strategies have evolved that have enabled sympatric congeneric species to survive in separate niche spaces (Slack 1997). The relatively recent diversification of bryophytes is correlated with the evolution of angiosperms, especially forest trees, providing new niches for both mosses and liverworts. There has also been much speciation of tropical liverworts and the evolution of adaptations of mosses for epiphytic life. In many groups of mosses, repeated reduction has occurred in the course of evolution of mosses for xeric and ephemeral habitats.

Acrocarpic mosses, those that produce sporophytes at the tips of usually erect main stems, were the earliest to evolve and most still live on terrestrial substrates, rocks and soil. A few groups are successful as epiphytes; for example, the Calymperaceae. Many pleurocarpous mosses, those that produce lateral rather than terminal sporophytes and are often prostrate and highly branched, are found in many habitats. They may be terrestrial, particularly on forest floors and in wet environments; others are aquatic. Many pleurocarps are epiphytic, particularly in the tropics. Virtually all are perennial, whereas some acrocarps are annual or ephemeral, as are some liverworts. All bryophytes are C_3 plants; anthocerotes alone have a carbon-concentrating mechanism.

Mosses have many specialized structures, some of which have physiological functions. For example, leaf axillary hairs secrete mucilage for juvenile leaves, thus preventing dehydration (Buck & Goffinet 2000). Paraphyllia, small leaflike or filiform structures on pleurocarpous moss stems, add photosynthetic surface area. Lamellae, sheets of cells with chlorophyll usually on the upper surface of the leaf in Polytrichaceae and some Pottiaceae, serve physiological functions. Not only do they restrict water loss, but probably can be viewed more importantly as an adaptation for increasing the area for CO_2 uptake when well supplied with water and thus photosynthetically active. One can estimate the ratio between CO_2 uptake and projected leaf area. There is a clear correlation between this value and 95% irradiance (Proctor 2009). The very great majority of moss leaves, including the greatly modified leaves of *Sphagnum* mosses, consist of one layer of cells, as do those of most liverworts apart from the Marchantiales. The dominant gametophyte stage of bryophytes is thus directly exposed to many environmental factors. Sexual reproduction in bryophytes, as in ferns, requires water for the transfer of swimming sperm to the egg. Antheridia and archegonia are often on different plants; the majority of mosses are dioicous. In some taxa male and female plants are widely separated geographically. It is thus not unexpected that bryophytes have evolved many types of asexual reproduction and dispersal. Fragments of the gametophyte can produce new plants, probably the main means of dispersal to new sites of many mosses, including *Sphagnum*. "Fragile" leaves, part of which regularly break off, have evolved in a variety of moss families. Specialized propagules such as gemmae are found very commonly in leafy liverworts and in many mosses. Flagella or flagellate branches occur in both acrocarps and pleurocarps. Bulbils on the rhizoids occur in some moss genera, particularly *Bryum*. These dispersal methods are very efficient. Leafy liverworts colonize almost every decorticated log in moist temperate forests in eastern North America, many by gemmae rather than by spores. *Tetraphis pellucida*, with its gemmae in specialized "splash cups," can be found on almost any rotten stump within its range. In addition, most moss sporophytes have capsules with peristome teeth that actively disperse spores. Peristomes have evolved, as for example in epiphytic mosses, to disperse spores efficiently in many different habitats. In the Splachnaceae, the dung mosses (coprophiles), spores are dispersed by insects, mainly by flies attracted to their special odors, with some coevolutionary parallels to insect pollination in angiosperms. The relationships of the species of *Splachnum* and other genera in the family to each other, to their substrates, and to environmental factors are complex (Marino 1997). The populations of some species have decreased in recent times, causing concern about land use practices and possibly climate change.

Bryophytes as Air Pollution Monitors

Bryophyte leaves, in contrast with those of vascular plants, do not have a thick cuticle. They are also, as noted above, ectohydric, obtaining their water and nutrients through the surface of the whole gametophyte. In addition, bryophytes are often exposed on rocks, tree bark, and soil. For all these reasons they are very close to, and in some respects at the mercy of, the environment. They have long been extensively used, as have lichens, as air pollution monitors, especially of sulphur dioxide and nitrogen oxides. Some species of mosses, especially epiphytes such as *Antitrichia curtipendula* and epiphytic species of *Orthotrichum*, are especially sensitive even to relatively low levels of pollutants compared with other mosses (Rao 1982; Porley & Hodgetts 2005). SO_2 in particular damages plasma membranes and causes chlorophyll to degrade. Some mosses were found to accumulate SO_2 one hundred times more efficiently than the leaves of vascular plants (Winner 1988). A great many studies of bryophytes as air pollution monitors have been carried out in many countries, including Britain, particularly from an early date (Burrell 1917). Some studies are ongoing, particularly in China and other areas heavily impacted by air pollution.

Desiccation tolerance

Desiccation tolerance has evolved in many bryophytes and is currently being intensively studied physiologically and by using techniques of molecular biology. It is important in terms of current climate change scenarios in which precipitation as well as temperature is predicted to change in the near future on a global scale. The mechanisms of true desiccation tolerance (as opposed to drought tolerance) are quite different from those of vascular plants; caution must therefore be used in making climate change predictions from vascular plant data alone.

A great variety of organisms show desiccation tolerance, including not only bryophytes and lichens but many other plants as well as animals and microorganisms. In bryophytes desiccation tolerance varies greatly among species, even for those in relatively moist environments, such as temperate forest epiphytes, and even for "similar" pleurocarps like *Hylocomium splendens* and *Rhytidiadelphus squarrosus* (Proctor 2000). Some mosses and leafy liverworts that live in continually moist or semi-aquatic habitats have not evolved (or have lost) desiccation tolerance. On the other hand, many epiphytes, such as *Ulota crispa*, which live on intermittently very dry bark even in generally moist forests, have evolved or retained desiccation tolerance.

Among desert mosses, desiccation tolerance is extremely well developed. Species of *Syntrichia* (*Tortula*) have been much studied. They can lose almost all of their water without disruption of cell structures; cell membranes and those of cell organelles remain intact. Experiments with *Syntrichia* (*Tortula*) *ruralis* show very rapid re-establishment of normal net photosynthesis in the renewed presence of water as was found in my early experiments on *Ulota crispa* (Tobiessen *et al.* 1979) and more recent ones (Tuba *et al.* 1996). Other experiments using chlorophyll fluorescence techniques show that recovery of photosynthesis is not affected by either chloroplast or cytoplasmic protein synthesis. It thus appears to be a matter of reassembly of pre-existing components rather than synthesis of new ones, also known as "constitutive" desiccation tolerance.

Most bryophytes live in conditions of intermittent water availability. They spend their lives either fully turgid during or after rain (especially desert bryophytes) or dry and metabolically inactive. They are actually drought-evaders rather than drought-resistors. The latter strategy is characteristic of most desert vascular plants, which have evolved storage organs, long roots, as well as transcription and protein synthesis after a slower drying period than that which bryophytes often experience.

Oliver (2009) discussed the development of true vegetative desiccation tolerance in bryophytes as a requirement for life on land, which very likely preceded the development of vascular tissue in tracheophytes. Vegetative desiccation tolerance is rare in vascular plants, but common, although as noted not universal, in bryophytes. Thus far only 158 species of moss, 51 species of liverwort, and one species of hornwort have been shown experimentally to have vegetative desiccation tolerance, but that number will likely increase rapidly with further experimental work. Most initial work was done on *Syntrichia ruralis*, but recent experiments have shown that the desert moss *Syntrichia caninervis* can actually survive rapid desiccation (within 30 minutes) to approximately –540 MMPa for up to six years, returning to normal metabolic activity upon rehydration (Oliver *et al.* 1993; Oliver 2009). Some bryophytes increase their level of desiccation tolerance after mild dehydration events prior to desiccation. Dehydration can be almost instantaneous and can also be a stressful cellular event.

In bryophytes it is not structural thickenings or other features of cell walls that are important in desiccation tolerance, as once thought, but inherent properties of the cellular components. Many bryophytes have mechanisms for cellular protection during the desiccation process, but this varies even within closely related species in one genus. In addition, the sporophyte generation may be less desiccation-tolerant than the highly tolerant gametophyte generation, as in *Tortula inermis* (Stark *et al.* 2007) Importantly, water loss in these mosses is too rapid for protein synthesis to occur; protein synthesis is very sensitive to loss of water from the cytoplasm and quickly ceases. Thus protein synthesis cannot account for cellular stability. Even if drying is slower, novel proteins are not transcripted during the drying process; the necessary proteins are already present in the cells in sufficient quantities. This contrasts with mechanisms in some of the few vegetatively desiccation-tolerant vascular plants, such as in "resurrection plants" (Selaginaceae). These involve a relatively slower drying process and the transcription of particular proteins by a range of dehydration-regulated genes. Oliver (2009) contends that there is a constitutive cellular protection mechanism that is "ready and waiting" to be challenged or activated by desiccation. Its effectiveness is at least clear in *Syntrichia ruralis* from the fact that the plasma membrane and cell organelle membranes remain intact during the drying process, as noted.

Desiccation tolerance is not yet fully understood in bryophytes but it does involve components present in the cells: sugars, largely sucrose, and protective proteins including antioxidants and enzymes involved in protection from the generation of reactive oxygen species (ROS). A genomic approach is currently being used to catalog genes whose products play a role in responses of bryophytes to desiccation and rehydration, but much remains to be resolved. The sequencing of the genome of *Physcomitrella patens* is an important tool. Even though *P. patens* is not a desiccation-tolerant species, researchers now have the ability to knock out and replace its genes, which will be a powerful tool for future work.

Future climate change, particularly in terms of changes in the geographic distribution of precipitation, is of course uncertain; many factors, particularly anthropogenic ones, are important. In some regions it is extremely likely that desiccation will become a problem; bryophytes may do better under such circumstances than vascular plants or animals.

Ecosystem functions of bryophytes

Bryophytes have important ecosystem functions that need to be considered. These functions of bryophytes have long been under study in many ecosystems in the temperate and boreal zones, in the tropics, and in the Arctic and Antarctic. These functions include high productivity and biomass accumulation in some ecosystems, as well as nitrogen fixation, nutrient cycling, food chains and animal interactions, colonization, vascular plant facilitation, mycorrhizal relationships (liverworts), and others. Many have been discussed extensively elsewhere (Rieley *et al.* 1979; Gerson 1982; Slack 1988; Bates 2000, 2009; O'Neill 2000; Duckett *et al.* 1991; Longton 1992; Sveinbjornsson & Oechel 1992; Porley & Hodgetts 2005; Rydin 2009; Vitt & Wieder 2009). Some are discussed in other chapters of this book for many of the ecosystems listed above.

Although in many environments vascular plants including forest trees are the dominant vegetation, in other environments, especially in the Arctic, the Antarctic, in alpine habitats in mountains above treeline and in bogs, fens, and larger peatlands, bryophytes are often the dominant plants in terms of both biomass and productivity. They also have important ecosystem functions in temperate rain forests as well as in wet high-elevation so-called "mossy forests," where, however, liverworts rather than mosses usually predominate. Much of the earth's boreal and arctic zones are covered by peatlands in which species of *Sphagnum* are dominant. These peatlands are very important as carbon sinks and are currently being impacted by climate change, a subject studied and discussed by several authors in this volume and previously by, e.g., O'Neill (2000) and Vitt & Wieder (2009). In addition to sequestering carbon, bryophytes in forests and elsewhere are important in water retention and nutrient cycling and also in relation to the vertebrate and invertebrate fauna (see references above).

Much information, some of it from recent studies (Bates 2009), is available on the functions of bryophytes in nutrient cycling, a function that is vital to vascular plants, both in terms of facilitation and competition. Nutrient cycling is important in a variety of ecosystems likely to be affected by continuing climate change, not only projected changes in temperature and precipitation, but also in increased atmospheric CO_2 and UVB. Bryophytes capture mineral nutrients by "facilitated diffusion," which involves ion channels and carrier proteins and depends on the existing gradients of concentration and electric charge across membranes. They frequently accumulate chemicals in much higher concentrations than in the ambient environment, one important reason for the use of mosses for biomonitoring of air pollution. There are a number of sources of nutrients (as well as other chemicals) that bryophytes accumulate with both wet and dry deposition. Experiments have shown that they also obtain nutrients from the substrate on which they are growing (e.g., Van Tooren *et al.* 1988). An early study (Tamm 1953) showed that *Hylocomium splendens* growing under a Norwegian forest canopy received most of its mineral nutrients from leachates from the tree canopy, i.e., wet deposition. This was also true in ombrotrophic bogs; wet deposition supplied mineral elements to *Sphagnum* (Malmer 1988). Woodin *et al.* (1985) showed that in a subarctic mire in Abisko, Sweden, *Sphagnum* captured NO_3 during both natural precipitation and experimental treatments more efficiently than rooted vascular plants, which presumably compete for nutrients in such ecosystems.

In addition, Oechel and Van Cleve (1986) found that bryophytes can be important in obtaining nutrients from wet deposition (precipitation) as well as from dust and litter before they can be taken up by rooted vascular plants. Experiments by van Tooren *et al.* (1990) showed that bryophytes in Dutch chalk grasslands absorb nutrients and grow during fall and winter while higher plants are inactive, and release nutrients by decomposition in spring and fall; these nutrients are then used by higher plants. Many more examples of nutrient cycling involving bryophytes could be cited. All of these relationships in part depend on ambient environmental factors currently in flux as a result of global climate change. Continued monitoring is needed.

Although most of the nutrient cycling studies to date involve mosses, studies of the liverwort *Blasia* are of interest. *Blasia pusilla* is very common in pioneer communities formed after deglaciation in Alaska (Slack & Horton 2010).

It contains nitrogen-fixing cyanobacteria (*Nostoc*). As reported by Bates (2009) from the work of J. C. Meeks and D. G. Adams, *Blasia* (as well as the hornwort *Anthoceros*, which also harbors *Nostoc*), when starved of nitrogen compounds, releases a chemical signal that induces the formation of short gliding filaments of the *Nostoc*, called hormogonia. These eventually move into the ventral auricles of *Blasia* and after developmental changes generate *Nostoc* filaments with a large number of N_2 -fixing heterocysts; about 80% of the fixed nitrogen is leaked to the host *Blasia* in the form of NH_3. Whatever the complex nature of this symbiosis, the molecular genetics of the switch in form of *Nostoc* is presently under study. In the Bering Strait region of Alaska, in this author's experience, *Blasia* is an early successional species. The plant dies, and mosses and vascular plant seedlings take part in the subsequent succession, presumably using the nitrogen leached from the *Nostoc*.

At the 2007 International Association of Bryologists (IAB) meeting in Kuala Lumpur, Malaysia, protocols were set up to be used internationally in monitoring bryophyte responses to various aspects of climate change. The responses of bryophytes are likely to interact in quite complex ways with climatic factors, often both earlier than and different from those of vascular plants. A great deal of recent research on bryophyte responses to all the above factors of climate change, both present and predicted, in diverse ecosystems and on several continents, is presented in succeeding chapters.

Although polar bears have recently been referred to, in relation to global warming, as the new "canaries in the coal mine," it is the bryophytes that deserve that title. They are sensitive not only to increasing global temperatures, but also to increasing carbon dioxide content of the atmosphere, to increasing UVB radiation, to decreasing precipitation in some regions, and to several factors affecting carbon storage, especially in peatlands. All of these effects have been studied by the authors of the succeeding chapters and by many others whose work they cite. Both monitoring and actual laboratory and field experiments are currently being conducted and their results, including those of long-term field experiments in a number of different ecosystems, from arctic and alpine to desert, are reported in this book.

References

Bates, J. W. (2000). Mineral nutrition, substratum ecology, and pollution. In *Bryophyte Biology*, ed. A. J. Shaw & B. Goffinet, pp. 248–311. Cambridge: Cambridge University Press.

Bates, J. W. (2009). Mineral nutrition and substratum ecology. In *Bryophyte Biology*, ed. B. Goffinet & A. J. Shaw, pp. 299–356. New York: Cambridge University Press.

Buck, W. R. & Goffinet, B. (2000). Morphology and classification of mosses. In *Bryophyte Biology*, ed. B. Goffinet & A. J. Shaw, pp. 71–123. Cambridge: Cambridge University Press.

Burrell, W. H. (1917). The mosses and liverworts of an industrial city. *The Naturalist* **42** (No. 499): 119–24.

Duckett, J. G., Renzaglia, K. S. & Pell, K. (1991). A light and electron microscope study of rhizoid-ascomycete associations and flagelliform axes in British hepatics. *New Phytologist* **118**: 233–55.

Gerson, U. (1982). Bryophytes and invertebrates. In *Bryophyte Ecology*, ed. A. J. E. Smith, pp. 291–331. London: Chapman & Hall.

Hébant, C. (1977). *The Conducting Tissues of Bryophytes*. Vaduz: J. Cramer.

Longton, R. E. (1992). The role of bryophytes and lichens in terrestrial ecosystems. In *Bryophytes and Lichens in a Changing Environment*, ed. J. W. Bates & A. M. Farmer, pp. 32–76. Oxford: Clarendon Press.

Malmer, N. (1988). Patterns in the growth and accumulation of inorganic constituents in the Sphagnum cover on ombrotrophic bogs in Scandinavia. *Oikos* **53**: 105–20.

Marino, P. C. (1997). Competition, dispersal, and coexistence of the Splachnaceae in patchy environments. *Advances in Bryology* **6**: 241– 63.

Oechel, W. C. & Van Cleve, K. (1986). The role of bryophytes in nutrient cycling in the taiga. In *Forest Ecosystems in the Alaskan Taiga*, ed. K. Van Cleeve, F. S Chapin III, P. W. Flannagan, L. A. Vierect & C. T. Dyrness, pp. 121–37. New York: Springer-Verlag.

Oliver, M. J. (2009). Biochemical and molecular mechanisms of desiccation tolerance in bryophytes. In *Bryophyte Biology*, ed. B. Goffinet & A. J. Shaw, pp. 269–97. New York: Cambridge University Press.

Oliver, M. J., Mishler, B. D. & Quisenberry, J. E. (1993). Comparative measures of desiccation-tolerance in the *Tortula ruralis* complex. I. Variation in damage control and repair. *American Journal of Botany* **80**: 127–36.

O'Neill, K. P. (2000). Role of bryophyte-dominated ecosystems in the global carbon budget. In *Bryophyte Biology*, ed. B. Goffinet & A. J. Shaw, pp. 344–68. Cambridge: Cambridge University Press.

Porley, R. & Hodgetts, N. (2005). *Mosses and Liverworts*. London: Collins.

Proctor, M. C. F. (2000). Physiological ecology. In *Bryophyte Biology*, ed. B. Goffinet & A. J. Shaw, pp. 225–47. Cambridge: Cambridge University Press.

Proctor, M. C. F. (2009). Physiological ecology. In *Bryophyte Biology*, ed. B. Goffinet & A. J. Shaw, pp. 237–68. New York: Cambridge University Press.

Rao, D. N. (1982). Responses of bryophytes to air pollution. In *Bryophyte Ecology*, ed. A. J. E. Smith, pp. 445–71. London: Chapman & Hall.

Rieley, J. O., Richards, P. W. & Bebbington, A. D. (1979). The ecological role of bryophytes in a North Wales woodland. *Journal of Ecology* **67**: 497–528.

Rydin, H. (2009). Population and community ecology of bryophytes. In *Bryophyte Biology*, ed. B. Goffinet & A. J. Shaw, pp. 393–444. New York: Cambridge University Press.

Slack, N. G. (1988). The ecological importance of lichens and bryophytes. In *Lichens and Bryophytes and Air Quality*, ed. T. W. Nash III & V. Wirth (*Bibliotheca Lichenologica* **30**), pp. 23–53. Berlin: J. Cramer.

Slack, N. G. (1997). Niche theory and practice. *Advances in Bryology* **6**: 169–204.

Slack, N. G. & Horton, D. G. (2010). Bryophytes and bryophyte ecology. In *Bering Glacier: Interdisciplinary Studies of Earth's Largest Temperate Surging Glacier*, ed. R. A. Schuchman & E. G. Josberger, pp. 365–71. Boulder, CO: Geological Society of America Press.

Stark, L. R., Oliver, M. J., Mishler, B. D. & McLetchie, D. N. (2007). Generational differences in response to desiccation stress in the desert moss, *Tortula inermis*. *Annals of Botany* **99**: 53–60.

Sveinbjornsson, B. & Oechel, W. C. (1992). Controls on growth and productivity of bryophytes: environmental limitations under current and anticipated conditions. In *Bryophytes and Lichens in a Changing Environment*, ed. J. W. Bates and A. M. Farmer, pp. 77–102. Oxford: Clarendon Press.

Tamm, C. O. (1953). Growth, yield, and nutrition in carpets of a forest moss (*Hylocomium splendens*). *Neddelanden från Statens Skogsforsknings Institut* **43**: 1–140.

Tobiessen, P., Slack, N. G. & Mott, K. A. (1979). Carbon balance in relation to drying in four epiphytes growing in different vertical ranges. *Canadian Journal of Botany* **57**: 1994–8.

Tuba, Z., Csintalan, Zs. & Proctor, M. C. F. (1996). Photosynthetic responses of a moss, *Tortula ruralis* ssp. *ruralis*, and the lichens *Cladonia convoluta* and *C. furcata* to water deficit and short periods of desiccation, and their ecophysiological significance: a baseline study at present-day CO_2 concentration. *New Phytologist* **133**: 353– 61.

van Tooren, B. F., den Hertog, J. & Verhaar, J. (1988). Cover, biomass and nutrient content of bryophytes in chalk grasslands. *Lindbergia* **14**: 47–58.

van Tooren, B. F., van Dam, D. & During, H. J. (1990). The relative importance of precipitation and soil as sources of nutrients for *Calliergonella cuspidata* in a chalk grassland. *Functional Ecology* **4**: 101–7.

Vitt, D. H. & Wieder, R. K. (2009). Structure and function of bryophyte-dominated peatlands. In *Bryophyte Biology*, ed. B. Goffinet & A. J. Shaw, pp. 357–91. New York: Cambridge University Press.

Winner, W. E. (1988). Responses of bryophytes to air pollution. In *Lichens, Bryophytes and Air Quality*, ed. T. W. Nash III & V. Wirth (*Bibliotheca Lichenologica* **30**), pp. 141–73. Berlin: J. Cramer.

Woodin, S., Press, M. C. & Lee, J. A. (1985). Nitrate reductase activity in *Sphagnum fuscum* in relation to wet deposition of nitrate from the atmosphere. *New Phytologist* **99**: 381–8.

2

Bryophyte Physiological Processes in a Changing Climate: an Overview

ZOLTÁN TUBA

Climate change as a result of global warming is predicted to be most pronounced at high latitudes. It is known from experimental studies that *Sphagnum* species respond to enhanced UV radiation by decreasing their growth (Huttunen *et al.* 2005). Some polar bryophytes reproduce sexually and form sporulating sporophytes, e.g., *Polytrichum hyperboreum* on the Svalbard tundra. Antarctic mosses tend to reproduce sexually more often at higher Antarctic latitudes (Lewis-Smith & Convey 2002). Simultaneously, increased emissions of nitrogenous air pollutants cause increased nitrogen deposition over the northern hemisphere (Bouwman *et al.* 2002).

Mires, i.e., wetlands actively accumulating organic material, are believed to play an important role in the global biogeochemical carbon cycle, potentially serving as major long-term carbon sinks. Vegetation structure, however, strongly influences the carbon sink capacity of mires (Malmer & Wallén 2005). In general, carbon accumulation is greater in *Sphagnum*-dominated mires than in sedge-dominated mires (Dorrepaal *et al.* 2005).

Graminoid dominance, however, leads to increased methane (CH_4) emissions, due both to increased root exudation of precursor compounds for CH_4 formation and to plant-mediated transport of CH_4 to the atmosphere through aerenchyma tissue, bypassing oxidation in the acrotelm. Because of such feedbacks on global carbon budgets, investigations of vegetation dynamics in response to ongoing pollution and warming are crucial for estimates of future scenarios of global change (Wiedermann *et al.* 2007).

There is great uncertainty in the estimations of the impacts and possible outcome of global climate changes. Only one of the global climate changes can be considered to be certain: the increase of atmospheric CO_2. By the end of the

Bryophyte Ecology and Climate Change, eds. Zoltán Tuba, Nancy G. Slack and Lloyd R. Stark. Published by Cambridge University Press.

twenty-first century it may reach about double its present concentration or even more (600–700 μmol mol^{-1}) (Houghton *et al.* 1990).

Even if, in the following century, CO_2 concentration of the atmosphere did not rise any higher, the impact of doubled CO_2 concentration would cause many further problems for a long period. Besides the climatic impacts of rising CO_2 concentration, it plays a significant role in the life of plants, as carbon is the basic element of their growth. Among its other impacts are changes in their vital processes, the quantity of their production, the composition of their yield and their reproduction and prevalence. Rising atmospheric CO_2 concentrations influence and modify vital processes of plants by changing their photosynthetic activity. By changing the physiology and ecology of plants, the base of the food chain, it affects and modifies all of life on earth. In the following discussion, recent aspects of bryophyte research are presented in terms of their climate change indicator capability.

Effects of temperature increase

Terrestrial bryophytes exhibit higher thermal optima and thermal tolerance than aquatic bryophytes (Carballeira *et al.* 1998). The acquisition of dry heat tolerance in bryophytes is expected to be more sharply selected than wet heat tolerance since under most conditions heat stress is encountered when gametophytes are desiccated (Hearnshaw & Proctor 1982; Tuba 1987; Meyer & Santarius 1998). This contrasts with vascular plants, the vast majority of which are incapable of complete desiccation and which exhibit the highest wet heat tolerance among land plants (Nobel *et al.* 1986).

High temperature results in structural changes of membranes, increased membrane fluidity and permeability (Deltoro *et al.* 1998), and finally to cell death. Heat and thermotolerance of *Funaria* moss sporophytes is a lesser known area and they are thought to be more sensitive to temperature than gametophytes. Nevertheless, sporophytic embryos of at least two desert mosses exhibit a phenological pattern of over-summering, thus tolerating long periods of thermal stress while desiccated (Stark 1997; Bonine 2004).

It was hypothesized (McLetchie & Stark 2006) that the sporophytes of *Microbryum starckeanum*, a species found at lower elevations in the Mojave Desert, would exhibit reduced thermotolerance relative to gametophytes. Sporophytic patches of *Crossidium, Tortula, Pterygoneurum, Funaria*, and *Microbryum* were investigated in southern Nevada, USA. Desiccated plants were cut to roughly 5 mm in length, and 30 plants were randomly assigned to each of four treatments: 35 °C (1 h), 55 °C (1 h), 75 °C (1 h), and 75 °C (3 h). All 120 measured gametophytes regenerated, with mean time to protonemal emergence ranging from 14.5 to

17.2 days across treatments. Sporophytes were more adversely affected by the thermal exposures than gametophytes. Plants that produced meiotic sporophytes took longer times to regenerate gametophytically when compared with those plants that aborted their sporophytes. This indicates generational differences in survival of *Microbryum starckeanum*. The 75 °C treatment resulted in the survival of maternal gametophytes, thus preserving the regeneration capability by protonemal production. However, embryonic sporophytes did not survive the same treatment. In arid-land regions, where patch hydration occurs only in the cooler months, short summer rainstorms may trigger sporophyte abortions (Stark 2002). Furthermore, sporophytes depend upon the maternal gametophytes that survive desiccation under both cool and hot temperatures, and thus sporophyte vigor depends on the gametophyte to which it is attached (Shaw & Beer 1997). Sporophyte thermotolerance can be considered as a combination of extrinsic stresses imposed on the sporophyte through temperature elevation and desiccation recovery, linked with dependence upon the physiological state of the maternal gametophyte (McLetchie & Stark 2006).

Nitrogen addition and increased temperature leads to decreased cover of mosses and an increased abundance of vascular plants in mire ecosystems (Heijmans *et al.* 2002), but the importance of trophic interactions and the relative importance of the two factors need further investigation. The effects of nitrogen addition on the species composition of vascular plants in other ecosystems can be mediated by fungi (Strengbom *et al.* 2002) or by plant–insect interactions (Throop & Lerdau 2004).

Results from warming experiments in natural vegetation suggest species-specific responses among natural pathogens (Roy *et al.* 2008), and it is anticipated that nitrogen-induced changes in plant tissue chemistry may be a key factor directing such interactions. Increased plant tissue nitrogen concentrations can increase plant susceptibility to damage from such organisms (Harvell *et al.* 2002; Strengbom *et al.* 2002).

In an area of northern Sweden with low background nitrogen deposition, Wiedermann *et al.* (2007) examined the data on vegetation responses, plant chemistry, and plant disease in order to elucidate the influence of enhanced temperature and increased nitrogen on trophic interactions that determine species composition of boreal mire vegetation. They provided results of a nine-year experiment with two levels of temperature and two levels of nitrogen and sulphur. To simulate climate change, warming chambers were used to raise mean air temperature by 3.6 °C, as measured 0.25 m above the bottom layer; this allowed evaluation of the direct effect of increased air and soil temperature. Their results show marked vegetation shifts in response to nitrogen addition and warming following nine years of continuous treatments. *Sphagnum* cover had drastically decreased, with average *Sphagnum* cover reduced from 100% to 63% in response to the nitrogen treatment and to 20% in the nitrogen plus temperature treatment. Temperature treatment alone did not impact *Sphagnum* cover. For *Sphagnum*, a negative effect of nitrogen was found on both the entire plots and the subplots, whereas a negative effect of temperature was recorded only when combined with increased nitrogen.

Disease incidence dropped significantly in the temperature-treated plots, namely from 24% to 0.5%, when averaged over temperature treatments. The vegetation showed only minor responses to the treatments during the first years of the experiment. *Sphagnum* growth was already reduced by the second year due to nitrogen and temperature treatment (Gunnarsson *et al.* 2004), but the cover remained constant. However, after nine years, these treatments had caused the initially closed *Sphagnum* carpet to nearly disappear, and the cover of vascular plant species had increased. This time lag in the response of the vascular species reflects the fact that *Sphagnum* mosses are efficient at absorbing applied nitrogen, and deep-rooted vascular plants growing in a dense *Sphagnum* carpet are thus, in the short term, prevented from having direct access to nitrogen added to the surface (Nordbakken *et al.* 2003). Eventually, *Sphagnum* nitrogen uptake results in excessive tissue nitrogen accumulation, which leads to decreased growth and moss decline (Limpens & Berendse 2003a) and increased growth of vascular species. Enhanced growth of the graminoid species (*Eriophorum vaginatum* L., *Carex* spp.) subsequently caused reduced growth of the *Sphagnum* mosses (Wiedermann *et al.* 2007).

Peat bogs play a large role in the global sequestration of carbon, and are often dominated by different *Sphagnum* species. Therefore, it is crucial to understand how *Sphagnum* vegetation in peat bogs will respond to global warming. In a greenhouse experiment, the growth of four *Sphagnum* species (*S. fuscum*, *S. balticum*, *S. magellanicum*, and *S. cuspidatum*) was studied (Breeuwer *et al.* 2008) in four temperature treatments (1.2, 14.7, 18.0 and 21.4°C). It was found found that the height, nitrogen content, and biomass production of all species increased with an increase in temperature, but their densities were lower at higher temperatures. Photosynthesis also increased with temperature up to an optimum of about 20–25 °C. As for their competitive abilities, *S. balticum* was the stronger competitor. However, under field conditions, this potential response may not be realized in instances of competition from vascular plants, drought stress, or extreme temperature increases.

The above findings indicate that production rates increase and that *Sphagnum* species abundances shift in response to global warming. Thus, hollow species such as *S. balticum* will lose competitive strength relative to hummock species such as *S. fuscum* and southern species such as *S. magellanicum*. The outcome for the C balance of bog ecosystems depends not only on the production rates, but also on decomposition, which also increases with temperature and increased N availability; hummock species decompose more slowly than hollow species (Limpens & Berendse 2003b).

Elevated carbon dioxide effects

The concentration of CO_2 in the atmosphere has been increasing over the past two centuries from about 280 ppm to a present value of 360 ppm, and is expected to reach more than twice its pre-industrial concentration in the next century (Houghton *et al.* 1990). The increasing CO_2 concentrations might be expected to have substantial impacts on plants and vegetation through changes caused to photosynthesis and other physiological processes (Körner & Bazzaz 1996). Over the past few years an enormous amount of experimental data has become available on the responses of non-DT (desiccation tolerant) plants to elevated concentrations of CO_2. Despite their importance (e.g., Oechel & Svejnbjörnsson 1978), the potential effect of climate change on bryophytes has received much less attention than effects on other plants such as trees, crops, and grasses (Körner & Bazzaz 1996).

Studying the effects of rising CO_2 concentration on natural vegetation of the earth without considering the responses of the bryophytes leaves a large (and sometimes crucial) lacuna in our understanding. Despite this and the volume of research on the physiology of desicccation tolerance (e.g., Tuba *et al.* 1998), there are many fewer reports that discuss in general how bryophytes would respond to a long-term high-CO_2 environment (Ötvös & Tuba 2005). There is, however, new research on this subject, reported in this volume.

Variations in atmospheric CO_2 concentration are nothing new: CO_2 concentrations were higher or lower during some earlier geological periods (Berner 1998). Bryophytes as ancient C_3 land plants experienced these changes in CO_2 concentrations. Results of research on bryophytes are comparable to those on desiccation-sensitive and evolutionarily young C_3 flowering plants, the latter representing the most widely investigated group in the field of global change.

It is not possible to draw parallels between the responses to CO_2 enrichment of DT bryophytes and non-DT plants without caution and qualification. There are some similarities. In both groups photosynthesis shows an immediate positive response to an elevated CO_2 environment, but in the longer term this tends to be negated to varying degrees by downward acclimation of photosynthesis or other limitations on production and growth. Most mosses live in nutrient-poor habitats and this exerts some limitation on their photosynthetic and production responses to elevated atmospheric CO_2 (Jauhiainen *et al.* 1998). Of the moss species studied, *Polytrichum formosum* comes closest to the non-DT homoiohydric plant pattern (Csintalan *et al.* 1995). Members of the Polytrichaceae have an internal conducting system, unusual among bryophytes. In general, mosses have limited source–sink differentiation. The great majority of bryophytes are poikilohydric and evergreen. This means that they often have to face suboptimal environmental factors and to react immediately to intermittent favorable periods (Tuba *et al.* 1998). There are indications that elevated CO_2 favors DT bryophytes most when their water content is too low or too high for positive net carbon assimilation at present CO_2 concentrations. There are also indications that they may generally (but not always) cope better with heavy metal pollution and other stresses at higher CO_2 (Tuba *et al.* 1999). The responses of bryophytes are thus likely to interact in quite complex ways with other climatic factors (Proctor 1990); no simple predictions can be made. Broad biogeochemical considerations predict that rising atmospheric CO_2 should result in faster net (photosynthetic) transfer of carbon from atmosphere to biosphere (Schlesinger 1997), but this has yet to be unequivocally demonstrated even for non-DT vascular plants. Much more experimental evidence from long-term experiments would be needed to make a confident forecast of the responses of desiccation-tolerant plants to the global environmental changes that may be expected later in this century.

Effects of increased light intensity and ozone depletion

Bryophytes, even those of open, exposed habitats, are regarded as shade plants with typically low chlorophyll a:b ratios. They show saturation of photosynthesis at rather low irradiances: around 20% of full sunlight, even in species of open, brightly lit habitats (Tuba 1987; Marschall & Proctor 2004). It is reasonable that poikilohydric photosynthetic organisms should be adapted to function at relatively low light levels since during periods of bright, dry sunny weather most bryophytes are in a desiccated, metabolically inactive state. Most of their photosynthesis takes place in rainy or cloudy weather, when irradiance may often be less than 20% of that in full sun.

For bryophytes of dry habitats, the major physiological need during transient exposures to bright sunshine, as the plant dries out, is likely to be for photoprotection rather than energy capture. The bryophytes that, prima facie, might be expected to be best adapted to photosynthesize under high light conditions are the species of mires, springs and other wet habitats, which usually remain constantly moist in full sun.

Marschall and Proctor (2004) examined the ratios of chlorophyll a to chlorophyll b, and of total chlorophylls to total carotenoids, in relation to bryophyte habitat and to the light-response of photosynthesis in a taxonomically wide range of material collected in southwest England, including most of the major groups of bryophytes.

Ozone in the stratosphere has altered the solar radiation reaching the surface of the earth at the UV-B wavelengths 280–315 nm, resulting in a 60% springtime increase in UV radiation at subarctic latitudes (Taalas *et al.* 1995). According to Rozema *et al.* (2002) the UV-B tolerance of mosses is probably due to the fact that they are evolutionarily older than flowering plants and may therefore have adapted to higher historic surface UV-B intensities than those occurring at present (Rozema *et al.* 2002) as the ozone layer developed gradually over time. UV protection and reflection of mosses depend on the type of moss: the ectohydric mosses have a thin lipid layer and no waxes on their surfaces for protection against irradiation. Some endohydric species have cuticles and wax deposits on their surfaces.

Bryophytes are vulnerable to UV-B damage because of their relatively undifferentiated simple structure (although there are exceptions like *Marchantia* and Polytrichaceae). The survival of Antarctic bryophytes under ozone depletion depends on their ability to acclimate to increasing UV-B radiation. UV-B absorbing pigments absorb UV-B radiation and convey photosynthetically active radiation. In polar and alpine bryophytes during physiologically inactive states, such as desiccation or freezing, the accumulation of these pigments is important because of their passive UV-B absorbing capacity, rather than repair mechanisms, which would necessitate an active metabolism. Amongst the photoprotective secondary metabolites present in mosses, flavonoids are the most important, consisting of flavone and flavanol aglycones and glycosides, anthocyanins and derivates, aurones, biflavonoids, dihydroflavonoids, isoflavones, and triflavones (Mues 2000). They are also found in the wall-bound fraction of some bryophytes, e.g., *Sphagnum* (Asakawa 1995). In *Mnium* and *Brachythecium*, *p*-coumaric acid and ferulic acid have also been identified (Davidson *et al.* 1989). Anthocyanins, especially polyacylated anthocyanins (Mori *et al.* 2005), are a type of flavonoid that absorbs in the visible region of the spectrum with a tail in the UV; it thus can offer protection from UV-B radiation. One piece of evidence for this is that they can reduce the rate of DNA photoproduct repair by filtering out the blue light needed to activate the photolyases that catalyse such repair (Hada *et al.* 2003), and by antioxidant activity, they may indirectly increase tolerance to UV-B radiation by neutralizing free radicals (Dunn & Robinson 2006).

Huttunen *et al.* (2005) studied crude methanol extract of UV-B-absorbing compounds from herbarium specimens to determine whether they reflect changes in the past radiation climate, and which subarctic species seem to be suitable to this analysis. The studied specimens were *Dicranum scoparium* Hedw., *Funaria hygrometrica* Hedw., *Hylocomium splendens* (Hedw.) Schimp., *Pleurozium schreberi* (Brid.) Mitt., *Polytrichastrum alpinum* (Hedw.) G.L.Sm., *Polytrichum commune* Hedw, *Sphagnum angustifolium* (Russow) C. Jens., *Sphagnum capillifolium* (Ehrh) Hedw., *Sphagnum fuscum* Klinggr., and *Sphagnum warnstorfii* Russow. From a comparison of different specimens, specific absorbance should be proportional to the radiation level to which the specimen surface was exposed, and the specimen surface area (SSA) that helped in characterizing the light response of the species. They found that the *Sphagnum* species had slightly decreasing contents of UV-B-absorbing compounds since the 1920s, but since the extraction of the cell wall fraction-bound UV-B-screening compounds such as *p*-coumaric acid, ferulic acid and sphagnorubins was not adequate, the methanol extraction thus does not indicate true changes in *Sphagnum* species. *S. capillifolium* was the only species with a significant decrease in absorbance over time; thus it might well be a suitable UV indicator species. In this case, *P. alpinum, F. hygrometrica*, and three of the four *Sphagnum* species reflected the past trends of global radiation and they will be studied further (Huttunen *et al.* 2005). These findings are in accordance with those of Taipale and Huttunen (2002), who concluded that UV-B enhancement induced larger variance in the concentration of UV-B-absorbing compounds in *H. splendens*.

From the limited number of UV-B field studies on polar bryophytes, it has been concluded that (partial) exclusion of solar UV-B in (sub)Antarctic areas with significantly reduced stratospheric ozone does not affect growth, photosynthesis, or DNA (Rozema *et al.* 2005). Lud *et al.* (2003) studied photosynthetic activity of the moss *Sanionia uncinata* (Hedw.) Loeske. It was investigated on Lé onie Island (67°35′S, 68°20′W, Antarctic Peninsula) in response to short-term changes in UV-B radiation. The researchers found that light response of relative electron transport rate through photosystem II (rel ETR=AF/ Fm′ × PPFD), net photosynthesis, and dark respiration remained unaffected by ambient summer levels of UV-B radiation. In addition, no DNA damage was measured as cyclobutane pyrimidine dimers (CPDs); only artificially enhanced UV-B radiation led to formation of 3–7 CPD (10^6 nucleotides)$^{-1}$. It was concluded that current ambient summer levels of UV-B radiation do not affect photosynthetic activity in *S. uncinata* (Lud *et al.* 2003).

It is known from experimental studies that *Sphagnum* species respond to enhanced UV radiation by decreasing their growth (Huttunen *et al.* 2005), but this tendency varies according to the investigated species and also the length of the investigation. Shoot length and shoot biomass of sub(Ant)Arctic *Sphagnum* species were slightly affected with enhanced UV-B, but biomass per unit ground area appeared to be unaffected (Searles *et al.* 2002; Robson *et al.* 2003). Robson *et al.* (2003) found that growth parameter changes were not significant within a year, but were when repeatedly measured over several years. The growth rate of *Sphagnum* was reduced by near-ambient UV-B over 3 years of study (Huttunen *et al.* 2005).

Dunn and Robinson (2006) measured concentrations of UV-B-absorbing pigments and anthocyanins in three Antarctic moss species over a summer growing season. The Antarctic endemic moss *Schistidium antarctici* was compared with the other two native mosses of the Windmill Island region of East Antarctica, *Bryum pseudotriquetrum* and *Ceratodon purpureus*, in order to determine whether the pigments were induced by UV-B radiation or other environmental parameters, as well as whether there were differences between species in their pigment response. They found that the two cosmopolitan species had higher concentrations of UV-B-absorbing compounds. *B. pseudotriquetrum* had the highest overall, *C. purpureus* intermediate levels, and the endemic *S. antarctici* showed the lowest concentration. *Bryum pseudotriquetrum* displayed a seasonal decline in UV-B-absorbing compounds.

As Robinson *et al.* (2005) reported earlier in their study of UV radiation screened from *S. antarctici* moss beds, UV radiation was found to cause photo-oxidation of chlorophyll and morphological damage but the accumulation of UV-B screening pigments did not occur. Among the three investigated species *S. antarctici* is the last species that is exposed to full sunlight once the snow melts (Wasley *et al.* 2006) and photo-oxidation arises as plants are exposed to full sunlight. They concluded that mosses do not behave as a consistent functional group with respect to UV-B radiation. The differences may be due to their different degrees of desiccation tolerance (Csintalan *et al.* 2001; Lud *et al.* 2002) since *C. purpureus* and *B. pseudotriquetrum* show higher desiccation tolerance than *S. antarctici* (Dunn & Robinson 2006).

UV-B-absorbing pigments and anthocyanins also responded to water availability and air temperature parameters in *B. pseudotriquetrum*, with the highest concentrations of pigments under low humidity and TWC (tissue water content), or low temperature, respectively. Accumulation of anthocyanins at low temperature is common in many plant species (Chalker-Scott 1999). Because in Antarctica temperature is an important factor for determining water availability, anthocyanin accumulation may be related to water availability. It is likely that UV-B-absorbing pigments play an antioxidant role, neutralizing free radicals produced by both excess photosynthetically active radiation and UV-B radiation, under desiccated or frozen conditions, a role that would enhance protection from UV-B radiation. Moss photosynthetic capacity is reduced at low water content, and excess light absorption becomes more problematic. The response of pigments to water availability in *B. pseudotriquetrum* suggests that they may play an important antioxidant role in this species (Dunn & Robinson 2006).

Effect of drying and desiccation tolerance

Desiccation causes several disruptions in plants, such as membrane leakage of ions and electrolytes, pH changes, solute crystallization, protein denaturation, and the formation of reactive oxygen species (Proctor & Tuba 2002; Mayaba & Beckett 2003). These disturbances result in growth reduction, chlorophyll degradation, and reduced regeneration potential (Barker *et al.* 2005). Desiccation tolerance (DT) is the ability to dry to balance with the ambient air and then recover and return to normal metabolism through remoistening (Proctor & Pence 2002). Plants are divided into two major groups in terms of desiccation tolerance: homoiochlorophyllous (HDT) and poikilochlorophyllous (PDT) types. Bryophytes are in the former group, in which plant species retain their chlorophyll during desiccation (Tuba 2008), thus suffering little or no damage following desiccation and essentially returning to a functional state within minutes (Proctor & Smirnoff 2000). Bryophytes have both constitutive cellular protection and a rehydration-induced repair–recovery system (Oliver *et al.* 2000). Bryophytes belong to the most prominent groups of desiccation-tolerant (DT) photosynthetic organisms. Recent synthetic phylogenetic analyses suggest that vegetative desiccation tolerance was primitively present in the bryophytes (the basal-most living clades of land plants), but was then lost in the evolution of tracheophytes (Oliver *et al.* 2000).

Desiccation damage to bryophytes is greatest when the period of hydration is short; this is a frequent case in moss habitats (Proctor 2004), involving a high frequency of rapid drying events. Bryophytes show wide degrees of DT, ranging from some species of *Sphagnum* that cannot tolerate a single drying event, to desert species of mosses that can recover maximal photosynthetic rates within one hour of rehydration after very long drying events. Desert mosses are also able to tolerate multiple wet/dry/wet periods (Stark *et al.* 2005). Desiccation tolerance of bryophytes has been widely studied via various methods including membrane permeability (Beckett et *al.* 2000), pigment concentrations (Newsham 2003), protein synthetic response (Oliver *et al.* 2004), and chlorophyll fluorescence (Takács *et al.* 2000).

Stark *et al.* (2007) studied the gametophyte and sporophyte of *Tortula inermis* in terms of their recovery after rehydration and showed a high level of recovery after rehydration in the gametophyte and sporophyte of controls (100% and 73%, respectively). After a single drying cycle, control gametophytes still had 100% recovery. However, after two cycles of rapid desiccation, recovery was compromised in gametophytes (63%). In sporophytes, a single drying cycle resulted in 23% recovery, whereas two drying cycles resulted in only 3% recovery. These results indicate that the sporophyte is more negatively affected by desiccation than the gametophyte. Reasons for the evolution of higher DT tolerance of the shoot (gametophyte) of bryophytes include the fact that they are perennials exposed to much variation in moisture, whereas the sporophytes with a shorter life span (1–2 years) do not experience long periods of desiccation during development. In addition, the development of sporophytes is strongly reduced in dry years owing to the requirement of free water for fertilization (Stark *et al.* 2007). Desiccation tolerance requires synthesis of an array of repair proteins, and is likely to be an energetically expensive process given the elevated respiration rates attending rehydration from a desiccated state (Tuba *et al.* 1996; Oliver *et al.* 2005). Inherent variations in the desiccation tolerance repair pathways may result in the generational difference in DT among bryophytes. Steady-state mRNA levels for Tr288 increase dramatically during slow drying, and Tr288 transcripts do not accumulate during rapid drying of moss gametophytes. On the other hand, during rehydration of rapidly dried tissue Tr288 transcript levels increase several-fold during the first 15 min after rehydration, whereas the relatively high amount of Tr288 mRNA sequestered in slowly dried material declines with time after the addition of water (Velten & Oliver 2001).

Lüttge *et al.* (2008) investigated three poikilohydric and homoiochlorophyllous moss species, *Campylopus savannarum* (C. Muell.) Mitt., *Racocarpus fontinaloides* (C. Muell.) Par. and *Ptychomitrium vaginatum* Besch., on sun-exposed rocks of a tropical inselberg (isolated outcrop or ridge) in Brazil subject to regular drying and wetting cycles. Their aim was to identify whether there were differences in the reduction of F' (ground chlorophyll fluorescence) in the desiccated state, in the photosynthetic capacity in the wetted state or in morphological traits among the species according to their distributional differences and niche occupation. The almost complete loss of ground chlorophyll fluorescence of photosystem II (PSII) F' in the desiccated state of all three mosses (Nabe *et al.* 2007; Lüttge *et al.* 2008) confirms an effective desiccation-induced energy dissipation in the reaction centers, notwithstanding their different niche occupation. The recovery of basic fluorescence upon rewatering had a very fast initial phase, with a drastic increase of F' within less than a minute followed by a more gradual increase regardless of the moss species.

This agrees with the idea that mosses, the homoiochlorophyllous desiccation-tolerant plants, maintain constituent compounds of the photosynthetic apparatus during desiccation (Tuba 2008) as a requirement for fast revitalization. This also means, however, that ground fluorescence is not useful for the characterization of either different species or microhabitats. $\Delta F/F'_m$ values of two species (*Campylopus savannarum* (C. Muell.) Mitt. and *Racocarpus fontinaloides* (C. Muell.)) in the belt around the vegetation islands, on the other hand, indicate a greater photosynthetic capacity compared with the species on the bare rock (*Ptychomitrium vaginatum* Besch.).

The rate of desiccation and rewetting depends on cushion form and size (Zotz *et al.* 2000). The small cushions of *P. vaginatum* under exposed conditions on the inselberg rocks experience short periods of carbon acquisition during and after rain, when carbon fixation is influenced by the diffusion barrier of excess water in the cushions. The other two species, *C. savannarum* and *R. fontinaloides*, produce larger cushions; water films do not develop and interfere with CO_2 diffusion. Further, water is available for longer after rain, so diffusion control of CO_2 uptake will contribute less and Rubisco relatively more to overall ^{13}C discrimination. This explains the more negative $\delta^{13}C$ values, although these are still less negative than expected if only gaseous CO_2 diffusion in air were effective for CO_2 supply to the mosses. The difference in water supply of the mosses in the microhabitats resulted in an altered diffusion limitation of carbon gain, and seems to differentiate the ecophysiological niches of the three bryophytes (Lüttge *et al.* 2008).

All the desiccation-tolerant species tested lost most of their PSII photochemical activity when photosynthetic electron transport was inhibited by air drying, while the PSI reaction center remained active under drying conditions in both sensitive and tolerant species. The activity, however, became undetectable in the light only in tolerant species due to deactivation of the cyclic electron flow around PSI and of the back reaction in PSI. Light-induced non-photochemical quenching (NPQ) was found to be induced not only by the xanthophyll cycle but also by a pH-induced, dithiothreitol-insensitive mechanism in desiccation-tolerant and -intolerant bryophytes (Nabe *et al.* 2007).

Forest ecosystems contain a major proportion of carbon stored on land as biomass and soil organic matter (SOM), and the long-term fate of carbon depends on whether it is stored in living biomass or soils. Forests take up CO_2 from the atmosphere through photosynthesis, and return large amounts through respiration by vegetation and decomposers in the soil. Soil carbon amounts exceed those in vegetation by 2:1 in northern temperate forests and by over 5:1 in boreal forests (Hyvönen *et al.* 2006). Thus changes in soil carbon stocks can be more important than changes in vegetation carbon stocks for forest carbon budgets (Medlyn *et al.* 2005). As a result of land-use change, deforestation, and fossil fuel burning, the global average atmospheric CO_2 concentration rose by approximately 35%, from 280 to 377 ppm, in 2004 (WMO 2006) of which approximately 60% is absorbed by oceans and terrestrial ecosystems. The average global temperature increased during the twentieth century by 0.6 °C, and projections are for an additional increase of 1.4–5.8 °C during the twenty-first century (IPCC 2001). Such changes will increase the rate of biochemical reactions and also lengthen growing seasons. Ecosystems are carbon sinks if CO_2 exceeds the losses of CO_2 in total ecosystem respiration ($R_E = R_A + R_H$) and volatile organic compounds. They are most probably carbon sinks in the daytime, carbon sources at night, carbon sinks in summer, and carbon sources in winter (Hyvönen *et al.* 2006).

Marschall and Beckett (2005) investigated inducible desiccation tolerance and its impact on photosynthesis on two bryophyte species, *Dumortiera hirsuta* and *Atrichum androgynum* (C. Müll.) as a result of hardening processes, such as partial dehydration or abscisic acid treatment as a substitute for partial dehydration. In *A. androgynum* hardening did not increase desiccation tolerance by sugar accumulation or enzyme activity induction, but rather by increased photoprotection by non-photochemical quenching (NPQ). Hardening treatments increased thylacoidal (qE), rather than reactions-based quenching (qI). However, the efficiency of PSII decreased, so they concluded that hardening shifts the photosynthetic apparatus from highly effective, to a somewhat lower, but more photoprotected, state. This explains the lower growth rate of the moss, but also its ability to survive stressed conditions (Marschall & Beckett 2005).

Desiccation tolerance cannot be considered as one mechanism, but rather as an interplay of several processes, such as the induction of protective carbohydrates, proteins, and also enzymatic and non-enzymatic antioxidants (Beckett *et al.* 2005). The tolerance mechanisms can be either constitutive or inducible, the latter of which has the advantage of not requiring much energy from growth and development (Marschall & Beckett 2005).

Conclusions

The effects of single factors causing global climate change will be discussed in separate chapters of this book. Many of the authors have conducted recent research and have reviewed the research of others in their special fields. Here I have provided an overview of the important processes that can be monitored with bryophyte research in a changing global climate.

Bryophytes are, in a sense, ideal plants for climate change indication, but the conclusions drawn from the experimental results should be carefully treated. There may be differences in greenhouse and field experiments, but long-term conclusions cannot be made about moss reactions to climate change from a single year of evaluation, as was seen in the outcomes of global warming experiments. Furthermore, bryophytes can successfully reflect UV-B radiation changes via the fact that their flavonoids are generally of the less oxidizable types, i.e., flavones and biflavones. This enables investigations of bryophyte herbarium specimens in retrospective studies at least on a regional scale, and the soluble UV-B-absorbing compounds of bryophytes may reflect the light climate.

We emphasize, and similarly for the effects of elevated CO_2, that it is not possible to draw parallels between the responses to global climate change of DT bryophytes and non-DT flowering plants without caution and qualifications. There are similarities, but most, though not all (e.g., limestone and rich fen), mosses live in nutrient-poor habitats and this exerts some limitation on their photosynthetic and production responses. In general, mosses have limited source–sink differentiation. Most bryophytes are poikilohydric and evergreen. This means that they often have to face suboptimal environmental factors and to react immediately to intermittent favorable periods (Tuba *et al.* 1998). There are also indications that bryophytes may generally (but not always) cope better with heavy metal pollution and other stresses at higher CO_2 (Tuba *et al.* 1999). The responses of bryophytes are thus likely to interact in quite complex ways with other climatic factors (Proctor 1990). No simple predictions can be made. More study is needed of all the processes discussed above and of comparisons with vascular, especially flowering, plants.

References

Asakawa, Y. (1995). In *Progress in the Chemistry of Organic Natural Products*, ed. W. Herz, G. W. Kirby, R. E. Moore, W. Steglich & C. Tamm, vol. 65, p. 1. Vienna: Springer.

Barker, D. H., Stark, L. R., Zimpfer, J. F., McLetchie, N. D. & Smith, S. D. (2005). Evidence of drought-induced stress on biotic crust moss in the Mojave Desert. *Plant, Cell and Environment* **28**: 939–47.

Beckett, R. P., Csintalan, Z. & Tuba, Z. (2000). ABA treatment increases both the desiccation tolerance of photosynthesis, and nonphotochemical quenching in the moss *Atrichum undulatum*. *Plant Ecology* **151**: 65–71.

Beckett, R. P., Marschall, M. & Laufer, Zs. (2005). Hardening enhances photoprotection in the moss *Atrichum androgynum* during rehydration by increasing fast rather than slow-relaxing quenching. *Journal of Bryology* **27**: 7–12.

Berner, R. A. (1998). The carbon cycle and CO_2 over Phanerozoic time: the role of land plants. *Philosophical Transactions of the Royal Society of London* B **353**: 75–82.

Bonine, M. L. (2004). Growth, reproductive phenology, and population structure in *Syntrichia caninervis*. M.S. thesis, University of Nevada, Las Vegas.

Bouwman, A. F., Van Vuuren, D. P., Derwent, R. G. & Posch, M. (2002). A global analysis of acidification and eutrophication of terrestrial ecosystems. *Water, Air, and Soil Pollution* **141**: 349–82.

Breeuwer A., Heijmans, M. M. P. D., Robroek, B. J. M. & Berendse, F. (2008). The effect of temperature on growth and competition between *Sphagnum* species. *Oecologia* **156**: 155–67.

Carballeira, A., Díaz, S., Vázquez, M. D. & López, J. (1998). Inertia and resilience in the response of the aquatic bryophyte *Fontinalis antipyretica* Hedw. to thermal stress. *Archives of Environmental Contamination and Toxicology* **34**: 343–9.

Chalker-Scott, L. (1999). Environmental significance of anthocyanins in plant stress responses. *Photochemistry & Photobiology* **70**: 1–9.

Crosby, M. R., Magill, R. E., Allen, B. & He, S. (2000). A checklist of the mosses. St. Louis, MO: Missouri Botanical Garden. www.mobot.org/MOBOT/tropicos/most/checklist.shtml.

Csintalan, Z., Tuba, Z. & Laitat, E. (1995). Slow chlorophyll fluorescence, net CO_2 assimilation and carbohydrate responses in *Polytrichum formosum* to elevated CO_2 concentrations. In *Photosynthesis from Light to Biosphere*, ed. P. Mathis, Vol. V, pp. 925–8. Dordrecht: Kluwer Academic Publishers.

Csintalan Z., Tuba, Z., Takács, Z. *et al.* (2001). Responses of nine bryophyte and one lichen species from different microhabitats to elevated UV-B radiation. *Photosynthetica* **39**: 317–20.

Davidson, A. J., Harborne, J. B. & Longton, R. E. (1989). Identification of hydroxycinnamic acid and phenolic acids in *Mnium hornum* and *Brachythecium rutabulum* and their possible role in protection against herbivory. *Journal of the Hattori Botanical Laboratory* **67**: 415–22.

Deltoro, V. I., Calatayud, A., Gimeno, C. & Barreno, E. (1998). Water relations, chlorophyll fluorescence, and membrane permeability during desiccation in bryophytes from xeric, mesic, and hydric environments. *Canadian Journal of Botany* **76**(11): 1923–9.

Dorrepaal, E., Cornelissen, J. H. C., Aerts, R., Wallen, B. & van Logtestijn, R. S. P. (2005). Are growth forms consistent predictors of leaf litter quality and decomposability across peatlands along a latitudinal gradient? *Journal of Ecology* **93**: 817–28.

Dunn, J. L. & Robinson, S. A. (2006). UV-B screening potential is higher in two cosmopolitan moss species than in a co-occurring Antarctic endemic moss – implications of continuing ozone depletion. *Global Change Biology* **12**(12): 2282–96.

Gunnarsson, U., Granberg, G. & Nilsson, M. B. (2004). Growth, production and interspecific competition in *Sphagnum*: effects of temperature, nitrogen and sulfur treatments on a boreal mire. *New Phytologist* **163**: 349–59.

Hada, H., Hidema, J., Maekawa, M. *et al.* (2003) Higher amounts of anthocyanins and UV absorbing compounds effectively lowered CPD photorepair in purple rice (*Oryza sativa* L.). *Plant Cell and Environment* **26**: 1691–1701.

Harvell, D., Mitchell, C. E., Ward, J. R. *et al.* (2002). Climate warming and disease risks for terrestrial and marine biota. *Science* **296**: 2158–62.

Hearnshaw, G. F. & Proctor, M. C. F. (1982). The effect of temperature on the survival of dry bryophytes. *New Phytologist* **90**: 221–8.

Heijmans, M. M. P., Klees, D. & Berendse, F. (2002).Competition between *Sphagnum magellanicum* and *Eriophorum angustifolium* as affected by raised CO_2 and increased N deposition. *Oikos* **97**: 415–25.

Houghton, J. T., Jenkins, G. J. & Ephramus, J. J. (1990). *Climate Change: The IPCC Scientific Assessment*. Cambridge: Cambridge University Press.

Huttunen, S., Lappalainen, N. M. & Turunen, J. (2005). UV-absorbing compounds in subarctic herbarium bryophytes. *Environmental Pollution* **133**: 303–14.

Hyvönen R., Ågren, G. I., Linder, S., *et al.* (2006). The likely impact of elevated [CO_2], nitrogen deposition, increased temperature and management on carbon sequestration in temperate and boreal forest ecosystems: a literature review. *New Phytologist* **173**: 463–80.

IPCC (2001). *Climate Change 2001: The Scientific Basis. Contribution of Working Group I to the Third Assessment Report of the Intergovernment Panel on Climate Change*, ed. J. T. Houghton, Y. Ding, D. J. Griggs *et al.* Cambridge and New York: Cambridge University Press.

Jauhiainen, J., Silvola, J. & Vasander, H. (1998). Effects of increased carbon dioxide and nitrogen supply on mosses. In *Bryology for the Twenty-first Century*, ed. J. W. Bates, N. W. Ashton & J. G. Duckett, pp. 343–60. Leeds: Maney Publishing and the British Bryological Society.

Körner, C. & Bazzaz, F. A. (1996). *Carbon Dioxide, Populations, and Communities*. San Diego, CA: Academic Press.

Lewis-Smith, R. I. & Convey, P. (2002). Enhanced sexual reproduction in bryophytes at high latitudes in the maritime Antarctic. *Journal of Bryology* **24**: 107–17.

Limpens, J. & Berendse, F. (2003a). Growth reduction of *Sphagnum magellanicum* subjected to high nitrogen deposition: the role of amino acid nitrogen concentration. *Oecologia* **135**: 339–45.

Limpens, J. & Berendse, F. (2003b). How litter quality affects mass loss and N loss from decomposing *Sphagnum*. *Oikos* **103**: 537–47.

Lud, D., Moerdijk, T. C. W., van der Poll, W. H. *et al.* (2002). DNA Damage and photosynthesis in Antarctic and Arctic *Sanionia uncinata* (Hedw.) Loeske under ambient and enhanced levels of UV-B radiation. *Plant, Cell and Environment* **25**: 1579–89.

Lud, D., Schlensog, M., Schroeter, B. & Huiskes, A. H. L. (2003). The influence of UV-B radiation on light-dependent photosynthetic performance in *Sanionia uncinata* (Hedw.) Loeske in Antarctica. *Polar Research* **26**: 225–32.

Lüttge, U., Meirelles, S. T. & Mattos, E. A. (2008). Strong quenching of chlorophyll fluorescence in the desiccated state in three poikilohydric and homoiochlorophyllous moss species indicates photo-oxidative protection on highly light-exposed rocks of a tropical inselberg. *Journal of Plant Physiology* **165**: 172–81.

Malmer, N. & Wallén, B. (2005). Nitrogen and phosphorus in mire plants: variation during 50 years in relation to supply rate and vegetation type. *Oikos* **109**: 539–54.

Marschall, M. & Beckett, R. P. (2005). Photosynthetic responses in the inducible mechanisms of desiccation tolerance of a liverwort and a moss. *Acta Biologica Szegediensis* **49** (1–2): 155–6.

Marschall, M. & Proctor, M. C. F. (2004). Are bryophytes shade plants? Photosynthetic light responses and proportions of chlorophyll a, chlorophyll b and total carotenoids. *Annals of Botany* doi:10.1093/aob/mch178, available online at www.aob.oupjournals.org.

Mayaba, N. & Beckett, R. P. (2003). Increased activities of superoxide dismutase and catalase are not the mechanism of desiccation tolerance induced by hardening in the moss *Atrichum androgynum*. *Journal of Bryology* **25**: 281–6.

McLetchie, D. N. & Stark, L. R. (2006). Sporophyte and gametophyte generations differ in their thermotolerance response in the moss *Microbryum*. *Annals of Botany* **97**: 505–11.

Medlyn, B. E., Berbigier, P., Clement, R. *et al.* (2005). The carbon balance of coniferous forests growing in contrasting climatic conditions: a model-based analysis. *Agricultural and Forest Meteorology* **131**: 97–124.

Meyer, H. & Santarius, K. A. (1998). Short-term thermal acclimation and heat tolerance of gametophytes of mosses. *Oecologia* **115**: 1–8.

Mori, M., Yoshida, K., Ishigaki, Y. *et al.* (2005). UV-B protective effect of a polyacylated anthocyanin, HBA, in flower petals of the blue morning glory, *Ipomoea tricolor* cv. Heavenly Blue. *Bioorganic & Medicinal Chemistry* **13**: 2015–20.

Mues, R. (2000). Chemical constituents and biochemistry. In *Bryophyte Biology*, ed. A. J. Shaw & B. Goffinet, pp. 150–81. Cambridge: Cambridge University Press.

Nabe, H., Funabiki, R., Kashino, Y., Koike, H. & Satoh, K. (2007). Responses to desiccation stress in bryophytes and an important role of dithiothreitol-insensitive non-photochemical quenching against photoinhibition in dehydrated states. *Plant and Cell Physiology* **48**: 1548–57.

Newsham, K. K. (2003). UV-B radiation arising from stratospheric ozone depletion influences the pigmentation of the Antarctic moss *Andreaea regularis*. *Oecologia* **135**: 327–31.

Nobel, P. S., Geller, G. N., Kee, Z. C. & Zimmerman, A. D. (1986). Temperatures and thermal tolerances for cacti exposed to high temperatures near the soil surface. *Plant, Cell and Environment* **9**: 279–87.

Nordbakken, J. F., Ohlson, M. & Högberg, P. (2003). Boreal bog plants: nitrogen sources and uptake of recently deposited nitrogen. *Environmental Pollution* **126**: 191–200.

Oechel, W. C. & Svejnbjörnsson, B. (1978). Primary production processes in arctic bryophytes at Barrow, Alaska. In *Vegetation and Production Ecology of an Alaskan Arctic Tundra*, ed. L. L. Tieszen, pp. 269–98. New York: Springer.

Oliver, M. J., Dowd, S. E., Zaragoza, J., Mauget, S. A. & Payton, P. R. (2004). The rehydration transcriptome of the desiccation-tolerant bryophyte *Tortula ruralis*: transcript classification and analysis. *BMC Genomics 2004* **5**: 89.

Oliver, M. J., Tuba, Z. & Mishler, B. D. (2000). The evolution of vegetative desiccation tolerance in land plants. *Plant Ecology* **151**: 85–100.

Oliver, M. J., Velten, J. J. & Mishler, B. D. (2005). Desiccation tolerance in bryophytes: a reflection of the primitive strategy for plant survival in dehydrating habitats? *Integrative and Comparative Biology* **45**: 788–99.

Ötvös, E. & Tuba, Z. (2005). Ecophysiology of mosses under elevated air CO_2 concentration: overview. *Physiology and Molecular Biology of Plants* **11**: 65–70.

Proctor, M. C. F. (1990). The physiological basis of bryophyte production. *Botanical Journal of the Linnean Society* **104**: 61–77.

Proctor, M. C. F. (2004). How long must a desiccation-tolerant moss tolerate desiccation? Some results of 2 years' data logging on *Grimmia pulvinata*. *Physiologia Plantarum* **122**: 21–7.

Proctor, M. C. F. & Pence, V. C. (2002). Vegetative tissues: bryophytes, vascular 'resurrection plants' and vegetative propagules. In *Desiccation and Plant Survival*, ed. H. Pritchard & M. Black, pp. 207–37. Wallingford, UK: CABI Publishing.

Proctor, M. C. F. & Smirnoff, N. (2000). Rapid recovery of photosystems on rewetting desiccation-tolerant mosses: chlorophyll fluorescence and inhibitor experiments. *Journal of Experimental Botany* **51**: 1695–1704.

Proctor, M. C. F. & Tuba, Z. (2002). Poikilohydry and homoihydry: antithesis or spectrum of possibilities? Tansley review no. 141. *New Phytologist* **156**: 27–49.

Robinson, S. A., Turnbull, J. D. & Lovelock, C. E. (2005). Impact of changes in natural ultraviolet radiation on pigment composition, physiological and morphological characteristics of the Antarctic moss, *Grimmia antarctici*. *Global Change Biology* **11**: 476–89.

Robson, M. T., Pancotto, V. A., Flint, S. D., *et al.* (2003). Six years of solar UV-B manipulations affect growth of *Sphagnum* and vascular plants in a Tierra del Fuego peatland. *New Phytologist* **160**: 379–89.

Roy, H. E., Brown, P. M. J., Rothery, P., Ware, R. L. & Majerus, M. E. N. (2008). Interactions between the fungal pathogen *Beauveria bassiana* and three species of coccinellid: *Harmonia axyridis*, *Coccinella septempunctata* and *Adalia bipunctata*. *BioControl* **53**: 265–76.

Rozema, J., Boelen, P. & Blokker, P. (2005). Depletion of stratospheric ozone over the Antarctic and Arctic: responses of plants of polar terrestrial ecosystems to enhanced UV-B, an overview. *Environmental Pollution* **137**: 428–42.

Rozema, J., Björn, L. O., Bornman, J. F. *et al.* (2002). The role of UV-B radiation in aquatic and terrestrial ecosystems – an experimental and functional analysis of the evolution of UV-absorbing compounds. *Journal of Photochemistry and Photobiology* B **66**: 2–12.

Schlesinger, W. H. (1997). *Biogeochemistry: an Analysis of Global Change*, 2nd edn. San Diego, CA: Academic Press.

Searles, P. S., Flint, S. D., Diaz, S. B. *et al.* (2002). Plant response to solar ultraviolet-B radiation in a southern South American *Sphagnum* peatland. *Journal of Ecology* **90**: 704–13.

Shaw, A. J. & Beer, S. C. (1997). Gametophyte–sporophyte variation and covariation in mosses. *Advances in Bryology* **6**: 35–63.

Stark, L. R. (1997). Phenology and reproductive biology of *Syntrichia inermis* (Bryopsida, Pottiaceae) in the Mojave Desert. *Bryologist* **100**: 13–27.

Stark, L. R. (2002). Skipped reproductive cycles and extensive sporophyte abortion in the desert moss *Tortula inermis* correspond to unusual rainfall patterns. *Canadian Journal of Botany* **80**: 533–42.

Stark, L. R., Nichols II, L., McLetchie, D. N. & Bonine, M. L. (2005). Do the sexes of the desert moss *Syntrichia caninervis* differ in desiccation tolerance? A leaf regeneration assay. *International Journal of Plant Sciences* **166**: 21–9.

Stark, L. R., Oliver, M. J., Mishler, B. D. & McLetchie, D. N. (2007). Generational differences in response to desiccation stress in the desert moss *Tortula inermis*. *Annals of Botany* **99**: 53–60.

Strengbom, J., Nordin, A., Näsholm, T. & L. Ericson, L. (2002). Parasitic fungus mediates change in nitrogen-exposed boreal forest vegetation. *Journal of Ecology* **90**: 61–7.

Taalas, P., Koskela, T., Kyro, E., Damski, J. & Supperi, A. (1995). *Ultraviolet Radiation in Finland*. In The Finnish Research Programme on Climate Change, Final Report, ed. J. Roos. *Publications of The Academy of Finland* **4**: 83–91.

Taipale, T. & Huttunen, S. (2002). Moss flavonoids and their ultrastructural localization under enhanced UV-B radiation. *Polar Research* **38**: 211–18.

Takács, Z., Lichtenthaler, H. K. & Tuba, Z. (2000). Fluorescence emission spectra of desiccation-tolerant cryptogamic plants during a rehydration – desiccation cycle. *Journal of Plant Physiology* **156**: 375–9.

Throop, H. L. & Lerdau, M. T. (2004). Effects of nitrogen deposition on insect herbivory: implications for community and ecosystem processes. *Ecosystems* **7**: 109–33.

Tuba, Z. (1987). Light, temperature and desiccation responses of CO_2-exchange in desiccation tolerant moss, *Tortula ruralis*. In Proceedings of the IAB Conference of Bryoecology, ed. T. Pócs, T. Simon, Z. Tuba & J. Podani, pp. 137–50. *Symp. Biol. Hung.* Vol. **35**, Part A. Budapest: Akadémiai Kiadó.

Tuba, Z. (2008). Notes on the poikilochlorophyllous desiccation-tolerant plants. *Acta Biologica Szegediensis* **52** (1): 111–13.

Tuba, Z., Csintalan, Zs. & Proctor, M. C. F. (1996). Photosynthetic responses of a moss, *Tortula ruralis* (Hedw.) Gaertn. *et al.* ssp. *ruralis*, and the lichens *Cladonia convoluta* (Lam.) P. Cout. and *C. furcata* (Huds.) Schrad. to water deficit and short periods of desiccation, and their eco-physiological significance: a baseline study at present-day CO_2 concentration. *New Phytologist* **133**: 353–61.

Tuba, Z., Csintalan, Zs., Szente, K., Nagy, Z. & Grace, J. (1998). Carbon gains by desiccation tolerant plants at elevated CO_2. *Functional Ecology* **12**: 39–44.

Tuba, Z., Proctor, M. C. F. & Takács, Z. (1999). Desiccation-tolerant plants under elevated air CO_2: a review. *Zeitschrift für Naturforschung* **54**: 788–96.

Velten, J. & Oliver, M. J. (2001). Tr288, a rehydrin with a dehydrin twist. *Plant Molecular Biology* **45**: 713–22.

Wasley, J., Robinson, S. A., Lovelock, C. E. *et al.* (2006). Some like it wet – an endemic Antarctic bryophyte likely to be threatened under climate change induced drying. *Functional Plant Biology* **33**: 443–55.

Wiedermann, M. M., Nordin, A., Gunnarsson, U., Nilsson, M. B. & Ericson, L. (2007). Global change shifts vegetation and plant – parasite interactions in a boreal mire. *Ecology* **88**: 454–64.

WMO. (2006). WMO Greenhouse Gas Bulletin 1. Geneva: World Meteorological Organization/Global Atmosphere Watch. http://www.wmo.ch/web/arep/gaw/ghg/ghg-bulletin-en-03–06.pdf

Zotz, G., Schweikert, A., Jetz, W. & Westerman, H. (2000). Water relations and carbon gain are closely related to cushion size in the moss *Grimmia pulvinata*. *New Phytologist* **148**: 59–67.

II ECOPHYSIOLOGY

3

Climatic Responses and Limits of Bryophytes: Comparisons and Contrasts with Vascular Plants

MICHAEL C. F. PROCTOR

Introduction

Bryophytes and flowering plants are very different in their strategies of adaptation to life on land. These differences are reflected in their responses to climate and their ecological and geographical distribution patterns. The difference is immediately obvious in any region for which both the vascular plant and the bryophyte flora are well known and well mapped. Britain and Ireland are good examples (Hill *et al.* 1991, 1992, 1994; Preston & Hill 1997; Hill & Preston 1998; Preston *et al.* 2002). Only around 10% of the vascular plants of Europe occur in the British Isles, and within Britain and Ireland the number of vascular plant species tends to decline from southeast to northwest. With bryophytes the trend is almost reversed. Britain and Ireland have one of the richest bryophyte floras of any comparably sized region of Europe (and a rich bryoflora even by world standards), and bryophyte diversity is heavily concentrated in upland regions and towards the west coast.

Strategies of adaptation to life on land

A planktonic green cell floating near the surface of a lake or the sea has all the necessities of life around it. It is surrounded on all sides by water, from which it can take up nutrients and exchange gases. Light is not a problem either, provided the cell does not sink too far below the surface of the water. All land plants face the problem that light for photosynthesis is only available in the aerial environment above ground, where the supply of water is erratic and limited. There are two possible strategies of adaptation to this problem (Proctor & Tuba 2002). The vascular plants evolved roots and water-conducting xylem to bring water from below ground – where it is (more or less) constantly

Bryophyte Ecology and Climate Change, eds. Zoltán Tuba, Nancy G. Slack and Lloyd R. Stark. Published by Cambridge University Press.

available – to the photosynthesizing foliage in the dry air above. Uptake of carbon dioxide from the air is inevitably accompanied by water loss, so an essential part of the vascular plant pattern of adaptation is a more or less waterproof cuticle over the aboveground parts, with stomata regulating water loss. This is the *homoiohydric* strategy.

The alternative *poikilohydric* strategy was to evolve desiccation tolerance and to metabolize and grow where and when water is available above ground, and to dry out and suspend metabolism when it is not. This pattern of adaptation is exemplified by the bryophytes, but by no means peculiar to them. Desiccation tolerance is probably exceedingly ancient in evolutionary terms, long pre-dating land plants as we know them. It is widespread in microorganisms, including cyanobacteria, almost universal in the spores of all later-evolved plant groups, and of course in the seeds of most flowering plants.

A corollary of the homoiohydric/poikilohydric distinction is that vascular plants are *endohydric*, but bryophytes are basically *ectohydric*. Vascular plants depend on internal conduction of water from the soil to the foliage. In bryophytes, the physiologically important free water is *outside* the plant body, held in capillary spaces around leaf bases, amongst paraphyllia or rhizoid tomentum, in the spaces between papillae on the leaf surface, or in the concavities of concave leaves. Many bryophytes can hold five or ten times their dry mass of water externally in this way. Many of the details of bryophyte leaf and shoot structure appear to be adaptations maximizing the capacity to store water (thus extending the time the cells are turgid and actively metabolizing) while minimizing interference with gas exchange.

Consequences of scale

Bryophytes are in general some two orders of magnitude smaller than flowering plants. The tallest trees (e.g., *Sequoia sempervirens*, *Eucalyptus regnans*) are around 100 m high. The tallest mosses (e.g., *Polytrichum commune*, *Dawsonia superba*) reach about 50 cm. At the lower end of the size scale there are many bryophytes a millimeter or so high, whereas very small flowering plants tend to be the exceptions that prove the rule, having all but forsaken the vascular pattern of adaptation (e.g., *Lemna minima*, *Anagallis minima*).

This difference in scale brings in its train major differences in physiology and responses to the environment. Other things being equal, a plant a tenth of the linear dimensions of another has a hundredth of the surface area, and a thousandth of the volume and mass (and its root system, if it had one, could exploit a thousandth of the volume of soil). Surface tension (which works at linear interfaces), is a trivial curiosity or occasional inconvenience at our own scale or the scale of a tree, but is a powerful force at the scale of many bryophyte structures. The force of gravity depends on mass (proportional to volume), so it is important at our scale and a limiting factor for tall trees, but trivial for bryophytes. A bryophyte, for simple reasons of scale, could not grow a root system big enough to support the rate of water loss from its aboveground surface area.

The vascular pattern of adaptation is unquestionably optimal for a large plant, but there is much reason to believe that the poikilohydric strategy is optimal for a small one – say, less than about a couple of centimeters high. Of the present major groups of land plants, the lycopods (represented by the modern clubmosses), equisetoids (including the modern horsetails) and the fern line (which later gave rise to the flowering plants), the mosses, and the liverworts, all have a fossil record going back to the Paleozoic, and give every evidence of having a history reaching back to the early differentiation of land plants. What is remarkable is that these groups have remained well differentiated and recognizable through some 350–400 million years of evolution, retaining essentially the same anatomy *and life strategy*. The (vascular) fern – gymnosperm – angiosperm line has dominated the earth's vegetation since the end of the Paleozoic, largely but not completely supplanting the equisetoids (themselves an early branch of the fern line) or the lycopods, two groups that were formerly much more diverse and dominated the Carboniferous forests. The (poikilohydric) mosses and liverworts have remained throughout as the major players in smaller-scale vegetation. They have been an evolutionary success story at their own scale, with some 20,000 species at the present day, distributed from the equator to the polar regions. There are few habitats from which bryophytes are completely absent, but they are particularly conspicuous in tropical-montane and oceanic-temperate forests, in the ground-layer of the boreal coniferous forest, in bogs and fens, and in alpine and subpolar fell-fields and tundra. It is significant that a very similar pattern of ecophysiological adaptation has evolved in a quite unrelated (and comparably successful) group of organisms, the lichens.

Climate and microclimate

"Beauty is in the eye of the beholder," and "climate" is in the perception of the organism that experiences it. Many climatic variables, such as temperature, humidity, and wind speed, show more or less steep gradients near the ground. Standard meteorological air temperature and humidity measurements are taken about 1.5 m above the ground, with the instruments freely exposed to the air but shaded from direct sun, as in a louvered Stevenson screen. That gives air ("shade") temperatures appropriate to our own scale, or to the scale of our farm livestock and tall crops or trees. We measure "grass minimum" temperature because many crops (and grasslands) are low-growing, and "ground frosts" can occur while the air temperature is still several degrees above freezing.

Solar radiation as a driver of climate and microclimate

What ultimately drives the earth's climate is the balance between radiation income from the sun, and heat loss to intergalactic space (Gates 1980; Monteith & Unsworth 1990; Barry & Chorley 2003). The intensity of solar radiation reaching the earth's upper atmosphere is about 1370 watts per square meter ($W\,m^{-2}$). Some of this incoming radiation is absorbed by the atmosphere, and some is reflected back into space by clouds, but much of the incoming energy is absorbed by the ground or by the oceans. The average maximum energy reaching the ground under a clear sky with the sun overhead is about 1000 $W\,m^{-2}$, but the air is seldom really clear, especially at low altitudes and in populated areas, and away from the tropics the sun is never fully overhead.

When solar radiation (light + near ultraviolet + near infrared, with peak energy at a wavelength of about 450 nm) is absorbed by a solid surface it is transformed into heat, which can leave the surface in only a limited number of ways. It may be *conducted* into the ground (or plant), it may heat the air close to the surface and be *convected* away by gaseous diffusion and air currents, or it may be *re-radiated* back to the environment as thermal infrared radiation (with peak energy far outside the visible spectrum at a wavelength of about 10,000 nm). If the surface is wet (or vegetated), heat will also be taken up to provide the *latent heat of evaporation* as water vapor is lost to the atmosphere. Under a clear sky at night, much of the daytime scenario is reversed, and outgoing radiation predominates. At surface temperatures of 0–25 °C, outgoing thermal infrared radiation is *c.* 300–450 $W\,m^{-2}$, but this is partially balanced by downward thermal infrared radiation from the gases in the atmosphere, so the net loss to a cloudless sky at night is commonly about 100 $W\,m^{-2}$ (these gases, mainly water vapor and CO_2, are responsible for the "greenhouse" effect). This loss of heat means that the ground (or plant) gets colder. The ground draws heat from the air; if the air is chilled below the dew point, water will condense as dew, yielding up its latent heat of evaporation. If the temperature of the surface falls below freezing point, the result is a ground frost.

These exchanges of energy at the ground or plant surface are much reduced in cloudy weather or under a forest canopy. The net loss of thermal infrared radiation under a cloudy sky is around half that under a clear sky. Under forest, the ground will generally be cooler than the canopy for much of the day, and warmer for much of the night, but the difference is seldom more than a few degrees, so the net gain or loss of thermal infrared radiation will be at most a few tens of $W\,m^{-2}$ and often much less.

Temperature and humidity gradients close to the surface

In the layer of air close to the ground, heat and water vapor diffuse away from a surface first by (relatively slow) molecular diffusion in the laminar boundary layer close to the surface, and then by (much faster) turbulent mixing of the air farther away from the surface. The temperature of the air (or any other gas) is an expression of the kinetic energy of its molecules, so heat diffuses by the diffusion of faster-moving molecules, and diffusion of heat, water vapor, and other gases follow essentially the same principles. The rate of diffusion of heat through a layer of air is proportional to the temperature difference, and inversely proportional to a diffusion resistance r_H; diffusion of water vapor is proportional to the concentration difference, and inversely to a diffusion resistance r_V (Fick's law of diffusion is analogous to Ohm's law in electricity). The two diffusion coefficients are numerically different, but related. The gradient of temperature and water vapor is steepest in the laminar boundary layer, becoming progressively less steep as turbulent mixing of the air becomes increasingly effective farther from the surface. Some of the effects of energy exchanges at the ground surface, the relation between air and bryophyte temperature wet and dry, and the influence of heat storage by the substratum, are illustrated in Fig. 3.1.

Wind speed and boundary layers

The flow around simple objects in an airstream, such as flat plates, cylinders, or spheres, is relatively well understood. Of course bryophytes and their substrata are not simple objects, but the relationships established for simpler objects provide concepts and orders of magnitude for thinking about bryophyte–atmosphere relationships. In general, the airflow very close to the surface of an object in an airstream is *laminar*, that is, the streamlines are even and parallel to the surface. Farther from the surface, the flow gradually becomes *turbulent*: the streamlines are irregular and form eddies which progressively increase in size and effectiveness at stirring and mixing the air. The average thickness of the laminar boundary layer of a small object is proportional to the square root of the size of the object, and inversely proportional to the square root of windspeed. For an object 1 cm in diameter at a windspeed of 1 m s^{-1} the calculated average thickness of the laminar boundary layer is *c.* 0.6 mm, increasing to *c.* 2 mm at 0.1 m s^{-1} (taken conventionally as "still air"). At still lower airspeeds, laminar conditions extend for several millimeters above the surface, and must prevail in much of the space amongst vegetation and in the vicinity of bryophytes in sheltered forest habitats. In these situations mass transfer of water vapor and carbon dioxide will be predominantly by (slow) molecular diffusion.

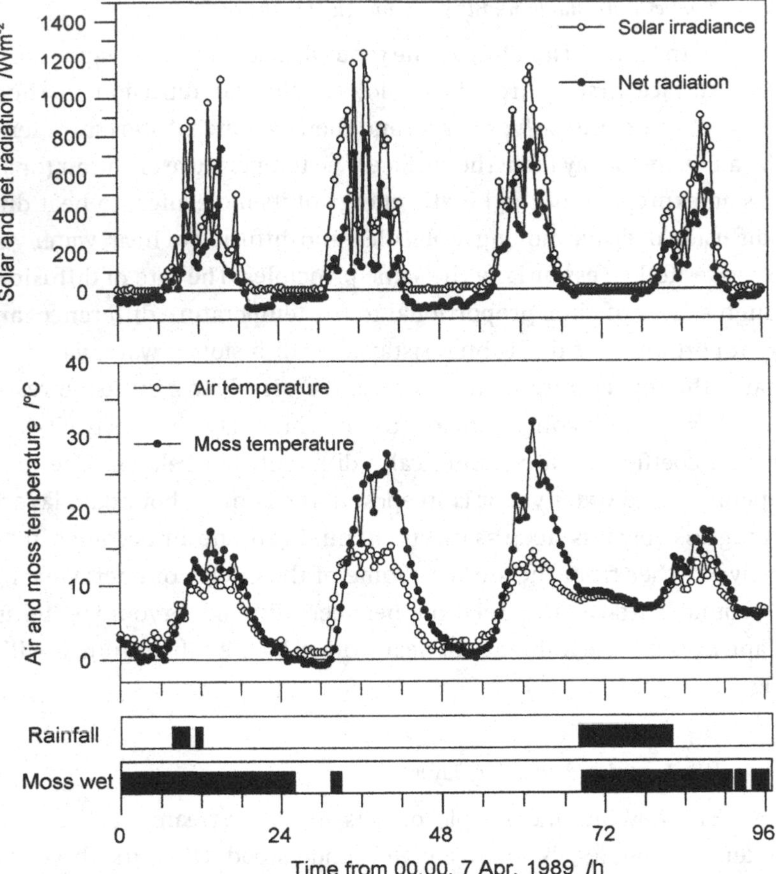

Fig. 3.1. Microclimatic measurements for four days during an unsettled spell of weather from 7–10 April, 1989, in a population of *Grimmia pulvinata* on top of a stone garden wall (*c.* 1.5 m high) at Morchard Bishop, Devon, UK. Clear sunshine and reflection from white clouds give periods when irradiance may substantially exceed 1000 $W\,m^{-2}$, alternating with much lower irradiance when the sun is obscured by clouds. Moss temperature tracks air temperature fairly closely when the moss is wet, a few degrees higher when radiation income is high (e.g., 7 and 10 April). When the moss is dry in bright sunshine its temperature may exceed that of the air by 10–15 °C or more, as on 8–9 April. On clear nights with substantial outgoing net radiation the moss may be several degrees cooler than the air; this led to a ground frost in the early hours of 8 April, with a brief "moss wet" period reflecting dewfall. Note the effect of heat storage by the wall, causing changes of moss temperature to lag behind changes of net radiation and air temperature.

Much of the morphological diversity expressed in bryophyte life forms (Gimingham & Robertson 1950; Gimingham & Birse 1957; Mägdefrau 1962; During 1992; Bates 1998) must reflect trade-offs between limiting water loss and acquiring sufficient carbon dioxide for growth. The complexity of form of bryophytes and their habitats is such that it is generally easier to measure the rate of water loss from a bryophyte under defined conditions, and to calculate the resistance to water loss (and hence effective boundary layer thickness) from that, than to attempt any more direct approach. In the relation of bryophytes to the atmosphere, their immediate surroundings are in effect an integral part of their physiology. Bryophyte shoots and leaves are often comparable in size with the thickness of laminar boundary layers. Consequently, bryophyte shoots and cushions tend to behave as single objects at low wind speeds (Proctor 1979). Beyond a critical windspeed the relation of water loss to velocity steepens; this may be partly due to the onset of turbulence, and partly to the fractally increasing area of the shoot or cushion surface as measured by a laminar boundary layer of progressively diminishing thickness. The diversity of surfaces presented by bryophytes to the atmosphere means that useful analogies may be found from sources as diverse as studies of the coats of animals and effects of wind on field crops. A good introduction to the physical principles involved is given by Gates (1980) and Monteith and Unsworth (1990).

Water loss and the heat budget

The evaporation of water is driven by the concentration difference of water vapor between the air in contact with the wet surface (saturated) and the ambient air. The saturated vapor pressure of water varies with temperature, and for water to evaporate the requisite latent heat must be supplied. What determines the temperature of the bryophyte? And where does the latent heat of evaporation come from? The answer in both cases depends on the balance between the components of the heat budget of the bryophyte. The latent heat of evaporation is supplied by some combination of radiative energy exchange at the surface, conduction from the substrate, and convective transfer from the air (Campbell 1977; Gates 1980; Monteith & Unsworth 1990; Jones 1992). Penman (1948), making some simplifying assumptions, derived an equation to estimate the rate of evaporation from a wet surface:

$$LE = [s(R_n - G) + \rho c_P(\chi_s - \chi)/r_H]/(\gamma^* + s)$$

where E is the rate of evaporation, L is the latent heat of evaporation, R_n is the net radiation balance of the surface, G is storage of heat by the substrate, s is the slope of the saturation vapor-density curve, ρ is the density of air, c_P is the specific heat of air, $\chi_s - \chi$ is the saturation deficit of the ambient air, and γ^* is the apparent psychrometer constant. The quantities r, c_P, s, and γ^* are "constants" which vary somewhat with temperature and can be looked up in tables. R_n, G, $(\chi_s - \chi)$ and r_H are variables which (at least in principle) can be measured.

Penman's equation provides a good starting point to consider the factors affecting water loss and their interactions. The left-hand term in the numerator reflects the supply of heat by radiation or conduction and the right-hand term heat drawn by convective transfer from the air. If the net radiation income is small and heat cannot be drawn from the substratum, the rate of evaporation will be determined mainly by the saturation deficit of the air and the boundary layer diffusion resistance of the bryophyte. The latent heat of evaporation will be drawn mostly from the air, and the bryophyte surface will be cooler than air temperature. In a humid sheltered situation in sun the position is reversed. The right-hand term is now small, and evaporation is determined mainly by the net radiation income; the bryophyte will now be warmer than the air. Evaporation will be at a minimum when net radiation income and saturation deficit are low, and boundary layer resistances are high (implying low windspeed). These conditions are realized *par excellence* in sheltered, shady, humid forests.

Responses of bryophytes to environmental factors

Water relations

The water relations of bryophyte cells do not differ greatly from those of other plants. The osmotic potential of the cell sap varies between about –0.5 MPa (thalloid liverworts of wet habitats) and –2.0 MPa. These figures are in the same range as for the general run of vascular plants (Larcher 1995). Metabolic activity declines as water is lost in much the same way as in vascular plants. Photosynthesis falls to low values at 30%–40% of the water content at full turgor (relative water content, RWC), but can be detected in some species down to a water potential of about –10 MPa (Dilks & Proctor 1979). Respiration continues to somewhat lower water contents. There are two major differences from vascular plants. First, most bryophytes carry greater or lesser amounts of external capillary water, which can vary widely without affecting the water content or water potential of the cells. Second, most bryophytes are at least to some degree desiccation-tolerant. A few big liverworts of wet, shady forests, such as *Monoclea forsteri* (Hooijmeijers 2008), provide a baseline. These are hardly or not at all desiccation-tolerant, and behave in very much the same way as vascular plant mesophyll cells, which are irreversibly damaged if their relative water content falls below *c.* 30%. This corresponds to a water potential of *c.* –5 to –7 MPa, and equilibrium with air at *c.* 95%–96% relative humidity. Some other bryophytes of permanently wet habitats are equally sensitive, but the great majority will stand drying to RWC 10%–15% (equilibrium with air at *c.* 75% RH) for at least a few days, and many are much more tolerant than that.

The external capillary water is exceedingly important to the ecophysiology of bryophytes. Water storage tissues in vascular plants are internal, and loss of stored water is immediately felt as a fall in the RWC and water potential of the cells. Water storage in bryophytes is external, in the capillary spaces of the plant's own structure and of its immediate surroundings. This external capillary water can vary within wide limits, with negligible changes in water potential; it is important in prolonging the time a bryophyte can photosynthesize after rain (Zotz *et al.* 2000; Proctor 2004). The internal cell water is generally only a small fraction of the water associated with a bryophyte. When the external capillary water is exhausted, the cells rapidly dry out and metabolic activity ceases. Consequently, for most of the time, bryophyte cells are *either* fully turgid, *or* dry and metabolically inactive. Water stress, with the cells metabolically active but below full turgor, is generally only transient.

External water storage potentially competes with area for gas exchange; many bryophyte structures seem to be adaptations to reconcile these two functions (Proctor 1979). Water is held around leaf bases and in the felt of rhizoids or other outgrowths around moss stems, leaving the laminae free for gas exchange. Some moss leaves are rough with papillae (e.g., *Encalypta* spp., *Anomodon* spp., *Thuidium* spp., many Pottiaceae). The spaces between these papillae can provide a network of channels for capillary conduction over the leaf surface, while the papilla tops are more or less water-repellent. It is noticeable how these species appear dark and waterlogged after heavy rain, quickly becoming opaque and bright green while retaining full turgor as excess water drains away. Many bryophytes have concave leaves, holding water in their concavities but leaving the convex surface free for gas exchange. Surface tension underlies all these mechanisms. Experimental evidence shows that excess superincumbent water depresses photosynthesis (e.g., Dilks & Proctor 1979; Meyer *et al.* 2008); in the field this is probably only a significant factor after heavy rain or in very wet habitats.

Desiccation tolerance

It cannot be too strongly emphasized that desiccation tolerance in bryophytes (also lichens and vascular "resurrection plants") is qualitatively different from "drought tolerance" as we normally think of it in vascular plants. Drought tolerance is the ability to maintain more or less normal metabolism under water stress in a range which extends at most down to about –7 to –9 MPa. Desiccation tolerance is the ability to survive loss of most of the water content of the cells (so that no liquid phase remains), to exhibit virtually total suspension of metabolic activity ("anhydrobiosis"), and to revive and resume normal metabolism on remoistening.

A great deal has been written about the desiccation tolerance of bryophytes; for reviews, see Proctor and Pence (2002) and Proctor *et al.* (2007b). Many species of desiccation-prone habitats, such as *Tortula (Syntrichia) ruralis*, *Grimmia pulvinata*, or *Schistidium apocarpum*, can withstand drying from full turgor to a water content of 5%–10% of their dry mass in half an hour or less, and after a few hours or days of desiccation recover positive net assimilation within a few minutes, and return to a positive net carbon balance within an hour or less. Recovery becomes slower (and ultimately less complete) with increasing desiccation time. Other species require slower drying, and always recover less rapidly. Some metabolic processes are slower to recover than respiration and photosynthesis. Full return of organelles to their pre-desiccation conformation and spatial relationships, and reassembly of the cytoskeleton, may take 12–24 h or more (Pressel *et al.* 2006, 2009; Proctor *et al.* 2007a). Re-establishment of the cell cycle takes a similar time (Mansour & Hallet 1981), so wet periods long enough to maintain a positive carbon balance may be insufficient for significant translocation or for cell division and growth. There is some evidence of drought hardening and dehardening in a number of species (Abel 1956; Schonbeck & Bewley 1981; Beckett 1999).

In the field, long periods of continuous dryness are surprisingly rare. In a two-year recording of *Grimmia pulvinata* cushions on a wall top in Devon, UK (Proctor 2004), the longest dry periods recorded were 367 h (15.3 d) in April – June 1989, and 268 h (11.2 d) in October – December the same year; the longest average duration of dry periods was in late summer (median 25.0 h, mid-point of logistic fit to data 52.9 h). Long "rainless" periods recorded by meteorological observations are not necessarily continuously dry for bryophytes, which may be moistened for significant times by rain events too small to be recorded, or by dew or cloudwater deposition. Bryophytes of open, sun-exposed sites need to withstand intense desiccation because they are likely to reach temperatures of 50–60 °C for significant periods during the day. In laboratory experiments they often survive dry for longest (like normal seeds) stored at 20%–30% RH. They also need to withstand frequent switches from dry to wet and back again. However, they may not be exposed to the longest periods dry, because unshaded sites are likely to be those of greatest dew deposition when humidity is high. Dewfall at night is rarely more than 0.4 mm, too small (and in the wrong place) to be recorded by routine meteorological measurements, but sufficient to keep at least some bryophytes at full turgor for an hour or two after sunrise (Csintalan *et al.* 2000). Forest species do not face the same intensity of desiccation, but they may need to tolerate long unbroken periods dry. They often survive best stored at 50%–70% RH. Canopy epiphytes in forests characterized by frequent cloud or mist occupy a special position. They experience frequent switches between wet and dry, but probably very rarely experience long dry periods, and they are never exposed to intense desiccation. This is the situation in the cloud forests of the tropical mountains (Wolf 1993; Leon-Vargas *et al.* 2006) and their extension into temperate oceanic regions, well seen at Pico del Ingles and elsewhere on the crest of the Sierra de Anaga, Tenerife (Islas Canarias, Spain).

For any desiccation-tolerant plant, long-term survival must depend critically on maintaining a positive net carbon balance over time. Respiration recovers more quickly than photosynthesis, and is stimulated in the early minutes of rewetting, so every desiccation event leads to an initial net carbon loss, which must be recouped as photosynthesis recovers (Dilks & Proctor 1976; Tuba *et al.* 1996). Alpert and Oechel (1985) and Alpert (1988) explored the significance of this in relation to the establishment and microdistribution of *Grimmia laevigata* on the surfaces of granite boulders, and Zotz *et al.* (2000) considered overall carbon balance in relation to cushion size (and water storage) in *Grimmia pulvinata*. Zotz and Rottenburger (2001) studied diel CO_2 balance through the year in *G. pulvinata*, *Schistidium apocarpum*, and *Syntrichia ruralis*, three common desiccation-tolerant mosses of walls and roofs.

The common supposition that bryophytes (and bryophyte diversity) are favored by high rainfall and high relative humidity *as such* is a misleading half truth. Species of wet habitats such as *Monoclea* spp., *Dumortiera hirsuta*, and *Jubula hutchinsiae* are indeed favored by high rainfall; a low saturation deficit will prolong the time for which bryophytes remain moist and metabolically active, and there is certainly a geographical correlation of high bryophyte diversity with high rainfall. However, for many bryophytes, what matters is *frequency* rather than *amount* of precipitation. This appears to be true of canopy epiphytes in at least some (perhaps all?) cloud forests, where the striking feature of microclimate measurements is not the constancy of high humidity, but the frequency of periods of 100% humidity from cloudwater deposition even in the season of least rainfall (Leon-Vargas *et al.* 2006). Dry periods (<80% RH, sometimes <70% RH), requiring at least some degree of desiccation tolerance, occur at all times of year.

Light and temperature

As with water relations, at the cell level, bryophytes respond to light and temperature in much the same way as vascular plants (C_3 plants in particular; all bryophytes have the C_3 pattern of photosynthesis). Net carbon fixation is the difference between CO_2 uptake by photosynthesis and CO_2 loss by respiration (Fig. 3.2). Respiration increases with temperature, and does so increasingly rapidly as temperature rises; dark respiration (loss of CO_2) is plotted at zero on the

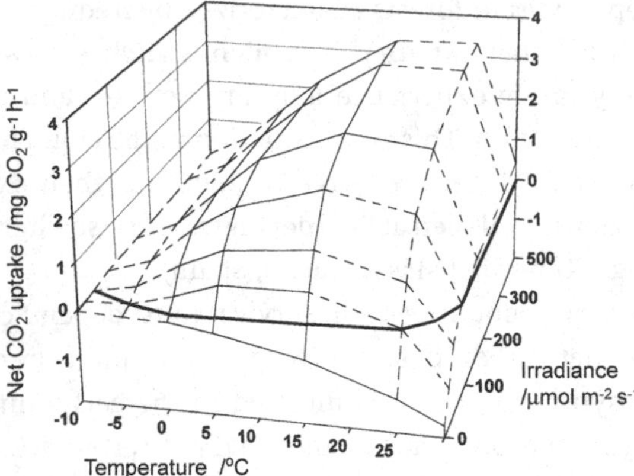

Fig. 3.2. Diagrammatic response of net CO_2 uptake to irradiance and temperature at ambient CO_2 concentration in a common temperate moss, based on the data of Stålfelt (1938) for *Hylocomium splendens*. Irradiances in the original measurements were expressed in kilolux, and have been roughly converted to equivalent quantum flux units. Broken lines are extrapolated beyond the range of the measurements. The bold line shows the compensation point, the line of zero net CO_2 exchange. Note that the temperature optimum for net CO_2 fixation tends to increase with irradiance, and the light saturation level rises with increasing temperature.

irradiance axis. At any particular temperature, as light increases, a point is reached where photosynthesis balances respiration, so there is no net exchange of CO_2. This is the *light compensation point*. Above this, the curve of net CO_2 uptake at first rises steeply, ultimately leveling off at *light saturation*. Both the irradiance at which saturation is reached, and the rate of net CO_2 uptake at saturation, are dependent on temperature. Many mosses maintain modest (sometimes substantial) net CO_2 uptake at or even a few degrees below 0 °C. Limitation of photosynthesis and respiration probably arises mainly from water loss as ice forms outside the cells. As temperature increases, net CO_2 uptake rises to a maximum, before declining again at the highest temperatures, ultimately reaching a point where photosynthesis fails to balance respiration and net CO_2 exchange again becomes negative. Thus there is an upper *temperature compensation point* which sets a limit to the highest temperature at which the bryophyte can grow. As Fig. 3.2 shows, the compensation point is in fact one line varying with irradiance and temperature.

We have very few good data on the light and temperature responses of bryophytes. Figure 3.2 is based on the data of Stålfelt (1938) for *Hylocomium splendens*; Russell (1990) and Proctor (1990) presented diagrams, based on the data of N. J. Collins for *Polytrichum alpestre* (*P. strictum*) and on those of Kallio and Heinonen (1975) for *Racomitrium lanuginosum*. These are both cool climate mosses, for which the data show temperature optima for net CO_2 uptake in the region of 5–10 °C; Ino (1990) found optimum net CO_2 uptake at *c.*10 °C in *Bryum pseudotriquetrum* in East Antarctica, and Davey and Rothery (1997) found optimum net uptake in the region of 10–20 °C in most of the 13 species they studied on the sub-Antarctic Signy Island. Fig. 3.2 is based on the data of Stålfelt (1938) for the common boreal forest floor moss *Hylocomium splendens*, with a temperature optimum near 20 °C, and is probably representative of temperate (and montane tropical) species. In controlled-environment growth experiments (at the low irradiance of 25 W m^{-2}, corresponding to *c.* 50 μmol m^{-2} s^{-1} PPFD), Furness and Grime (1982a, b) found optimum relative growth rate (RGR) in the common pleurocarpous moss *Brachythecium rutabulum* at *c.* 19 °C, but at 5 °C the reduction in RGR was less than 40%. In 40 species of varied ecology, mostly from the surroundings of Sheffield, UK, optimum growth was generally seen between 15 and 25 °C, and in most species there was considerable gain in mass at temperatures below 10 °C. At higher temperatures, performance declined sharply; many species died when kept continuously at temperatures above 30 °C. Many bryophytes can survive short periods at temperatures higher than this, and lethal limits are in the same range as for C_3 vascular plants, *c.* 40–50 °C (Larcher 1995; Liu *et al.* 2004).

Desiccation-tolerant bryophytes can withstand very much higher temperatures dry than wet. Time of survival dry is dependent on temperature. Desiccation-tolerant species such as *Syntrichia ruralis* or *Racomitrium lanuginosum* survive only for minutes at 100 °C, but for many months at 0 °C (Hearnshaw & Proctor 1983). Probably the very long survival times recorded for many species reflect not so much any need for long survival at normal ambient (shade) temperatures, as a response to the intermittent much higher temperatures experienced by dry bryophytes in sun-exposed places, where they may reach 50–60 °C on sunny days.

Light, photosynthesis and CO_2

Bryophytes have been regarded as "shade plants" because they typically show light saturation of photosynthesis at relatively modest irradiance (Marschall & Proctor 2004). Most bryophytes have unistratose leaves; calculation suggests that at current ambient CO_2 concentrations, and assuming a reasonable value for the liquid-phase diffusion resistance between the moist leaf surface and the chloroplasts, the maximum rate at which CO_2 can diffuse into a single surface could be matched by the energy available from a photon irradiance of *c.* 250 μmol m^{-2} s^{-1} – around 14% of full sunlight (Proctor 2005). A unistratose leaf has two surfaces, and leaves generally overlap on bryophyte shoots, so that the "leaf area index" may be *c.* 6 in *Syntrichia ruralis* and *c.* 20 in species with more densely leafy shoots. However, not all of that area is available for gas exchange, and in practice photosynthesis is saturated below 500 μmol m^{-2} s^{-1} in most mosses, and the saturation irradiance exceeds 1000 µmol m^{-2} s^{-1} for only few. This may not be a serious constraint in poikilohydric organisms that mainly photosynthesize in rainy or overcast weather.

However, two groups of bryophytes have evolved ventilated photosynthetic systems analogous to vascular plant leaves. The Polytrichaceae have densely-set photosynthetic lamellae on the upper leaf surface, and in many species photosynthesis saturates at irradiances approaching or even exceeding full sunlight; saturation irradiance is correlated with the effective "mesophyll" area (Proctor 2005). The thick thalli of Marchantiales have systems of chambers housing the photosynthetic tissues (often filaments growing from the floor of the chamber), opening to the dorsal surface by pores. They too include species which reach saturation at irradiances far above the general run of liverworts with unistratose leaves or simple thalli. The Anthocerotae possess, uniquely among archegoniate land plants, a biochemical carbon-concentrating mechanism, which reduces the diffusion limitation on CO_2 uptake, but it cannot rival a ventilated photosynthetic system like a *Marchantia* thallus or a *Polytrichum* or angiosperm leaf in CO_2 acquisition per unit area (Meyer *et al.* 2008).

If CO_2 diffusion limits photosynthesis at high irradiance in many bryophytes, their CO_2 uptake would be expected to respond positively to enhanced CO_2 concentration. The data of Silvola (1985) for mire and forest bryophytes in Finland show that this is indeed so (Fig. 3.3). Other studies (Sveinbjörnsson & Oechel 1992; Jauhiainen *et al.* 1994; van der Heijden *et al.* 2000) show that this is

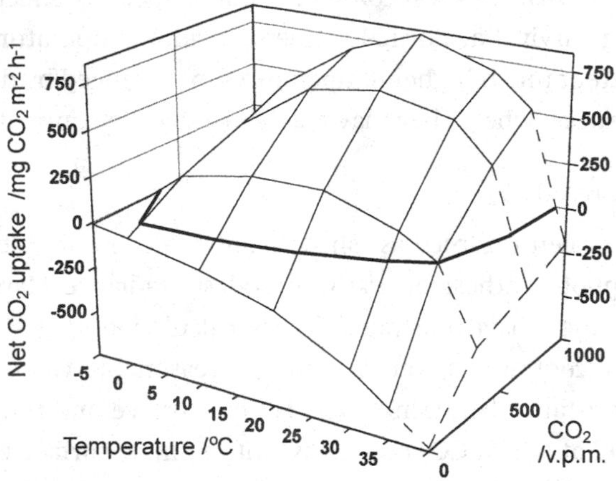

Fig. 3.3. Diagrammatic response of net CO_2 uptake to temperature and CO_2 concentration, at an irradiance (photon flux) of 500 µmol m^{-2} s^{-1}, in a boreal forest moss, based on the data of Silvola (1985) for *Dicranum majus*. Broken lines are extrapolated beyond the range of the measurements. The bold line shows the compensation point. Note that the temperature optimum for net CO_2 fixation tends to increase with rising ambient CO_2.

not translated straightforwardly into increased growth. Other regulatory factors must be at work too, including nutrient and biochemical limitations. As Fig. 3.3 shows, temperature and CO_2 concentration interact in their effect on net CO_2 uptake. At moderate temperatures we may speak of a *CO_2 compensation point*, which for bryophytes is commonly around 100 v.p.m.; at moderate to high ambient CO_2 we can again speak of an upper temperature compensation point. In fact, net CO_2 uptake by a bryophyte follows a single four-dimensional response surface to temperature, irradiance, and CO_2 concentration, and the compensation point is a single line following a course through that 4-D diagram. Unfortunately that is beyond easy visualization for most of us!

Desiccation-tolerant bryophytes in sun-exposed situations must be exposed while turgid and metabolically active to light greatly in excess of that needed for carbon fixation, at least for short intervals. In many (perhaps all) of these species photosynthetic electron flow inferred from chlorophyll fluorescence does not saturate, but continues to rise (often near-linearly) at high irradiance (Marschall & Proctor 2004; Proctor 2009). Apparently similar non-saturating electron flow has been reported by Pannewitz *et al.* (2003) from *Hennediella heimii* in the Antarctic. In the species that have been investigated, CO_2 and O_2 act as interchangeable electron sinks, and the non-saturating component of electron flow is to photoreduction of oxygen. It is always associated with high levels of non-photochemical quenching, and appears to be important in photoprotection (Proctor 2009; M. C. F. Proctor & N. Smirnoff, unpublished data).

Conclusions

Climatic change is nothing new. Many of our familiar bryophyte genera predate the breakup of Gondwanaland late in the Mesozoic. They have lived through immense geographic and climatic changes. In Cretaceous times atmospheric CO_2 concentrations were several times their present level, and temperatures were warmer with less difference between tropical and polar regions. Deciduous forest extended to polar latitudes in both the northern and southern hemispheres, including Antarctica, which was at that time connected to both South America and Australasia. The distribution of land and sea, temperature, and atmospheric CO_2 concentrations have all changed greatly over the past 100 million years, which have been, broadly, a period of declining CO_2 and declining temperature. Atmospheric CO_2 probably reached its lowest ever concentrations (*c.* 200 v.p.m.) at the maxima of the Pleistocene glaciations. At the end of the last glaciation, climatic warming was very rapid at the onset of the Allerød interstadial about 13,000 years ago, and particularly so at the close of the cold Younger Dryas period that followed it. The evidence suggests that at that time (about 10,000 years ago), temperature rose by some 10 °C over a century or two, accompanied by a rapid rise in CO_2 to its post-glacial (pre-industrial) level of *c.* 280 v.p.m. (Wilson *et al.* 2000).

Climatic change did not end with the close of the last glaciation. In Bronze Age northwest Europe, summer temperatures were *c.* 2 °C higher than now, becoming cooler and wetter in Roman, and somewhat warmer and drier in medieval times. From the sixteenth to the nineteenth centuries the climate was cooler again, and this period (the "Little Ice Age") saw the maximum post-glacial extent of the glaciers of the Alps. The only certainty is change itself. We all tend to see the world in which we grew up as "normal", and all that went before as a prelude to it. This is an illusion; the present has no privileged status. It is simply one period, one moment among many in an ever-changing scene.

Identifying the causes of apparent change in bryophyte distributions is never easy, and climatic causes have to be sought against a background of change due to changes in land use and other causes (Söderström 1992). Further, interpreting bryophyte distributional limits in terms of standard climatic data is fraught with problems. The ability of the lichen *Parmotrema endosulphureum* to survive rising temperatures in lowland tropical forests has been questioned (Zotz *et al.* 2003), and the same question might be asked of bryophytes and C_3 vascular plants. This underlines the need for much more data on the temperature responses of tropical bryophytes, and much more field data on tropical forest (and other) microclimates in which these bryophytes grow. The ground layer bryophytes of mesic forests and mires may pose the fewest difficulties in predicting effects of climatic change, because their temperatures are likely to track air (shade) temperatures reasonably closely. Even for these, microclimatic considerations may be important. Microclimate can cut surprisingly far across conventional climatic zones. *Bryum argenteum* and *Ceratodon purpureus* find congenial niches from the Arctic to the Antarctic, and most places in between. Davey and Rothery (1997) measured midday temperatures of 11–12 °C at 10 mm beneath *Chorisodontium aciphyllum* in the sub-Antarctic; many of the temperate bryophytes of Furness and Grime (1982b) spend most of their time photosynthesizing at temperatures little higher that this. The more restrictive limits of many less common species are likely to be set by interactions of water availability and the light and temperature climate, and these will demand a subtler approach than simplistic correlations with standard climatic data.

References

Abel, W. O. (1956). Die Austrocknungsresistenz der Laubmoose. *Österreichische Akademie der Wissenschaften, Mathematisch-Naturwissenschaftliche Klasse, Sitzungsberichte*, Abteilung 1, **165**: 619–707.

Alpert, P. (1988). Survival of a desiccation-tolerant moss, *Grimmia laevigata*, beyond its observed microdistributional limit. *Journal of Bryology* **15**: 219–27.

Alpert, P. & Oechel, W. C. (1985). Carbon balance limits microdistribution of *Grimmia laevigata*, a desiccation-tolerant plant. *Ecology* **66**: 660–9.

Barry, R. G. & Chorley, R. J. (2003). *Atmosphere, Weather and Climate*, 8th edn. London: Routledge.

Bates, J. W. (1998). Is life-form a useful concept in bryophyte ecology? *Oikos* **82**: 223–37.

Beckett, R. P. (1999). Partial dehydration and ABA induce tolerance to desiccation-induced ion leakage in the moss *Atrichum androgynum*. *South African Journal of Botany* **65**: 1–6.

Campbell, G. S. (1977). *An Introduction to Environmental Biophysics*. New York: Springer-Verlag.

Csintalan, Zs., Takács, Z., Proctor, M. C. F., Nagy, Z. & Tuba, Z. (2000). Early morning photosynthesis of the moss *Tortula ruralis* following summer dew fall in a Hungarian temperate dry sandy grassland. *Plant Ecology* **151**: 51–4.

Davey, M. C. & Rothery, P. (1997). Interspecific variation in respiratory and photosynthetic parameters in Antarctic bryophytes. *New Phytologist* **137**: 231–40.

Dilks, T. J. K. & Proctor, M. C. F. (1976). Effects of intermittent desiccation on bryophytes. *Journal of Bryology* **9**: 249–64.

Dilks, T. J. K. & Proctor, M. C. F. (1979). Photosynthesis, respiration and water content in bryophytes. *New Phytologist* **82**: 97–114.

During, H. J. (1992). Ecological classifications of bryophytes and lichens. In *Bryophytes and Lichens in a Changing Environment*, ed. J. W. Bates & A. M. Farmer, pp. 1–31. Oxford: Clarendon.

Furness, S. E. & Grime, J. P. (1982a). Growth rates and temperature responses in bryophytes. I. An investigation of *Brachythecium rutabulum*. *Journal of Ecology* **70**: 513–23.

Furness, S. E. & Grime, J. P. (1982b). Growth rates and temperature responses in bryophytes. II. A comparative study of species of contrasted ecology. *Journal of Ecology* **70**: 525–36.

Gates, D. M. (1980). *Biophysical Ecology*. New York: Springer-Verlag.

Gimingham, C. H. & Birse, E. M. (1957). Ecological studies on growth-form in bryophytes. I. Correlations between growth-form and habitat. *Journal of Ecology* **45**: 533–45.

Gimingham, C. H. & Robertson, E. T. (1950). Preliminary investigations on the structure of bryophyte communities. *Transactions of the British Bryological Society* **1**: 330–44.

Hearnshaw, G. F. & Proctor, M. C. F. (1983).The effect of temperature on the survival of dry bryophytes. *New Phytologist* **90**: 221–8.

Hill, M. O. & Preston, C. D. (1998). The geographical relationships of British and Irish bryophytes. *Journal of Bryology* **20**: 127–226.

Hill, M. O., Preston, C. D. & Smith, A. J. E. (1991–4). *Atlas of the Bryophytes of Britain and Ireland*, Vol.1 (1991), Vol.2 (1992), Vol.3 (1994). Colchester: Harley Books.

Hooijmeijers, C. (2008). Membrane integrity, oxidative damage and chlorophyll fluorescence during dehydration of the thalloid liverwort *Monoclea forsteri* Hook. *Journal of Bryology* **30**: 217–22.

Ino, Y. (1990). Field measurement of net photosynthesis of mosses at Langhovde, East Antarctica. *Ecological Research* **5**: 195–205.

Jauhiainen, J., Vasander, H. & Silvola, J. (1994). Response of *Sphagnum fuscum* to N deposition and increased CO_2. *Journal of Bryology* **18**: 83–96.

Jones, H. G. (1992). *Plants and Microclimate*, 2nd edn. Cambridge: Cambridge University Press.

Kallio, P. & Heinonen, S. (1975). CO_2 exchange and growth of *Rhacomitrium lanuginosum* and *Dicranum elongatum*. In *Fennoscandian Tundra Ecosystems, Vol. 1, Plants and Microorganisms*, ed. F. E. Wiegolaski, pp. 138–48. New York, Heidelberg, Berlin: Springer-Verlag.

Larcher, W. (1995). *Physiological Plant Ecology*, 3rd edn. New York: Springer-Verlag.

Leon-Vargas, Y., Engwald, S. & Proctor, M. C. F. (2006). Microclimate, light-adaptation and desiccation tolerance of epiphytic bryophytes in two Venezuelan cloud forests. *Journal of Biogeography* **33**: 901–13.

Liu, Y., Li, Z., Cao, T. & Glime, J. M. (2004). The influence of high temperature on cell damage and shoot survival rates of *Plagiomnium acutum*. *Journal of Bryology* **26**: 265–71.

Mägdefrau, K. (1962). Life-forms of bryophytes. In *Bryophyte Ecology*, ed. A. J. E. Smith, pp. 45–58. London: Chapman & Hall.

Mansour, K. S. & Hallet, J. N. (1981). Effects of desiccation on DNA synthesis and the cell cycle of the moss *Polytrichum formosum*. *New Phytologist* **87**: 315–24.

Marschall, M. & Proctor, M. C. F. (2004). Are bryophytes shade plants? Photosynthetic light responses and proportions of chlorophyll-*a*, chlorophyll-*b* and total carotenoids. *Annals of Botany* **94**: 593–603.

Meyer, M., Seibt, U. & Griffiths, H. (2008). To concentrate or ventilate? Carbon acquisition, isotope discrimination and physiological ecology of early land plant life forms. *Philosophical Transactions of the Royal Society of London B*. doi: 10.1098/rtb.2008.0039.

Monteith, J. L. & Unsworth, M. H. (1990). *Principles of Environmental Physics*, 2nd edn. Cambridge: Cambridge University Press.

Pannewitz, S., Green, T. G. A., Scheidegger, C., Schlensog, M. & Schroeter, B. (2003). Activity pattern of the moss *Hennediella heimii* (Hedw.) Zand. in the Dry Valleys, Southern Victoria Land, Antarctica during the mid-austral summer. *Polar Biology* **26**: 545–50.

Penman, H. L. (1948). Natural evaporation from open water, bare soil and grass. *Proceedings of the Royal Society of London A* **194**: 120–45.

Pressel, S., Ligrone, R. & Duckett, J. G. (2006). Effects of de- and rehydration on food-conducting cells in the moss *Polytrichum formosum*: a cytological study. *Annals of Botany* **98**: 67–76.

Pressel S., Duckett, J. G., Ligrone, R. & Proctor, M. C. F. (2009). Effects of de- and rehydration in desiccation-tolerant liverworts: a physiological and cytological study. *International Journal of Plant Science* **170**: 182–99.

Preston, C. D. & Hill, M. O. (1997). The geographic relationships of British and Irish vascular plants. *Botanical Journal of the Linnean Society* **124**: 1–120.

Preston, C. D., Pearman, D. A. & Dines, T. D. (2002). *New Atlas of the British and Irish Flora*. Oxford: Oxford University Press.

Proctor, M. C. F. (1979). Structure and ecophysiological adaptation in bryophytes. In *Bryophyte Systematics*, ed. G. C. S. Clarke & J. G. Duckett, pp. 479–509. London: Academic Press.

Proctor, M. C. F. (1981). Diffusion resistances in bryophytes. In *Plants and their Atmospheric Environment*, ed. J. Grace, E. D. Ford & P. G. Jarvis, pp. 219–29. Oxford: Blackwell.

Proctor, M. C. F. (1982). Physiological ecology: water relations, light and temperature responses, carbon balance. In *Bryophyte Ecology*, ed. A. J. E. Smith, pp. 333–81. London: Chapman & Hall.

Proctor, M. C. F. (1990). The physiological basis of bryophyte production. *Botanical Journal of the Linnean Society* **104**: 61–77.

Proctor, M. C. F. (2003). Experiments on the effects of different intensities of desiccation on bryophyte survival, using chlorophyll fluorescence as an index of recovery. *Journal of Bryology* **25**: 215–24.

Proctor, M. C. F. (2004). How long must a desiccation-tolerant moss tolerate desiccation? Some results of two years' data-logging on *Grimmia pulvinata*. *Physiologia Plantarum* **122**: 21–7.

Proctor, M. C. F. (2005). Why do Polytrichaceae have lamellae? *Journal of Bryology* **27**: 221–9.

Proctor, M. C. F. (2009). Physiological ecology. In *Bryophyte Biology*, ed. B. Goffinet & A. J. Shaw, 2nd edn, pp. 239–68. Cambridge: Cambridge University Press.

Proctor, M. C. F., Duckett, J. G. & Ligrone, R. (2007a). Desiccation tolerance in the moss *Polytrichum formosum* Hedw.: physiological and fine-structural changes during desiccation and recovery. *Annals of Botany* **99**: 75–93.

Proctor, M. C. F., Oliver, M. J., Wood, A. J. *et al.* (2007b). Desiccation tolerance in bryophytes: a review. *Bryologist* **110**: 595–621.

Proctor, M. C. F. & Pence, V. C. (2002). Vegetative tissues: bryophytes, vascular resurrection plants and vegetative propagules. In *Desiccation and Survival in Plants: Drying without Dying*, ed. M. Black & H. W. Pritchard, pp. 207–37. Wallingford, UK: CABI Publishing.

Proctor, M. C. F. & Tuba, Z. (2002). Poikilohydry and homoiohydry: antithesis or spectrum of possibilities? *New Phytologist* **156**: 327–49.

Russell, S. (1990). Bryophyte production and decomposition in tundra ecosystems. *Botanical Journal of the Linnean Society* **104**: 3–22.

Schonbeck, M. W. & Bewley, J. D. (1981). Responses of the moss *Tortula ruralis* to desiccation treatments. II. Variation in desiccation tolerance. *Canadian Journal of Botany* **59**: 2707–12.

Silvola, J. (1985). CO_2 dependence of photosynthesis in certain forest and peat mosses and simulated photosynthesis at various actual and hypothetical CO_2 concentrations. *Lindbergia* **11**: 86–93.

Söderström, L. (1992). Invasions and range expansions and contractions of bryophytes. In *Bryophytes and Lichens in a Changing Environment*, ed. J. W. Bates & A. M. Farmer, pp. 131–58. Oxford: Clarendon.

Stålfelt, M. G. (1938). Der Gasaustauch der Moose. *Planta* **27**: 30–60.

Sveinbjörnsson, B. & Oechel, W. C. (1992). Controls on growth and productivity of bryophytes: environmental limitations under current and anticipated conditions. In *Bryophytes and Lichens in a Changing Environment*, ed. J. W. Bates & A. M. Farmer, pp. 77–102. Oxford: Clarendon.

Tuba, Z., Csintalan, Zs. & Proctor, M. C. F. (1996). Photosynthetic responses of a moss, *Tortula ruralis* ssp. *ruralis* and the lichens *Cladonia convoluta* and *C. furcata* to water deficit and short periods of desiccation, and their ecophysiological significance: a baseline study at present CO_2 concentration. *New Phytologist* **133**: 353–61.

van der Heijden, E., Jauhiainen, J. & Silvola, J. (2000). Effects of elevated atmospheric CO_2 concentration and increased nitrogen deposition on growth and chemical composition of ombrotrophic *Sphagnum balticum* and oligo-mesotrophic *Sphagnum papillosum*. *Journal of Bryology* **22**: 175–82.

Wilson, R. C. L., Drury, S. A. & Chapman, J. L. (2000). *The Great Ice Age*. London: Routledge.

Wolf, J. H. D. (1993). Epiphyte communities of tropical montane rain forest in the northern Andes. I Lower montane communities. *Phytocoenologia* **22**: 1–52.

Zotz, G. & Rottenburger, S. (2001). Seasonal changes in diel CO_2 exchange of three Central European moss species: a one-year field study. *Plant Biology* **3**: 661–9.

Zotz, G., Schultz, S. & Rottenburger, S. (2003). Are tropical lowlands a marginal habitat for macrolichens? Evidence from a field study with *Parmotrema endosulphureum* in Panama. *Flora* **198**: 71–7.

Zotz, G., Schweikert, A., Jetz, W. & Westerman, H. (2000). Water relations and carbon gain in relation to cushion size in the moss *Grimmia pulvinata* (Hedw.) Sm. *New Phytologist* **148**: 59–67.

4

Effects of Elevated Air CO_2 Concentration on Bryophytes: a Review

ZOLTÁN TUBA, EDIT ÖTVÖS, AND ILDIKÓ JÓCSÁK

Introduction

The concentration of CO_2 in the atmosphere has been increasing over the past two centuries from about 280 ppm to a present value of 360 ppm, and is expected to reach more than twice the pre-industrial concentration in this century (Houghton *et al.* 1990). Variations in atmospheric CO_2 concentration are nothing new; CO_2 concentrations were higher or lower during some earlier geological periods. Bryophytes as ancient C_3 land plants experienced these changes of CO_2 concentrations in air. What is new is that the present increase is faster than most changes that have taken place in the geologically recent past. Results of research on bryophytes are compared with those on desiccation-sensitive and evolutionarily younger vascular C_3 plants, the most widely investigated group in the field of global change.

Desiccation-tolerant (DT) bryophytes are an important component of the photosynthesizing biomass, including arctic and alpine tundras, temperate, mediterranean and sub/tropical grasslands, and non-arborescent communities of arid and semi-arid habitats (Kappen 1973; Smith 1982; Hawksworth & Ritchie 1993). For example, *Sphagnum* species are globally important owing to their considerable peat-forming ability and their potential impact on global climatic cycles (Gorham 1991; Franzén 1994).

Globally, peatlands are estimated to cover between 3.8 and 4.1 million square kilometers (Charman 2002), equivalent to about 3% of the land surface. Peat accumulation over thousands of years has resulted in a vast store of 450 × 1015 g C (Gorham 1991), which is at least 20% of the global carbon store in terrestrial ecosystems. Despite this, the potential effects of elevated air CO_2 on mosses

Bryophyte Ecology and Climate Change, eds. Zoltán Tuba, Nancy G. Slack and Lloyd R. Stark. Published by Cambridge University Press.

have received much less attention (Tuba *et al.* 1999) than effects on vascular plants (e.g., Körner & Bazzaz 1966; Drake *et al.* 1997).

The importance of mosses lies not only in their biomass production but also in their ability to modify their environment in terms of microclimate, water economy, soil characterics, and decomposition rate (Oechel & Lawrence 1985). Most mosses exist in nutrient-poor habitats, but the mineral demand of most mosses is relatively modest and moss production is commonly not nutrient-limited. In general, mosses have limited source–sink differentiation and they are poikilohydric and evergreen. This means that they often have to face rather limiting environmental factors and to respond immediately to intermittent favorable periods. The present ambient CO_2 concentration can be suboptimal for bryophytes (Tuba *et al.* 1998).

Direct effects of elevated CO_2 may benefit moss species more than vascular plants. Long-term elevated air CO_2 experiments on bryophytes are rare. In general, a warmer, moister, and CO_2-rich future environment may enhance bryophyte growth and result in thicker and denser bryophyte mats on forest floors. However, the net effect of an extended growing season on carbon dynamics is uncertain, since both decomposition and carbon uptake are expected to increase.

In bryophytes, enhanced short-term photosynthesis may or may not be reflected in increased production. Acclimation may result in changes in production, reproductive success, and competitive ability, which in turn can cause a compositional change in the bryophyte communities. Mosses show differences in their acclimation to elevated CO_2 concentration, which may vary from marked down-regulation to upward regulation (Jarvis 1993). Usually, they give an immediate positive response of photosynthesis to elevated CO_2, but the longer-term effect is usually reduced by down-regulation or feedback inhibition of photosynthesis, or other limitations on production and growth (Drake *et al.* 1997).

With regard to current temperature limitations on productivity in boreal ecosystems, it seems certain that soil warming will change the distribution of vegetation including bryophytes in northern latitudes. Differential responses to changing environmental conditions may change the competitive relationships between species.

The relationships between productivity and geographical position of *Sphagnum*-dominated wetlands and climatic parameters were investigated by Gunnarsson (2005). There were interspecific differences in productivity, which could be explained by both phylogeny and microhabitat preferences. The wetter microhabitat carpets and lawns had higher productivity than the drier hummocks. Climatic conditions (mean annual temperature and precipitation), together with geographical factors, were able to explain 40% of the variation in productivity. The most important single factor explaining productivity on a global scale in *Sphagnum*-dominated wetlands was the mean annual temperature. Climatic parameters, together with geographical position, are important for estimating the global patterns of *Sphagnum* productivity, and can be used to estimate productivity changes in *Sphagnum*-dominated wetlands under climatic warming scenarios.

Over the past decade an enormous number of experimental data have become available on the responses of plants to elevated concentrations of CO_2, but investigations on mosses are few. This chapter reviews present knowledge of the ecophysiological responses to elevated air CO_2 of different species of bryophytes and summarizes the results of previous research.

Immediate, short- and medium-term effects of elevated air CO_2

Effects of elevated air CO_2 on mosses have been examined in short- (1–3 months), medium- (4–10 months) and long- (>10 months) term studies. The photosynthesis of DT (desiccation-tolerant) mosses shows an immediate positive response to an elevated CO_2 environment, but in the longer term this tends to be negated to varying degrees by downward acclimation of photosynthesis or other limitations on production and growth (Tuba *et al.* 1999). Since most mosses exist in nutrient-poor habitats, as noted above, this exerts some limitation on their photosynthetic and production responses to elevated atmospheric CO_2 (e.g., Jauhiainen *et al.* 1998). Desiccation tolerance is well developed in mosses (Proctor 1981), which retain their chlorophyll content and photosynthetic apparatus during desiccation (Bewley 1979). Tuba *et al.* (1998) found that plants of *Syntrichia ruralis* exposed to 700 ppm CO_2 responded by increasing CO_2 uptake by more than 30%.

In general, the present ambient CO_2 concentration is suboptimal for mosses; results of short-term laboratory experiments show a positive response to elevated CO_2. The immediate effect of exposure to high CO_2 levels is to increase net photosynthesis of mosses dramatically, and this may more than compensate for low photosynthetically active radiation (Silvola 1985).

When exposure to elevated CO_2 lasts for 20–40 days, the primary advantageous effects start to decline (Jauhiainen & Silvola 1996) and signs of negative, downward acclimation of photosynthetic system in mosses appear (Drake *et al.* 1997). Moss production is commonly not nutrient-limited (Oechel & Svejnbjörnsson 1978). Despite this, N addition was the only controlling factor in the biomass production of various *Sphagna* in the experiments of Jauhiainen *et al.* (1998); simultaneous exposure to elevated CO_2 had a negligible effect. Baker and Boatman (1990) detected synergism between nutrient addition (P, K, N) and high CO_2 concentration in *Sphagnum cuspidatum*. Overdieck (1993) described phosphorus as the main limiting nutrient for higher plants under elevated air CO_2 concentration. In this view, these plants can tolerate some decrease in concentrations of all minerals but phosphorus. More phosphorus is required to enable increased flux of carbon through the photosynthetic carbon reduction cycle. *Sphagna* generally receive their nutrients from atmospheric sources; in the study of Jauhiainen *et al.* (1998), *Sphagnum* failed to respond positively to extra N. It would be worthwhile to test other, more minerotrophic bryophyte species.

In the study of Van der Heijden *et al.* (2000a), the peat moss *Sphagnum recurvum* was grown for six months in controlled environments fumigated with air containing 350 and 700 ppm CO_2. Elevated CO_2 consistently reduced rates of dark respiration by 40%–60%, thus providing the CO_2-enriched plants with greater carbon supplies to support their growth and other physiological processes, relative to control plants exposed to ambient air. However, elevated CO_2 only stimulated dry mass production in plants that were simultaneously subjected to the lowest of three nitrogen treatments. In a similar four-month study conducted by Van der Heijden *et al.* (2000b), elevated CO_2 increased total plant biomass in *Sphagnum papillosum* by 70% and elevated nitrogen increased it by 53%. However, neither elevated CO_2 nor elevated nitrogen induced a growth response in *Sphagnum balticum*.

The rate of net photosynthesis in *Sphagnum fuscum* was measured during 50–122 days, and subsequently during short-term (1/2 h) exposure to 350, 700, 1000, or 2000 ppm CO_2 concentrations (Jauhiainen & Silvola 1999). Raised CO_2 concentrations caused a general increase in the rate of net photosynthesis, increasing the rate of photosynthesis at light saturation and causing a given rate of net CO_2 exchange to be reached at lower light fluxes. The relative increase in the rate of net photosynthesis by increasing radiation intensity was independent of the CO_2 treatment. The rates of net photosynthesis at enhanced CO_2 concentrations gradually decreased compared with rates found with the 350 ppm treatment; this acclimation was also noticed during short-term exposure to all four CO_2 concentrations. At 2000 ppm CO_2, the depression of net photosynthesis at high water contents (found at lower CO_2 concentrations) was removed. Observed rates of net photosynthesis indicated that water-use efficiency of *Sphagnum* was not coupled with constant long-term CO_2 concentrations.

Sphagnum recurvum var. *mucronatum* was grown at 360 (ambient) and 700 ppm (elevated) atmospheric CO_2 in combination with different nitrogen deposition rates (6, 15, 23 g N m^{-2} per year), in a short- and long-term growth chamber experiment (Van der Heijden *et al.* 2000a). After 6 months, elevated atmospheric CO_2 in combination with the lowest nitrogen deposition rate increased plant dry mass by 17%. In combination with a high nitrogen deposition rate, biomass production was not significantly stimulated. At the start of the experiment, photosynthesis was stimulated by elevated atmospheric CO_2, but was down-regulated to control levels after three days of exposure. Elevated CO_2 substantially reduced dark respiration, which resulted in a continuous increase in soluble sugar content in capitula. Doubling atmospheric CO_2 reduced total nitrogen content in capitula but not in stems at all nitrogen deposition rates. Reduction in total nitrogen content coincided with a decrease in amino acids, but soluble protein levels remained unaffected. Elevated CO_2 induced a substantial shift in the partitioning of nitrogen compounds in capitula. Soluble sugar concentration was negatively correlated with total nitrogen content, which implied that the reduction in amino acid content in capitula exposed to elevated CO_2 might be caused by the accumulation of soluble sugars. Growth was not stimulated by increased nitrogen deposition. High nitrogen deposition, resulting in a capitulum nitrogen content in excess of 15 mg g^{-1} dry mass, was detrimental to photosynthesis, reduced water content, and induced necrosis.

Decrease in Rubisco content as a response to CO_2 enrichment can explain well the fall of the light-saturated rate of photosynthesis. As to the mosses, there was no correlation between photosynthesis and growth responses in the experiments of Jauhiainen *et al.* (1998). The increased amount of photosynthates enlarged the non-structural carbohydrate pool in *Sphagnum papillosum* and *S. balticum* at elevated CO_2 concentration. This phenomenon is a well-established indication of insufficient sink strength (Drake *et al.* 1997). Mosses can be characterized by little assimilate translocation and little sink differentiation. However, the control by end-product or sink-demand cannot be generalized to all bryophytes, as Svejnbjörnsson and Oechel (1992) found a positive correlation between total non-structural carbohydrate pool and maximum photosynthetic rates in mosses in Alaska.

The effects of raised CO_2 and increased atmospheric N deposition on growth of *Sphagnum* and other plants were studied in bogs at four sites across Western Europe (Berendse *et al.* 2001). Contrary to expectations, elevated CO_2 did not significantly affect *Sphagnum* biomass growth. Increased N deposition reduced *Sphagnum* mass growth because it increased the cover of vascular plants and the tall moss *Polytrichum strictum*. Such changes in plant species composition may decrease carbon sequestration in *Sphagnum*-dominated bog ecosystems.

Chlorophyll fluorescence, CO_2 assimilation, and dark respiration were studied by Takács *et al.* (2004) in heavy metal-treated (Cd, Pb) *Syntrichia ruralis* growing at elevated CO_2 (700 ppm) and present-day CO_2 (350 ppm) concentrations in OTCs (open top chambers) for one month. Photosynthetic quantum yield of PSII (F_v/F_m) decreased in reponse to both heavy metals. The effect of the Cd treatment was statistically significant only at the high CO_2 level. Chlorophyll "yield" values decreased in reponse to Pb and mixed (Cd, Pb) treatment, and reached zero in the Cd treatment. There were no significant differences between CO_2 treatments in any of the fluorescence parameters. CO_2 gas exchange measurements revealed that the elimination of photosynthetic activity due to heavy metal treatment was independent of the CO_2 level. According to Takács *et al.* (2004), future high CO_2 concentrations might be beneficial for CO_2 assimilation by the desiccation-tolerant mosses with physiological levels of heavy metals by improving their carbon balance. Elevated CO_2 concentrations could partly ameliorate the deleterious effects of heavy metal stress as well, but plant responses in the high CO_2 concentration to the heavy metals were very diverse.

In *Syntrichia ruralis*, during desiccation from the fully wetted state, net maximum assimilation rate was achieved at a tissue water content of about 60%, and thereafter the assimilation fell to zero after 140 minutes at present CO_2 and 160 minutes at elevated CO_2 (Tuba *et al.* 1998). The rate of assimilation at elevated CO_2 was 25%–35% enhanced at 60% water content, and during the whole desiccation period the assimilation was increased by elevated CO_2 from 83 to 140 mmol kg^{-1} dry mass in the moss.

In another experiment, Z. Tuba *et al.* (unpublished) studied net CO_2 assimilation, slow chlorophyll fluorescence (Kautsky effect) and carbohydrate responses to six months' exposure to high CO_2 (467, 583, 700 μmol mol^{-1}) in *Polytrichum formosum*. The initial slope of the $A/c_{internal}$ curve was lower in all of the high CO_2 treatments compared with the controls. This reduction in slope, indicating a decrease in Rubisco capacity, was higher in the 583 and 700 μmol mol^{-1} treatments than in the 467 μmol mol^{-1} treatment. At a c_i corresponding to about 700 μmol mol^{-1} CO_2, net CO_2 assimilation was higher in the 467 μmol mol^{-1} plants than in the ambient plants; there was no change in the 583 μmol mol^{-1} treatment and assimilation was reduced in the 700 μmol mol^{-1} treatment.

Photochemical activity (variable fluorescence decrease ratio, Rfd, at 690 nm and 735 nm) was significantly reduced in the 583 and 700 μmol mol^{-1} treatments. The reduction was probably caused by the downward acclimation of the light processes. There was no change in the chlorophyll a+b content in the elevated CO_2 treatments. The starch content of the mosses increased in the CO_2 treatments in the order of 467, 583, and 700 μmol mol^{-1}. The acclimation of photosynthesis was downward in all three treatments: the least so in the plants grown at 467 μmol mol^{-1} and increasingly so at 583 and 700 μmol mol^{-1}

CO_2. The responses to long-term elevated CO_2 in the moss *P. formosum* did not differ from those reported for vascular plants. This suggests that the role of the stomata in acclimation is negligible.

Pannewitz *et al.* (2005) provided a comprehensive matrix for photosynthesis and major environmental parameters for three dominant Antarctic moss species (*Bryum subrotundifolium, B. pseudotriquetrum* and *Ceratodon purpureus*). They determined their short-term photosynthetic responses to as much as a 5.5-fold increase in atmospheric CO_2 concentration (a multi-step increase from *c.*360 ppm to *c.*2000 ppm) at various light intensities and air temperatures. These measurements revealed that the net photosynthetic rate of these moss species showed a large response to increase in CO_2 concentration and this rose with increase in temperature. In *B. subrotundifolium*, net photosynthesis saturated above *c.*1000 ppm CO_2, but *B. pseudotriquetrum* showed no saturation up to 2000 ppm, particularly at 20 °C, which was the highest temperature studied. More specifically, Pannewitz *et al.* (2005) reported that at 2000 ppm CO_2 net photosynthetic rates for *B. subrotundifolium* were 60%–80% higher than at the accepted ambient level of 360 ppm, and net photosynthesis of *B. pseudotriquetrum* was more than doubled.

Long-term responses

For long exposure of plants to higher than present-day CO_2 concentration, five techniques have been developed: open top chamber (OTC; Last 1986), closed chamber (Körner & Arnone 1992; Payer *et al.* 1993) free air CO_2 enrichment (FACE; Rogers *et al.* 1992), solardome (Rafael *et al.* 1995) and wind tunnel (Soussana & Loiseau 1997).

The use of open top chambers (OTCs), to be placed in the field to enclose portions of ecosystems, was a first step toward less artifactual experiments. OTCs (Ashenden *et al.* 1992; Leadley & Drake 1993) were mainly developed for air pollution research. Chambers are usually cylindrical or polygonal metallic frames, fixed to the soil and with supporting transparent plastic coverings. They enclose a volume of air; a frustum at the chamber top may reduce the entering of external air. Air is blown in the chambers by a fan, feeding a perforated pipe that encircles the base of the chambers. CO_2 is added to the air flow, using different tools to improve the mixing of pure CO_2 in the air flow (Tuba *et al.* 2003). Tuba *et al.* (2002) developed a cryptogam (lichens and mosses) fumigation system from the OTC system of Ashenden *et al.* (1992), which was successfully used in grassland experiments. The main difference between cryptogams and higher plants is that cryptogams are poikilohydric, and this needs to be taken into consideration when exposing cryptogams to elevated atmospheric CO_2 (Tuba *et al.* 2002). Tuba *et al.* (2002) placed cryptogams into OTCs with their original substratum. As to the FACE system, it eliminated the problems arising from the altered microclimate within the chamber, since CO_2 was released by valves or by perforated plastic tubes into the air around the experimental plots. Without walls, however, the vertical profile of the CO_2 concentration was much steeper. Therefore, in case of epiphytic cryptogams OTCs were preferred. As wind velocity and CO_2 concentration above the plot could change from second to second, this control system had to react much faster.

In the long run, photosynthetic apparatus seems to be even more influenced by elevated CO_2. *Polytrichum formosum* was exposed to four different CO_2 levels for 11 months in open top chambers (Csintalan *et al.* 1995). In this case, even photochemical activity, as reflected by relative fluorescence decrease ratio (Rfd), was negatively affected by the two highest CO_2 concentrations (700 and 683 ppm). The higher the CO_2 concentration applied, the greater the decrease in chlorophyll a+b content, as indicated by the ratio of fluorescence intensity F690:F730. Obviously, the more common signs of downward regulation could also be detected: lowered Rubisco capacity and elevated content of soluble sugars and starch.

Most surprisingly, the ectohydric moss *Syntrichia ruralis* did not display any signs of acclimation in photosynthesis after four months of exposure to 700 ppm CO_2 (Csintalan *et al.* 1997). In addition to this, elevated CO_2 concentration resulted in a slightly but significantly higher net photosynthesis rate from the fifteenth minute onwards in the rehydration period in the moss. The difference between the responses of *P. formosum* and *S. ruralis* is difficult to explain, but may be related to the more elaborate morphology of the *Polytrichum*. This discrepancy between the behavior of *P. formosum* and that of *S. ruralis* is a warning against uncritically treating all mosses as a uniform group.

Although OTCs minimize the influence of experimental conditions on the vegetation and ensure both a sufficient control of CO_2 concentration at an acceptable investment and running cost, they still suffer from artifacts. FACE systems have proven to be suitable for exposing plants to elevated CO_2 concentrations with minimal disturbance of their natural environment and have been successfully used (Miglietta *et al.* 2001).

MiniFACE systems represent a further symplification of the FACE concept, and were built for the fumigation of (for example) grasslands (Miglietta *et al.* 2001). Each miniFACE ring consists of a horizontal and circular plenum resting on the soil; air is injected by a blower, and is vented through small holes from the plenum itself, or from vertical pipes (Tuba *et al.* 2003). CO_2 injection is regulated by a series of flow controllers (one per ring), operated by a control algorithm using both CO_2 concentrations measured in the center of the ring and wind speed data (Lewin *et al.* 1992).

Miglietta *et al.* (2001) investigated the effects of elevated CO_2 on the net exchange of CO_2 between bogs and the atmosphere, and on the biodiversity of bog communities at five climatically different sites across Europe, using a miniFACE design. A major challenge to investigate the effects of elevated CO_2 on ecosystems is to apply elevated CO_2 concentrations to growing vegetation without changing the physical conditions such as climate and radiation. Most available CO_2 enrichment methods disturb the natural conditions to some degree, as for instance closed chambers or OTCs. Their results showed that increased wind speeds improved the miniFACE system temporal performance. Spatial analyses showed no apparent CO_2 gradients across a ring during a 4 day period and the mean differences between each sampling point and the center of the ring did not exceed 10%. Observations made during a windy day, causing a CO_2 concentration gradient, and observations made during a calm day, indicated that short-term gradients tend to average out over longer periods of time. On a day with unidirectional strong winds, CO_2 concentrations at the upwind side of the ring center were higher than those made at the center and at the downwind side of the ring center, but the bell-shaped distribution was found to be basically the same for the center and the four surrounding measurement points, implying that the short-term (1 s) variability of CO_2 concentrations across the miniFACE ring is almost the same at any point in the ring. Based on gas dispersion simulations and measured CO_2 concentration profiles, the possible interference between CO_2-enriched and control rings was found to be negligible beyond a center-to-center ring distance of 6 m.

In a miniFACE study, Heijmans *et al.* (2001) maintained peat moss monoliths (dominated by *Sphagnum magellanicum*) in 1 m diameter circular plots at atmospheric CO_2 concentrations of 360 and 560 ppm for three growing seasons. After three years, elevated CO_2 was observed to have increased the height and dry mass of this *Sphagnum* species by 36% and 17%, respectively. In contrast, Mitchell *et al.* (2002) established miniFACE plots in a cutover bog dominated by *Polytrichum strictum* and *Sphagnum fallax* and reported that a 210 ppm increase in the air's CO_2 concentration reduced the total biomass of these species by 17% and 14%, respectively. These miniFACE experiments suggest that atmospheric CO_2 enrichment can increase or decrease the growth of certain moss species. Moreover, in a set of miniFACE experiments conducted in bog environments located in four European countries, no CO_2-induced growth effects on non-vascular plants were reported (Berendse *et al.* 2001; Hoosbeek *et al.* 2001), in spite of the fact that some of the tested bog communities contained the CO_2-responsive *Sphagnum papillosum* and *Sphagnum magellanicum* species. However, elevated nitrogen deposition reduced *Sphagnum* growth at two of the study sites (Berendse *et al.* 2001), while it enhanced the production of *Polytrichum* in one study site (Mitchell *et al.* 2002).

The competition between *Sphagnum magellanicum* and a vascular plant (*Eriophorum angustifolium*) as affected by raised CO_2 and increased N deposition was studied in a greenhouse experiment by exposing whole peat sections with monocultures and mixtures of *Sphagnum* and *Eriophorum* to ambient (350 ppm) or raised (560 ppm) atmospheric CO_2 concentrations, combined with low or high N deposition (Heijmans *et al.* 2002). Growth of the two species was monitored for three growing seasons. Raised CO_2 and/or increased N deposition did not change the competitive relationships between *Sphagnum* and *Eriophorum*, but rather had independent effects. Raised CO_2 had a positive effect both on *Sphagnum* and *Eriophorum* biomass, although with *Eriophorum* the effect was transient, probably because of P limitation.

Heijmans *et al.* (2008) developed a bog ecosystem model that includes vegetation, carbon, nitrogen and water dynamics. Two groups of vascular plant species and three groups of *Sphagnum* species competed with each other for light and nitrogen. The model was tested by comparing the outcome with long-term historic vegetation changes in peat cores. A climate scenario was used to analyze the future effects of atmospheric CO_2, temperature, and precipitation. The main changes in the species composition since 1766 were simulated by the model. Simulations for a future warmer, and slightly wetter, climate with doubling CO_2 concentration suggested that little future change in species composition will occur. This prediction was due to the contrasting effects of increasing temperatures (favoring vascular plants) and CO_2 (favoring *Sphagnum*). Further analysis of the effects of temperature showed that simulated carbon sequestration is negatively related to vascular plant expansion. Model results showed that increasing temperatures may still increase carbon accumulation at cool, low N deposition sites, but will decrease carbon accumulation at high N deposition sites. The effects of temperature, precipitation, N deposition and atmospheric CO_2 are not straightforward, but interactions between these components of global change exist. These interactions are the result of changes in vegetation composition. When analyzing long-term effects of global change, vegetation changes should be taken into account and predictions should not be based on temperature increase alone.

Natural CO_2 springs enable studies of CO_2 effects in plants and plant communities to investigate long-term and evolutionary effects of elevated CO_2 on plant physiology (Raschi *et al.* 1997). In geothermal areas, natural CO_2 vents are not uncommon; in some cases, they have exposed portions of terrestrial ecosystems to elevated CO_2 for centuries or even millennia. This offers the chance to study long-term plant adaptations as well as interspecies competition and changes in ecosystem biogeochemistry. It cannot be denied that CO_2 springs are not perfect experimental tools, in consequence of fluctuations in CO_2 concentration (Tuba *et al.* 2003). Moreover, in most of the gas vents the presence of sulphur pollutants can strongly affect plant response to elevated CO_2.

Csintalan *et al.* (2005) measured the net CO_2 assimilation rates of five moss species growing in the vicinity of a natural CO_2-emitting spring near Laiatico (Toscana, Italy). Apart from CO_2, the vents emitted small amounts of H_2S, but its concentration never exceeded 0.04 μmol mol^{-1}, which was not considered harmful. Mosses around the CO_2 spring are exposed to daytime CO_2 concentration of about 700 ppm throughout the year with short-term variations between 500 and 1000 μmol mol^{-1} depending on wind speed and convective turbulance. The researchers made their measurements at a CO_2 concentration of 350 ppm, in order to determine the nature of any photosynthetic acclimation. This work revealed that the CO_2 assimilation was higher in the native CO_2 vent species. The net CO_2 assimilation rates of the plants that had been exposed to the nominally doubled atmospheric CO_2 concentration for their entire lives (but that were measured at a CO_2 concentration of 350 ppm) were 42% greater (*Ctenidium molluscum*), 44% greater (*Hypnum cupressiforme*), 49% greater (*Pseudoscleropodium purum*), 80% greater (*Pleurochaete squarrosa*), and 85% greater (*Platygyrium repens*) than the net CO_2 assimilation rates of plants of the same species that were both grown and measured at 350 ppm CO_2. Contrary to the results of many shorter-term elevated-CO_2-exposure experiments, they determined that native mosses showed upward acclimation, indicative of an extremely positive long-term photosynthetic adjustment of the five species to life-long atmospheric CO_2 enrichment. The results suggested that the long-term adjustment of photosynthetic capacity to elevated atmospheric CO_2 may be diverse and strongly species-dependent. Their data demonstrated that the photochemical reactions of photosynthesis probably play a less important role in the acclimation of mosses to elevated CO_2; however, the CO_2 assimilation responds more sensitively.

E. Ötvös *et al.* (unpublished) investigated the heavy metal content of mosses growing under elevated air CO_2 concentrations in OTCs and around natural CO_2 vents. Five moss species were collected near a natural CO_2 source (about 700–1000 ppm) in 'Slovenske gorice' (Slovenia), in Laiatico (Italy), and from control sites, where CO_2 concentration was measured at 350 ppm. Ten moss species were transplanted into OTCs, where CO_2 concentration was maintained at 700 ppm. Control moss species were grown at 360 ppm CO_2. In most cases an increased photosynthesis at elevated CO_2 caused increased dry matter production, which was indicated by the increased carbon content.

This had an influence on heavy metal contents in the investigated mosses. As the production increased, heavy metal concentration decreased in 11 moss species. As a consequence it could be concluded that in the future, the increased air CO_2 concentrations may strongly influence the elemental contents in mosses, as heavy metal concentrations can be relatively decreased and "diluted" because of the increased dry matter production.

Conclusions

In mosses, photosynthesis shows an immediate positive response to elevated CO_2, but in the longer term this tends to be negated to downward acclimation of photosynthesis or other limitations on production and growth. Most mosses live in nutrient-poor habitats and this exerts some limitation on their photosynthetic and production responses to elevated atmospheric CO_2. As mosses usually have limited source–sink differentiation and they are poikilohydric and evergreen, they often have to face suboptimal environmental factors and to react immediately to intermittent favorable periods. There are indications that elevated CO_2 favors mosses most when their water content is too low or too high for positive net carbon assimilation at present CO_2. Broad biogeochemical considerations predict that rising atmospheric CO_2 should result in faster net (photosynthetic) transfer of carbon from the atmosphere to the biosphere (Schlesinger 1997). Some results show that the effects of temperature, precipitation, N deposition, and atmospheric CO_2 are not straightforward, but interactions between these components of global change exist. These interactions are the result of changes in vegetation composition. Short-term laboratory experiments show a positive response to elevated CO_2. When exposure to elevated CO_2 lasts for a prolonged time (20–40 days), the primary advantageous effects start to decline and signs of negative, downward acclimation of photosynthetic system in mosses can also be detected. A downward regulation occurs if plants grown under elevated CO_2 concentration assimilate less CO_2 than controls brought promptly under high CO_2. Decreases in maximum assimilation capacity are usually due to decreases in Rubisco level and/or activity.

When analyzing long-term effects of global change, vegetation changes should be taken into account and predictions should not be based on temperature increase alone (Heijmans *et al.* 2008). The different acclimation may result in different production, reproductive success, and competitive ability, which in turn can cause a compositional change in the bryophyte communities. For example, the downward acclimation in *Syntrichia ruralis* in the semidesert temperate grasslands indicates that its dominance in a future high-CO_2 climate may be reduced and other, less frequent species may increase through their upward acclimation (Takács *et al.* 2004).

The long-term adjustment of photosynthetic capacity to elevated atmospheric CO_2 may be diverse among species with similar characteristics and be strongly species dependent. The photochemical reactions of photosynthesis probably play less important roles in the acclimation of mosses to elevated CO_2; however, CO_2 assimilation responds more sensitively. Moss species may apparently respond in a different way from vascular plants to CO_2 enrichment (Csintalan *et al.* 2005).

Small terricolous bryophytes may experience variable and probably somewhat elevated levels of CO_2 close to the soil surface derived from root and microbial respiration. DT bryophytes photosynthesize and grow only when they are wet. Since the desiccation–rehydration cycle itself causes profound changes in the photosynthetic activity of these plants (Tuba *et al.* 1994), the effects of increased atmospheric CO_2 concentration on their carbon balance and growth need not necessarily parallel those of vacular plants (Csintalan *et al.* 1997).

Much more experimental evidence from long-term experiments would be needed to make a confident forecast of the responses of mosses that may be expected by the end of the century with expected even higher levels of CO_2 in the atmosphere.

Acknowledgements

The support of the Hungarian Széchenyi Agroecology (OM-3B/0057/2002), the Italian–Hungarian (Project No. 1–71/99), South-African–Hungarian (DAF-9/1998; DAF-11/2001), Portuguese–Hungarian (PORT-12/2001), Slovenian–Hungarian (SLO-10/2001), and Indian–Hungarian (IND-11/2001) Bilateral Intergovernmental Science & Technology Co-operation Projects is gratefully acknowledged.

References

Ashenden, T. W., Baxter, R. & Rafarel, C. R. (1992). An inexpensive system for exposing plants in the field to elevated concentrations of CO_2. *Plant Cell and Environment* **15**: 365–72.

Baker, R. G. & Boatman, D. J. (1990). Some effects of nitrogen, phosphorus, potassium and carbon dioxide concentration on the morphology and vegetative reproduction of *Sphagnum cuspidatum* Ehrh. *New Phytologist* **116**: 605–11.

Berendse, F., Van Breemen, N., Rydin, H. *et al.* (2001). Raised atmospheric CO_2 levels and increased N deposition cause shifts in plant species composition and production in *Sphagnum* bogs. *Global Change Biology* **7**: 591–8.

Bewley, D. J. (1979). Physiological aspects of desiccation tolerance. *Annual Review of Plant Physiology* **30**: 195–238.

Charman, D. (2002). *Peatlands and Environmental Change*. Chichester: John Wiley & Sons.

Csintalan, Zs., Tuba, Z. & Laitat, E. (1995). Slow chlorophyll fluorescence, net CO_2 assimilation and carbohydrate responses in *Polytrichum formosum* to elevated CO_2 concentrations. In *Photosynthesis from Light to Biosphere*, ed. P. Mathis, Vol. V, pp. 925–8. Dordrecht: Kluwer Academic Publishers.

Csintalan, Zs., Takács, Z., Tuba, Z. *et al.* (1997). Some ecophysiological responses of a desiccation tolerant grassland lichen and moss under elevated CO_2: preliminary findings. *Abstracta Botanica* **21**: 309–15.

Csintalan, Z., Juhász, A., Benkő, Z., Raschi, A. & Tuba, Z. (2005). Photosynthetic responses of forest-floor moss species to elevated CO_2 level by a natural CO_2 vent. *Cereal Research Communications* **33**: 177–80.

Drake, B. G., González-Meler, M. A. & Long, S. P. (1997). More efficient plants, a consequence of rising tropospheric CO_2? *Annual Review of Plant Physiology and Plant Molecular Biology* **48**: 607–37.

Franzén, L. G. (1994). Are wetlands the key to the ice-age cycle enigma? *Ambio* **23**: 300–8.

Gorham, E. (1991). Northern peatlands: role in the carbon cycle and probable responses to climatic warming. *Ecological Applications* **1**: 182–95.

Gunnarsson, U. (2005). Global patterns of *Sphagnum* productivity. *Journal of Bryology* **27**: 269–79.

Hawksworth, D. L. & Ritchie, J. M. (1993). *Biodiversity and Biosystematic Priorities: Microorganisms and Invertebrates*. Wallingford: CAB International.

Heijmans, M. M. P. D., Berendse, F., Arp, W. J. *et al.* (2001). Effects of elevated carbon dioxide and increased nitrogen deposition on bog vegetation in the Netherlands. *Journal of Ecology* **89**: 268–79.

Heijmans, M. M. P. D., Klees, H. & Berendse, F. (2002). Competition between *Sphagnum magellanicum* and *Eriophorum angustifolium* as affected by raised CO_2 and increased N deposition. *Oikos* **97**: 415–25.

Heijmans, M. M. P. D., Mauquoy, D., van Geel, B. & Berendse, F. (2008). Long-term effects of climate change on vegetation and carbon dynamics in peat bogs. *Journal of Vegetation Science* **19**: 307–20.

Hoosbeek, M. R., van Breeman, N., Berendse, F. *et al.* (2001). Limited effect of increased atmospheric CO_2 concentration on ombrotrophic bog vegetation. *New Phytologist* **150**: 459–63.

Houghton, J. T., Jenkins, G. J. & Ephramus, J. J. (1990). *Climate Change: The IPCC Scientific Assessment*. Cambridge: Cambridge University Press.

Jarvis, P. G. (1993). Global change and plant water relations. In: *Water Transport in Plants under Climatic Stress*, ed. M. Borghetti, J. Grace & A. Raschi, pp. 1–13. Cambridge: Cambridge University Press.

Jauhiainen, J. & Silvola, J. (1996). The effect of elevated CO_2 concentration on photosynthesis of *Sphagnum fuscum*. In: *Northern Peatlands in Global Climatic Change*, ed. R. Laiho, J. Laine & H. Vasander, pp. 23–9. Helsinki: Academy of Finland.

Jauhiainen, J., Silvola, J. & Vasander, H. (1998). Effects of increased carbon dioxide and nitrogen supply on mosses. In *Bryology for the Twenty-first Century*, ed. J. W. Bates, N. W. Ashton & J. G. Duckett, pp. 343–60. Leeds: Maney Publishing and the British Bryological Society.

Jauhiainen, J. & Silvola, J. (1999). Photosynthesis of *Sphagnum fuscum* at long-term raised CO_2 concentrations. *Annales Botanici Fennici* **36**: 11–19.

Kappen, L. (1973). Response to extreme environments. In *The Lichens*, ed. V. Ahmadjian & M. E. Hale, pp. 311–80. New York, London: Academic Press.

Körner, C. A. & Arnone III, J. A. (1992). Responses to elevated carbon dioxide in artificial tropical ecosystems. *Science* **257**: 1672–5.

Körner, C. & Bazzaz, F. A. (1966). *Carbon Dioxide, Populations, and Communities*. San Diego, CA: Academic Press.

Last, F. T. (ed.) (1986). Microclimate and plant growth in open top chambers. *CEC, Air Pollution Research Report* **5**. EUR 11257.

Leadley, P. W. & Drake, B. G. (1993). Open top chambers for exposing plant canopies to elevated CO_2 concentration and for measuring net gas exchange. *Vegetatio* **104/105**: 3–16.

Lewin, K. F., Hendrey, G. & Kolber, Z. (1992). Brookhaven National Laboratory Free-Air Carbon dioxide Enrichment facility. *Critical Reviews in Plant Sciences* **11**: 135–41.

Miglietta, F., Hoosbeek, M. R., Foot, J. *et al.* (2001). Spatial and temporal performance of the miniFACE (free air CO_2 enrichment) system on bog ecosystems in Northern and Central Europe. *Environmental Monitoring and Assessment* **66**: 107–27.

Mitchell, E. A. D., Butler, A., Grosvernier, P. *et al.* (2002). Contrasted effects of increased N and CO_2 supply on two keystone species in peatland restoration and implications for global change. *Journal of Ecology* **90**: 529–33.

Oechel, W. C. & Svejnbjörnsson, B. (1978). Primary production processes in arctic bryophytes at Barrow, Alaska. In *Vegetation and Production Ecology of an Alaskan Arctic Tundra*, ed. L. L. Tieszen, pp. 269–98. New York: Springer.

Oechel, W. C. & Lawrence, W. T. (1985). Taiga. In *Physiological Ecology of North American Plant Communities*, ed. B. F. Chabot & H. A. Mooney, pp. 66–94. New York: Chapman and Hall.

Overdieck, D. (1993). Elevated CO_2 and mineral content of herbaceous and woody plants. *Vegetatio* **104/105**: 403–11.

Pannewitz, S., Green, T. G. A., Maysek, K. *et al.* (2005). Photosynthetic responses of three common mosses from continental Antarctica. *Antarctic Science* **17**: 341–52.

Payer, H. D., Blodow, P., Koefferlein, M. *et al.* (1993). Controlled environment chambers for experimental studies on plant responses to CO_2 and interactions with pollutants. In *Design and Execution of Experiments on CO_2 Enrichment*, ed. D. Schulze & H. Mooney, *Commission of the European Communities, Ecosystem Research Report*, **6**: 127–47.

Proctor, M. C. F. (1981). Physiological ecology of bryophytes. In *Advances in Bryology*, ed. W. Schultze-Motel, Vol. 1, pp. 79–166. Vaduz: Cramer.

Rafael, C. R., Ashenden, T. W. & Roberts, T. M. (1995). An improved solardome system for exposing plants to elevated CO_2 and temperature. *New Phytologist* **131**: 481–90.

Raschi, A., Miglietta, F., Tognetti, R. & van Gardingen, P. R. (1997). *Plant Responses to Elevated CO_2: Evidence from Natural Springs*. Cambridge: Cambridge University Press.

Rogers, H. H., Prior, S. A. & O'Neill, E. G. (1992). Cotton root and rhizosphere responses to free-air CO_2 enrichment. *Critical Reviews in Plant Sciences* **11**: 251–63.

Silvola, J. (1985). CO_2 dependence of photosynthesis in certain forest and peat mosses and simulated photosynthesis at various actual and hypothetical CO_2 concentrations. *Lindbergia* **11**: 86–93.

Schlesinger, W. H. (1997). *Biogeochemistry: an Analysis of Global Change*, 2nd edn. San Diego, CA: Academic Press.

Smith, A. J. E. (1982). *Bryophyte Ecology*. London: Chapman & Hall.

Soussana, J. F. & Loiseau, P. (1997). Temperate grass swards and climatic changes. The role of plant-soil interactions in elevated CO_2. *Abstracta Botanica* **21**: 223–34.

Sveinbjörnsson, B. & Oechel, W. C. (1992). Controls on growth and productivity of bryophytes: environmental limitations under current and anticipated conditions. In *Bryophytes and Lichens in a Changing Environment*, ed. J. W. Bates & A. M. Farmer, pp. 77–102. Oxford: Clarendon Press.

Takács, Z., Ötvös, E., Lichtenthaler, H. K. & Tuba, Z. (2004). Chlorophyll fluorescence and CO_2 exchange of the heavy metal-treated moss, *Tortula ruralis* under elevated CO_2 concentration. *Physiology and Molecular Biology of Plants* **10**: 291–6.

Tuba, Z., Lichtenthaler, H. K., Csintalan, Zs., Nagy, Z. & Szente, K. (1994). Reconstitution of chlorophylls and photosynthetic CO_2 assimilation in the desiccated poikilochlorophyllous plant *Xerophyta scabrida* upon rehydration. *Planta* **192**: 414–20.

Tuba, Z., Csintalan, Zs., Szente, K., Nagy, Z. & Grace, J. (1998). Carbon gains by desiccation tolerant plants at elevated CO_2. *Functional Ecology* **12**: 39–44.

Tuba, Z., Proctor, M. C. F. & Takács, Z. (1999). Dessication-tolerant plants under elevated air CO_2: a review. *Zeitschrift für Naturforschung* **54c**: 788–96.

Tuba, Z., Ötvös, E. & Sóvári, A. (2002). Studying the effects of elevated concentrations of carbon dioxide on lichens using Open Top Chambers. In *Protocols in Lichenology*, ed. I. Kranner, R. P. Beckett & A. K. Varma, pp. 212–23. Berlin: Springer.

Tuba, Z., Raschi, A., Lannini, G. M. *et al.* (2003). Vegetations with various environmental constraints under elevated atmospheric CO_2 concentrations. In *Abiotic Stresses in Plants*, ed. L. S. di Toppi & B. Pawlik-Skowronska, pp. 157–204. Dordrecht: Kluwer Academic Publishers.

Van der Heijden, E., Verbeek, S. K. & Kuiper, P. J. C. (2000a). Elevated atmospheric CO_2 and increased nitrogen deposition: effects on C and N metabolism and growth of the peat moss *Sphagnum recurvum* P. Beauv. var. *mucronatum* (Russ.) Warnst. *Global Change Biology* **6**: 201–12.

Van der Heijden, E., Jauhiainen, J., Silvola, J., Vasander, H. & Kuiper, P. J. C. (2000b). Effects of elevated atmospheric CO_2 concentration and increased nitrogen deposition on growth and chemical composition of ombrotrophic *Sphagnum balticum* and oligo-mesotrophic *Sphagnum papillosum*. *Journal of Bryology* **22**: 175–82.

5

Seasonal and Interannual Variability of Light and UV Acclimation in Mosses

NIINA M. LAPPALAINEN, ANNA HYYRYLÄINEN, AND SATU HUTTUNEN

Introduction

Anthropogenic ozone depletion in the stratosphere causes enhanced ultraviolet-B (UV-B) radiation on the Earth's surface (Taalas *et al.* 2000; ACIA 2005). Ozone layer thickness and ozone depletion vary with season and latitude. At present, the ozone layer has had measurable reductions at mid-latitudes, and is most vulnerable near the poles. The ozone hole over Antarctica has occurred consistently since the early 1980s; over the years, it has varied in depth and size. Harmful UV-B radiation is partly absorbed by the stratospheric ozone layer, but the ozone layer has no attenuating effect on UV-A radiation.

The intensity of solar UV radiation incident on organisms and ecosystems is influenced by a range of factors, making it a highly dynamic component of the environment. Solar elevation contributes to latitudinal, seasonal, and diurnal variations in UV; these variations are more pronounced for UV-B than for UV-A. The increase in UV-A penetration with altitude might be little more than that for total irradiance, but penetration of UV-B is higher (Paul & Gwynn-Jones 2003). Clouds, albedo, and aerosols also influence the diurnal, seasonal, and interannual variation of UV radiation (Taalas *et al.* 2000). The largest relative increase in UV-B caused by ozone depletion has occurred at high latitudes. In the Northern Hemisphere, Arctic areas of Scandinavia are expected to be affected by the largest UV changes and steepest ozone depletion (Björn *et al.* 1998; Taalas *et al.* 2000). The greatest increase in UV-B radiation at high latitudes occurs in the spring. According to calculations, springtime enhancement of erythemal UV doses for the period 2010–2020, relative to 1979–1992, could be up to 90% in the 60–90° N region, and 100% in the 60–90° S region (Taalas *et al.* 2000). The corresponding annual maximum increases in UV doses could be up to 14% in

Bryophyte Ecology and Climate Change, eds. Zoltán Tuba, Nancy G. Slack and Lloyd R. Stark. Published by Cambridge University Press.

the north and 40% in the south in 2010–2020, and 2% and 27% in 2040–2050, respectively (Taalas *et al.* 2000). The Arctic summer, with continuous sunlight lasting from a few days to several months, creates special light climate conditions.

Since natural doses of UV-B radiation had previously been rather low at high latitudes owing to low solar angles and a naturally thick ozone layer, it is presumed that plants from these areas are adapted to lower radiation levels compared with plants from low latitudes and thus are susceptible to increasing UV-B radiation (Caldwell *et al.* 1980; Robberecht *et al.* 1980). In high-latitude areas, where drought and excess light are combined with low temperatures, mosses are abundant ground layer plants. Mosses experience varying light exposure, depending upon habitat conditions, from open mires to closed forests.

Mosses are a key element of these ecosystems, since they mediate the exchange of water, heat, and trace gases between rhizosphere and atmosphere (Oechel & Van Cleve 1986). Desiccation-tolerance, leaves of one cell layer, thin cuticles, and absence of roots are predominant features in mosses (Proctor 1982; Proctor *et al.* 2007). The specific organization of mosses and their adaptation to low doses of UV-B has led to the conclusion that these plants are vulnerable to increasing UV-B radiation (Gehrke *et al.* 1996). However, some studies have reported no response from mosses exposed to enhanced UV radiation (see Boelen *et al.* 2006). Besides UV-B radiation, UV-A radiation can have harmful effects on plants (Day *et al.* 1999; Rozema *et al.* 2001; Albert *et al.* 2007).

The objective of this chapter is to characterize the magnitude and variability of seasonal and interannual UV-induced responses of different moss species from different regions *in situ*. There has only been a limited number of long-term outdoor studies where UV-B was considered. Also, studies with repeated sampling during a year are scarce. However, in some cases, the responses of different populations of the same species are available. Most of the studies have used UV-B lamps to simulate a decrease in stratospheric ozone concentration. Some studies have monitored the effects of ambient UV-B, or used screens to reduce solar UV-B. Our own experience is based on studying seasonal responses of young active gametophyte segments in subarctic and boreal environments. The most common variables used to measure impact of changes in light environment on mosses are photosynthesis, growth, and secondary metabolites.

Seasonal variability

In mosses, photosynthesis and growth are dependent on favorable conditions. Bryophytes have the strategy of photosynthesizing and growing actively when water and light are available, and, in case these factors are limited, suspending metabolism while drying (Vitt 1990). In bryophytes, photosynthesis takes place mostly in cloudy weather, with irradiance often less than 20% of that in the full sun (Marschall & Proctor 2004). Bryophytes of dry habitats utilize photoprotection mechanisms rather than energy capture when experiencing exposure to bright sunshine (Marschall & Proctor 2004). However, mosses show no seasonal variation in their capacity for rapid recovery from unfavorable environmental conditions (Valanne 1984).

Under northern conditions, studies on moss photosynthesis and growth reveal clear seasonality (Callaghan *et al.* 1978) with increasing photosynthetic activity towards the end of the growing season (Huttunen *et al.* 1981). Photosynthesis is more dependent on light than on temperature, and the ability of mosses to use dew is an obvious advantage (Kellomäki *et al.* 1977; Csintalan *et al.* 2000). The light compensation point is temperature dependent. Generally, measurable net photosynthesis takes place at 0 °C and the lower temperature compensation point is reached between –5 °C and – 10 °C (Proctor 1982). The maximal net assimilation rate depends on the size of the young, photosynthetically active segment of the shoot, which varies according to season (Skre & Oechel 1981).

In Antarctica, mosses reveal seasonal variations in photosynthetic parameters, with clear maxima in summer, dependence on humidity and no dependence on temperature (Davey & Rothery 1996). A positive correlation has been found between the water supply in the moss habitat and the degree of seasonal variation in respiration and photosynthesis in mosses. Photosynthesis was found to be higher in *Brachythecium austro-salebrosum* than in the other species from less hydric habitats (Davey & Rothery 1996).

The growing season in mid-latitudes is much longer than in high latitudes. In southern New Mexico, new growth of *Syntrichia ruralis* starts in midwinter and stops the following winter (Mishler & Oliver 1991). In some species of Middle England, and in *Syntrichia ruralis* of southern New Mexico, a rapid growth phase has been mostly observed in late summer and early autumn, continuing with gains in dry matter during late winter and spring (Rincon & Grime 1989; Mishler & Oliver 1991). This coincides with high atmospheric humidity and low temperatures. In contrast, some species (colony forming or endohydric) have a rapid growth phase in spring (Rincon & Grime 1989). These mid- and low-latitudinal seasonal growth patterns in mosses cannot be directly applied to growth at high latitudes, owing to snow cover and lower temperatures during winter months.

In addition to primary photosynthetic assimilates, mosses have a specific lipid metabolism (Mikami & Hartmann 2004). Seasonality in lipid metabolism in mosses enables their acclimation to cold, and it is one of the reasons why bryophytes dominate the primary production at high polar latitudes and in alpine mountain habitats (Aro & Karunen 1979).

Bryophytes tend to invest in secondary metabolites; those with antioxidant activity form an important chemical defense mechanism in mosses (Basile *et al.* 1999). Flavonoids and other phenolic compounds are abundant in mosses, functioning as an effective UV screen (Mues 2000; see Grace 2005). Flavonoid biosynthesis is known to be induced by many environmental factors, not just light, but often by low temperatures. The amphipathic character of flavonoids affects their behavior in cold temperatures (Ollila *et al.* 2002). During cold acclimation and freezing, flavonoids act as scavengers for reactive oxygen species (Korn *et al.* 2008).

Shortcuts have to be applied to assess the potential effects of UV environment on secondary chemistry of bryophyte species (Cornelissen *et al.* 2007). Methanol (MeOH)-soluble UV-absorbing compounds have been widely used to evaluate UV-B-induced effects in mosses. This method does not necessarily reveal the whole protective capacity of the UV-absorbing compounds, since portions of the compounds are bound in cell walls (Taipale & Huttunen 2002; Clarke & Robinson 2008; Lappalainen 2010). The cell wall-bound UV-absorbing compounds have recently been argued to have a significant role in UV-protection, at least in some moss species (Table 5.1; Clarke & Robinson 2008). A study of

Table 5.1. *Seasonal and interannual variability of moss responses under ambient UV radiation in Antarctica*

Abbreviations: ↑ = increase in the variable, ↓ = decrease in the variable, UAC = methanol-extractable UV-B-absorbing compounds, WUAC = cell wall-bound UV-absorbing compounds, Chl = chlorophylls, Car = carotenoids, Anth=anthocyanins, Fv/Fm = maximum quantum yield of PSII, O_3 = ozone layer depth.

Species	1998 / 1999	1999 / 2000	2002 / 2003	Ref
Andreaea regularis	UAC: ↑ Car: ↑ Chl: no			1
Bryum pseudotriquetrum		UAC: ↑, high in spring Anth: ↑	UAC : WUAC almost 1:1	2, 3
Ceratodon purpureus		UAC: no Anth: ↓, with ↑ O_3	UAC : WUAC 1:6	2, 3
Schistidium antarctici		UAC: no Anth: no	WUAC dominate	2, 3
Sanionia uncinata	UAC: ↑ Car: ↑ Chl: no	UAC: ↑ Car: ↑ Chl: no Fv/Fm: no		4

References: (1) Newsham 2003; (2) Dunn & Robinson 2006; (3) Clarke & Robinson 2008; (4) Newsham *et al.* 2002.

three Antarctic moss species showed the most (*Ceratodon purpureus*) and the least (*Schistidium antarctici*) UV-B-tolerant species to have the majority of their UV-absorbing compounds bound in cell walls, and the least UV-tolerant species to have only half of the combined soluble and cell-wall-bound compounds of the other two species. *Bryum pseudotriquetrum*, which showed an increase in the amount of methanol-soluble UV-absorbing compounds under enhanced UV-B radiation, was characterized by high content of both soluble and cell-wall-bound compounds present in the plant in almost equal parts (Dunn & Robinson 2006; Clarke & Robinson 2008; Otero Labarta 2008).

In mosses, seasonal variability in secondary metabolites appears to be species-specific. It has been suggested that the differences between species – for example, in UV-B tolerance – is related to their desiccation tolerance (Bornman & Teramura 1993; Csintalan *et al.* 2001; Lud *et al.* 2002; Dunn & Robinson 2006) or their tolerance to other environmental factors.

Boreal and sub-arctic (Finland and Sweden)

We collected samples of several moss species – *Dicranum polysetum*, *Dicranum scoparium*, *Hylocomium splendens*, *Pleurozium schreberi*, *Pohlia nutans*, *Polytrichum juniperinum*, *Polytrichum piliferum* and *Racomitrium canescens* – from different locations in Finland between the years 2000 and 2005 (Fig. 5.1). Samplings were done several times in a season from June to October. The sampling sites were in or at the edge of a forest, or at an open site, in Oulu (65° N, 27° E), in a dry pine forest in Sodankylä (67° N, 26° E), and on a seashore on Hailuoto Island (65° N, 24° E). Most mosses were sampled at their natural sites, but some of them were transplanted into common garden experiments before sampling (Fig. 5.1). Samplings were also done at different altitudes at Laukukero Mountain in Northern Finland.

The extractable UV-B-absorbing compounds were determined by extracting *c.* 5 mg of dry green moss gametophyte into acidified methanol (Lappalainen *et al.* 2008a). Absorbance of the extract was measured with a spectrophotometer at wavelengths 280 and 300 nm, summed, and divided by the specific leaf area (SLA, $mm^2\ mg^{-1}$) of the sample. The method used is suitable for classifying the principal UV-absorbing molecules present and revealing changes in their concentration (Seo *et al.* 2008).

Seasonal trends of methanol-extractable UV-B-absorbing compounds varied in different boreal species studied (Fig. 5.1). In *Pleurozium schreberi*, *Hylocomium splendens*, and *Dicranum polysetum*, the compound levels tend to decrease from early summer towards September (Fig. 5.1A, B). Retrospective analyses of *H. splendens* revealed highest UV-B absorbances in June (Huttunen *et al.* 2005b).

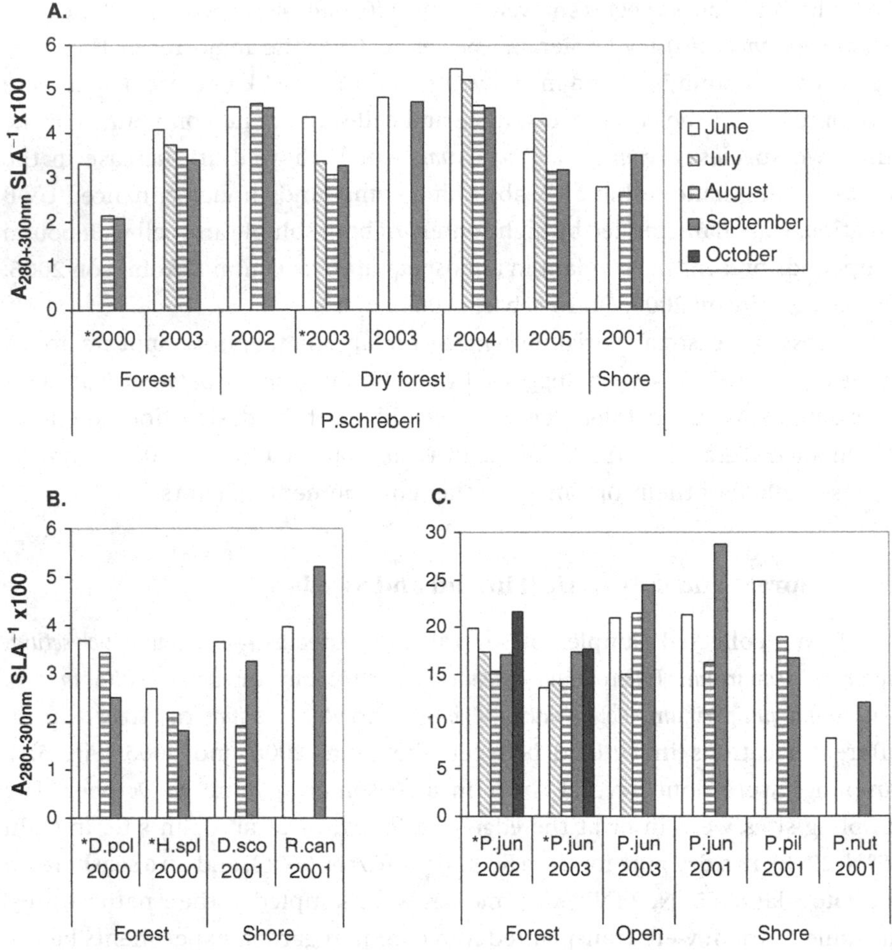

Fig. 5.1. Seasonal variation in methanol-extractable UV-B-absorbing compounds per specific leaf area (SLA) in moss samples from Finland. The samples of (A) *Pleurozium schreberi*, (B) *Dicranum polysetum*, *Hylocomium splendens*, *D. scoparium*, *Racomitrium canescens*, (C) *Polytrichum juniperinum*, *P. piliferum*, and *Pohlia nutans* were collected from Oulu, Sodankylä, and Hailuoto Island. Transplanted mosses are marked with an asterisk. From Lappalainen *et al.* 2007, 2008a and unpublished data.

Nevertheless, there are differences between years in seasonal behavior of *P. schreberi* (Fig. 5.1A), and the retrospective analyses of *P. schreberi* showed highest UV-B absorbances in September (Huttunen *et al.* 2005b).

Compared with *Pleurozium*, *Hylocomium*, *Dicranum*, and *Racomitrium* (Fig. 5.1A, B), endohydric *Polytrichum* species had considerably higher concentrations of methanol-extractable UV-B-absorbing compounds (Fig. 5.1C). *Polytrichum juniperinum* exhibited a seasonal pattern, with an increase in the amount of UV-B-absorbing compounds towards September and October (Fig. 5.1C). No increase was observed by September in another endohydric species, *Polytrichum piliferum*.

Interestingly, the site affects the seasonality. The results of our studies in Finland show that the majority of mosses from the seashore site exhibit maximum compound production in September (Fig. 5.1). At the experimental sites located in or at the edge of the forest, the content of the UV-B-absorbing compounds increases in September only in *Polytrichum juniperinum*. The seashore site is more open, the albedo is higher owing to reflectance from the sea, and it is normally warmer in autumn.

Transplanted mosses (marked with an asterisk in Fig. 5.1) reveal seasonality in UV-absorbing compounds, although it may be assumed that additional stress caused by transplantation masks the effect of UV-B enhancement under adaptation processes during the first few treatment years (Lappalainen *et al.* 2007). Under enhanced UV-B radiation in a one-year transplantation study, boreal *Sphagnum* species have shown both accumulation (*S. papillosum*) and decrease (*S. angustifolium*) in UV-absorbing compounds during the highest UV radiation period at midsummer (Niemi *et al.* 2002b). *S. angustifolium* also showed enhanced membrane permeability under enhanced UV-B, and *S. papillosum* increased chlorophyll and carotenoid production under UV-A, at the time of highest UV radiation during midsummer.

The content of methanol-extractable UV-B-absorbing compounds in mosses seems to correlate with altitude. In June 2001, mosses growing *in situ* above the timberline (415 m a.s.l.) of Laukukero Mountain exhibited decreasing specific UV absorbance with increasing altitude (unpublished data).

Length increment and biomass in experiments from high latitudes are often measured once in a season, usually at the end of the growing season. In a study with several measurements, length increment of *Dicranum elongatum* and *Sphagnum fuscum* in a moderately wet habitat occurred mainly in the latter part of July (Sonesson *et al.* 2002). The same trend was also notable in a moderately dry habitat in *Sphagnum*, but not in *Dicranum*.

Antarctic (Antarctica and Tierra del Fuego)

Antarctic studies report results from ambient (Table 5.1), enhanced (Lud *et al.* 2002, 2003; Boelen *et al.* 2006) and reduced (Searles *et al.* 1999; Huiskes *et al.* 2001; Lud *et al.* 2002, 2003; Robinson *et al.* 2005) UV-B radiation experiments. In Antarctic conditions the studies tend to be very short, which makes tracing the seasonal and interannual variability difficult.

Under ambient UV, *Andreaea regularis*, *Bryum pseudotriquetrum*, and *Sanionia uncinata* showed an increase in methanol-extractable UV-B-absorbing compounds with increasing UV radiation (Table 5.1), the level being highest during austral spring in December. *Ceratodon purpureus* and *Schistidium antarctici*, which, according to Dunn and Robinson (2006), rely more on cell wall-bound UV-absorbing compounds, did not show seasonal variation in methanol-extractable compounds. Carotenoids increased with increasing ambient UV in *Andreaea regularis* and *Sanionia uncinata* (Table 5.1). Seasonal variations of anthocyanin levels were found to be species-dependent. Study of UV-A protection *in situ* shows resistance to UV-A in Antarctic *Ceratodon purpureus* and *Bryum subrotundifolium*, with an ability to rapidly alter the level of UV-A protection under changed irradiation conditions in the latter species (Green *et al.* 2005).

In experiments with enhanced or reduced UV-B, Antarctic mosses showed little or no effects, and seasonality in the variables cannot be estimated. *Grimmia antarctici* showed differences in the amount of chlorophylls, turf reflectance, and morphology, and *Sanionia uncinata* in branching, under reduced and ambient UV (Lud *et al.* 2002; Robinson *et al.* 2005). Enhanced UV-B induced DNA damage in *Sanionia uncinata* in transplantation experiments of two days or fewer, but it had no effect in an experiment of a few weeks duration *in situ* (Lud *et al.* 2002, 2003; Boelen *et al.* 2006). No information about seasonality is reported for the two longer-term UV-B exclusion studies of 14 months and 2 years (Lud *et al.* 2002; Robinson *et al.* 2005).

Laboratory and greenhouse

We have focused on *in situ* studies in this review, but some results of studies performed in laboratories and greenhouses are listed briefly below as well. Short-term UV-enhancement studies have lasted from six days to three months. The aquatic moss *Fontinalis antipyretica* showed a decrease in photosynthetic pigments, photosynthesis, and growth, and an increase in dark respiration rate and schlerophylly, with shade samples being more sensitive (Martínez-Abaigar *et al.* 2003; Núñez-Olivera *et al.* 2004, 2005). In *Hylocomium splendens*, phenological development was accelerated and growth increased (Johanson *et al.* 1995). *Leucobryum glaucum, Mnium hornum, Plagiomnium undulatum*, and *Plagiothecium undulatum* showed decreases in fluorescence (Takács *et al.* 1999; Csintalan *et al.* 2001). In *Polytrichum commune*, a 15% ozone depletion decreased photosynthetic pigments and UV-absorbing compounds, and increased sucrose and glucose synthesis, but a 25% ozone depletion induced an increase in photosynthetic pigments, a reduction in sucrose, and a degradation of cellular organelles, having no effect on UV-absorbing compounds (Barsig *et al.* 1998). Fluorescence was temporarily affected in *Polytrichum formosum*, decreased in *Sphagnum capillifolium*, and stayed unaffected in *Syntrichia ruralis* (Takács *et al.* 1999; Csintalan *et al.* 2001). In *Dicranum scoparium*, fluorescence parameters showed no or only transient responses (Takács *et al.* 1999; Csintalan *et al.* 2001). Of these species, *Dicranum scoparium, Hylocomium splendens*, and *Polytrichum commune* have also been studied *in situ* (Fig. 5.1; Table 5.2).

Table 5.2. *Interannual variability of UV-B enhancement effects at northern latitudes in Finland, Sweden, and Svalbard*

Abbreviations: UAC, methanol-extractable UV-absorbing compounds (SD = standard deviation); transplanted, experiments with transplanted moss material; Biomass, includes height, dry mass, shoot morphology, and male gametophyte length growth; Chl, Car, chlorophylls a and/or b, carotenoids, ratio between Chl and Car; UV-R, ultraviolet radiation; H_2O, increased precipitation; °C, increased temperature, *, change in the UAC compared with UV-A control.

Species	Duration of enhanced UV experiments *in situ* (years)							Ref
	1st	2nd	3rd	4th	5th	6th	7th	
Aulacomnium turgidum (1996–1997)		Biomass: no, with °C ↓						1
Dicranum elongatum (1973)	UAC: no Biomass: no Chl: no							2
(1997–1998)	UAC: no Biomass: no Chl: no	UAC: no Biomass: no Chl: no						2
Dicranum polysetum (2000, transplanted)	UAC: no Biomass: no							3
Hylocomium splendens (1993–1995)	UAC: SD ↑ Biomass: ↓	Biomass: ↓	UAC: SD ↑ Biomass: ↓ Chl, Car: ↓					4
(1993–1999)	Biomass: no, with H_2O ↑	UAC: no Biomass: ↓, with H_2O ↑ Chl, Car: no	Biomass: ↓, with H_2O ↑	Biomass: no, with H_2O ↑ (during 4th to 7th year)				5, 6
(2000, transplanted)	UAC: ↑ Biomass: no							7

Table 5.2. (*cont.*)

Species	Duration of enhanced UV experiments *in situ* (years)							Ref
	1st	2nd	3rd	4th	5th	6th	7th	
Pleurozium schreberi (2002–2006)	UAC vs. UV-R: positive correlation (during 1st to 4th)				UAC: ↑, SD ↑			8, 9
	UAC: ↑ * Biomass: no Chl, Car: no	UAC: no Biomass: no Chl, Car: no	UAC: no Biomass: ↓ Chl, Car: no	UAC: no Biomass: no Chl, Car: no				
(2000, transplanted)	UAC: no Biomass: no							7
(2003, transplanted)	UAC: no							10
Polytrichum commune (1993–1995)	UAC: no Biomass: no	Biomass: no	UAC: ↓ Biomass: ↓ Chl, Car: ↓					4
Polytrichum hyperboreum (1996–1997)		Biomass: no, with °C ↑						1
(1996–2002)							Biomass: ↓	11
Polytrichum juniperinum (2002–2007)						UAC: SD↑ Biomass: ↓		12, 13

References: (1) Björn *et al.* 1998; (2) Sonesson *et al.* 2002; (3) T. Taipale *et al.* unpublished; (4) Gehrke 1999; (5) Gehrke *et al.* 1996; (6) Phoenix *et al.* 2001; (7) Taipale & Huttunen 2002; (8) Lappalainen *et al.* 2008a; (9) Lappalainen *et al.* 2008b; (10) Lappalainen *et al.* 2007; (11) Rozema *et al.* 2006; (12) Lappalainen *et al.* 2008c; (13) Lappalainen *et al.* 2010.

Interannual variability

Studies of interannual comparisons are scarce, and the observation period varies between two and seven years (Table 5.2). Studies often concentrate on the treatment effect and the interannual variability may be ignored.

The possibility of using herbarium specimens to reconstruct past variations in UV-B radiation has been explored with several plant species. Flavonoid concentration of *Bryum argenteum* correlates with historical ozone levels for the period 1960–1990 (Markham *et al.* 1990). In herbarium specimens of ten subarctic bryophyte species, the interannual variability is remarkable (Huttunen *et al.* 2005a). Variable numbers of available herbarium specimens per species make it difficult to draw conclusions, but some trends can be observed. With increasing summertime total radiation, *Pleurozium schreberi* and *Polytrichastrum alpinum* showed a decreasing amount of methanol-extractable UV-absorbing compounds per unit surface area for the period 1926–1996, but *Sphagnum capillifolium* revealed an increase (Huttunen *et al.* 2005a). A study carried out in northern Spain showed fewer variations between years than within one season in physiological parameters of three aquatic bryophytes (Otero Labarta 2008). However, interannual variance was observed in bulk UV-absorbance of *Bryum pseudotriquetrum* between three years of enhanced UV-B (Otero Labarta 2008). The two phenolic UV-B-absorbing compound components of sporopollenin of the outer wall of spores of clubmoss (Lycophyta), *p*-coumaric and ferulic acid, were shown to be strongly regulated by the historical variations in UV-B radiation (Lomax *et al.* 2008). In a study of the liverwort *Jungermannia exsertifolia* subsp. *cordifolia*, the variability of *p*-coumaroylmalic acid was explained by a model based on collection year, collection month, latitude, and altitude (Otero *et al.* 2009).

Boreal and Arctic (Finland, Sweden, and Svalbard)

To our knowledge, results of enhanced UV experiments *in situ* of five years or more have been reported for four moss species (Table 5.2). In two cases, for *Hylocomium splendens* and *Pleurozium schreberi*, results have been reported annually (Gehrke *et al.* 1996; Phoenix *et al.* 2001; Lappalainen *et al.* 2008a, b). In *H. splendens*, reduction in biomass was observed during the second and third year out of seven treatment years. No response was seen in secondary metabolites during the second year (Table 5.2).

We exposed *Pleurozium schreberi* and *Polytrichum juniperinum* to several years of enhanced UV-B radiation treatment *in situ* in northern Finland (Lappalainen *et al.* 2008a; Lappalainen *et al.* 2010). The treatments were performed with UV lamps

Fig. 5.2. The enhanced UV-B experiment in Sodankylä, northern Finland.

covered with proper filters, and a modulated system simulated about 20% ozone depletion (Fig. 5.2). The methanol-extractable UV-absorbing compounds and the annual growth were determined. In *Pleurozium schreberi*, the concentration of the methanol-soluble UV-absorbing compounds correlated positively with the amount of UV radiation received during four years of the experiment (Table 5.2). An increase of the content of UV-B absorbing compounds under enhanced UV-B was observed during the first (when compared to UV-A control), and again at the fifth year of the study. The height increment and dry mass production were stimulated by UV-A during the second year, but interestingly, dry mass production decreased under both enhanced UV-B and UV-A during the third year of the research. During the fourth year, no treatment effects were detected. These results indicate that even though *P. schreberi* seems to acclimate to the enhanced UV-B radiation after a few years, the effects can reappear.

The timing of detectable effects of UV enhancement *in situ* varies among moss species. In Boreal and Arctic studies (Table 5.2), enhanced UV-B radiation has caused decreased height increment and dry mass production in several species (*Hylocomium splendens*, *Pleurozium schreberi*, *Polytrichum commune*, *P. hyperboreum*, *P. juniperinum*, some *Sphagnum* species; see Table 5.2 for references; Gehrke *et al.* 1996; Gehrke 1998; Niemi *et al.* 2002a). Detectable effects have appeared after one, two, or three years of UV-B treatment, and have been continuous or transient. Treatment effects on biomass have also been observed after the sixth and seventh treatment year (Table 5.2). Desiccation tolerance of moss species has also been observed to be linked to tolerance of UV-B radiation (Takács *et al.* 1999). No treatment effects have been observed in *Dicranum* species (1- and 2-year studies) or in the majority of the one-year studies with transplanted moss material. However, one transplantation study (Table 5.2) showed increasing amounts of UV-B-absorbing compounds in *Hylocomium splendens* after only one season of treatment. *Sphagnum* species, which grow in moist conditions, have also shown UV-induced effects in several variables during the first year of treatment (Niemi *et al.* 2002a, b; Sonesson *et al.* 2002). Negative effects on biomass indicate susceptibility of some *Sphagnum* species to enhanced UV-B radiation.

In three out of fourteen Boreal and Arctic studies, mosses have shown UV-B induced responses during the first enhancement year (Table 5.2). Responses to increased UV-B are likely to be affected by simultaneous changes in other environmental factors. Ectohydric mosses do not have underground storage organs to buffer climatic fluctuations between years (Callaghan *et al.* 1978), therefore their buffering capacity against weather changes may be lower than that of higher plants (Callaghan *et al.* 1997). For example, in *Hylocomium splendens*, temperatures during early summer influence the annual growth and cause variance between years (Callaghan *et al.* 1997). The ectohydric *Sphagnum* species (Gehrke *et al.* 1996; Gehrke 1998; Niemi *et al.* 2002a, b; Sonesson *et al.* 2002), *H. splendens* and *P. schreberi* show UV-B induced responses during the first treatment year, but the desiccation tolerant endohydric *Polytrichum* species show responses only after a few years, although our knowledge of *Polytrichum* species is very limited (Table 5.2). Predominantly ectohydric *Dicranum* species have also shown no response during the first two years (Table 5.2). Desiccation tolerance of moss species has also been observed to be linked to tolerance of UV-B radiation (Takács *et al.* 1999).

Endohydric mosses have well differentiated internal water conducting tissues, cuticles, and rhizoids (Buck & Goffinet 2000). In *Polytrichum juniperinum*, the content of soluble UV-B-absorbing compounds increased towards autumn and was relatively high under snow cover, which indicates the importance of anticipated seasonality in overwintering conditions (Lappalainen *et al.* 2010). In the endohydric moss *Polytrichastrum alpinum*, benzonaphtoxanthenones (Seo *et al.* 2008) and epicuticular waxes give additional protection depending on latitude and season (Huttunen *et al.* 2004; N. Lappalainen *et al.*, unpublished data).

Changes in precipitation and temperature can affect bryophyte responses to enhanced UV-B radiation. In *Hylocomium splendens*, increases in precipitation with UV-B radiation inverted the negative effect on biomass of UV-B radiation alone (Gehrke *et al.* 1996; Phoenix *et al.* 2001). The positive effect of increased water availability on biomass was measured during the entire seven-year experiment. Increases in carbon dioxide simultaneously with UV-B also enhanced growth (Gehrke *et al.* 1996). Increased temperature and precipitation enhanced growth of *Dicranum elongatum* in a two-year study, whereas enhanced UV-B had no effect

(Sonesson *et al.* 2002). The bog species *Sphagnum fuscum* showed increased growth with increased precipitation and temperature; however, increased temperature with increased UV-B had the opposite effect (Sonesson *et al.* 2002).

In 1995 in three simultaneous studies with two species (*H. splendens* and *P. commune*), detectable changes in most variables were observed under enhanced UV-B radiation (Table 5.2). Both species experienced a decrease in dry mass production and *H. splendens* decreased in stem length compared with earlier years (Gehrke *et al.* 1996; Gehrke 1999). These studies were performed in Abisko (68.35°N, 18.82°E), northern Sweden.

Stunted shoots and decreased spatial shoot density were observed in *Sphagnum fuscum* after two years of enhanced UV-B (Gehrke 1998). Stunted shoots in *Polytrichum commune* and decreased shoot segment area in *Hylocomium splendens* were observed after the third treatment year in Sweden (Gehrke 1999). In Svalbard, no effect on coverage of *Polytrichum hyperboreum* or *Sanionia uncinata* was observed after 2, 6, or 7 years of enhanced UV-B radiation (Rozema *et al.* 2006). Length of male gametophytes of *P. hyperboreum* was reduced after the seventh year of enhanced UV-B (Rozema *et al.* 2006).

According to a recent meta-analysis of polar bryophytes and angiosperms, UV-B radiation induces the production of UV-B-absorbing compounds, with a mean increase of 7.4 % on dry mass basis, but the increase was observed under UV-B exclusion (screens) and under natural UV-B radiation, not under enhanced UV-B radiation (UV lamps) (Newsham & Robinson 2009). This was partly explained by the unstable outputs from the lamps at low polar temperatures. The meta-analysis revealed negative UV-B induced effects on biomass and growth as well, but no effect on photosynthetic pigments or photosynthetic parameters (Newsham & Robinson 2009). When interpreting the results from the bryophyte point of view, one has to keep in mind that the data are compromised by combining bryophytes and angiosperms (Newsham & Robinson 2009).

Antarctic (Antarctica and Tierra del Fuego)

The duration of Antarctic studies varies from several hours to two years; studies were conducted between years 1997 and 2003 (Table 5.1). The majority of studies lasted between one and four months. With this limited amount of information, no conclusions about the interannual variation in UV response of Antarctic mosses can be drawn.

Conclusions

Seasonality of UV response exists in mosses, and it varies according to species, location, weather, and irradiation conditions. Since large changes in climate are predicted to occur in northern Europe, it can be assumed that this will affect seasonal and interannual patterns of moss responses to solar radiation. In this review, we have concentrated mainly on terrestrial moss species. So far it is difficult to compile a complete overview of interannual variability of moss responses in a changing light environment. However, on the basis of existing data we can state that transient UV effects or their absence during the first experimental years do not exclude the possibility of effects appearing in long-term experiments. Since the effects of UV-B radiation and the seasonal patterns vary between moss species, more species must be studied.

References

ACIA (2005). *Arctic Climate Impact Assessment*. Cambridge: Cambridge University Press.

Albert, A., Ernst, D., Heller, W. *et al.* (2007). Biological dose functions of plants in the UV-B and UV-A range determined from experiments in a sun simulator. Proceedings of the UV Conference "One Century of UV Radiation Research", 18–20 September, Davos, Switzerland.

Aro, E.-M. & Karunen, P. (1979). Fatty acid composition of polar lipids in *Ceratodon purpureus* and *Pleurozium schreberi*. *Physiologia Plantarum* **45**: 265–9.

Barsig, M., Schneider, K. & Gehrke, C. (1998). Effects of UV-B radiation on fine structure, carbohydrates, and pigments in *Polytrichum commune*. *Bryologist* **101**: 357–65.

Basile, A., Giordano, S., López-Sáez, J. A. & Cobianchi, R. C. (1999). Antibacterial activity of pure flavonoids isolated from mosses. *Phytochemistry* **52**: 1479–82.

Björn, L. O., Callaghan, T. V., Gehrke, C. *et al.* (1998). The problem of ozone depletion in Northern Europe. *Ambio* **27**: 275–9.

Boelen, P., de Boer, M. K., de Bakker, N. V. J. & Rozema, J. (2006). Outdoor studies on the effects of solar UV-B on bryophytes: overview and methodology. *Plant Ecology* **182**: 137–52.

Bornman, J. F. & Teramura, A. H. (1993). Effects of ultraviolet-B radiation on terrestrial plants. In *Environmental UV Photobiology*, ed. A. R. Young, L. O. Björn, J. Moan & W. Nultsch, pp. 427–70. New York: Plenum Press.

Buck, W. R. & Goffinet, B. (2000). Morphology and classification of mosses. In *Bryophyte Biology*, ed. A. J. Shaw & B. Goffinet, pp. 71–123. Cambridge: Cambridge University Press.

Caldwell, M. M., Robberecht, R. & Billings, W. D. (1980). A steep latitudinal gradient of solar ultraviolet-B radiation in the arctic-alpine life zone. *Ecology* **61**: 600–11.

Callaghan, T. V., Carlsson, B. Å., Sonesson, M. & Temesváry, A. (1997). Between-year variation in climate-related growth of circumarctic polulations of the moss *Hylocomium splendens*. *Functional Ecology* **11**: 157–65.

Callaghan, T. V., Collins, N. J. & Callaghan, C. H. (1978). Photosynthesis, growth and reproduction of *Hylocomium splendens* and *Polytrichum commune* in Swedish Lapland. *Oikos* **31**: 73–88.

Clarke, L. J. & Robinson, S. A. (2008). Cell wall-bound ultraviolet-screening compounds explain the high ultraviolet tolerance of the Antarctic moss, *Ceratodon purpureus*. *New Phytologist* **179**: 776–83.

Cornelissen, J. H. C., Lang, S. I., Soudzilovskaia, N. A. & During, H. J. (2007). Comparative cryptogam ecology: a review of bryophyte and lichen traits that drive biogeochemistry. *Annals of Botany* **99**: 987–1001.

Csintalan, Z., Takács, Z., Proctor, M. C. F., Nagy, Z. & Tuba, Z. (2000). Early morning photosynthesis of the moss *Tortula ruralis* following summer dew fall in a Hungarian temperate dry sandy grassland. *Plant Ecology* **151**: 51–4.

Csintalan, Z., Tuba, Z., Takács, Z. & Laitat, E. (2001). Responses of nine bryophyte and one lichen species from different microhabitats to elevated UV-B radiation. *Photosynthetica* **39**: 317–20.

Davey, M. C. & Rothery, P. (1996). Seasonal variation in respiratory and photosynthetic parameters in three mosses from the maritime Antarctic. *Annals of Botany* **78**: 719–28.

Day, T. A., Ruhland, C. T., Grobe, C. W. & Xiong, F. (1999). Growth and reproduction of Antarctic vascular plants in response to warming and UV radiation reductions in the field. *Oecologia* **119**: 24–35.

Dunn, J. L. & Robinson, S. A. (2006). Ultraviolet B screening potential is higher in two cosmopolitan moss species than in a co-occurring Antarctic endemic moss: implications of continuing ozone depletion. *Global Change Biology* **12**: 2282–96.

Gehrke, C. (1999). Impacts of enhanced ultraviolet-B radiation on mosses in a subarctic heath ecosystem. *Ecology* **80**: 1844–51.

Gehrke, C. (1998). Effects of enhanced UV-B radiation on production-related properties of a *Sphagnum fuscum* dominated subarctic bog. *Functional Ecology* **12**: 940–7.

Gehrke, C., Johanson, U., Gwynn-Jones, D. *et al.* (1996). Effects of enhanced ultraviolet-B radiation on terrestrial subarctic ecosystems and implications for interactions with increased atmospheric CO_2. *Ecological Bulletins* **45**: 192–203.

Grace, S. C. (2005). Phenolics as antioxidants. In *Antioxidants and Reactive Oxygen Species in Plants*, ed. N. Smirnoff, pp. 141–168. Oxford: Blackwell Publishing Ltd.

Green, T. G. A., Kulle, D., Pannewitz, S., Sancho, L. G. & Schroeter, B. (2005). UV-A protection in mosses growing in continental Antarctica. *Polar Biology* **28**: 822–7.

Huiskes, A. H. L., Lud, D. & Moerdijk-Poortvliet, T. C. W. (2001). Field research on the effects of UV-B filters on terrestrial Antarctic vegetation. *Plant Ecology* **154**: 77–86.

Huttunen, S., Karhu, M. & Kallio, S. (1981). The effect of air pollution on transplanted mosses. *Silva Fennica* **15**: 495–504.

Huttunen, S. & Virtanen, V. (2004). Wax and UV-B-absorbing compounds of *Polytrichastrum alpinum* from different climatic regions. In Proceedings of the IUFRO meeting, *Forests under Changing Climate, Enhanced UV and Air Pollution*, pp. 17–21, August 25–30, Oulu, Finland.

Huttunen, S., Lappalainen, N. M. & Turunen, J. (2005a). UV-absorbing compounds in subarctic herbarium bryophytes. *Environmental Pollution* **133**: 303–14.

Huttunen, S., Taipale, T., Lappalainen, N. M. *et al.* (2005b). Environmental specimen bank samples of *Pleurozium schreberi* and *Hylocomium splendens* as indicators of the radiation environment at the surface. *Environmental Pollution* **133**: 315–26.

Johanson, U., Gehrke, C., Björn, L. O., Callaghan, T. V. & Sonesson, M. (1995). The effects of enhanced UV-B radiation on a subarctic heath ecosystem. *Ambio* **24**: 106–11.

Kellomäki, S., Hari, P. & Koponen, T. (1977). Ecology of photosynthesis in *Dicranum* and its taxonomic significance. Congres International der Bryologie Bordeaux 21–28 Novembre 1977. *Bryophytorum Bibliotheca* **13**: 485–507.

Korn, M., Peterek, S., Mock, H.-P., Heyer, A. G. & Hincha, D. K. (2008). Heterosis in the freezing tolerance, and sugar and flavonoid contents of crosses between *Arabidopsis thaliana* accessions of widely varying freezing tolerance. *Plant, Cell and Environment* **31**: 813–27.

Lappalainen, N. M. (2010). The responses of ectohydric and endohydric mosses under ambient and enhanced ultraviolet radiation. *Acta Universitatis Ouluensis*, doctoral thesis (A558). Tampere, Finland: Juvenes Print.

Lappalainen, N. M., Huttunen, S. & Suokanerva, H. (2007). Red-stemmed feather moss *Pleurozium schreberi* (Britt.) Mitt. – a bioindicator of UV radiation? *Proceedings of the ISEST Conference, November 13–16, Beijing, China*, pp. 14–18.

Lappalainen, N. M., Huttunen, S. & Suokanerva, H. (2008a). Acclimation of a pleurocarpous moss *Pleurozium schreberi* (Britt.) Mitt. to enhanced ultraviolet radiation *in situ*. *Global Change Biology* **14**: 321–33.

Lappalainen, N. M., Huttunen, S. & Suokanerva, H. (2008b). The ectohydric moss *Pleurozium schreberi* (Britt.) Mitt. after 5 years of enhanced UV-B radiation *in situ*. In *Proceedings of the Moss meeting, August 15–18, 2008, Tampere, Finland. Physiologia Plantarum* **133**: P04–024.

Lappalainen, N. M., Huttunen, S., Suokanerva, H. & Lakkala, K. (2008c). Acclimation of an endohydric moss *Polytrichum juniperinum* Hedw. to light and ultraviolet radiation. In *Proceedings of the 23rd IUFRO Conference, September 7–12, 2008, Murten, Switzerland*, p. 126.

Lappalainen, N. M., Huttunen, S., Suokanerva, H. & Lakkala, K. (2010). Seasonal acclimation of the moss *Polytrichum juniperinum* Hedw. to natural and enhanced ultraviolet radiation. *Environmental Pollution* **158**: 891–900.

Lomax, B. H., Fraser, W. T., Sephton, M. A. *et al.* (2008). Plant spore walls as a record of long-term changes in ultraviolet-B radiation. *Nature Geoscience* **1**: 592–6.

Lud, D., Moerdijk, T. C. W., van de Poll, W. H., Buma, A. G. J. & Huiskes, A. H. L. (2002). DNA damage and photosynthesis in Antarctic and Arctic *Sanionia uncinata* (Hedw.) Loeske under ambient and enhanced levels of UV-B radiation. *Plant, Cell and Environment* **25**: 1579–89.

Lud, D., Schlensog, M., Schroeter, B. & Huiskes, A. H. L. (2003). The influence of UV-B radiation on light-dependent photosynthetic performance in *Sanionia uncinata* (Hedw.) Loeske in Antarctica. *Polar Biology* **26**: 225–32.

Markham, K. R., Franke, A., Given, D. R. & Brownsey, P. (1990). Historical Antarctic ozone level trends from herbarium specimen flavonoids. *Bulletin de Liaison Groupe de Polyphenols* **15**: 230–5.

Marschall, M. & Proctor, M. C. F. (2004). Are bryophytes shade plants? Photosynthetic light responses and proportions of chlorophyll *a*, chlorophyll *b* and total carotenoids. *Annals of Botany* **94**: 593–603.

Martínez-Abaigar, J., Núñez-Olivera, E., Beaucourt, N. *et al.* (2003). Different physiological responses of two aquatic bryophytes to enhanced ultraviolet-B radiation. *Journal of Bryology* **25**: 17–30.

Mikami, K. & Hartmann, E. (2004). Lipid metabolism in mosses. In *New Frontiers in Bryology: Physiology, Molecular Biology and Functional Genomics*, ed. A. J. Wood, M. J. Oliver & D. J. Cove, pp. 133–155. Dordrecht: Kluwer Academic Publishers.

Mishler, B. D. & Oliver, M. J. (1991). Gametophytic phenology of *Tortula ruralis*, a desiccation-tolerant moss, in the Organ Mountains of southern New Mexico. *Bryologist* **94**: 143–53.

Mues, R. (2000). Chemical constituents and biochemistry. In *Bryophyte Biology*, ed. A. J. Shaw & B. Goffinet, pp. 150–81. Cambridge: Cambridge University Press.

Newsham, K. K. (2003). UV-B radiation arising from stratospheric ozone depletion influences the pigmentation of the Antarctic moss *Andreaea regularis*. *Oecologia* **135**: 327–31.

Newsham, K. K., Hodgson, D. A., Murray, A. W. A., Peat, H. J. & Lewis Smith, R. I. (2002). Response of two Antarctic bryophytes to stratospheric ozone depletion. *Global Change Biology* **8**: 972–83.

Newsham, K. K. & Robinson, S. A. (2009). Responses of plants in polar regions to UVB exposure: a meta-analysis. *Global Change Biology* **15**: 2574–89.

Niemi, R., Martikainen, P. J., Silvola, J. *et al.* (2002a). Responses of two *Sphagnum* moss species and *Eriophorum vaginatum* to enhanced UV-B in a summer of low UV intensity. *New Phytologist* **156**: 509–15.

Niemi, R., Martikainen, P. J., Silvola, J. *et al.* (2002b). Elevated UV-B radiation alters fluxes of methane and carbon dioxide in peatland microcosms. *Global Change Biology* **8**: 361–71.

Núñez-Olivera, E., Martínez-Abaigar, J., Tomás, R., Beaucourt, N. & Arróniz-Crespo, M. (2004). Influence of temperature on the effects of artificially enhanced UV-B radiation on aquatic bryophytes under laboratory conditions. *Photosynthetica* **42**: 201–12.

Núñez-Olivera, E., Arróniz-Crespo, M., Martínez-Abaigar, J., Tomás, R. & Beaucourt, N. (2005). Assessing the UV-B tolerance of sun and shade samples of two aquatic bryophytes using short-term tests. *Bryologist* **108**: 435–48.

Oechel, W. C. & Van Cleve, K. (1986). The role of bryophytes in nutrient cycling in the taiga. In *Forest Ecosystems in the Alaskan Taiga*, ed. K. Van Cleve, F. S. Chapin III, P. W. Flanagan, L. A. Viereck & C. T. Dyrness, pp. 121–137. Berlin: Springer-Verlag.

Ollila, F., Halling, K., Vuorela, P., Vuorela, H. & Slotte, J. P. (2002). Characterization of flavonoid biomembrane interactions. *Archives of Biochemistry and Biophysics* **399**: 103–8.

Otero Labarta, S. (2008). Ecophysiological bases for the use of aquatic bryophytes as bioindicators of ultraviolet radiation. Doctoral Thesis. Universidad de la Rioja, Departament de Agricultura y alimentacion, pp. 147–76.

Otero, S., Núñez-Olivera, E., Martínez-Abaigar, J., Tomás, R. & Huttunen, S. (2009). Retrospective bioindication of total ozone and ultraviolet radiation using hydroxycinnamic acid derivatives of herbarium samples of an aquatic liverwort. *Environmental Pollution* **157**: 2335–44.

Paul, N. D. & Gwynn-Jones, D. (2003). Ecological roles of solar UV radiation: towards an integrated approach. *Trends in Ecology and Evolution* **18**: 48–55.

Phoenix, G. K., Gwynn-Jones, D., Callaghan, T. V., Sleep, D. & Lee, J. A. (2001). Effects of global change on a sub-Arctic heath: effects of enhanced UV-B radiation and increased summer precipitation. *Journal of Ecology* **89**: 256–67.

Proctor, M. C. F. (1982). Physiological ecology: water relations, light and temperature responses, carbon balance. In *Bryophyte Ecology*, ed. A. J. E. Smith, pp. 333–81. London: Chapman and Hall.

Proctor, M. C. F., Ligrone, R. & Duckett, J. G. (2007). Desiccation tolerance in the moss *Polytrichum formosum*: physiological and fine-structural changes during desiccation and recovery. *Annals of Botany* **99**: 75–93.

Rincon, E. & Grime, J. P. (1989). An analysis of seasonal patterns of bryophyte growth in a natural habitat. *Journal of Ecology* **77**: 447–55.

Robberecht, R., Caldwell, M. M. & Billings, W. D. (1980). Leaf ultraviolet optical properties along a latitudinal gradient in the arctic-alpine life zone. *Ecology* **61**: 612–19.

Robinson, S. A., Turnbull, J. D. & Lovelock, C. E. (2005). Impact of changes in natural ultraviolet radiation on pigment composition, physiological and morphological characteristics of the Antarctic moss, *Grimmia antarctici*. *Global Change Biology* **11**: 476–89.

Rozema, J., Boelen, P., Solheim, B., *et al.* (2006). Stratospheric ozone depletion: high arctic tundra plant growth on Svalbard is not affected by enhanced UV-B after 7 years of UV-B supplementation in the field. *Plant Ecology* **182**: 121–35.

Rozema, J., Noordijk, A. J., Broekman, R. A., *et al.* (2001). (Poly)phenolic compounds in pollen and spores of Antarctic plants as indicators of solar UV-B. *Plant Ecology* **154**: 11–26.

Searles, P. S., Flint, S. D., Díaz, S. B. *et al.* (1999). Solar ultraviolet-B radiation influence on *Sphagnum* bog and *Carex* fen ecosystems: first field season findings in Tierra del Fuego, Argentina. *Global Change Biology* **5**: 225–34.

Seo, C., Choi, Y.-H., Sohn, J. H. *et al.* (2008). Ohioensins F and G: protein tyrosine phosphatase 1B inhibitory benzonaphthoxanthenones from the Antarctic moss *Polytrichastrum alpinum*. *Bioorganic and Medicinal Chemistry Letters* **18**: 772–5.

Skre, O. & Oechel, W. C. (1981). Moss functioning in different taiga ecosystems in interior Alaska. I. Seasonal, phenotypic, and drought effects on photosynthesis and response patterns. *Oecologia* **48**: 50–9.

Sonesson, M., Carlsson, B. Å., Callaghan, T. V. *et al.* (2002). Growth of two peat-forming mosses in subarctic mires: species interactions and effects of simulated climate change. *Oikos* **99**: 151–60.

Taalas, P., Kaurola, J., Kylling, A. *et al.* (2000). The impact of greenhouse gases and halogenated species on future solar UV radiation doses. *Geophysical Research Letters* **27**: 1127–30.

Taipale, T. & Huttunen, S. (2002). Moss flavonoids and their ultrastructural localization under enhanced UV-B radiation. *Polar Record* **38**: 211–18.

Takács, Z., Csintalan, Z., Sass, L. *et al.* (1999). UV-B tolerance of bryophyte species with different degrees of desiccation tolerance. *Journal of Photochemistry and Photobiology B: Biology* **48**: 210–15.

Valanne, N. (1984). Photosynthesis and photosynthetic products in mosses. In *Experimental Biology of Bryophytes*, ed. A. F. Dyer & J. G. Duckett, pp. 257–73. London: Academic Press.

Vitt, D. H. (1990). Growth and production dynamics of boreal mosses over climatic, chemical and topographic gradients. *Botanical Journal of the Linnean Society* **104**: 35–59.

III AQUATIC BRYOPHYTES

6

Ecological and Physiological Effects of Changing Climate on Aquatic Bryophytes

JANICE M. GLIME

Introduction

If you go hunting for bryophytes in the tropics, you soon learn that streambeds are depauperate and searching is futile. Witness the absence of such aquatic taxa as *Fontinalis*, *Hygroamblysteglum*, and *Rhynchostegium ripar ioides*, so common in temperate mountainous areas. Ruttner (1955) reports that *Fontinalis* is especially common at 10–15 m depth in alpine lakes, but that in the tropics it is nowhere. And consider the paucity of bryophytes in exposed, warm valley and flatland temperate streams. Ward (1986) described the altitudinal zonation in a Rocky Mountain, USA, stream and noted that bryophytes had the greatest biomass in headwaters, whereas tracheophytes were absent from higher elevations. Suren (1996), in studying 118 streams on South Island, New Zealand, reported that sites with no bryophytes had a lower mean elevation than did sites with bryophytes. Hence rising temperatures are likely to force aquatic bryophytes into higher elevations or more northern locations.

Furthermore, factors that correlate with warmer temperatures may alter bryophyte distributions. Cappelletti and Bowden (2006) suggest that global warming will increase the soluble reactive phosphorus, water temperature, and discharge of Arctic rivers, hence changing other factors that might favor tracheophytes over bryophytes or change the species composition of the aquatic bryophyte communities. Elevated temperatures can be expected to change nutrients, CO_2 concentrations, flow rates, flooding depth and frequency, competing primary producers, light penetration, and seasonal coordination.

Bryophyte Ecology and Climate Change, eds. Zoltán Tuba, Nancy G. Slack and Lloyd R. Stark. Published by Cambridge University Press.

Temperature optima and variation

Bryophytes in streams are often negatively correlated with temperature (Szarek 1994). They are C_3 plants and therefore have low temperature compensation points and low temperature maxima (Dilks & Proctor 1975; Johnson 1978; Glime & Acton 1979; Fornwall & Glime 1982; Furness & Grime 1982; Glime 1982, 1984, 1987a, b, c; Claveri & Mouvet 1995; Carballeira *et al.* 1998).

Dilks and Proctor (1975) demonstrated that aquatic bryophyte species tended to be different from terrestrial species in their temperature optima, with most terrestrial species in their study having a photosynthetic optimum around 25–30 °C and an upper level temperature compensation point of 35–40 °C. Aquatic species, on the other hand, exhibited an optimum at 20 °C (*Nardia compressa*) and 15–20 °C (*Fontinalis squamosa*). Harder (1925) reported optima of 8–20 °C for *Fontinalis antipyretica*, with optima varying with season. Saitoh *et al.* (1970) found that the optimum temperature for photosynthesis in *F. hypnoides* was 20 °C, whereas for the tracheophyte species *Ceratophyllum demersum*, *Potamogeton crispus*, and *Cabomba caroliniana* the optimum was 30 °C (Fig. 6.1). *Fontinalis hypnoides* continued to have a net gain in photosynthesis with a slow decline above 20 °C, whereas the tracheophytes had a rapid decline above their optimum. Furthermore, Carballeira *et al.* (1998) showed that *F. antipyretica* initially experienced respiratory increase with temperature, but after about ten days, respiration declined markedly.

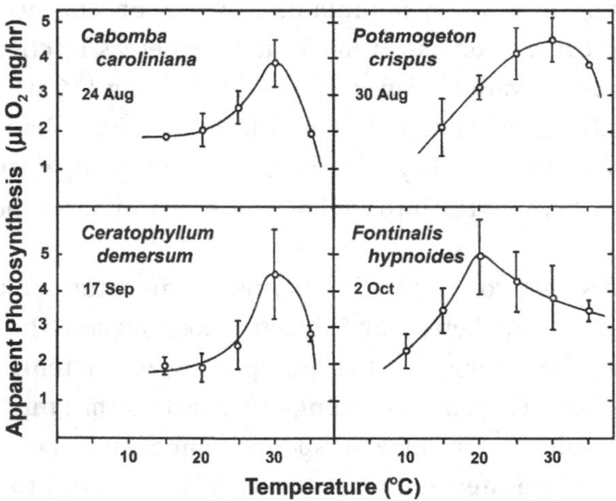

Fig. 6.1. Apparent photosynthesis of *Fontinalis hypnoides* compared to that of three aquatic tracheophytes. Redrawn from Saitoh *et al.* (1970).

Even these comparatively low optima may be artificially high because of the short duration of measurement. Glime and Acton (1979) and Fornwall and Glime (1982) found that exposure to temperatures above 15 °C for three weeks caused a depression in photosynthetic activity, placing the optimum temperature for growth at about 15 °C for most species of *Fontinalis* (Furness & Grime 1982; Glime 1987a, b, c). Glime and Acton (1979) found that if *Fontinalis duriaei* was collected after a hot period of three or more weeks in the field, its productivity was reduced and its ability to maintain high photosynthetic levels or growth was reduced compared to populations collected after a prolonged cool period.

Bryophytes have little ability to acclimate to temperature changes, at least in the short term (Antropova 1974). Carballeira *et al.* (1998) stressed *Fontinalis antipyretica* for two, four, and ten days at 30 °C, then transferred it to 16 °C for 40 days. The moss experienced 50% recovery within ten days, but those that had been collected from a river with abnormally high temperatures due to hot spring input did not fare any better than those collected from a "normal" river site. Pigment ratios returned to normal more quickly than did gross and net photosynthesis or respiration. On the other hand, Fornwall and Glime (1982) determined that field-acclimated *Fontinalis duriaei* had a cold response (0–1 °C acclimation, November – early March) and a warm response (9–16 °C, April – October) that had significantly different Arrhenius slopes. This relationship corresponded with low net assimilation in November through March and higher net assimilation in April through October. Nevertheless, the moss appeared to be most healthy (deep green) in March through June compared to August through January (olive green).

Spring often accounts for the greatest growth, but at least some occurs in winter. In the northern Pennines of England, Kelly and Whitton (1987) found maximum growth in spring in *Rhynchostegium riparioides* (= *Eurhynchium riparioides*), with a smaller peak in autumn, and minimal growth in winter, suggesting that this moss may even benefit from elevated temperatures earlier in spring and later in autumn growing seasons. Johnson (1978) found that new growth in a north Swedish river started as soon as there were ice-free areas such as riffles, although the water temperature was still 0 °C. Nevertheless, the highest growth rate occurred just before the highest stream temperature occurred. But in a North Carolina, USA, stream, *Fontinalis* became dominant over the red alga *Lemanea australis* only when it was warmer (Everitt & Burkholder 1991).

Global warming models predict great variation in weather in the temperate zone, at least in the short term, and thus some aquatic systems may likewise have more variable temperatures. Vanderpoorten *et al.* (1999) found that

Hygroamblystegium tenax, *Chiloscyphus pallescens*, and *Pellia endiviifolia* decreased in frequency as the standard deviation of temperature increased and were absent in streams where the standard deviation reached more than 4 °C. Other species seem to require a greater span of temperature. Vanderpoorten *et al.* (1999) found that *Cinclidotus danubicus* was absent in streams with a standard deviation of less than 4 °C, and that *Leptodictyum riparium* and *R. riparioides* increased in frequency with increasing standard deviation.

But even the relatively heat-tolerant *L. riparium* (Tremp 2003) in Massachusetts, USA, had an optimum temperature of only 23 °C and died at 33 °C (Sanford 1979). At 30 °C, it produced numerous branch initials (*c.* 0.5 mm), accounting for a larger mean rate of branch growth compared with stem growth, but many failed to develop (Sanford 1979). Like most of the aquatic species in experiments by Dilks and Proctor (1975; *Anthelia julacea*, *Calliergon trifarium*, *Fontinalis squamosa*, *Racomitrium aquaticum*, *Nardia compressa*) and by Glime and co-workers (Glime & Acton 1979; Glime 1982, 1987a, b, c; *F. duriaei*, *F. hypnoides*, *F. novae-angliae*, *F. antipyretica* var. *gigantea*, *F. neomexicana*), growth decreased rapidly at temperatures above optimum, especially at 20 °C and higher. Even *Hygrohypnum ochraceum*, which can live on wet rocks and in the splash of waterfalls (Crum & Anderson 1981), died after four weeks at 30 °C (Sanford *et al.* 1974). Its optimum temperature range for growth was 18–21 °C.

Changes in climate will affect the environment surrounding the streams and lakes, and this will, in turn, affect the conditions of these aquatic habitats. Westlake (1981) found that in chalk streams, mosses were the only plants in denser shade. Higher temperatures and greater drought could decrease the canopy cover of streams, thus facilitating a rise in water temperature and increased exposure to UV-B radiation (discussed elsewhere in this volume).

Despite an inability to photosynthesize and grow at high temperatures, some aquatic mosses have a remarkable ability to survive. *Fontinalis novae-angliae* survived two weeks of intermittent boiling in the laboratory, but all its leaves died (Glime & Carr 1974). Researchers found it one year later where it had been replaced in its native stream, sporting a new green leaf!

Effects on development

Normal development is the product of a coordinated suite of responses to temperature, coupled with other necessary conditions in the environment (water availability, light intensity, photoperiod, nutrient availability). Sexual reproduction is often keyed to temperature, but may also require specific photoperiods. Male and female reproductive organs may respond to different environmental cues and become out of sync.

Several *Fontinalis* species have sexual reproduction in autumn and produce capsules in winter, with spring abrasion releasing the spores and removing all evidence of capsules (Glime 1982, 1984; Glime & Knoop 1986). For germination to be successful, temperature and photoperiod must be coordinated (Kinugawa & Nakao 1965). In *Fontinalis squamosa* large, green spores germinated in as little as 5 days at 14 and 20 °C, whereas those at 3 °C required 15 days or did not germinate at all and no further development occurred after distension of the spore (Glime & Knoop 1986). The best protonemal growth occurred at 14 °C, and only those protonemata cultured at 14 °C produced buds. Hence, in this species, if the spring season is shortened so that summer water temperatures of 20 °C are reached before buds are produced, the protonemata are likely to fail. This can be further complicated by early arrival of leaves on canopy trees, reducing available light.

The good news is that aquatic bryophytes seem rarely to rely on sexual reproduction, but rather depend on vegetative reproduction, including fragmentation (e.g., Glime *et al.* 1979). Nevertheless, vegetative reproduction may be unsuccessful owing to lack of coordination of stream conditions with branching, growth, and rhizoid development (Glime 1982, 1984). During periods of high flow, anchor ice, and surface ice, breakage is frequent (e.g., Steinman & Boston 1993). Subsequent high flow distributes the broken pieces, some of which lodge on substrata where they can attach and spread (Conboy & Glime 1971; Glime *et al.* 1979). *Hygroamblystegium fluviatile* and *Fontinalis duriaei* took 9–10 weeks to attach when held firmly against a rock (Glime *et al.* 1979). Rhizoidal attachment was greatest at 15 °C (Fig. 6.2), a temperature typically higher than that for optimum growth, and sometimes plummeted at 20 °C. Furthermore, rhizoidal growth under pool conditions was generally less than that in flowing water. Glime and Raeymaekers (1987) demonstrated that temperature effects on branching in five species of *Fontinalis* tended to follow a curve similar to that of growth (which included branch length), but that rhizoid development occurred only in a narrow range of temperatures (Fig. 6.2).

Elevated carbon dioxide effects

Gases quickly dissipate in warm water. For photosynthetic plants, this means that CO_2 quickly becomes limiting (Burr 1941). Many higher aquatic plants are known to use bicarbonates as their source of carbon (Allen & Spence 1981), and others such as *Isoetes* are CAM plants (Keeley & Bowes 1982). But aquatic bryophytes in general do not seem to have either of these mechanisms (James 1928; Ruttner 1947; Steemann Nielsen 1947; Bain & Proctor 1980; Allen & Spence 1981; Osmond *et al.* 1981; Prins & Elzenga 1989; Madsen *et al.*

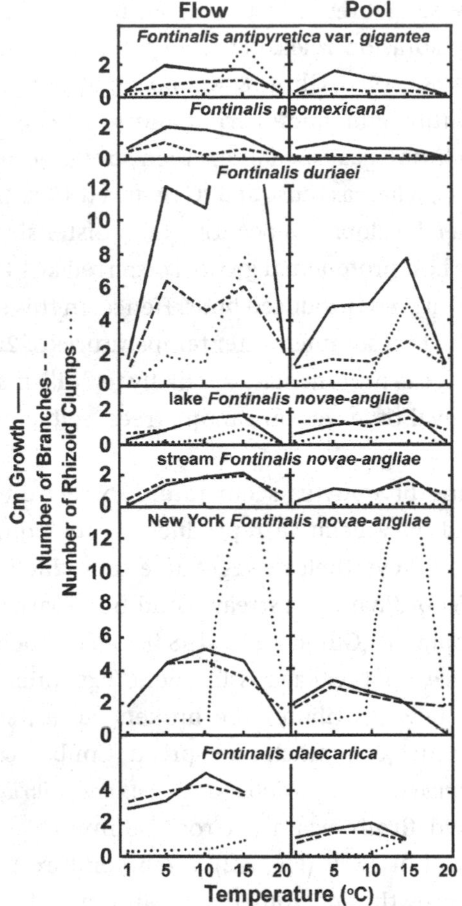

Fig. 6.2. Comparison of number of rhizoid clumps (dotted line), number of branches (dashed line), and mean growth (thick line) of 2–3 cm stems for five species of *Fontinalis* in flowing water and pool conditions. Graphs redrawn from Glime & Raeymaekers (1987) and Glime (2007).

1993; Ballesteros *et al.* 1998; Raven *et al.* 1998) and lack CO_2-concentrating mechanisms (Raven 1991) such as pyrenoids. Some evidence suggests that bicarbonates play a role in attaining photosynthetic carbon (James 1928; Burr 1941; Peñuelas 1985; Vitt *et al.* 1986). Raven *et al.* (1998) suggest that *Fissidens* cf. *mahatonensis* and *Fontinalis antipyretica* may have some sort of CO_2-concentrating mechanism.

Aquatic bryophytes have low Rubisco activity (Proctor 1981; Bowes 1985; Farmer *et al.* 1986), resulting in low rates of photosynthesis (Glime & Vitt 1984; Mouvet 1986). Furthermore, respiration increases rapidly as the temperature rises up to 20 °C (Sommer & Winkler 1982). In *Fontinalis antipyretica*, the Q_{10} in the range of 5–15 °C was 2.0 for maximum gross photosynthesis and 2.1 for respiration, but in the range of 10–20 °C, it was 1.4 and 2.0, respectively (Maberly 1985a). C_3 plants, including mosses and liverworts, typically have a low CO_2 compensation point (Andersen & Pedersen 2002), permitting them to take advantage of the cold temperatures.

Tremp (2003) states that limitations of aquatic bryophyte growth are due to their lack of aerenchyma and the low CO_2 diffusion coefficient in water [approximately four orders of magnitude less than in air (Proctor 1984)]. Sanford *et al.* (1974) demonstrated the increase in growth of *Hygrohypnum ochraceum* with increased availability of CO_2 up to the field level of 6 mg l^{-1}. Jenkins and Proctor (1985) demonstrated that boundary-layer resistance in flow rates less than 0.01 mm s^{-1} limits the photosynthetic ability of *Fontinalis* because of the difficulty of obtaining CO_2. Proctor (1984) found little change in the photosynthetic rate of *F. antipyretica* down to flow rates of *c.*1 cm s^{-1}, but Madsen *et al.* (1993) found that net photosynthesis decreased significantly as flow rate increased from 1 to 8.6 cm s^{-1}, while dark respiration increased. Maberly (1985a) showed that the slope of photosynthesis vs. CO_2 in *F. antipyretica* increased linearly with increase in temperature, suggesting that the relationship was due to boundary layer resistance. When *Fontinalis* forms streamers that separate branches, this maximizes its surface area under limiting boundary-layer conditions. The layering of branches is less problematic for mat-forming species like *Nardia compressa* and *Scapania undulata* that have a high leaf area index (Jenkins & Proctor 1985). Nevertheless, these two species exhibit a steep photosynthetic decline as the flow rate decreases below about 10 cm s^{-1}, demonstrating the importance of using turbulence to get CO_2 into the small spaces between the leaves (Proctor 1984).

Effects of altered flow

In regions where flow rates decrease owing to reduced rainfall (a common consequence of global warming), the availability of CO_2, typically captured briefly by turbulent waters, will decrease (Madsen *et al.* 1993). Bain and Proctor (1980) suggested that some bryophytes may prefer waterfalls and turbulent water because of their available atmospheric CO_2. Raven *et al.* (1998) suggested that CO_2 is rarely limiting in these habitats. Suren (1996) found that in streams of South Island, New Zealand, waterfalls supported the highest bryophyte cover. The conditions would permit them to be constantly wet, yet obtain atmospheric CO_2. Riffles, runs, and pools had considerably less cover than these turbulent areas. Odland *et al.* (1991) found that a 92% reduction in the mean annual discharge of a river reduced the

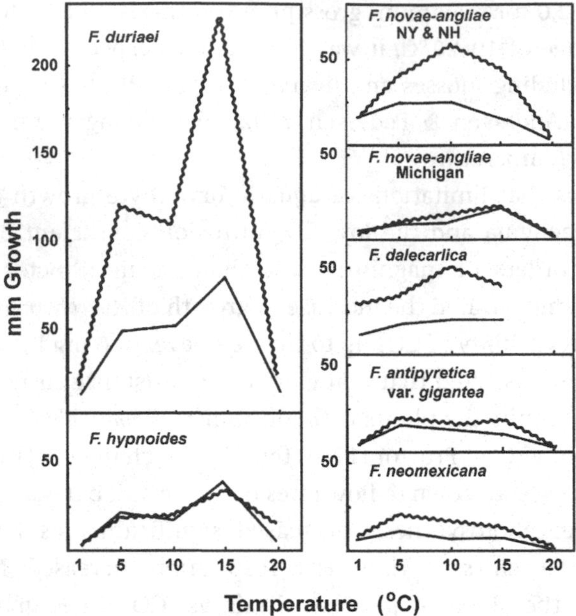

Fig. 6.3. Comparison of temperature effects and flowing water (wavy line) vs pool (thin line) conditions for combined branch and stem growth of *Fontinalis* species from the Keweenaw Peninsula of Michigan (except as noted) for 15 weeks. From Glime (2007), redrawn from Glime (1987a).

spray from its waterfall by 98%–100%, resulting in a great decline in water-loving bryophytes.

Glime (1987a) demonstrated reduced growth of *Fontinalis* in the laboratory when pool conditions were provided instead of flowing water conditions (Fig. 6.3). Reduction in flow from regulated rivers suggests that similar reductions from climate change could negatively impact bryophytes (Englund *et al.* 1997). At sites in Swedish rivers with reduced flow, species richness was reduced by 22% and at sites with regulated but unreduced flow by 26%. Large, submersed species (*F. antipyretica*, *F. dalecarlica*) diminished, and smaller, substrate-hugging *Blindia acuta* and *Schistidium agassizii* increased in abundance.

On South Island, New Zealand, Suren (1996) found that streams with no bryophytes likewise had the lowest catchment yield, longest periods of low-flow events, and highest coefficient of variation of flow, all events likely to be consequences of global warming. On the other hand, frequency and duration of floods seemed to have no significant effect on bryophyte abundance. In Nepal, monsoonal floods and resultant movement of streambed rocks increased the importance of streambed stability (Suren & Ormerod 1999). As streambeds shifted from high stability to moderate stability at higher altitudes, bryophytes shifted from *Isopterygium* spp., *Philonotis* spp., *Rhizomnium punctatum*, and Lejeuneaceae to *Eurhynchium praelongum*, *Rhynchostegium* spp., *Fissidens grandifrons*, and *Hygroamblystegium* spp. Biggs and Saltveit (1966) reported that bryophytes are often restricted to areas with stable substrates but high velocities, where tracheophyte colonization is low. Changes to the substrate stability could have a major impact on bryophyte community structure.

Desiccation due to higher temperatures and reduced rainfall

For aquatic bryophytes, desiccation may result from drying of streambeds and lowering of water levels that result from higher temperatures and reduced rainfall. This will strand plants on rocks above water. Proctor (1982) found that aquatic mosses were particularly susceptible to death by drying out and cited *Philonotis calcarea*, *Fontinalis antipyretica*, *Scorpidium scorpioides*, and *Cladopodiella fluitans* as always dying upon even "slight drying." Steere (1976) reported that *F. squamosa* from the Alaskan Arctic did not survive one week of air drying. On the other hand, Glime (1971) isolated *F. novae-angliae* and *F. dalecarlica* on the stream bank out of water for one year, demonstrating that they could survive (but not grow) if the stream temporarily dried up. In fact, plants that were removed in September and replaced in the water in April (shortly after snowmelt) regained their normal green color within a few days. Even plants that had been isolated for a full year and returned to the stream water in September developed new branches and bright green tips after one month. Although the leaves died, the stems held the resiliency needed to produce new leaves and shoots. Irmscher (1912) reported that leaves of air-dried *F. antipyretica* and *F. squamosa* died in seven days, but both subsequently grew new sprouts. *Fontinalis flaccida* regained its green color and grew after air-drying on a herbarium sheet in the laboratory for three months (J. Glime, unpublished data). Nevertheless, truly aquatic bryophytes like *F. antipyretica* and *Nardia compressa* have little or no growth when above water (Jenkins 1982).

Dhindsa (1985) suggests that immediate availability of NADPH may be important in repair of damaged membranes and other cellular components following desiccation. The xerophytes resume dark CO_2 fixation immediately upon rehydration, but the aquatic *Cratoneuron filicinum* does not. Bewley (1979) explains that the ability of some aquatic mosses to survive desiccation may result from their capacity to retain polysomes and renew protein synthesis upon rehydration. Loss of polysomes occurs in rapidly dried *Bryum pseudotriquetrum* and *Cratoneuron filicinum*. In slowly dried *C. filicinum*, some cells survive and limited photosynthesis resumes. In a warming environment, frequency of drying is likely to increase, causing a greater number of cell deaths. If this drying is followed by a short period of hydration, then another dry period, repeatedly, the mosses will not have enough time to rebuild membranes and repair damage before they once again are dried (Peterson & Mayo 1975; Dilks & Proctor 1976; Proctor 1981). Repetition of this short-term cycle is a common cause of moss death (Bewley 1974; Dilks & Proctor 1976; Gupta 1977; Oliver & Bewley 1984).

Disturbance

Changes in rainfall regimes due to climate change will create unprecedented disturbances in some areas. Suren (1996) found that some bryophytes may disappear for a long time, whereas others, such as *Fissidens rigidulus* in New Zealand, may be able to recolonize quickly after a major disturbance. This species' "flexible morphology" permits it to persist in small clumps in tiny cracks and to colonize huge areas, forming turfs that cover most of a boulder. It appears to rely on vegetative propagation, perhaps including fragmentation, with capsules appearing rarely. In northern Finland, Muotka and Virtanen (1995) found that *Blindia acuta* and *Hygrohypnum luridum*, both relatively small for stream mosses, were characteristic of disturbed sites, whereas larger species such as *Fontinalis* spp. and *Hygrohypnum ochraceum* required more stable environments, displaying poor recolonization after 14 months (Englund 1991). If stream bryophytes require 9–10 weeks to attach, an intervening flood could disrupt that attachment. On the other hand, in New Zealand, flooding did not hamper the bryophyte cover, and streams that flooded on a regular basis often had the most "luxuriant" growths (Suren 1996). By contrast, streams with frequent low flow periods displayed negative effects on bryophytes, most likely owing to their low desiccation tolerance (Suren 1996).

Temperature effects on bryophytes in lakes

Bryophytes are known for their ability to live below the depths of tracheophytes in lakes, growing slowly in low light (Fogg 1977; Chambers & Kalff 1985; Riis & Sand-Jensen 1997), but gaining the advantage of higher CO_2 levels and nutrients from decomposition and lower temperatures on the bottom (Osmond *et al.* 1981; Maberly 1985b; Riis & Sand-Jensen 1997). Mountain lakes of Scotland have only bryophytes, owing to the long period of ice cover (Light 1975). It is not unusual to find the upper portions of the hypolimnion of cold-water lakes such as those in the Canadian High Arctic (Hawes *et al.* 2002), Crater Lake, Oregon, USA, or Lake Tahoe on the California–Nevada border, USA (McIntire *et al.* 1994), to be almost completely covered by bryophytes. In such clear, cold-bottomed lakes, they can occupy depths as low as 140 m (McIntire *et al.* 1994).

Like stream bryophytes, most lake bryophytes are restricted to higher latitudes (> 55°N; Chambers & Kalff 1985). This may be a direct function of temperature and respiratory loss of CO_2 at warmer temperatures, but it is also likely a function of competition from both tracheophytes and algae for both light and CO_2, as well as for physical space and nutrients at warmer temperatures (Chambers & Kalff 1985). In Ontario lakes, mosses grew down to 50 m when decreased rainfall and increased evaporation, due to global warming, resulted in fewer organics being washed into the lakes (Felley 2003). This resulted in loss of phytoplankton and deeper penetration of light. When temperatures rise in late summer, competition for CO_2 from phytoplankton limits productivity of *Fontinalis antipyretica* in North Bay, British Isles (Maberly 1985b), suggesting that temperatures that depress the hypolimnion could eliminate the bryophytes.

Increased phosphorus effects

Cappelletti and Bowden (2006) predict that global warming will increase the soluble reactive phosphorus in Arctic rivers and streams. In their experiments, Finlay and Bowden (1994) determined that such nutrient changes would change the dominance of bryophyte species, with *Hygrohypnum* spp. and *Fontinalis neomexicana* becoming abundant in P-fertilized portions of the river. After 13 years of P fertilization, little effect was shown on *Schistidium agassizii*, but *Hygrohypnum ochraceum* and *H. alpinum*, formerly rare in the river, exhibited extensive growth (Arscott *et al.* 2000).

Interaction with pollutants

Rising temperatures will undoubtedly be coupled with other pollutants, and temperature affects the rate of chemical reactions. Claveri and Mouvet (1995) tested this relationship in *Rhynchostegium riparioides*, using copper pollution. After 12 days of contamination by a copper solution containing 80 $\mu g\,L^{-1}$ of copper, followed by a 10-day decontamination period, they observed a decrease in moss vitality as the temperature increased from 7 to 19 °C. Both controls and contaminated mosses experienced denaturation of chlorophyll at 29 °C, indicating temperature damage. However, the accumulation rate of copper was seemingly unaffected by temperature during 22 days of exposure.

The effect of temperature on the adsorption of heavy metals depends on the metal, and possibly on the moss as well. Cadmium adsorption on dry *Fontinalis antipyretica* seems unaffected by temperature, but the same moss experienced an increase in adsorption of zinc from 11.5 mg g^{-1} at 5 °C to 14.7 mg g^{-1} at 30 °C (Martins *et al.* 2004).

Some forms of pollution may actually favor the expansion of bryophyte populations. Tracheophytes are often affected detrimentally by acidification, whereas bryophytes can be favored by resulting higher levels of free CO_2 (Sand-Jensen & Rasmussen 1978).

Ecosystem effects

Changes in bryophyte cover, whether by increase or decrease, will affect the other organisms in that aquatic system. For example, loss of submerged bryophyte cover due to a drop in water level created problems at spawning time for the fish *Rutilus rutilus* (roach) that normally distributed its eggs among the branches of *Fontinalis antipyretica* (Mills 1981). Some bryophytes provide shelter for both young and older fish (Heggenes & Saltveit 2002).

Bryophytes serve as habitats for a multitude of micro- and macrofauna in both streams and lakes. The distribution of these organisms is often dependent on the distribution of bryophytes, providing a variety of flow, detrital, light, cover, and food organism niches (Cowie & Winterbourn 1979; Suren 1988, 1991; Brittain & Saltveit 1989). Mosses collect detrital matter that affects the invertebrates that are able to live and feed there (Johnson 1978). Englund (1991) showed that disturbance causing moss-covered rocks to be overturned reduced the density of 13 out of 16 invertebrate taxa. Chantha *et al.* (2000) demonstrated that in a stream in Quebec, Canada, dense growths of mosses had more but smaller invertebrates with little net difference in biomass, but in all cases both epiphyte and invertebrate densities were much greater on the mosses than on nearby rocks. In fact, *Fontinalis antipyretica* was introduced to South Africa for the expressed purpose of increasing invertebrate biomass (Richards 1946). It backfired because the moss-adapted species were not present and the rock-adapted species were unable to live among the mosses that now occupied their substrate.

Numerous reports on stream invertebrates indicate the importance of bryophytes in providing a different kind of niche that contributes to the diversity of species. For example, moss-inhabiting species in New Zealand alpine streams were predominantly Chironomidae (midges), stoneflies *Zelandoperla* and *Zelandobius*, Dorylaimoidea nematodes, Oibatei (mites), Hydracarinae (water mites), Copepoda, and Ostracoda (Suren 1988). These were 5–15 times as abundant as the mayflies (*Deleatidium* and *Nesameletus*) that were most abundant in the rocky areas. Bryophytes contribute to the accessible food by accumulating periphyton and organic matter (Suren 1991). Cowie and Winterbourn (1979) reported a different suite of insects occupying the bryophytes in such places as torrential mid-channel, waterfall spray zone, and outer spray zone. Brittain and Saltveit (1989) reported that the increased growth of mosses and periphyton below many dams increases the numbers of mayfly (Ephemerellidae), but may greatly restrict or eliminate the rock-dwelling Heptageniidae.

Periphyton productivity itself is an important contribution of stream bryophytes. In Greenland (Johansson 1980) and Signy Island in the Antarctic (Broady 1977), bryophytes are important algal substrates. In a stream in Yorkshire, UK, diatoms on *Palustriella commutata* and *Rhynchostegium riparioides* accounted for more than 70% of the periphyton volume of the stream (Pentecost 1991). Pentecost estimated the productivity to be 190 g C m^{-2} for *P. commutata*, 260 g C m^{-2} for *R. riparioides*, and 1–40 g C m^{-2} for the algae periphytic on the mosses. Chantha *et al.* (2000) reported that 43% of the spatial variation in algal biomass could be explained by the moss biomass, but that the algae tended to be at the surface and therefore did not increase linearly with moss biomass increase. The increase in moss biomass resulted in an increase in number of invertebrates, but they were smaller at the larger moss biomass levels, presumably due to the increased difficulty of navigating within the moss bed. Here, likewise, the algal biomass was much greater (5–10-fold) than on rocks.

New distributions

Many bryophytes will extend to higher latitudes and altitudes, as suggested by these relationships in several *Fontinalis* species (Fig. 6.4). Alpine

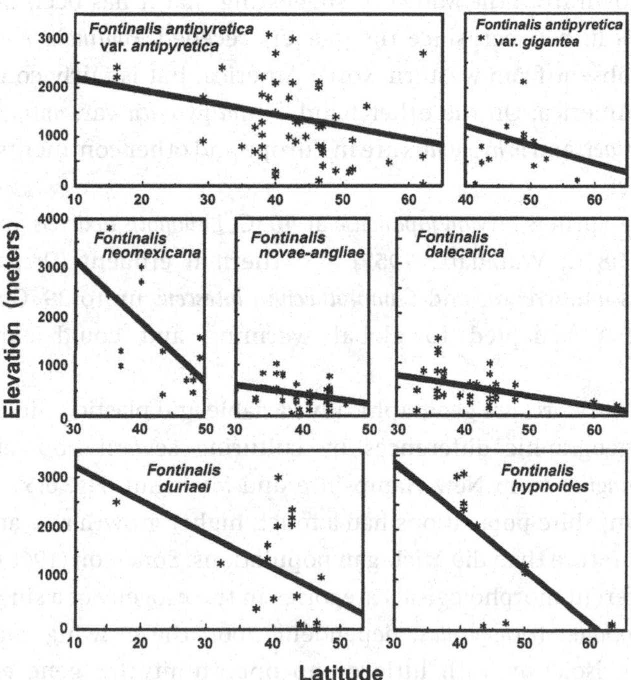

Fig. 6.4. Relationship of elevation to latitude of occurrence in six species of *Fontinalis*, based on collections cited in Welch (1960).

streams are typically dominated by bryophytes (Geissler 1982), where competition by tracheophytes is reduced by the low temperature, short growing season, and often very fast flow.

The fossil record in Kråkenes Lake in Norway suggests that the response of mosses to climatic changes is rapid, occurring within a decade of the temperature change (Birks *et al.* 2000)! Jonsgard and Birks (1995) and Birks *et al.* (1998) suggest that mosses are able to colonize newly available habitats as rapidly as tracheophytes. Although bryophyte spores are capable of long distance transport (van Zanten 1978a, b; van Zanten & Gradstein 1987), those from wet or aquatic habitats are resistant to drying for shorter periods of time than are those from dry habitats. Even the rainforest bryophytes typically can survive being dry for only half a year or less (van Zanten 1984), and liverworts have less resistance than mosses (van Zanten & Gradstein 1988).

More typically, it appears that attaining a new distribution is slow. *Fontinalis squamosa* has remained on the European side of the Atlantic Ocean and *F. novae-angliae* on the North American side. *Fontinalis novae-angliae*, which would seemingly do quite well in Iceland and Greenland, has not yet reached there. *Fontinalis neomexicana* has remained in western North America with only two populations known from the Midwest, suggesting that it has been unable to recolonize areas in between since the glaciers receded. *Fontinalis antipyretica* var. *gigantea* is absent from western North America, but is fairly common in eastern North America. On the other hand, *F. antipyretica* var. *antipyretica, F. dalecarlica*, *F. duriaei*, and *F. hypnoides* are in Europe and other continents, as well as North America.

Mosses in hot springs (*Bryum japanense* at 40 °C, *Philonotis laxiretis* and *Bryum cyclophyllum* at 38 °C; Watanabe 1957) and thermal effluent (*Drepanocladus kneiffii, Fontinalis antipyretica*, and *Camptothecium lutescens* up to 39 °C; Schiller 1938) may be pre-adapted to global warming and could expand in distribution.

Mosses and liverworts are geographically variable and plastic. Glime (1987c) demonstrated geographic differences by culturing several populations of *Fontinalis novae-angliae* from New Hampshire and Michigan (Fig. 6.5). After 15 weeks, New Hampshire populations had a much higher growth rate and lower optimum temperature than did Michigan populations. Sonesson (1966) demonstrated four different morphological categories in the progeny of a single specimen of *Drepanocladus trichophyllus*, dependent upon the growing conditions. Because of long isolation with little or no opportunity for gene exchange between aquatic populations, it is likely that many physiological races have developed.

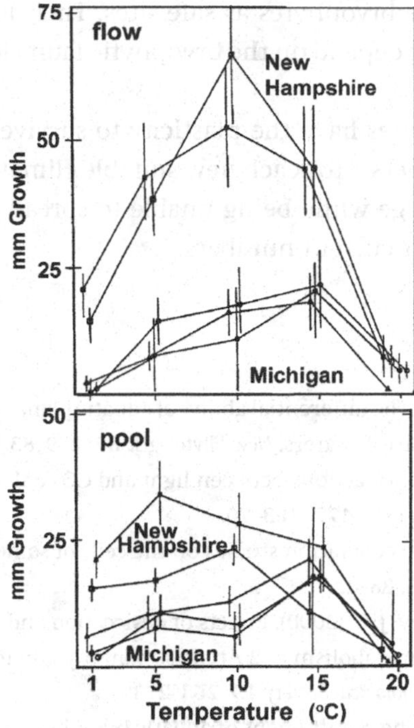

Fig. 6.5. Comparison of the temperature responses of two New Hampshire populations of *Fontinalis novae-angliae* with three populations from the Upper Peninsula of Michigan, USA, after 15 weeks in artificial streams ($n = 40$). Redrawn from Glime (1987c).

Conclusions

With climate change, relative abundances of bryophyte species will change, as demonstrated in the Kuparuk River, Alaska, USA (Arscott *et al.* 2000). Frahm (1997) suggests that the extended distribution of *Fissidens fontanus* and *F. rivularis* in the Rhine is due to raised water temperatures. Pentecost and Zhang (2002) attributed the more northerly distribution of *Palustriella* (*Cratoneuron*) *commutata* compared with *Eucladium verticillatum* to the influence of temperature. Higher respiration rates, decreased photosynthetic rates, less available CO_2, altered flow rates, more desiccation events, altered nutrient availability, and greater competition from algae and tracheophytes will force bryophytes to higher latitudes and altitudes. Changed seasonal cues, changes in frequency of flooding and dry periods, and reduced canopy light will all impact the productivity and alter the species composition in the bryophyte communities.

Loss of bryophytes in the warmer parts of the world will reduce periphyton and the invertebrates that depend not only on the periphyton and collected detritus but which also depend on the bryophytes as safe sites. Their loss will affect fish and larger invertebrates that depend on the bryophytic fauna for food and spawning sites.

The prognosis is that some bryophytes have the plasticity to survive where they are, some will have sufficient dispersal to reach new suitable climates, and others will lose part or all of their range while being unable to spread to new areas quickly enough to maintain their current numbers.

References

Allen, E. D. & Spence, D. H. N. (1981). The differential ability of aquatic plants to utilize the inorganic carbon supply in fresh waters. *New Phytologist* **87**: 269–83.

Andersen, T. & Pedersen, O. (2002). Interactions between light and CO_2 enhance the growth of *Riccia fluitans*. *Hydrobiologia* **477**: 163–70.

Antropova, T. A. (1974). Temperature adaptation studies on the cells of some bryophyte species. *Tsitologiya* **16**: 38–42.

Arscott, D. B., Bowden, W. B. & Finlay, J. C. (2000). Effects of desiccation and temperature/irradiance on the metabolism of 2 Arctic stream bryophyte taxa. *Journal of the North American Benthological Society* **19**: 263–273.

Bain, J. T. & Proctor, M. C. F. (1980). The requirement of aquatic bryophytes for free CO_2 as an inorganic carbon source: some experimental evidence. *New Phytologist* **86**: 393–400.

Ballesteros, D., García-Sánchez, M. J., Heredia, M. A., Felle, H. & Fernández, J. A. (1998). Inorganic carbon acquisition in *Riccia fluitans* L. *Journal of Experimental Botany* **49**: 1741–7.

Bewley, J. D. (1974). Protein synthesis and polyribosome stability upon desiccation of the aquatic moss *Hygrohypnum luridum*. *Canadian Journal of Botany* **52**: 423–7.

Bewley, J. D. (1979). Physiological aspects of desiccation tolerance. *Annual Review of Plant Physiology* **30**: 195–238.

Biggs, B. J. F. & Saltveit, S. J. (eds.). (1996). Hydraulic habitat of plants in streams. *Regulated Rivers: Research and Management* **12**(2–3): 131–144.

Birks, H. H., Battarbee, R. W., Birks, H. J. B. *et al.* (2000). The development of the aquatic ecosystem at Kraekenes Lake, western Norway, during the late-glacial and early-Holocene – a synthesis. *Journal of Paleolimnology* **23**: 91–114.

Birks, H. J. B., Heegaard, E., Birks, H. H. & Jonsgard, B. (1998). Quantifying bryophyte – environment relationships. In *Bryology for the Twenty-first Century*, ed. J. W. Bates, N. W. Ashton & J. G. Duckett, pp. 305–319. Maney Publishing and the British Bryological Society, UK.

Bowes, G. (1985). Pathways of CO_2 fixation by aquatic organisms. In *Inorganic Carbon Uptake by Aquatic Photosynthetic Organisms*, ed. W. J. Lucas & J. A. Berry, pp. 187–210. Rockville, MD: American Society of Plant Physiologists.

Brittain, J. & Saltveit, S. J. (1989). A review of the effect of river regulation on mayflies (Ephemeroptera). *Regulated Rivers: Research & Management* **3**: 191–204.

Broady, P. A. (1977). The Signy Island terrestrial reference sites: VII. The ecology of the algae of site 1, a moss turf. *Bulletin of the British Antarctic Survey* **45**: 47–62.

Burr, G. O. (1941). Photosynthesis of algae and other aquatic plants. In *Symposium on Hydrobiology*, ed. J. G. Needham, pp. 163–81. Madison, WI: University of Wisconsin Press.

Cappelletti, C. K. & Bowden, W. B. (2006). Implications of global warming on photosynthesis and respiration in an Arctic tundra river: Consequences to the C cycle. NABStracts, North American Benthological Society Meeting, 2006.

Carballeira, A., Díaz, S., Vázquez, M. D. & López, J. (1998). Inertia and resilience in the responses of the aquatic bryophyte *Fontinalis antipyretica* Hedw. to thermal stress. *Archives of Environmental Contamination and Toxicology* **34**: 343–9.

Chambers, P. A. & Kalff, J. (1985). Depth distribution and biomass of submersed aquatic macrophyte communities in relation to Secchi depth. *Canadian Journal of Fisheries and Aquatic Science* **42**: 701–9.

Chantha, S.-C., Cloutier, L. & Cattaneo, A. (2000). Epiphytic algae and invertebrates on aquatic mosses in a Quebec stream. *Archives of Hydrobiology* **147**: 143–60.

Claveri, B. & Mouvet, C. (1995). Temperature effects on copper uptake and CO_2 assimilation by the aquatic moss *Rhynchostegium riparioides*. *Archives of Environmental Contamination and Toxicology* **28**: 314–20.

Conboy, D. A. & Glime, J. M. (1971). Effects of drift abrasives on *Fontinalis novae-angliae* Sull. *Castanea* **36**: 111–14.

Cowie, B. & Winterbourn, M. J. (1979). Biota of a subalpine springbrook in the Southern Alps. *New Zealand Journal of Marine and Freshwater Research* **13**: 295–301.

Crum, H. A. & Anderson, L. E. (1981). *Mosses of Eastern North America*, 2 vols. New York: Columbia University Press.

Dhindsa, R. S. (1985). Non-autotrophic CO_2 fixation and drought tolerance in mosses. *Journal of Experimental Botany* **36**: 980–8.

Dilks, T. J. K. & Proctor, M. C. F. (1975). Comparative experiments on temperature responses of bryophytes: assimilation, respiration and freezing damage. *Journal of Bryology* **8**: 317–36.

Dilks, T. J. K. & Proctor, M. C. F. (1976). Effects of intermittent desiccation on bryophytes. *Journal of Bryology* **9**: 249–64.

Englund, G. (1991). Effects of disturbance on stream moss and invertebrate community structure. *Journal of the North American Benthological Society* **10**: 143–53.

Englund, G., Jonsson, B. G. & Malmqvist, B. (1997). Effects of flow regulation on bryophytes in North Swedish rivers. *Biological Conservation* **79**: 79–86.

Everitt, D. T. & Burkholder, J. M. (1991). Seasonal dynamics of macrophyte communities from a stream flowing over granite flatrock in North Carolina, USA. *Hydrobiologia* **222**: 159–72.

Farmer, A. M., Maberly, S. C. & Bowes, G. (1986). Activities of carboxylation enzymes in freshwater macrophytes. *Journal of Experimental Botany* **37**: 1568–1573.

Felley, J. (2003). Climate change undermines recovering lakes. *Environmental Science and Technology* **37**: 346a–347a.

Finlay, J. C. & Bowden, W. B. (1994). Controls on production of bryophytes in an Arctic tundra stream. *Freshwater Biology* **32**: 455–66.

Fogg, G. E. (1977). Aquatic primary production in the Antarctic. *Philosophical Transactions of the Royal Society of London B* **279**: 27–38.

Fornwall, M. D. & Glime, J. M. (1982). Cold and warm-adapted phases in *Fontinalis duriaei* Schimp. as evidenced by net assimilatory and respiratory responses to temperature. *Aquatic Botany* **13**: 165–77.

Frahm, J.-P. (1997). Zur Ausbreitung von Wassermoosen am Rhein (Deutschland) und an seinen Nebenflüssen seit dem letzten Jahrhundert. [The spreading of aquatic bryophytes along the Rhine (Germany) and its tributaries since the last century.] *Limnologica* **27**: 251–61.

Furness, S. B. & Grime, J. P. (1982). Growth rate and temperature responses in bryophytes. *Journal of Ecology* **70**: 513–23.

Geissler, P. (1982). Alpine communities. In *Bryophyte Ecology*, ed. A. J. E. Smith, pp. 167–90. New York: Chapman and Hall.

Glime, J. M. (1971). Response of two species of *Fontinalis* to field isolation from stream water. *Bryologist* **74**: 383–6.

Glime, J. M. (1982). Response of *Fontinalis hypnoides* to seasonal temperature variations. *Journal of the Hattori Botanical Laboratory* **53**: 181–93.

Glime, J. M. (1984). Physio-ecological factors relating to reproduction and phenology in *Fontinalis dalecarlica. Bryologist* **87**: 17–23.

Glime, J. M. (1987a). Phytogeographic implications of a *Fontinalis* (Bryopsida) growth model based on temperature and flow conditions for six species. *Memoirs of the New York Botanic Garden* **45**: 154–70.

Glime, J. M. (1987b). Growth model for *Fontinalis duriaei* based on temperature and flow conditions. *Journal of the Hattori Botanical Laboratory* **62**: 101–9.

Glime, J. M. (1987c). Temperature optima of *Fontinalis novae-angliae*: Implications for its distribution. *Symposia Biologica Hungarica* **35**: 569–76.

Glime, J. M. (2007). *Bryophyte Ecology*. Volume 1. *Physiological Ecology*. Ebook sponsored by Michigan Technological University and the International Association of Bryologists. <http://www.bryoecol.mtu.edu/>.

Glime, J. M. & Acton, D. W. (1979). Temperature effects on assimilation and respiration in the *Fontinalis duriaei* – periphyton association. *Bryologist* **82**: 382–92.

Glime, J. M. & Carr, R. E. (1974). Temperature survival of *Fontinalis novae-angliae* Sull. *Bryologist* **77**: 17–22.

Glime, J. M. & Knoop, B. C. (1986). Spore germination and spore development of *Fontinalis squamosa*. *Journal of the Hattori Botanical Laboratory* **61**: 487–97.

Glime, J. M. & Raeymaekers, G. (1987). Temperature effects on branch and rhizoid production in six species of *Fontinalis*. *Journal of Bryology* **14**: 779–90.

Glime, J. M. & Vitt, D. H. (1984). The physiological adaptations of aquatic Musci. *Lindbergia* **10**: 41–52.

Glime, J. M., Nissila, P. C., Trynoski, S. E. & Fornwall, M. D. (1979). A model for attachment of aquatic mosses. *Journal of Bryology* **10**: 313–20.

Gupta, R. K. (1977). A study of photosynthesis and leakage of solutes in relation to the desiccation effects in bryophytes. *Canadian Journal of Botany* **55**: 1186–94.

Harder, R. (1925). Über die Assimilation von Kalte- und Warmeindividuen der Gleichen Pflanzenspecies. *Jahrbuch für Wissenschaftlichen Botanik* **64**: 169–200.

Hawes, I., Andersen, D. T. & Pollard, W. H. (2002). Submerged aquatic bryophytes in Colour Lake, a naturally acidic polar lake with occasional year-round ice-cover. *Arctic* **55**: 380–8.

Heggenes, J. & Saltveit, S. J. (2002). Effect of aquatic mosses on juvenile fish density and habitat use in the regulated River Suldalslågen, western Norway. *River Research and Applications* **18**: 249–64.

Irmscher, E. (1912). Über die Resistenz der Laubmoose gegen Austrocknung und Kalte. *Jahrbuch für Wissenschaftlichen Botanik* **50**: 387–449.

James, W. O. (1928). Experimental researches on vegetable assimilation and respiration. XIX. The effect of variations of carbon dioxide supply upon the rate of assimilation of submerged water plants. *Proceedings of the Royal Society of London* B **103**: 1–42.

Jenkins, J. T. (1982). Effects of flowrate on the ecology of aquatic bryophytes. Ph.D. Thesis, University of Exeter, UK.

Jenkins, J. T. & Proctor, M. C. F. (1985). Water velocity, growth-form and diffusion resistances to photosynthetic CO_2 uptake in aquatic bryophytes. *Plant, Cell & Environment* **8**: 317–23.

Johansson, C. (1980). Attached algal vegetation in some streams from the Narssaq area, south Greenland. *Acta Phytogeographica Suecica* **68**: 89–96.

Johnson, T. (1978). Aquatic mosses and stream metabolism in a north Swedish river. *Internationale Vereinigung für Theoretische und Angewandte Limnologie* **20**: 1471–7.

Jonsgard, B. & Birks, H. H. (1995). Late-glacial mosses and environmental reconstructions at Kråkenes, western Norway. *Lindbergia* **20**: 64–82.

Keeley, J. E. & Bowes, G. (1982). Gas exchange characteristics of the submerged aquatic crassulacean acid metabolism plant, *Isoetes howellii*. *Plant Physiology* **70**: 1455–8.

Kelly, M. G. & Whitton, B. A. (1987). Growth rate of the aquatic moss *Rhynchostegium riparioides* in northern England. *Freshwater Biology* **18**: 461–8.

Kinugawa, K. & Nakao, S. (1965). Spore germination and protonemal growth of *Bryum pseudo-triquetrum* as a long-day plant. *Miscellanea Bryologica et Lichenologica* **3**: 136–7.

Light, J. J. (1975). Clear lakes and aquatic bryophytes in the mountains of Scotland. *Journal of Ecology* **63**: 937–43.

Maberly, S. C. (1985a). Photosynthesis by *Fontinalis antipyretica*. I. Interactions between photon irradiance, concentration of carbon dioxide and temperature. *New Phytologist* **100**: 127–40.

Maberly, S. C. (1985b). Photosynthesis by *Fontinalis antipyretica*. II. Assessment of environmental factors limiting photosynthesis and production. *New Phytologist* **100**: 141–55.

Madsen, T. V., Enevoldsen, H. O. & Jorgensen, T. B. (1993). Effects of water velocity on photosynthesis and dark respiration in submerged stream macrophytes. *Plant, Cell & Environment* **16**: 317–22.

Martins, R. J. E., Pardo, R. & Boaventura, R. A. R. (2004). Cadmium(II) and zinc(II) adsorption by the aquatic moss *Fontinalis antipyretica*: effect of temperature, pH and water hardness. *Water Research* **38**: 693–9.

McIntire, C. D., Phinney, H. K., Larson, G. L. & Buktenica, M. (1994). Vertical distribution of a deep-water moss and associated epiphytes in Crater Lake, Oregon. *Northwest Science* **68**(1): 11–21.

Mills, C. A. (1981). The spawning of roach *Rutilus rutilus* (L.) in a chalk stream. *Fisheries Management* **12**: 49–54.

Mouvet, C. (1986). Métaux lourds et mousses aquatiques. *Synthese Methodologique*. Metz: Agence de l'Eau, Rhone-Med-Corse.

Muotka, T. & Virtanen, R. (1995). The stream as a habitat templet for bryophytes: species' distributions along gradients in disturbance and substratum heterogeneity. *Freshwater Biology* **33**: 141–60.

Odland, A., Birks, H. H., Botnen, A., Tønsberg, T. & Vevle, O. (1991). Vegetation change in the spray zone of a waterfall following river regulation in Aurland, western Norway. *Regulated Rivers: Research & Management* **6**: 147–62.

Oliver, M. J. & Bewley, J. D. (1984). Plant desiccation and protein synthesis. IV. RNA synthesis, stability, and recruitment of RNA into protein synthesis during desiccation and rehydration of the desiccation-tolerant moss, *Tortula ruralis*. *Plant Physiology* **74**: 21–5.

Osmond, C. B., Valanne, N., Haslam, S. M., Uotila, P. & Roksandic, Z. (1981). Comparisons of $\delta^{13}C$ values in leaves of aquatic macrophytes from different habitats in Britain and Finland; some implications for photosynthetic processes in aquatic plants. *Oecologia* **50**: 117–24.

Pentecost, A. (1991). Algal and bryophyte flora of a Yorkshire (U.K.) hill stream: a comparative approach using biovolume estimations. *Archives of Hydrobiology* **121**: 181–201.

Pentecost, A. & Zhang, Z. (2002). Bryophytes from some travertine-depositing sites in France and the U.K.: relationships with climate and water chemistry. *Journal of Bryology* **24**: 233–41.

Peñuelas, J. (1985). HCO_3 as an exogenous carbon source for aquatic bryophytes *Fontinalis antipyretica* and *Fissidens grandifrons*. *Journal of Experimental Botany* **36**: 441–8.

Peterson, W. L. & Mayo, J. M. (1975). Moisture stress and its effect on photosynthesis in *Dicranum polysetum*. *Canadian Journal of Botany* **53**: 2897–900.

Prins, H. B. A. & Elzenga, J. T. M. (1989). Bicarbonate utilization: function and mechanism. *Aquatic Botany* **34**: 59–83.

Proctor, M. C. F. (1981). Physiological ecology of bryophytes. In *Advances in Bryology*, ed. W. Schultze-Motel, pp. 79–166. Vaduz: J. Cramer.

Proctor, M. C. F. (1982). Physiological ecology: water relations, light and temperature responses, carbon balance. In *Bryophyte Ecology*, ed. A. J. E. Smith, pp. 333–82. London: Chapman & Hall.

Proctor, M. C. F. (1984). Structure and ecological adaptation. In *The Experimental Biology of Bryophytes*, ed. A. F. Dyer & J. G. Duckett, pp. 9–37. New York: Academic Press.

Raven, J. A. (1991). Implications of inorganic carbon utilization: ecology, evolution, and geochemistry. *Canadian Journal of Botany* **69**: 908–24.

Raven, J. A., Griffiths, H., Smith, E. C. & Vaughn, K. C. (1998). New perspectives in the biophysics and physiology of bryophytes. In *Bryology for the Twenty-first Century*, ed. J. W. Bates, N. W. Ashton & J. G. Duckett, pp. 261–75. Leeds: Maney Publishing and the British Bryological Society.

Richards, P. W. (1946). The introduction of *Fontinalis antipyretica* Hedw. into South Africa and its biological effects. *Transactions of the British Bryological Society* **1**: 16.

Riis, T. & Sand-Jensen, K. (1997). Growth reconstruction and photosynthesis of aquatic mosses: influence of light, temperature and carbon dioxide at depth. *Journal of Ecology* **85**: 359–72.

Ruttner, F. (1947). Zur Frage der Karbonatassimilation der Wasserpflanzen. I. Teil: Die beiden Haupttypen der Kohlenstaufnahme. *Österreichische Botanische Zeitschrift* **94**: 265–94.

Ruttner, F. (1955). Zur Ökologie tropischer Wassermoose. *Archives of Hydrobiology*, Suppl. **21**: 343–81.

Saitoh, M. K., Narita, K. & Isikawa, S. (1970). Photosynthetic nature of some aquatic plants in relation to temperature. *Botanical Magazine* **83**: 10–12.

Sand-Jensen, K. & Rasmussen, L. (1978). Macrophytes and chemistry of acidic streams from lignite mining areas. *Botanisk Tidsskrift* **72**: 105–11.

Sanford, G. R. (1979). Temperature related growth patterns in *Amblystegium riparium*. *Bryologist* **82**: 525–32.

Sanford, G. R., Bayer, D. E. & Knight, A. W. (1974). *An evaluation of environmental factors affecting the distribution of two aquatic mosses in the Sacramento River near Anderson, California*. University of California Departments of Botany, Water Science, and Engineering.

Schiller, J. (1938). Die Förderung des Wachstums von Moosen im Gasteiner Thermalwasser. *Osterreichische Botanische Zeitschrift* **87**: 114–18.

Sommer, C. & Winkler, S. (1982). Reaktionen im Gaswechsel von *Fontinalis antipyretica* Hedw. nach experimentellen Belastungen mit Schwermetallverbindungen. *Archiv für Hydrobiologie* **93**: 503–14.

Sonesson, M. (1966). On *Drepanocladus trichophyllus* in the Tornetrask area. *Botaniska Notiser* **119**: 379–400.

Steemann Nielsen, E. (1947). Photosynthesis of aquatic plants with special reference to the carbon sources. *Dansk Botanisk Arkiv* **12**: 5–71.

Steere, W. C. (1976). Ecology, phytogeography and floristics of Arctic Alaskan bryophytes. *Journal of the Hattori Botanical Laboratory* **41**: 47–72.

Steinman, A. D. & Boston, H. L. (1993). The ecological role of aquatic bryophytes in a heterotrophic woodland stream. *Journal of the North American Benthological Society* **12**: 17–26.

Suren, A. M. (1988). Ecological role of bryophytes in high alpine streams of New Zealand. *Internationale Vereinigung für Theoretische und Angewandte Limnologie* **23**: 1412–16.

Suren, A. M. (1991). Bryophytes as invertebrate habitat in two New Zealand alpine streams. *Freshwater Biology* **26**: 399–418.

Suren, A. M. (1996). Bryophyte distribution patterns in relation to macro-, meso-, and micro-scale variables in South Island, New Zealand streams. *New Zealand Journal of Marine and Freshwater Research* **30**: 501–23.

Suren, A. M. & Ormerod, S. J. (1999). Aquatic bryophytes in Himalayan streams: testing a distribution model in a highly heterogeneous environment. *Freshwater Biology* **40**: 697–716.

Szarek, E. (1994). The effect of abiotic factors on chlorophyll a in attached algae and mosses in the Sucha Woda stream (High Tatra Mts., southern Poland). *Acta Hydrobiologica (Cracow)* **36**: 309–22.

Tremp, H. (2003). Ecological traits of aquatic bryophytes and bioindication. Accessed on 31 October 2003 at <http://www.uni-hohenheim.de/www320/german/homepages/horst/image/pdfs/aquatic_bryophytes.pdf>.

Vanderpoorten, A., Klein, J.-P., Stieperaere, H. & Trémolières, M. (1999). Variations of aquatic bryophyte assemblages in the Rhine Rift related to water quality. I. The Alsatian Rhine floodplain. *Journal of Bryology* **21**: 17–23.

Vitt, D. H., Glime, J. M. & LaFarge-England, C. (1986). Bryophyte vegetation and habitat gradients of montane streams in western Canada. *Hikobia* **9**: 367–85.

Ward, J. V. (1986). Altitudinal zonation in a Rocky Mountain stream. *Archiv für Hydrobiologie Supplement* **74**: 133–99.

Watanabe, R. (1957). On some mosses growing in hotspring water. *Miscellanea Bryologica et Lichenologica* **13**: 2.

Welch, W. H. (1960). *A Monograph of the Fontinalaceae*. The Hague: Martinus Nijhoff.

Westlake, D. F. (1981). Temporal changes in aquatic macrophytes and their environment. In *Dynamique de Populations et Qualite de l'Eau* (Actes du Symposium de l'Institute d'Ecologie du Bassin de la Somme, Chantilly, 1979), ed. H. Hoestlandt, pp. 110–38. Paris: Gauthier-Villars.

Zanten, B. O. van (1978a). Experimental studies on trans-oceanic long-range dispersal of moss spores in the Southern Hemisphere. In *Congres International de Bryologie, Bryophytorum Bibliotheca* **13**: 715–33.

Zanten, B. O. van (1978b). Experimental studies on trans-oceanic long-range dispersal of moss spores in the Southern Hemisphere. *Journal of the Hattori Botanical Laboratory* **44**: 455–82.

Zanten, B. O. van (1984). Some considerations on the feasibility of long-distance transport in bryophytes. *Acta Botanica Neerlandica* **33**: 231–2.

Zanten, B. O. van & Gradstein, S. R. (1987). Feasibility of long-distance transport in Colombian hepatics, preliminary report. *Symposia Biologica Hungarica* **35**: 315–22.

Zanten, B. O. van & Gradstein, S. R. (1988). Experimental dispersal geography of neotropical liverworts. *Nova Hedwigia*, suppl. **90**: 41–94.

7

Aquatic Bryophytes under Ultraviolet Radiation

JAVIER MARTÍNEZ-ABAIGAR AND ENCARNACIÓN NÚÑEZ-OLIVERA

Introduction

Ultraviolet radiation (UVR) has many effects on photosynthetic organisms. It is a minority component (about 6%) of solar radiation in comparison with the dominant visible/photosynthetic and infrared bands. However, UVR is a natural environmental factor that has been involved in the appearance of diverse adaptive changes in organisms through the development of life on Earth (Cockell & Knowland 1999). UVR induces a number of biological processes in all living organisms, including humans, and many of them are harmful. In this respect, among the three wavelength categories into which UVR is divided by the CIE (Commission Internationale d'Eclairage), the most damaging UV-C (< 280 nm) is not relevant at the present time because of its complete absorption by stratospheric oxygen and ozone, but both UV-B (280–315 nm) and UV-A (315–400 nm) penetrate the biosphere and have significant biological effects. These effects are highly dependent on wavelength, and different biological weighting functions have been conceived to calculate the biologically effective UV (UV_{BE}). UV_{BE} encompasses UV-A and UV-B. However, given the logarithmic increase in effectiveness with decreasing wavelength, UV_{BE} is dominated by UV-B, especially at shorter wavelengths. Therefore, most studies on the effects of UVR have dealt with UV-B. This has been especially true since the discovery of the anthropogenic stratospheric ozone reduction, because UV-B (and not UV-A) is absorbed by stratospheric ozone, and thus ozone reduction leads to an increase in surface UV-B levels. Nevertheless, the present tendency is to also pay attention to UV-A in the development of biological weighting functions (Flint *et al.* 2003), considering that the wavelength limit between UV-B and UV-A is somewhat diffuse.

Bryophyte Ecology and Climate Change, eds. Zoltán Tuba, Nancy G. Slack and Lloyd R. Stark. Published by Cambridge University Press.

UVR is a delicate meteorological variable to measure; for example, there are discrepancies between ground-based measurements and satellite estimates of surface UV irradiance (Seckmeyer *et al.* 2008). UV irradiance at ground level depends on a number of factors, such as latitude, season, time of day, altitude, presence of clouds or aerosols, surface reflectivity, and ozone levels (McKenzie *et al.* 2007). The ozone loss as a result of anthropogenic emissions of halogenated carbon compounds has been most dramatic over the Antarctic continent, although reductions at Arctic and mid-latitudes have also been observed. At mid-latitudes, the resultant increase in solar UV-B has been estimated at 6%–12% above the 1980 levels (McKenzie *et al.* 2003), although this increase may be masked by large seasonal changes and geographic differences (Häder *et al.* 2007). Elevated UVR is expected to continue for several decades because of the lack of a full recovery of ozone. Consequently, studies on the effects of ambient and enhanced UV levels on organisms and ecosystems are becoming increasingly important, also taking into account that interactions between ozone depletion and climate change may occur through variations in cloudiness, aerosols, and surface reflectivity (McKenzie *et al.* 2007).

In humans, an excessive UV exposure (mainly to UV-B, but also to UV-A) causes acute and chronic damage to eyes and skin, including sunburn and cancer, and down-regulates immune responses, although some beneficial effects of UVR are also clear (Norval *et al.* 2007). Excessive exposures seem to be caused by irresponsible habits rather than by stratospheric ozone reduction. In photosynthetic organisms, increased UV may cause diverse damage to the photosynthetic apparatus: pigment degradation, photoinhibition, and decreases in quantum yield, photosynthetic rates, and the activity of the Calvin cycle enzymes (Jansen *et al.* 1998). In addition, DNA alterations, oxidative damage, and changes in mineral absorption can occur. This may lead to alterations in growth and development. However, some controversy about the ecological relevance of these effects still persists, because much of the early work concerning the UV effects was conducted indoors using unrealistically high UV doses and UV/PAR proportions (PAR is the photosynthetically active radiation, i.e., the part of the solar spectrum driving photosynthesis in plants, which extends from 400 to 700 nm). Thus, the extrapolation of these results to field conditions seems difficult (Searles *et al.* 2001a; Day & Neale 2002). At the ecosystem level, UV can affect litter decomposition, nutrient cycling, trophic interactions, and the competitive balance between species (Caldwell *et al.* 2007). Photosynthetic organisms may develop a number of protection and repair mechanisms against the adverse effects of UV (Jansen *et al.* 1998): UV-absorbing compounds (flavonoids, phenyl-propanoids, mycosporine-like amino acids, etc.), antioxidant and photoprotective mechanisms, and repairing or turnover of damaged biomolecules such as DNA and proteins.

The effects of UVR on photosynthetic organisms have been studied mainly in terrestrial plants, especially crops (Caldwell *et al.* 2007), and in marine phytoplankton and macroalgae (Day & Neale 2002; Häder *et al.* 2007). Effects of UVR on photosynthetic organisms from freshwater ecosystems have been little investigated, in line with their minor contribution to the global biomass and primary production. However, lakes and rivers have outstanding ecological importance as local systems and it would be desirable to know the effects of UVR on the organisms inhabiting them and evaluate their potential vulnerability. In lakes, the penetration of UVR and its effects on phytoplankton have been the most studied topics, under both field and laboratory conditions (Häder *et al.* 2007). In flowing waters (rivers and streams), most data come from laboratory studies and not much fieldwork has been done (Rader & Belish 1997a, b; Kelly *et al.* 2003), probably owing to intrinsic methodological problems derived from the strongly dynamic and changeable environmental conditions (discharge, water velocity, bed morphology, depth, etc.). Mires, including bogs, represent another type of freshwater ecosystem, with particular importance at high latitudes and altitudes, in which significant fieldwork has been done in relation to UVR.

Bryophytes and ultraviolet radiation

Bryophytes are acquiring increasing importance in the context of UVR research and are usually mentioned in the most recent reviews on this topic (Björn & Mckenzie 2007; Caldwell *et al.* 2007; Häder *et al.* 2007; Newsham & Robinson 2009). This is a logical consequence of the increasing work on the effects of UVR on bryophytes that has been performed mainly in the past decade. To our knowledge, about 75 papers containing original data have been published on this topic (Table 7.1). The research has focused mainly on terrestrial and semiaquatic bryophytes from Antarctic habitats and circumpolar heathlands and peatlands. Most of the species incorporated are mosses: several *Sphagnum* species, *Hylocomium splendens* (typical from forest soils), and *Polytrichum commune* (typical from a wide range of acid habitats in damp to wet situations). Liverworts have been notably less studied than mosses, except the aquatic *Jungermannia exsertifolia* subsp. *cordifolia* and the terrestrial *Cephaloziella varians*; no hornwort has been investigated in this regard.

Diverse methodological approaches have been applied. Studies under both field and controlled conditions have been conducted, and in the latter case both laboratory and greenhouse studies were employed. Manipulative experiments have included the two main experimental options in the context of UVR research: UVR supplementation using lamps to address the ozone depletion issue (38 studies), and UVR exclusion using filters to assess the effects of current UVR levels (17 studies). When using lamps, most results have been obtained

Table 7.1. *Original papers on the effects of UV radiation on bryophytes*

"Species Used": L, liverwort; M, moss. "Ambient": T, terrestrial; P, peatlands; A, aquatic; R, rivers or streams; L, lakes. "Type of Experiment": F, Field; G, greenhouse; L, laboratory; E, exclusion of UV-B radiation; S, supplement of UV-B radiation; N, samples exposed to natural levels of solar radiation; VSh, very short duration (less than 1 day); Sh, short duration (1–30 days); M, medium duration (longer than one month and shorter than 6 months); Lo, long duration (6 months – 1 year); VLo, very long duration (longer than 1 year); ?, undetermined duration; H, historical study (comparison of samples over a prolonged period). "Variables used": A, alterations in DNA; C, cover; Fl, chlorophyll fluorescence; FlS, fluorescence spectra; G, growth; H, hydric relations; M, morphology; Mt1, primary metabolites (glucids, proteins, lipids); Mt2, secondary metabolites, including UV-absorbing compounds; N, mineral nutrients; Ox, variables of oxidative stress (peroxide content, lipid peroxidation, ascorbate, superoxide dismutase, peroxidase, catalase); P, photosynthesis; Ph, phenology; PP, photosynthetic pigments; PS1 and PS2, activity of photosystems I and II, respectively; R, respiration; Rf, reflectance indices; Sc, sclerophylly; U, ultrastructure; Z, other variables.

Reference	Species used	Ambient	Type of experiment	Variables used
Arróniz-Crespo *et al.* (2004)	*Chiloscyphus polyanthos* (L), *Jungermannia exsertifolia* subsp. *cordifolia* (L), *Marsupella sphacelata* (L), *Scapania undulata* (L), *Brachythecium rivulare* (M), *Bryum alpinum* (M), *Bryum pseudotriquetrum* (M), *Fontinalis antipyretica* (M), *Palustriella commutata* (M), *Philonotis seriata* (M), *Polytrichum commune* (M), *Racomitrium aciculare* (M), *Rhynchostegium riparioides* (M), *Sphagnum flexuosum* (M)	A (R)	F, N	Mt2, Sc
Arróniz-Crespo *et al.* (2006)	*Jungermannia exsertifolia* subsp. *cordifolia* (L)	A (R)	F, N	Fl, Mt2, P, PP, R, Sc
Arróniz-Crespo *et al.* (2008a)	*Jungermannia exsertifolia* subsp. *cordifolia* (L)	A (R)	L, S, M	Fl, Mt2, PP, Sc
Arróniz-Crespo *et al.* (2008b)	*Jungermannia exsertifolia* subsp. *cordifolia* (L)	A (R)	L, S, M	Fl, Mt2, PP, Sc
Ballaré *et al.* (2001)	*Sphagnum magellanicum* (M)	P	F, E, VLo	G, Mt2
Barsig *et al.* (1998)	*Polytrichum commune* (M)	P	G, S, M	Mt1, Mt2, PP, U
Björn *et al.* (1998)	*Aulacomnium turgidum* (M), *Dicranum elongatum* (M), *Hylocomium splendens* (M), *Polytrichum commune* (M), *P. hyperboreum* (M), *Sphagnum fuscum* (M)	T, P	F, S, M-VLo	G, H

Blokker *et al.* (2006)	*Ceratodon purpureus* (M), *Polytrichum commune* (M)	—	?	Mt2
Boelen *et al.* (2006)	*Chorisodontium aciphyllum* (M), *Polytrichum strictum* (M), *Sanionia uncinata* (M), *Warnstorfia sarmentosa* (M)	T, P	F, S, Sh, M	A, Mt2
Clarke & Robinson (2008)	*Bryum pseudotriquetrum* (M), *Ceratodon purpureus* (M), *Schistidium antarctici* (M)	T	F, N, M	Mt2
Conde-Álvarez *et al.* (2002)	*Riella helicophylla* (L)	A (L)	L, E, VSh	Fl, Mt2, P, PP, R
Csintalan *et al.* (2001)	*Dicranum scoparium* (M), *Leucobryum glaucum* (M), *Mnium hornum* (M), *Pellia epiphylla* (L), *Plagiomnium undulatum* (M), *Plagiothecium undulatum* (M), *Polytrichum formosum* (M), *Sphagnum capillifolium* (M), *Tortula ruralis* (M)	T	L, S, Sh-M	Fl, FlS, Mt2
Dunn & Robinson (2006)	*Bryum pseudotriquetrum* (M), *Ceratodon purpureus* (M), *Schistidium antarctici* (M)	T	F, N, M	Mt2
Gehrke (1998)	*Sphagnum fuscum* (M)	P	F, S, VLo	G, M, Mt2, P, PP, R
Gehrke (1999)	*Hylocomium splendens* (M), *Polytrichum commune* (M)	T, P	F, S, VLo	G, M, Mt2, PP
Gehrke *et al.* (1996)	*Hylocomium splendens* (M), *Sphagnum fuscum* (M)	T, P	F, S, VLo	G, H, Mt2, PP
Green *et al.* (2000)	*Bryum argenteum* (M)	T	F, E, VSh	Fl, P
Green *et al.* (2005)	*Bryum subrotundifolium* (M), *Ceratodon purpureus* (M)	T	F, N, Sh	M, Mt2
Harris (2009)	*Plagiomnium spp.* (M)	T	N	Mt2
Hooijmaijers & Gould (2007)	*Isotachis lyallii* (L), *Jamesoniella colorata* (L)	T	F, N, ?	Fl, Mt2, PP
Hughes *et al.* (2006)	*Drepanocladus* sp. (M)	T	F, N, Sh	Z
Huiskes *et al.* (1999)	*Sanionia uncinata* (M)	T	—	—
Huiskes *et al.* (2001)	*Sanionia uncinata* (M)	T	F, E, Sh	Fl
Huttunen *et al.* (1998)	*Dicranum* sp. (M), *Hylocomium splendens* (M), *Polytrichum commune* (M)	T, P	G, S, ?	M
Huttunen *et al.* (2005a)	*Dicranum scoparium* (M), *Funaria hygrometrica* (M), *Hylocomium splendens* (M), *Pleurozium schreberi* (M), *Polytrichum commune* (M), *Polytrichastrum alpinum* (M), *Sphagnum angustifolium* (M), *S. capillifolium* (M), *S. fuscum* (M), *S. warnstorfii* (M)	T, P	N, H	M, Mt2

Table 7.1. (*cont.*)

Reference	Species used	Ambient	Type of experiment	Variables used
Huttunen *et al.* (2005b)	*Hylocomium splendens* (M), *Pleurozium schreberi* (M)	T	N, H	M, Mt2
Ihle (1997)	*Conocephalum conicum* (L)	T	L, S, VSh	Mt1
Ihle & Laasch (1996)	*Conocephalum conicum* (L)	T	L, S, VSh-Sh	Fl, Mt1, Mt2, P
Johanson *et al.* (1995)	*Hylocomium splendens* (M)	T	G, S, ?	G, Ph
Kato-Noguchi & Kobayashi (2009)	*Hypnum plumaeforme* (M)	T	L, S, Sh	Mt2
Lappalainen *et al.* (2008)	*Pleurozium schreberi* (M)	T	F, S, VLo	G, Mt2, PP, Sc
Lewis Smith (1999)	*Bryum argenteum* (M), *Bryum pseudotriquetrum* (M), *Ceratodon purpureus* (M)	T	F, E, M	G
Lovelock & Robinson (2002)	*Bryum pseudotriquetrum* (M), *Ceratodon purpureus* (M), *Grimmia antarctici* (M)	T	F, N, ?	Mt2, PP, Rf
Lud *et al.* (2002)	*Sanionia uncinata* (M)	T	F, L, E, S, VSh-VLo	A, G, Fl, M, P, Mt2, PP
Lud *et al.* (2003)	*Sanionia uncinata* (M)	T	F, E, S, VSh-Sh	A, Fl, Mt2, P, PP, R
Markham *et al.* (1990)	*Bryum argenteum* (M)	T	N, H	Mt2
Markham *et al.* (1998)	*Marchantia polymorpha* (L)	T	G, S, M	G, M, Mt2, Ph
Martínez-Abaigar *et al.* (2003)	*Jungermannia exsertifolia* subsp. *cordifolia* (L), *Fontinalis antipyretica* (M)	A (R)	L, S, M	Fl, Mt2, P, PP, R, Sc
Martínez-Abaigar *et al.* (2004)	*Jungermannia exsertifolia* subsp. *cordifolia* (L), *Fontinalis antipyretica* (M)	A (R)	L, S, M	G, M
Martínez-Abaigar *et al.* (2008)	*Jungermannia exsertifolia* subsp. *cordifolia* (L), *Fontinalis antipyretica* (M)	A (R)	L, S, M	Fl, Mt2, P, PP, R, Sc
Martínez-Abaigar *et al.* (2009)	*Jungermannia exsertifolia* subsp. *cordifolia* (L), *Marsupella sphacelata* (L), *Scapania undulata* (L), *Brachythecium rivulare* (M), *Bryum pseudotriquetrum* (M), *Racomitrium aciculare* (M)	A (R)	L, S, Sh	Fl, G, Mt2, PP, Sc

Montiel *et al.* (1999)	*Sanionia uncinata* (M)	T	F, S, ?	Fl
Newsham (2003)	*Andreaea regularis* (M)	T	F, N, M	Mt2, PP
Newsham *et al.* (2002)	*Sanionia uncinata* (M), *Cephaloziella varians* (L)	T	F, N, Sh-M	Fl, Mt2, PP
Newsham *et al.* (2005)	*Cephaloziella varians* (L)	T	F, N, E, M	Mt2, PP
Niemi *et al.* (2002a)	*Sphagnum angustifolium* (M), *S. papillosum* (M), *S. magellanicum* (M)	P	F, S, M	G, Mt2, PP
Niemi *et al.* (2002b)	*Sphagnum balticum* (M), *Sphagnum papillosum* (M)	P	F, S, M	G, Mt2, PP
Núñez-Olivera *et al.* (2004)	*Jungermannia exsertifolia* subsp. *cordifolia* (L), *Fontinalis antipyretica* (M)	A (R)	L, S, M	Fl, G, Mt1, Mt2, P, PP, R, Sc
Núñez-Olivera *et al.* (2005)	*Jungermannia exsertifolia* subsp. *cordifolia* (L), *Fontinalis antipyretica* (M)	A (R)	L, S, Sh	Fl, Mt1, Mt2, P, PP, R, Sc
Núñez-Olivera *et al.* (2009)	*Jungermannia exsertifolia* subsp. *cordifolia* (L)	A (R)	F, N, VLo	A, Fl, Mt2, Sc
Otero *et al.* (2006)	*Jungermannia exsertifolia* subsp. *cordifolia* (L)	A (R)	L, S, Sh	A, Fl, Mt2, P, PP, R, Sc
Otero *et al.* (2008)	*Clasmatocolea vermicularis* (L), *Noteroclada confluens* (L), *Pachyglossa dissitifolia* (L), *Pseudolepicolea quadrilaciniata* (L), *Triandrophyllum subtrifidum* (L), *Breutelia dumosa* (M), *Bryum laevigatum* (M), *Pohlia wahlenbergii* (M), *Racomitrium lamprocarpum* (M), *Scorpidium revolvens* (M), *Scouleria patagonica* (M), *Sphagnum fimbriatum* (M), *Vittia pachyloma* (M), *Warnstorfia exannulata* (M), *Warnstorfia sarmentosa* (M)	A (R)	F, N	Mt2, Sc
Otero *et al.* (2009)	*Jungermannia exsertifolia* subsp. *cordifolia* (L)	A (R)	N, H	M, Mt2
Phoenix *et al.* (2001)	*Hylocomium splendens* (M)	T	F, S, VLo	G, H
Post & Vesk (1992)	*Cephaloziella exiliflora* (L)	T	F, N, Sh	M, Mt2, P, PP, U
Prasad *et al.* (2004)	*Riccia sp.* (L)	T	L, S, VSh	Ox, PP, PS1, PS2
Rader & Belish (1997b)	*Fontinalis neomexicana* (M)	A (R)	F, E-S, M	G
Robinson *et al.* (2005)	*Grimmia antarctici* (M)	T	F, E, VLo	Fl, H, M, Mt2, P, PP, Rf
Robson *et al.* (2003)	*Sphagnum magellanicum* (M)	P	F, E, VLo	G, M

Table 7.1. (*cont.*)

Reference	Species used	Ambient	Type of experiment	Variables used
Robson *et al.* (2004)	*Sphagnum magellanicum* (M)	P	F, E, VLo	G, M
Rozema *et al.* (2002)	*Tortula ruralis* (M)	T	F, E, ?	G, Mt2
Rozema *et al.* (2006)	*Polytrichum hyperboreum* (M), *Sanionia uncinata* (M)	T	F, S, VLo	C, G
Ryan *et al.* (2009)	*Bryum argenteum* (M)	T	N, H	Mt2
Schipperges & Gehrke (1996)	*Hylocomium splendens* (M), *Sphagnum fuscum* (M)	T, P	F-L, S, M- VLo	G, H, P
Searles *et al.* (1999)	*Sphagnum magellanicum* (M)	P	F, E, Lo	G, Mt2, PP
Searles *et al.* (2001b)	*Sphagnum magellanicum* (M)	P	F, E, VLo	G, M, Mt2
Searles *et al.* (2002)	*Sphagnum magellanicum* (M)	P	F, E, VLo	G, M, Mt2, PP
Snell *et al.* (2007)	*Cephaloziella varians* (L)	T	F, N, M	Fl, Mt2, PP
Snell *et al.* (2009)	*Cephaloziella varians* (L)	T	F, E, M	Fl, Mt2, P, PP
Sonesson *et al.* (1996)	*Hylocomium splendens* (M)	T	L, S, M	G, P
Sonesson *et al.* (2002)	*Dicranum elongatum* (M), *Sphagnum fuscum* (M)	P	F, S, VLo	G, H
Taipale & Huttunen (2002)	*Hylocomium splendens* (M), *Pleurozium schreberi* (M)	T	F, S, M	Mt2
Takács *et al.* (1999)	*Dicranum scoparium* (M), *Leucobryum glaucum* (M), *Mnium hornum* (M), *Pellia epiphylla* (L), *Plagiothecium undulatum* (M), *Polytrichum formosum* (M), *Tortula ruralis* (M)	T	G, S, Sh-M	Fl
Turnbull & Robinson (2009)	*Bryum pseudotriquetrum* (M), *Ceratodon purpureus* (M), *Grimmia antarctici* (M)	T	F, N, M	A
Turnbull *et al.* (2009)	*Bryum pseudotriquetrum* (M), *Ceratodon purpureus* (M), *Grimmia antarctici* (M)	T	L, S, VSh	A

under unrealistic conditions of UV irradiance, UVR daily pattern, proportions of UV-B, UV-A and PAR, etc. If filters are used, above-ambient UV-B levels associated with ozone reduction cannot be provided, and filters may modify the micro-environmental conditions of the samples. All of these methodological problems may limit the ecological relevance of the results obtained. A third way to assess UVR effects exploits natural gradients of UVR, such as temporal variations in the Antarctic during the occurrence of an "ozone hole", or spatial variations with water depth or altitude. In bryophytes, 22 studies of this kind have been carried out. The advantage of such studies is that no environmental circumstance of the plants is modified, something greatly needed in the context of UVR works. However, these experiments cannot reproduce the conditions of potential ozone reductions, require an accurate measurement of the UVR gradient, and should differentiate the influence of interacting factors that could obscure the specific effects of UVR. The duration of the experiments has been diverse, from a few hours of UV exposure (usually under controlled conditions) to several years (under field conditions), and the bryophyte responses have been assessed using morphological and, especially, physiological variables (Table 7.1).

The results obtained are diverse, since UVR has been found to stimulate, depress, or have no effect on the bryophyte performance. To a certain extent, this may have been caused by the diversity of species and experimental conditions used in the different studies. Several investigations have found a growth reduction in bryophytes exposed to UVR, but this effect seems to depend on the species considered, the experimental design, and other additional factors such as temperature, water availability, and CO_2 concentration. Other harmful effects (chlorophyll degradation, reduction in photosynthesis rates and F_v/F_m) are even less clear, since contradictory results have been reported. The increase in UV-absorbing compounds, the most usual response of vascular plants to enhanced UV (Searles *et al.* 2001a), has been manifested less frequently in bryophytes. Beneficial effects of UVR on bryophyte growth have also been reported (Johanson *et al.* 1995; Björn *et al.* 1998; Searles *et al.* 1999; Phoenix *et al.* 2001), which further complicates the global interpretation of the results. This controversy contrasts with intuitive thoughts that bryophytes would be strongly sensitive to UVR because of their structural simplicity and the consequent lack of defenses commonly found in higher plants: hairs, epicuticular waxes, thick cuticles, multilayered epidermis, etc. Given that bryophyte leaves are mostly unistratose and lack air spaces, the molecular targets of UVR could be reached much more easily.

Globally, the responses of bryophytes to UVR and their protective systems are still poorly characterized, and thus further study is required under both controlled and field conditions. In particular, long-term field studies under realistic enhancements of UVR (see Lappalainen *et al.* 2008; see also Chapter 5, this volume) are badly needed. However, it is already clear that bryophytes as a group are not as strongly UVR-sensitive as could be anticipated, taking into consideration their structural limitations, and that many of the species studied can acclimate well to high levels of UVR.

Aquatic bryophytes and ultraviolet radiation

The concept of "aquatic bryophyte"

Aquatic bryophytes are integrated in a frame defined by two different environmental gradients: current–wave action and water level fluctuation (Vitt & Glime 1984). With respect to the first factor, aquatic bryophytes may be limnophilous (living in standing waters) to rheophilous (living in running waters). With respect to the second gradient, they range from obligate aquatics to facultative aquatics and semiaquatic emergents. So defined, a great diversity of bryophytes can be considered as aquatics, and they prevail in different environments, such as mountain streams, deep lakes, and certain wetlands, where they play a relevant ecological role in primary production, nutrient cycles, and food webs. Depending on the system where they occur, they can support periphyton and provide a refuge, and occasionally provide food, for protozoans, micro- and macro-invertebrates, amphibians, and fish. Bryophyte domination in those environments is primarily based on their tolerance to adverse environmental factors.

The key stresses and disturbances in streams are abrasion by turbulent water and suspended solids, substratum movement, cold water, seasonal desiccation, nutrient limitation in soft waters, CO_2 limitation in the stagnant parts of alkaline streams, high photosynthetic and UV irradiances in unshaded high-altitude streams, and diaspore difficulties in attaching to new substrates. In lakes, the adverse factors are waterlogging, cold water, low irradiances in the deep zones, high hydrostatic pressure, and abrasion along the shores. Finally, in mires, waterlogging, fluctuation of the water table level, high irradiances, and particular mineral stresses dependent on the diverse chemical composition of water in the different systems, are typical unfavorable factors.

Effects of UVR on lake bryophytes

In an important study on the effects of UVR on lake bryophytes (Conde-Álvarez *et al.* 2002), samples of the thalloid liverwort *Riella helicophylla* from a shallow saline lake were collected and cultivated for one day, throughout a natural daily light cycle, under two radiation treatments: full solar radiation (UVR + PAR) and solar radiation deprived of UVR (PAR treatment). There were significant differences between the two treatments in the maximum quantum yield of PSII (F_v/F_m), the effective quantum yield of photosynthetic energy conversion of PSII (Φ_{PSII}), the electron transport rate (ETR), and the initial slope of ETR vs. irradiance curve (all higher in PAR plants than in UV+PAR plants throughout the day), photosynthetic capacity (higher in PAR plants only at noon), chlorophyll a (lower in UV+PAR only at 11.00), and phenolic compounds (higher in UV+PAR only at 13.30). No differences between treatments were found in dark respiration, photochemical quenching and carotenoid concentration, and only slight ones in non-photochemical quenching (higher in UV +PAR only in the morning). In conclusion, although solar UVR caused some transitory damage to photosynthesis, recovery took place in the afternoon, and thus no irreversible damage occurred in the short term.

Effects of UVR on bog bryophytes

Several studies have been conducted on bryophytes from circumpolar bogs, most of them under long-term field conditions (Table 7.1), which emphasize the significance of the results obtained. In southern latitudes, the effects of UVR exclusion have been studied in one bryophyte, *Sphagnum magellanicum*, in a bog in Tierra del Fuego. In a field experiment several months long, height growth of this species was not affected by either near-ambient or reduced solar UV-B (Searles *et al.* 1999). However, in a more prolonged experiment (three years), Searles *et al.* (2001b) found that height growth was less under near-ambient UV-B than under reduced UV-B, whereas mass per unit of height growth was greater under near-ambient UV-B. The increased height growth under reduced UV-B was counteracted by an increased volumetric density under near-ambient UV-B, so that biomass production was not influenced by the different treatments during the three years of the study (Searles *et al.* 2002). Chlorophylls, carotenoids, chlorophyll a/b ratio, and UV-absorbing compounds were not affected by UV-B manipulation (Searles *et al.* 1999, 2002). Robson *et al.* (2003) continued the experiment for another three years and corroborated previously reported results. Thus, at least in Tierra del Fuego, ambient levels of UVR seemed to affect morphogenic rather than production processes in *Sphagnum magellanicum*.

In Arctic bogs, UVR supplements have been applied in the field to simulate realistic ozone reductions (15%–20%). The height increment of *Sphagnum fuscum* was reduced by 20% in the first year of exposure to enhanced UVR (25% enhancement over the controls: Gehrke *et al.* 1996). In a two-year experiment, height increment, spatial shoot density, and dark respiration of the same species decreased under enhanced UVR, but dry mass per unit length increased, and thus biomass production did not change over the course of the experiment

(Gehrke 1998). In the same study, the integrity of the photosynthetic apparatus was somewhat affected by enhanced UVR, since the concentration of chlorophyll a and carotenoids decreased, but net photosynthesis, chlorophyll a/b ratio and the levels of UV-absorbing compounds hardly changed. As with *Sphagnum magellanicum* under ambient levels of UVR, biomass production of *Sphagnum fuscum* was not modified by enhanced UVR, although height growth was affected. In addition, Gehrke (1998) pointed out that the great variability in productivity among microsites probably masked any effect of enhanced UVR.

Other studies in Arctic bogs used modulated systems that provide UV supplements proportional to ambient UV levels so that the simulation of UVR enhancement is more realistic. Using this system, two different studies of three months' duration (one growing season) were conducted by Niemi *et al.* (2002a,b). In the first one, three *Sphagnum* species (*S. angustifolium*, *S. papillosum*, and *S. magellanicum*) were used and the effects of UV-A and UV-B radiation were differentiated. Membrane damage, chlorophyll and carotenoid concentrations, and UV-absorbing compounds did not show clear changes under UV-A or UV-B enhancements (30% over ambient levels), and no significant differences between the treatments in either capitulum or stem dry mass of *S. angustifolium* (the only species tested in this respect) were observed. In their second experiment, Niemi *et al.* (2002b) found more clear effects on *Sphagnum balticum* and *S. papillosum*, as both species showed significantly higher membrane permeability under enhanced UVR. However, the remaining variables measured (capitulum and stem biomass, chlorophyll, carotenoids, carbon isotope discrimination, and UV-absorbing compounds) did not show significant changes between samples exposed to current and enhanced UVR levels, except a surprising increase in chlorophyll, carotenoids, and chlorophyll a/b ratio under enhanced UVR in *S. balticum*.

In a more complex experimental design, Sonesson *et al.* (2002) tested the effect of increased UV-B, temperature, and irrigation during two consecutive seasons in *Sphagnum fuscum* and *Dicranum elongatum*. Increased UV-B had no statistically significant overall effect in length growth, chlorophyll and flavonoid contents, nor in the interaction between the species. However, the growth of *S. fuscum* responded negatively to increased UV-B under increased temperature at the peak of the growing season, probably because of water deficit. Different responses of *S. fuscum* in this study and previous ones (Gehrke *et al.* 1996; Gehrke 1998) were attributed to different weather conditions during the experimental periods.

Effects of UVR on mountain stream bryophytes

Mountain streams might be particularly exposed to the effects of UVR, since (1) the biologically active UVR increases between 5% and 20% per 1000 m altitudinal increase (Björn *et al.* 1998); (2) many organisms live emersed and fully exposed to UVR, or immersed at relatively low depths, where UVR can also reach them because it easily penetrates into the typically occurring oligotrophic shallow waters (Frost *et al.* 2005), and (3) the low temperatures prevailing during most of the year may limit the development of metabolically dependent mitigating mechanisms, such as the synthesis of UV-absorbing compounds and the action of antioxidant and DNA-repairing systems.

In a pioneer study on the effects of UVR on stream bryophytes, Rader and Belish (1997b) carried out a ten-week field experiment in which samples of the moss *Fontinalis neomexicana* were transplanted from a reference site to both a shaded and an open section of a mountain stream, and were irradiated with enhanced levels of UV-B radiation. The UV-exposed transplants from the open site showed an important reduction in dry biomass with respect to those under ambient conditions, whereas the transplants from the shaded site were not affected either by ambient or enhanced UV. The moss failed to grow at any site and under any treatment condition, and there was a loss of material in all samples from the beginning to the conclusion of the experiment. These facts show clearly the difficulties in implementing field manipulative experiments (using filters or lamps) in streams, especially in the long term.

Much more manipulative work has been done under laboratory conditions (Table 7.1). This is useful to characterize UVR responses while preventing the interference of other environmental factors. However, laboratory results are not directly comparable to those obtained under field conditions, and in particular, should not be used to predict the consequences of a potential UVR increase due to ozone reduction. In the laboratory, aquatic bryophytes have been cultivated under enhanced UVR simulating a realistic ozone reduction (20%) during periods ranging from 3 to 82 days. In two studies, the effects of UV-A and UV-B were measured separately, but contrasting results were obtained because in one case samples exposed to UV-A showed a behavior similar to that of those exposed only to PAR (Martínez-Abaigar *et al.* 2003), whereas in the other study the responses of UV-A samples were more similar to those of samples exposed to UV-A plus UV-B (Otero *et al.* 2006). Results of both studies are probably little comparable because the PAR level applied was notably different (100 vs. 500 $\mu mol\ m^{-2}\ s^{-1}$), and thus UV-A effects are still little characterized. In *Sphagnum*, UV-A radiation has been demonstrated to have little biological effect (Niemi *et al.* 2002a).

Some field studies using natural gradients of UVR have also been carried out (Table 7.1). In the following sections, we will detail the results obtained on mountain stream bryophytes in both laboratory experiments using enhanced UVR and in field studies using natural UVR gradients.

Laboratory studies on mountain stream bryophytes: responses of the variables used

Bryophyte responses have been analyzed in terms of diverse physiological variables: the photosynthetic pigment composition (chlorophyll and total carotenoid concentration, chlorophyll a/b ratio, chlorophylls/phaeopigments and chlorophylls/carotenoids ratios), xanthophyll index, and the relationship ([antheraxanthin + zeaxanthin]/chlorophyll a); the rates of net photosynthesis; some variables of chlorophyll fluorescence (the maximum quantum yield of PSII, F_v/F_m; the effective quantum yield of photosynthetic energy conversion of PSII, Φ_{PSII}; the apparent electron transport rate through PSII, ETR; and the quenching due to non-photochemical dissipation of absorbed light energy, NPQ); the rates of dark respiration; sclerophylly index (the quotient between the dry mass and the shoot area); the level of UV-absorbing compounds (analyzed both globally by spectrophotometry and individually by HPLC); DNA damage (appearance of thymine dimers); protein concentration; length growth, and morphological symptoms (both macro- and microscopic).

Not all the variables used were equally UV-responsive. UV stress may be preferentially indicated by a decrease in F_v/F_m and chlorophylls/phaeopigments ratios, and, to a lesser extent, by a decrease in chlorophyll a/b ratio and net photosynthesis rates. The fact that all these variables are directly related to the photosynthetic process is not surprising, because several of its components are recognized molecular targets of UVR (Jansen *et al.* 1998). In addition, these variables have been frequently used as indices of physiological vitality because they decline under stress conditions caused by diverse harmful factors, such as cold, high light, water deficit, or pollutants (Martínez-Abaigar & Núñez-Olivera 1998; Maxwell & Johnson 2000; DeEll & Toivonen 2003).

Dark-adapted values of F_v/F_m provide information about the intactness and potential efficiency of PSII, and are used as a sensitive indicator of plant photosynthetic performance (Maxwell & Johnson 2000). A decrease in F_v/F_m in particular indicates the phenomenon of photoinhibition, which destroys the central core protein D1 of PSII; D1 is a well-known target of UVR (Jansen *et al.* 1998). Stream bryophytes are more easily photoinhibited under enhanced UVR than under PAR only (Martínez-Abaigar *et al.* 2003).

The brown-colored phaeopigments are products of chlorophyll breakdown. Chlorophyll/phaeopigment ratios may be useful to evaluate the physiological state of a plant because they indicate the proportion of intact chlorophylls and decrease under stress conditions (Martínez-Abaigar & Núñez-Olivera 1998). The first ratio is based on the fact that the blue maximum of chlorophylls at 430–435 nm shifts to 410–415 nm in the respective phaeopigments, and the quotient between the absorbances at 430 and 410 nm, or between the absorbances at 435 and 415 nm, indicates the relative proportion of chlorophylls and phaeopigments. The second ratio requires the acidification of the pigment extract, which totally transforms chlorophylls into phaeopigments, and then the ratio between the absorbance at 665 nm before and after acidification is calculated. As the red maximum of phaeopigments at 665 nm is lower than those of their respective chlorophylls, this ratio will again indicate the relative proportion of chlorophylls and phaeopigments.

The chlorophyll a/b quotient has a notable ecophysiological relevance and typically decreases in plants that are experiencing senescence or are under stress conditions (Martínez-Abaigar & Núñez-Olivera 1998), since the degradation of the light-harvesting complexes of the photosystems, relatively enriched in chlorophyll b, is slower than that of the core complexes, in which only chlorophyll a occurs. The chlorophyll a/b quotient also decreases in shade plants.

The alterations of key components of the photosystems (D1 protein, chlorophylls) may lead to a general decrease of photosynthesis rates under enhanced UV, which can be intensified by UV-driven damage in photosynthetic enzymes (Jansen *et al.* 1998). However, photosynthetic rates may remain unaffected under field conditions (Allen *et al.* 1998). Even in the laboratory, damage to photosynthetic pigments, proteins or enzymes does not necessarily imply a decrease in photosynthetic rates.

Other physiological variables that have been shown to be UV-responsive in bryophytes from mountain streams, although to a lower extent than those described above, are the sclerophylly index, chlorophyll and carotenoid concentrations, and growth. In cormophytes (ferns and seed plants), leaf sclerophylly increases with the development of non-photosynthetic structures such as epidermal cells, hairs, cuticles, epicuticular materials, cell walls, and vascular and supporting tissues, as well as with increasing contents of organic and inorganic solutes. In bryophytes, shoot sclerophylly may depend on the proportion of leaves (phyllidia) and stems (caulidia), the leaf architecture, the ratio of photosynthetic and non-photosynthetic (vascular, supporting-mechanical) tissues, the leaf thickness and papillosity, the proportion between cell walls and protoplasts, and the organic and inorganic contents. The higher values of the sclerophylly index reported under enhanced UV in bryophytes may be due to lower elongation, which could lead to the production of less soft tissue. Hence, this index may indicate morphogenic changes and could be an indirect measurement of growth. The sclerophylly index could be interpreted in a similar way to the mass per unit length ratio used in other bryophytes (Gehrke 1998; Searles *et al.* 2001b).

Chlorophylls and carotenoids are UV targets (Jansen *et al.* 1998) and thus both pigments would be expected to decrease under enhanced UVR. However, they may increase, decrease, or remain unaltered, and thus the effects of UVR on chlorophylls and carotenoids in bryophytes are still obscure, probably owing to interspecific and experimental differences. Usually, the concentrations of total chlorophylls and total carotenoids fluctuate in the same direction in response to enhanced UVR, because chlorophylls and carotenoids are components of photosystems and, when these are constructed or degraded, both types of pigments increase or decrease, respectively. However, their functions are not the same: chlorophylls are specialized for light absorption, whereas carotenoids are involved in either photoprotection or light absorption. Thus, if carotenoids involved in photoprotection were individually analyzed, their changes would probably be independent of those of chlorophylls. Photoprotecting carotenoids have been analyzed only once in stream bryophytes; the ratio (antheraxanthin + zeaxanthin)/chlorophyll a increased under enhanced UVR, showing a higher photoprotection or a preferential degradation of chlorophyll a (Otero *et al.* 2006).

With respect to growth measurements, only length growth has been tested in response to enhanced UVR in bryophytes from mountain streams. A reduction in length growth under enhanced UVR was found only in *Jungermannia exsertifolia* subsp. *cordifolia* among the six species tested, but otherwise this liverwort was UV-tolerant on the basis of most physiological variables. This suggests that the reduction in length growth would be a morphogenic rather than a damage response. It should be also taken into account that growth measurement in bryophytes is not an easy task to do, and the high variability of data can obscure the interpretation.

Damages caused by enhanced UVR in stream bryophytes, as indicated by the variables described above, are quite similar to those found in other photosynthetic organisms, from phytoplankton to flowering plants. Those variables are indicative of the basic physiological state of the plant, and the responses found are non-specific and may be caused by other adverse factors than enhanced UVR. Thus, the interpretation of results must be cautious. Nevertheless, when a high number of experimental variables is used, multivariate analyses may help demonstrate a "UV general syndrome" suffered by the plants exposed to enhanced UVR and indicated by a combination of variables (Núñez-Olivera *et al.* 2004; Otero *et al.* 2006).

A separate comment should be made about DNA damage. This kind of damage has been rarely measured in bryophytes, and it generally increased in samples exposed to enhanced UVR under laboratory conditions, but not in the field (Lud *et al.* 2002, 2003; Boelen *et al.* 2006; Otero *et al.* 2006; Núñez-Olivera *et al.* 2009; Turnbull & Robinson 2009; Turnbull *et al.* 2009). In the only aquatic species tested (the liverwort *Jungermannia exsertifolia* subsp. *cordifolia*), DNA damage appeared exclusively in the samples exposed to enhanced UVR (mainly consisting of

UV-B) in the laboratory. Thus, this response seemed to be largely UV-B-specific, congruent with the target character of DNA with respect to UV-B. Effects of UVR on respiration remain obscure because contrasting responses have been found.

Finally, morphological responses to enhanced UVR were studied in two species. The moss *Fontinalis antipyretica* showed some damage, manifested in brown coloration, development of the central fibrillar body in the cells, chloroplast disappearance, and the presence of protoplasts progressively vesiculose, vacuolized, and finally hyaline (Martínez-Abaigar *et al.* 2004). These symptoms are somewhat unspecific and have been described in several pleurocarpous mosses as a response to diverse processes of senescence and stress (both natural and anthropogenic) (Glime & Keen 1984; Gimeno & Puche 1999). The only specific response of the moss to enhanced UVR was a color change in the cell walls, from yellow to orange-brown.

Laboratory studies on mountain stream bryophytes: responses of UV-absorbing compounds

While the variables described above may represent differentially sensitive markers to detect UVR damage, other variables may rather show acclimation processes. This is the case of the accumulation of UV-absorbing compounds, which is the most common response of vascular plants to enhanced UVR, with an average increase around 10% (Searles *et al.* 2001a). There is no simple direct relationship between this accumulation and UVR tolerance, but UV-absorbing compounds could reduce UVR penetration and, consequently, damage to the potential targets. They may offer additional protection through free-radical scavenging activity (Sroka 2005).

The most used variable to quantify the levels of UV-absorbing compounds, both in bryophytes and other photosynthetic organisms, has been the bulk UV absorbance. It should be noted that this measurement is based on a simple methanol extraction, and thus the UV-absorbing compounds bound to the cell wall may be underextracted or not extracted at all, leading possibly to the underestimation of the total UV absorption capacity. An adequate evaluation of the contribution of methanol-extractable and non-extractable UV-absorbing compounds to the bulk UV-absorption capacity of bryophytes, both mosses and liverworts, remains to be done. UV-absorbing compounds have been demonstrated to be located in cell walls, where they could contribute decisively to UV protection (Clarke & Robinson 2008), but the fraction more easily extractable in methanol is probably located in vacuoles, as in vascular plants.

In bryophytes, an increase in UV absorbance with increasing UVR has been much more frequently found in liverworts (in ten out of 14 experiments, conducted on six species) than in mosses (in seven out of around 45 experiments, in which about 20 species were used). Similar results have been found in bryophytes from mountain streams. In addition, liverworts show much higher values of UV absorbance than mosses under natural conditions, and the absorption spectra in the UV band are notably different in both groups (Arróniz-Crespo *et al.* 2004; Otero *et al.* 2008). These facts suggest that (1) the accumulation of methanol-extractable UV-absorbing compounds represents a frequent and constitutive protecting mechanism against UVR in liverworts, but not in mosses; and (2) this mechanism is inducible by UVR much more frequently in liverworts than in mosses. The apparently different behavior of both bryophyte groups with respect to the accumulation of methanol-extractable UV-absorbing compounds could be related to the fact that mosses and liverworts are presently considered to be more phylogenetically separated than previously thought (Qiu *et al.* 2006). Given the important role of liverworts in the water-to-land transition (Qiu *et al.* 2006), their efficient accumulation of UV-absorbing compounds could have been one of the factors favoring their success in the colonization of land, an environment with higher UV levels than the aquatic habitat.

Another critical point regarding UV-absorbing compounds in bryophytes is that the measurement of the bulk UV absorbance might be insufficient to assess the effects of UVR because each individual compound may respond to UVR in a different manner. Thus, it would be better to use HPLC methods allowing the chemical characterization of the individual compounds. This approach has been progressively more frequently used in bryophytes (Markham *et al.* 1998; Blokker *et al.* 2006; Harris 2009; Kato-Noguchi & Kobayashi 2009; Snell *et al.* 2009), especially through the measurement of hydroxycinnamic acid derivatives in the aquatic liverwort *Jungermannia exsertifolia* subsp. *cordifolia* (Arróniz-Crespo *et al.* 2006, 2008a, b; Otero *et al.* 2006, 2009; Martínez-Abaigar *et al.* 2008; Núñez-Olivera *et al.* 2009). These studies make clear that, although all the individual compounds contribute to the bulk UV absorbance, their UVR responses can be different: some of them may increase under enhanced UVR, whereas others may decrease or remain unaltered. These contrasting responses, together with the limited capacity of methanol to extract cell wall-bound compounds, could be reasons for the relatively frequent failure to demonstrate an increase in the bulk UV absorbance of bryophytes (especially in mosses) under enhanced UVR.

Laboratory studies on mountain stream bryophytes: factors influencing UVR responses

The factors influencing the UVR responses of aquatic bryophytes can be divided into internal and environmental ones. With respect to internal factors, UVR responses may depend primarily on the species, as in the case of vascular plants, and bryophytes should not be grouped as a single functional type regarding UVR effects. The aquatic bryophytes most used in laboratory studies have been the foliose liverwort *Jungermannia exsertifolia* subsp. *cordifolia*, together with the moss *Fontinalis antipyretica*, but five additional species, two liverworts and three mosses, have been used (Table 7.1). Overall, effects of enhanced UVR are not dramatic for most species under laboratory conditions, except for some particularly sensitive species, such as *Fontinalis antipyretica*. The UVR sensitivity of this moss may not be relevant under field conditions, because some populations can be found withstanding high UVR levels above the upper limit of the forest. Interspecific differences in UVR tolerance may be based on the degree of development of repairing and protecting mechanisms, such as the accumulation of UV-absorbing compounds (Martínez-Abaigar *et al.* 2003), but the influence of structural characteristics cannot be discarded.

A direct relationship between the UV tolerance and the desiccation tolerance of the different species has been suggested in bryophytes (Takács *et al.* 1999; Csintalan *et al.* 2001; Turnbull *et al.* 2009). In the present chapter, aquatic bryophytes are considered as a very diverse group ranging from obligate aquatics to semiaquatic emergents (Vitt & Glime 1984), and thus their desiccation tolerance may be equally diverse. One of the obligate aquatics (*Fontinalis antipyretica*) has been revealed as the most sensitive species among those tested, supporting the above-mentioned hypothesis.

Another internal factor is the tissue age. In UV-exposed samples of the aquatic liverwort *Jungermannia exsertifolia* subsp. *cordifolia*, five UV-absorbing compounds (hydroxycinnamic acid derivatives) were differentially distributed between apical and basal segments of the shoots, thus providing presumably different levels of UV protection (Arróniz-Crespo *et al.* 2008b).

Among environmental factors, a low culture temperature (2 °C) has been shown to enhance the adverse effects of UVR in the UV-sensitive *Fontinalis antipyretica*, but not in the more UV-tolerant *Jungermannia exsertifolia* subsp. *cordifolia* (Núñez-Olivera *et al.* 2004). Thus, the adverse effects of cold and UVR were apparently additive in the moss, whereas this additiveness was lacking in the liverwort. The combined effect of cold and UVR would probably be due to the limitation of the development of metabolically related protection and repair mechanisms against UVR. Other culture conditions apart from temperature, such as the PAR level and the proportions UVR/PAR, may influence the UVR responses.

UVR responses are also influenced by the previous light acclimation of the different samples to sun or shade conditions (Núñez-Olivera *et al.* 2005). Shade samples were more susceptible to enhanced UVR than sun samples, whose protection against UVR would be more efficient. Again, this effect was found only in a UVR-sensitive species (*Fontinalis antipyretica*), but not in a UVR-tolerant one (*Jungermannia exsertifolia* subsp. *cordifolia*). In the latter, sun and shade samples responded to enhanced UVR in a similar way, and both types of sample would be efficiently protected. Another example of acclimation to high UVR levels was suggested for samples of aquatic bryophytes collected in high-altitude sites and/or on dates near the summer solstice (Martínez-Abaigar *et al.* 2009). These samples were more tolerant of enhanced UVR than samples of the same species collected from lower altitudes or in periods of the year with lower UVR levels. Thus, the responses to UVR could be influenced by the collection place and collection date of the samples.

Exposure to heavy metals, such as cadmium, may exacerbate the damaging consequences of enhanced UVR, since both agents are harmful to plant metabolism through similar mechanisms, such as the production of reactive oxygen species, DNA damage, or alterations in the photosynthetic machinery (Otero *et al.* 2006). Cadmium and UVR may interact differently depending on the variable considered. In the aquatic liverwort *Jungermannia exsertifolia* subsp. *cordifolia*, both cadmium and enhanced UVR caused chlorophyll degradation and a strong inhibition of PSII activity, together with an increase in the mechanisms of non-photochemical dissipation of energy (increase in the xanthophyll index). Both adverse factors may affect, in a different manner, PSII activity and the photosynthetic machinery in general: UVR radiation inactivates PSII reaction centers, whereas cadmium acts mainly on the oxygen-evolving complex but also on several other photosynthetic sites and processes. Consequently, both factors may have additive effects on those variables. However, cadmium caused a diminution in photosynthesis rates, probably because the activity of Rubisco and other enzymes of the Calvin cycle were affected, whereas UVR did not. UVR increased the levels of UV-absorbing compounds, such as *p*-coumaroylmalic acid, and cadmium did the same, especially with phaselic and feruloylmalic acids. These responses could be related to both the more efficient absorption of harmful UVR and the enhanced protection against oxidative stress. DNA damage was almost specifically elicited by UV-B radiation, whereas cadmium itself had modest effects in this respect. Nevertheless, the strongest DNA damage was recorded in the presence of both UV-B and cadmium, which could act synergistically through two different mechanisms: UV-B would primarily induce the formation of thymine dimers and cadmium would impair the enzymatic repair mechanisms of DNA. In conclusion, UVR and cadmium seemed to operate additively on some physiological processes, whereas other responses were probably due to either factor alone. The samples exposed to both UVR and cadmium showed the most intense damage.

The interaction between UVR and mineral availability has been little investigated in bryophytes, in spite of their important peculiarities regarding mineral nutrition. Among mineral nutrients, phosphorus frequently exerts a limiting effect on aquatic phototrophs, because of its importance in ATP production and thus in overall cell metabolism (in particular, in the repairing of UV-damaged proteins and DNA). Some evidence from microalgae suggests that phosphorus limitation increases UV sensitivity and that phosphorus enrichment minimizes UV damage. However, in two bryophytes from mountain streams, an improvement in the availability of phosphorus, with the consequence of a 1.7- to 3.7-fold increase in phosphorus tissue concentration, did not modify the responses of diverse physiological variables to enhanced UVR (Martínez-Abaigar *et al.* 2008). This was probably due to the fact that the bryophytes studied had low nutritional requirements, and their natural tissue phosphorus concentrations would be sufficient for their metabolic activities. Thus, increasing phosphorus would result only in luxury consumption, without any improvement in either the photosynthetic performance or the mechanisms of UVR protection and repairing.

Field studies using natural gradients of UVR: stream bryophytes as UVR bioindicators

Two field studies using natural gradients of UVR have been conducted on stream bryophytes. In the first one, Arróniz-Crespo *et al.* (2006) demonstrated a direct association between the increased UVR levels along an altitudinal gradient and both the bulk UV absorbance and the concentrations of several hydroxycinnamic acid derivatives (in particular, two coumarins) in *Jungermannia exsertifolia* subsp. *cordifolia*. These compounds could play a role as UVR protectors. In the second study, Núñez-Olivera *et al.* (2009) found that, for three consecutive years, the concentration of *p*-coumaroylmalic acid in *Jungermannia exsertifolia* subsp. *cordifolia* was higher in summer–autumn than in winter–spring, being positively correlated with UVR levels. In addition, the best model explaining UV-B radiation took into consideration the concentration of this compound and the ozone level. Ozone maximum was reached in early spring (March–April), UV-B in summer (June–July), and *p*-coumaroylmalic acid in autumn (September–October, although its values were high from June). Thus, ozone and *p*-coumaroylmalic acid would compensate for each other in UV-B modeling (Fig. 7.1).

The relationships found between the concentrations of hydroxycinnamic acid derivatives in *Jungermannia exsertifolia* subsp. *cordifolia* and the spatial and temporal changes in UVR show that there is potential in the study of using stream bryophytes as UVR bioindicators. This may take advantage of the well-known bioindication ability of aquatic bryophytes in a number of pollution processes and environmental changes (Glime 1992). In addition, the concentration of *p*-coumaroylmalic acid in herbarium samples of *Jungermannia exsertifolia* subsp. *cordifolia* was useful in the

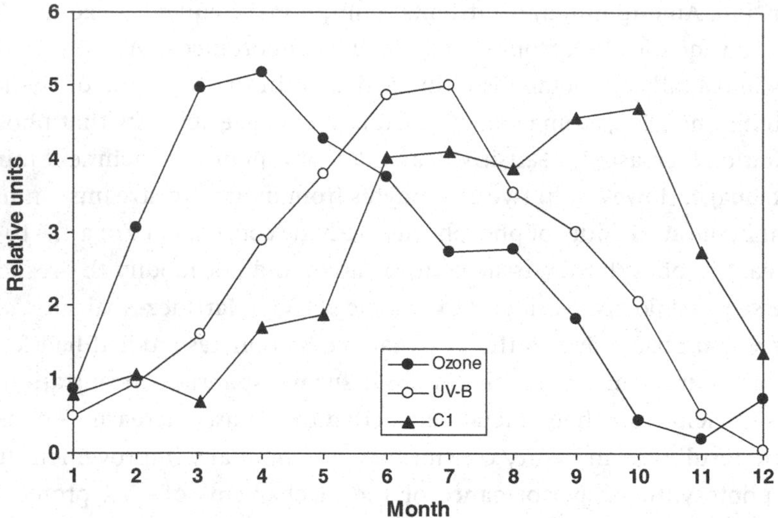

Fig. 7.1. Monthly changes of total ozone in the site in which samples of the aquatic liverwort *Jungermannia exsertifolia* subsp. *cordifolia* were collected, together with the UV-B daily doses at a meteorological station near the sampling site, and the concentrations of *p*-coumaroylmalic acid (C1) in the liverwort. Mean values of the three years studied are shown in relative units, to allow an easy comparison of the three variables.

reconstruction of past ozone and UVR levels in northern Europe (Otero *et al.* 2009). This reconstruction showed that there was no significant temporal trend in UVR in the period 1918–2006, which agrees with previous UVR reconstructions based on both purely climatic models and biological proxies. These studies, together with those of Ryan *et al.* (2009) and Snell *et al.* (2009) in the Antarctic, are the first ones using individual UV-absorbing compounds of bryophytes in UVR bioindication, given that all the previous studies were based on the bulk UV absorbance of methanolic extracts (Newsham *et al.* 2002, 2005; Newsham 2003; Huttunen *et al.* 2005a, b; Robinson *et al.* 2005).

For UVR bioindication purposes, an adequate selection of both variables and species must be carried out. We propose the use of hydroxycinnamic acid derivatives in the aquatic liverwort *Jungermannia exsertifolia* subsp. *cordifolia* for this aim, for several reasons: (1) liverworts are known for their great variety of secondary metabolites, such as many different phenolic constituents; (2) UV-absorbing compounds of liverworts in general, and of this species in particular, are usually more UVR-responsive than those of mosses; (3) the selected species is large enough (2–10 cm long) to allow for easy manipulation, and it forms extensive and frequently unmixed masses that provide plenty of biomass; (4) given the long-lived perennial character of this species, healthy biomass may be available throughout the year; (5) if permanently submerged populations are selected, the interference of the typical transitory desiccation of bryophytes is prevented, and thus the responses may be easier to interpret and model; (6) this species has a wide distribution range over the northern hemisphere, which facilitates studies across wide geographical scales; (7) one of its hydroxycinnamic acid derivatives, *p*-coumaroylmalic acid, is specifically induced by enhanced UVR in laboratory studies under different PAR levels and correlates well with the temporal changes in UVR under field conditions; and (8) a chemically similar UV-absorbing compound (*p*-coumaric acid) was used in the Antarctic graminoid *Deschampsia antarctica* as a UV-B indicator (Ruhland *et al.* 2005) and has been repeatedly recommended for the reconstruction of past UV-B levels (Rozema *et al.* 2001; Blokker *et al.* 2006; Lomax *et al.* 2008).

Conclusions

Ultraviolet radiation (UVR) is an ecological factor that has accompanied life since its origins and whose study has increased greatly since the discovery of the anthropogenic stratospheric ozone reduction; this reduction leads to an increase in UVR (specifically UV-B) at ground level.

UVR increase may cause diverse damage to photosynthetic organisms, and UVR tolerance is a complex process depending on the interaction between this damage, protection and repair mechanisms against it, and acclimation.

Laboratory studies are useful to characterize responses to UVR. However, long-term realistic field manipulative experiments using either filters or lamps, together with experiments exploiting natural UVR gradients, are needed to assess the effects of current UVR levels, and to predict the effects of potentially higher UVR levels derived from the stratospheric ozone reduction.

The noteworthy variability of the effects of UVR on bryophytes may be due to the diversity of species, variables, and experimental conditions used in the different studies, together with the influence of environmental factors. Thus, it is necessary to take into account the methodological approaches to appropriately interpret the results obtained. Nevertheless, and under a global perspective, UVR responses in bryophytes do not seem to be very different from those found in other photosynthetic organisms, from phytoplankton to flowering plants.

Given that aquatic habitats frequently dominated by bryophytes, such as bogs and mountain streams, may be particularly affected by UVR, we have detailed here the effects of UVR on bryophytes from diverse aquatic environments. These effects have also been put in the context of global UVR research.

Under laboratory conditions, effects of enhanced UVR are not dramatic for bryophytes from mountain streams, except for particularly sensitive species, such as *Fontinalis antipyretica*. UVR stress may be indicated by a decrease in the maximum quantum yield of photosystem II (F_v/F_m), chlorophyll/phaeopigment ratios, chlorophyll a/b quotient, and net photosynthesis rates. However, no variable responds in the same manner in every species, and this still limits our global comprehension of UVR effects. Some species can be protected from UVR damage by the accumulation of protective methanol-extractable UV-absorbing compounds.

Field studies suggest that biomass production of peatland bryophytes is not at risk at current levels of UVR, nor at the potentially higher levels derived from predicted ozone reductions. Nevertheless, the complex interaction of ozone reduction (and concomitant UVR increase) with factors of climate change, such as increasing temperature and changes in cloudiness and water availability, makes predictions uncertain.

The effects of UVR on aquatic bryophytes depend not only on specific genetic factors (the species), but also on environmental factors (temperature, presence of heavy metals) and the previous acclimation of the samples to sun or shade conditions (light history), low or high altitude, etc.

Liverworts seem to have higher amounts of both constitutive and inducible methanol-extractable UV-absorbing compounds than mosses. This difference could be related to the presently recognized phylogenetic distance between both groups, and also to the importance of liverworts in the water-to-land transition.

The responses of aquatic bryophytes to UVR are still poorly understood, especially on the topics of their acclimation capacity and protection mechanisms: antioxidant systems, DNA repairing, and the accumulation of UV-absorbing compounds. Regarding this last particular topic, some methodological problems should be solved, such as the extraction of compounds from cell walls and vacuoles. In addition, the use of individual UV-absorbing compounds in the quantification of the UV absorption capacity of bryophytes (further than the already used bulk UV absorbance of methanolic extracts) should be encouraged.

Aquatic bryophytes have been used as bioindicators of numerous pollution processes and environmental changes. We propose their use as UVR bioindicators, which would require an adequate selection of both variables and species. Promising variables regarding this point are F_v/F_m, because of its sensitivity to UVR, and the concentration of certain individual UV-absorbing compounds (such as *p*-coumaroylmalic acid), owing to its remarkable specificity of response. The liverwort *Jungermannia exsertifolia* subsp. *cordifolia* seems to be a good candidate species because of its responsiveness to UVR, availability of abundant healthy biomass throughout the year, and relatively wide distribution range.

Acknowledgements

We are grateful to the Ministerio de Ciencia e Innovación of Spain (Project CGL2008–04450) for financial support.

References

Allen, D. J., Nogués, S. & Baker, N. R. (1998). Ozone depletion and increased UV-B radiation: is there a real threat to photosynthesis? *Journal of Experimental Botany* **49**: 1775–88.

Arróniz-Crespo, M., Núñez-Olivera, E., Martínez-Abaigar, J. & Tomás, R. (2004). A survey of the distribution of UV-absorbing compounds in aquatic bryophytes from a mountain stream. *Bryologist* **107**: 202–8.

Arróniz-Crespo, M., Núñez-Olivera, E., Martínez-Abaigar, J., *et al.* (2006). Physiological changes and UV protection in the aquatic liverwort *Jungermannia exsertifolia* subsp. *cordifolia* along an altitudinal gradient of UV-B radiation. *Functional Plant Biology* **33**: 1025–36.

Arróniz-Crespo, M., Núñez-Olivera, E. & Martínez-Abaigar, J. (2008a). Hydroxycinnamic acid derivatives in an aquatic liverwort as possible bioindicators of enhanced UV radiation. *Environmental Pollution* **151**: 8–16.

Arróniz-Crespo, M., Phoenix, G., Núñez-Olivera, E. & Martínez-Abaigar, J. (2008b). Age-specific physiological responses to UV radiation in the aquatic liverwort *Jungermannia exsertifolia* subsp. *cordifolia*. *Cryptogamie Bryologie* **29**: 115–26.

Ballaré, C. L., Rousseaux, M. C., Searles, P. S., *et al.* (2001). Impacts of solar ultraviolet-B radiation on terrestrial ecosystems of Tierra del Fuego (southern Argentina). An overview of recent progress. *Journal of Photochemistry and Photobiology B: Biology* **62**: 67–77.

Barsig, M., Schneider, K. & Gehrke, C. (1998). Effects of UV-B radiation on fine structure, carbohydrates, and pigments in *Polytrichum commune*. *Bryologist* **101**: 357–65.

Björn, L. O. & McKenzie, R. L. (2007). Attempts to probe the ozone layer and the ultraviolet-B levels of the past. *Ambio* **36**: 366–71.

Björn, L. O., Callaghan, T. V., Gehrke, C., *et al.* (1998). The problem of ozone depletion in northern Europe. *Ambio* **27**: 275–9.

Blokker, P., Boelen, P., Broekman, R. & Rozema, J. (2006). The occurrence of *p*-coumaric acid and ferulic acid in fossil plant materials and their use as UV-proxy. *Plant Ecology* **182**: 197–207.

Boelen, P., De Boer, M. K., De Bakker, N. V. J. & Rozema, J. (2006). Outdoor studies on the effects of solar UV-B on bryophytes: overview and methodology. *Plant Ecology* **182**: 137–52.

Caldwell, M. M., Bornman, J. F., Ballaré, C. L., Flint, S. D. & Kulandaivelu, G. (2007). Terrestrial ecosystems, increased solar ultraviolet radiation, and interactions with other climate change factors. *Photochemical and Photobiological Sciences* **6**: 252–66.

Clarke, L. J. & Robinson, S. A. (2008). Cell wall-bound ultraviolet-screening compounds explain the high ultraviolet tolerance of the Antarctic moss, *Ceratodon purpureus*. *New Phytologist* **179**: 776–83.

Cockell, C. S. & Knowland, J. (1999). Ultraviolet radiation screening compounds. *Biological Review* **74**: 311–45.

Conde-Álvarez, R. M., Pérez-Rodríguez, E., Altamirano, M., *et al.* (2002). Photosynthetic performance and pigment content in the aquatic liverwort *Riella helicophylla* under natural solar irradiance and solar irradiance without ultraviolet light. *Aquatic Botany* **73**: 47–61.

Csintalan, Z., Tuba, Z., Takács, Z. & Laitat, E. (2001). Responses of nine bryophyte and one lichen species from different microhabitats to elevated UV-B radiation. *Photosynthetica* **39**: 317–20.

Day, T. A. & Neale, P. J. (2002). Effects of UV-B radiation on terrestrial and aquatic primary producers. *Annual Review of Ecology and Systematics* **33**: 371–96.

DeEll, J. R. & Toivonen, P. M. A. (2003). *Practical Applications of Chlorophyll Fluorescence in Plant Biology*. Boston, MA: Kluwer.

Dunn, J. L. & Robinson, S. A. (2006). Ultraviolet B screening potential is higher in two cosmopolitan moss species than in a co-occurring Antarctic endemic moss: implications of continuing ozone depletion. *Global Change Biology* **12**: 2282–96.

Flint, S. D., Ryel, R. J. & Caldwell, M. M. (2003). Ecosystem UV-B experiments in terrestrial communities: a review of recent findings and methodologies. *Agricultural and Forest Meteorology* **120**: 177–89.

Frost, P. C., Larson, J. H., Kinsman, L. E., Lamberti, G. A. & Bridgham, S. D. (2005). Attenuation of ultraviolet radiation in streams of northern Michigan. *Journal of the North American Benthological Society* **24**: 246–55.

Gehrke, C. (1998). Effects of enhanced UV-B radiation on production related properties of a *Sphagnum fuscum* dominated subarctic bog. *Functional Ecology* **12**: 940–7.

Gehrke, C. (1999). Impacts of enhanced ultraviolet-B radiation on mosses in a subarctic heath ecosystem. *Ecology* **80**: 1844–51.

Gehrke, C., Johanson, U., Gwynn-Jones, D. *et al.* (1996). Effects of enhanced ultraviolet-B radiation on terrestrial subarctic ecosystems and implications for interactions with increased atmospheric CO_2. *Ecological Bulletins* **45**: 192–203.

Gimeno, C. & Puche, F. (1999). Chlorophyll content and morphological changes in cellular structure of *Rhynchostegium riparioides* (Hew.) Card. (Brachytheciaceae, Musci) and *Fontinalis hypnoides* Hartm. (Fontinalaceae, Musci) in response to water pollution and transplant containers on Palancia river (East Spain). *Nova Hedwigia* **68**: 197–216.

Glime, J. M. (1992). Effects of pollutants on aquatic species. In *Bryophytes and Lichens in a Changing Environment*, ed. J. W. Bates & A. M. Farmer, pp. 333–61. Oxford: Clarendon Press.

Glime, J. M. & Keen, R. E. (1984). The importance of bryophytes in a man-centered world. *Journal of the Hattori Botanical Laboratory* **55**: 133–46.

Green, T. G. A., Schroeter, B. & Seppelt, R. D. (2000). Effect of temperature, light and ambient UV on the photosynthesis of the moss *Bryum argenteum* Hedw. in continental Antarctica. In *Antarctic Ecosystems: Models for Wider Ecological Understanding*, ed. W. Davison, C. Howard-Williams & P. Broady, pp. 165–70. Christchurch: The Caxton Press.

Green, T. G. A., Kulle, D., Pannewitz, S., Sancho, L. G. & Schroeter, B. (2005). UV-A protection in mosses growing in continental Antarctica. *Polar Biology* **28**: 822–7.

Häder, D. P., Kumar, H. D., Smith, R. C. & Worrest, R. C. (2007). Effects of solar UV radiation on aquatic ecosystems and interactions with climate change. *Photochemical and Photobiological Sciences* **6**: 267–85.

Harris, E. S. J. (2009). Phylogenetic and environmental lability of flavonoids in a medicinal moss. *Biochemical Systematics and Ecology* **37**: 180–92.

Hooijmaijers, C. A. M. & Gould, K. S. (2007). Photoprotective pigments in red and green gametophytes of two New Zealand liverworts. *New Zealand Journal of Botany* **45**: 451–61.

Hughes, K. A., Scherer, K., Svenoe, T. *et al.* (2006). Tundra plants protect the soil surface from UV. *Soil Biology and Biochemistry* **38**: 1488–90.

Huiskes, A. H. L., Lud, D., Moerdijk-Poortvliet, T. C. W. & Rozema, J. (1999). Impact of UV-B radiation on Antarctic terrestrial vegetation. In *Stratospheric Ozone Depletion: The Effects of Enhanced UV-B Radiation on Terrestrial Ecosystems*, ed. J. Rozema, pp. 313–37. Leiden: Backhuys Publishers.

Huiskes, A. H. L., Lud, D. & Moerdijk-Poortvliet, T. C. W. (2001). Field research on the effects of UV-B filters on terrestrial Antarctic vegetation. *Plant Ecology* **154**: 77–86.

Huttunen, S., Kinnunen, H. & Laakso, K. (1998). Impact of increased UV-B on plant ecosystems. *Chemosphere* **36**: 829–33.

Huttunen, S., Lappalainen, N. M. & Turunen, J. (2005a). UV-absorbing compounds in subarctic herbarium bryophytes. *Environmental Pollution* **133**: 303–14.

Huttunen, S., Taipale, T., Lappalainen, N. M. *et al.* (2005b). Environmental specimen bank samples of *Pleurozium schreberi* and *Hylocomium splendens* as indicators of the radiation environment at the surface. *Environmental Pollution* **133**: 315–26.

Ihle, C. (1997). Degradation and release from the thylakoid membrane of Photosystem II subunits after UV-B irradiation of the liverwort *Conocephalum conicum*. *Photosynthesis Research* **54**: 73–8.

Ihle, C. & Laasch, H. (1996). Inhibition of photosystem II by UV-B radiation and the conditions for recovery in the liverwort *Conocephalum conicum* Dum. *Botanica Acta* **109**: 199–205.

Jansen, M. A. K., Gaba, V. & Greenberg, B. M. (1998). Higher plants and UV-B radiation: balancing damage, repair and acclimation. *Trends in Plant Sciences* **3**: 131–5.

Johanson, U., Gehrke, C., Björn, L. O., Callaghan, T. V. & Sonesson, M. (1995). The effects of enhanced UV-B radiation on a subarctic heath ecosystem. *Ambio* **24**: 106–11.

Kato-Noguchi, H. & Kobayashi, K. (2009). Jasmonic acid, protein phosphatase inhibitor, metals and UV-irradiation increased momilactone A and B concentrations in the moss *Hypnum plumaeforme*. *Journal of Plant Physiology* **166**: 1118–22.

Kelly, D. J., Bothwell, M. L. & Schindler, D. W. (2003). Effects of solar ultraviolet radiation on stream benthic communities: an intersite comparison. *Ecology* **84**: 2724–40.

Lappalainen, N. M., Huttunen, S. & Suokanerva, H. (2008). Acclimation of a pleurocarpous moss *Pleurozium schreberi* (Britt.) Mitt. to enhanced ultraviolet radiation *in situ*. *Global Change Biology* **14**: 321–33.

Lewis Smith, R. I. (1999). Biological and environmental characteristics of three cosmopolitan mosses dominant in continental Antarctica. *Journal of Vegetation Science* **10**: 231–42.

Lomax, B. H., Fraser, W. T., Sephton, M. A. *et al.* (2008). Plant spore walls as a record of long-term changes in ultraviolet-B radiation. *Nature Geoscience* **1**: 592–6.

Lovelock, C. E. & Robinson, S. A. (2002). Surface reflectance properties of Antarctic moss and their relationship to plant species, pigment composition and photosynthetic function. *Plant, Cell & Environment* **25**: 1239–50.

Lud, D., Moerdijk, T. C. W., Van de Poll, W. H., Buma, A. G. J. & Huiskes, A. H. L. (2002). DNA damage and photosynthesis in Antarctic and Arctic *Sanionia uncinata* (Hedw.) Loeske under ambient and enhanced levels of UV-B radiation. *Plant, Cell & Environment* **25**: 1579–89.

Lud, D., Schlensog, M., Schroeter, B. & Huiskes, A. H. L. (2003). The influence of UV-B radiation on light-dependent photosynthetic performance in *Sanionia uncinata* (Hedw.) Loeske in Antarctica. *Polar Biology* **26**: 225–32.

Markham, K. R., Franke, A., Given, D. R. & Brownsey, P. (1990). Historical Antarctic ozone level trends from herbarium specimen flavonoids. *Bulletin de Liaison du Groupe Polyphenols* **15**: 230–5.

Markham, K. R., Ryan, K. G., Bloor, S. J. & Mitchell, K. A. (1998). An increase in the luteolin:apigenin ratio in *Marchantia polymorpha* on UV-B enhancement. *Phytochemistry* **48**: 791–4.

Martínez-Abaigar, J. & Núñez-Olivera, E. (1998). Ecophysiology of photosynthetic pigments in aquatic bryophytes. In *Bryology for the Twenty-first Century*, ed. J. W. Bates, N. W. Ashton & J. G. Duckett, pp. 277–92. Leeds: Maney Publishing and the British Bryological Society.

Martínez-Abaigar, J., Núñez-Olivera, E., Beaucourt, N. *et al.* (2003). Different physiological responses of two aquatic bryophytes to enhanced ultraviolet-B radiation. *Journal of Bryology* **25**: 17–30.

Martínez-Abaigar, J., Núñez-Olivera, E., Tomás, R. *et al.* (2004). Daños macroscópicos y microscópicos causados por un aumento de la radiación ultravioleta-B en dos briófitos acuáticos del Parque Natural de Sierra Cebollera (La Rioja, norte de España). *Zubía* **22**: 143–63.

Martínez-Abaigar, J., Otero, S., Tomas, R. & Núñez-Olivera, E. (2008). High-level phosphate addition does not modify UV effects in two aquatic bryophytes. *Bryologist* **111**: 444–54.

Martínez-Abaigar, J., Otero, S., Tomas, R. & Núñez-Olivera, E. (2009). Effects of enhanced ultraviolet radiation on six aquatic bryophytes. *Cryptogamie-Bryologie* **30**: 157–75.

Maxwell, K. & Johnson, G. N. (2000). Chlorophyll fluorescence – a practical guide. *Journal of Experimental Botany* **51**: 659–68.

McKenzie, R. L., Björn, L. O., Bais, A. & Ilyasd, M. (2003). Changes in biologically active ultraviolet radiation reaching the Earth's surface. *Photochemical and Photobiological Sciences* **2**: 5–15.

McKenzie, R. L., Aucamp, P. J., Bais, A. F., Björn, L. O. & Hyas, M. (2007). Changes in biologically-active ultraviolet radiation reaching the Earth's surface. *Photochemical and Photobiological Sciences* **6**: 218–31.

Montiel, P., Smith, A. & Keiller, D. (1999). Photosynthetic responses of selected Antarctic plants to solar radiation in the southern maritime Antarctic. *Polar Research* **18**: 229–35.

Newsham, K. K. (2003). UV-B radiation arising from stratospheric ozone depletion influences the pigmentation of the Antarctic moss *Andreaea regularis*. *Oecologia* **135**: 327–31.

Newsham, K. K., Hodgson, D. A., Murray, A. W. A., Peat, H. J. & Lewis Smith, R. I. (2002). Response of two Antarctic bryophytes to stratospheric ozone depletion. *Global Change Biology* **8**: 972–83.

Newsham, K. K., Geissler, P., Nicolson, M., Peat, H. J. & Lewis-Smith, R. I. (2005). Sequential reduction of UV-B radiation in the field alters the pigmentation of an Antarctic leafy liverwort. *Environmental and Experimental Botany* **54**: 22–32.

Newsham, K. K., & Robinson, S. A. (2009). Responses of plants in polar regions to UVB exposure: a meta-analysis. *Global Change Biology* **15**: 2574–89.

Niemi, R., Martikainen, P. J., Silvola, J. *et al.* (2002a). Elevated UV-B radiation alters fluxes of methane and carbon dioxide in peatland microcosms. *Global Change Biology* **8**: 361–71.

Niemi, R., Martikainen, P. J., Silvola, J. *et al.* (2002b). Responses of two *Sphagnum* moss species and *Eriophorum vaginatum* to enhanced UV-B in a summer of low UV intensity. *New Phytologist* **156**: 509–15.

Norval, M., Cullen, A. P., De Gruijl, F. R., *et al.* (2007). The effects on human health from stratospheric ozone depletion and its interactions with climate change. *Photochemical and Photobiological Sciences* **6**: 232–51.

Núñez-Olivera, E., Martínez-Abaigar, J., Tomás, R., Beaucourt, N. & Arróniz-Crespo, M. (2004). Influence of temperature on the effects of artificially enhanced UV-B radiation on aquatic bryophytes under laboratory conditions. *Photosynthetica* **42**: 201–12.

Núñez-Olivera, E., Arróniz-Crespo, M., Martínez-Abaigar, J., Tomás, R. & Beaucourt, N. (2005). Assessing the UV-B tolerance of sun and shade samples of two aquatic bryophytes using short-term tests. *Bryologist* **108**: 435–48.

Núñez-Olivera, E., Otero, S., Tomás, R. & Martínez-Abaigar, J. (2009). Seasonal variations in UV-absorbing compounds and physiological characteristics in the aquatic liverwort *Jungermannia exsertifolia* subsp. *cordifolia* over a three-year period. *Physiologia Plantarum* **136**: 73–85.

Otero, S., Núñez-Olivera, E., Martínez-Abaigar, J. *et al.* (2006). Effects of cadmium and enhanced UV radiation on the physiology and the concentration of UV-absorbing compounds of the aquatic liverwort *Jungermannia exsertifolia* subsp. *cordifolia*. *Photochemical and Photobiological Sciences* **5**: 760–9.

Otero, S., Cezón, K., Martínez-Abaigar, J. & Núñez-Olivera, E. (2008). Ultraviolet-absorbing capacity of aquatic bryophytes from Tierra del Fuego (Argentina). *Journal of Bryology* **30**: 290–6.

Otero, S., Núñez-Olivera, E., Martínez-Abaigar, J., Tomás, R. & Huttunen, S. (2009). Retrospective bioindication of stratospheric ozone and ultraviolet radiation using hydroxycinnamic acid derivatives of herbarium samples of an aquatic liverwort. *Environmental Pollution* **157**: 2335–44.

Phoenix, G. K., Gwynn-Jones, D., Callaghan, T. V., Sleep, D. & Lee, J. A. (2001). Effects of global change on a sub-Arctic heath: effects of enhanced UV-B radiation and increased summer precipitation. *Journal of Ecology* **89**: 256–67.

Post, A. & Vesk, M. (1992). Photosynthesis, pigments, and chloroplast ultrastructure of an Antarctic liverwort from sun-exposed and shaded sites. *Canadian Journal of Botany* **70**: 2259–64.

Prasad, S. M., Dwivedi, R., Zeeshan, M. & Singh, R. (2004). UV-B and cadmium induced changes in pigments, photosynthetic electron transport activity, antioxidant levels and antioxidative enzyme activities of *Riccia sp. Acta Physiologiae Plantarum* **26**: 423–30.

Qiu, Y. L., Li, L., Wang, B. *et al.* (2006). The deepest divergences in land plants inferred from phylogenomic evidence. *Proceedings of the National Academy of Sciences of the United States of America* **103**: 15511–16.

Rader, R. B. & Belish, T. A. (1997a). Effects of ambient and enhanced UV-B radiation on periphyton in a mountain stream. *Journal of Freshwater Ecology* **12**: 615–28.

Rader, R. B. & Belish, T. A. (1997b). Short-term effects of ambient and enhanced UV-B on moss (*Fontinalis neomexicana*) in a mountain stream. *Journal of Freshwater Ecology* **12**: 395–403.

Robinson, S. A., Turnbull, J. D. & Lovelock, C. E. (2005). Impact of changes in natural ultraviolet radiation on pigment composition, physiological and morphological characteristics of the Antarctic moss, *Grimmia antarctici*. *Global Change Biology* **11**: 476–89.

Robson, T. M., Pancotto, V. A., Flint, S. D., *et al.* (2003). Six years of solar UV-B manipulations affect growth of *Sphagnum* and vascular plants in a Tierra del Fuego peatland. *New Phytologist* **160**: 379–89.

Robson, T. M., Pancotto, V. A., Ballaré, C. L. *et al.* (2004). Reduction of solar UV-B mediates changes in the *Sphagnum* capitulum microenvironment and the peatland microfungal community. *Oecologia* **140**: 480–90.

Rozema, J., Noordijk, A. J., Broekman, R. A., *et al.* (2001). (Poly)phenolic compounds in pollen and spores of Antarctic plants as indicators of solar UV-B. A new proxy for the reconstruction of past solar UV-B? *Plant Ecology* **154**: 11–26.

Rozema, J., Björn, L. O., Bornman, J. F., *et al.* (2002). The role of UV-B radiation in aquatic and terrestrial ecosystems – an experimental and functional analysis of the evolution of UV-absorbing compounds. *Journal of Photochemistry and Photobiology B: Biology* **66**: 2–12.

Rozema, J., Boelen, P., Solheim, B., *et al.* (2006). Stratospheric ozone depletion: high arctic tundra plant growth on Svalbard is not affected by enhanced UV-B after 7 years of UV-B supplementation in the field. *Plant Ecology* **182**: 121–35.

Ruhland, C. T., Xiong, F. S., Clark, W. D. & Day, T. A. (2005). The influence of ultraviolet-B radiation on growth, hydroxycinnamic acids and flavonoids of *Deschampsia antarctica* during springtime ozone depletion in Antarctica. *Photochemistry and Photobiology* **81**: 1086–93.

Ryan, K. G., Burne, A. & Seppelt, R. D. (2009). Historical ozone concentrations and flavonoid levels in herbarium specimens of the Antarctic moss *Bryum argenteum*. *Global Change Biology* **15**: 1694–702.

Schipperges, B. & Gehrke, C. (1996). Photosynthetic characteristics of subarctic mosses and lichens. *Ecological Bulletins* **45**: 121–6.

Searles, P. S., Flint, S. D., Díaz, S. B. *et al.* (1999). Solar ultraviolet-B radiation influence on *Sphagnum* bog and *Carex* fen ecosystems: first field season findings in Tierra del Fuego, Argentina. *Global Change Biology* **5**: 225–34.

Searles, P. S., Flint, S. D. & Caldwell, M. M. (2001a). A meta-analysis of plant field studies simulating stratospheric ozone depletion. *Oecologia* **127**: 1–10.

Searles, P. S., Kropp, B. R., Flint, S. D. & Caldwell, M. M. (2001b). Influence of solar UV-B radiation on peatland microbial communities of southern Argentina. *New Phytologist* **152**: 213–21.

Searles, P. S., Flint, S. D., Díaz, S. B. *et al.* (2002). Plant response to solar ultraviolet-B radiation in a southern South American *Sphagnum* peatland. *Journal of Ecology* **90**: 704–13.

Seckmeyer, G., Pissulla, D., Glandorf, M., *et al.* (2008). Variability of UV irradiance in Europe. *Photochemistry and Photobiology* **84**: 172–9.

Snell, K. R. S., Convey, P. & Newsham, K. K. (2007). Metabolic recovery of the Antarctic liverwort *Cephaloziella varians* during spring snowmelt. *Polar Biology* **30**: 1115–22.

Snell, K. R. S., Kokubun, T., Griffiths, H. *et al.* (2009). Quantifying the metabolic cost to an Antarctic liverwort of responding to an abrupt increase in UVB radiation exposure. *Global Change Biology* **15**: 2563–73.

Sonesson, M., Callaghan, T. V. & Carlsson, B. A. (1996). Effects of enhanced ultraviolet radiation and carbon dioxide concentration on the moss *Hylocomium splendens*. *Global Change Biology* **2**: 67–73.

Sonesson, M., Carlsson, B. A., Callaghan, T. V., *et al.* (2002). Growth of two peat-forming mosses in subarctic mires: species interactions and effects of simulated climate change. *Oikos* **99**: 151–60.

Sroka, Z. (2005). Antioxidative and antiradical properties of plant phenolics. *Zeitschrift für Naturforschung Section C – Journal of Biosciences* **60**: 833–43.

Taipale, T. & Huttunen, S. (2002). Moss flavonoids and their ultrastructural localization under enhanced UV-B radiation. *Polar Record* **38**: 211–18.

Takács, Z., Csintalan, Z., Sass, L. *et al.* (1999). UV-B tolerance of bryophyte species with different degrees of desiccation tolerance. *Journal of Photochemistry and Photobiology B: Biology* **48**: 210–15.

Turnbull, J. D. & Robinson, S. A. (2009). Accumulation of DNA damage in Antarctic mosses: correlations with ultraviolet-B radiation, temperature and turf water content vary among species. *Global Change Biology* **15**: 319–29.

Turnbull, J. D., Leslie, S. J. & Robinson, S. A. (2009). Desiccation protects two Antarctic mosses from ultraviolet-B induced DNA damage. *Functional Plant Biology* **36**: 214–21.

Vitt, D. H. & Glime, J. M. (1984). The structural adaptations of aquatic Musci. *Lindbergia* **10**: 95–110.

IV DESERT AND TROPICAL ECOSYSTEMS

8

Responses of a Biological Crust Moss to Increased Monsoon Precipitation and Nitrogen Deposition in the Mojave Desert

LLOYD R. STARK, D. NICHOLAS MCLETCHIE, STANLEY D. SMITH, AND MELVIN J. OLIVER

Introduction

Global climate change in the Mojave Desert will likely result in a greater intensity of summer (monsoon) rain events and greater N deposition. The nitrogen cycle has already been significantly altered by human activities to the extent that anthropogenically released N now equals natural terrestrial biological fixation (Vitousek *et al.* 1997; Galloway 1998). Because most bryophytes receive the bulk of their nutrients from direct atmospheric deposition (Bates 2000), this influx of N can affect the productivity of individual species and thus may alter bryophyte community structure and function. In addition to N deposition, global change models for the southwestern USA predict significant increases in summer precipitation in the northern Mojave Desert (Taylor & Penner 1994; Higgins & Shi 2001). The interaction between increased N deposition and an increased monsoon effect on bryophytes in the arid southwestern USA is largely unknown. Although growth rates of desert bryophytes are relatively low compared with bryophytes in more mesic ecosystems, the contribution of biological soil crusts (a community of cyanobacteria, mosses, lichens, algae, and fungi) to the global cycling of trace gases can be significant in regard to global budgets (Zaady *et al.* 2000).

Most field studies have found a rapid negative effect of N fertilization on the growth and productivity of mosses, with nutrient uptake a function of desiccation regime, temperature, and light. For several bryophyte species, high experimental N deposition rates decreased biomass production except in a widely

Bryophyte Ecology and Climate Change, eds. Zoltán Tuba, Nancy G. Slack and Lloyd R. Stark. Published by Cambridge University Press.

tolerant species of *Sphagnum* (Jauhiainen *et al.* 1998). Field applications of NPK did not stimulate productivity in the mosses *Calliergonella cuspidata* (van Tooren *et al.* 1990) or *Dicranum majus* (Bakken 1995), and in the latter study the authors attributed this to drought. Similarly, shoot extension rate was significantly depressed from controls in fertilized and fertilized + high temperature treatments in *Hylocomium splendens* (Jägerbrand *et al.* 2003). Moreover, all five species in the latter study responded negatively to fertilization and combined temperature and fertilization treatments, confirming previous results from the Alaskan tundra (Chapin *et al.* 1995) and from a shrub heath community (Press *et al.* 1998). Nutrient uptake is affected by desiccation regime, with productivity decreased in intermittently hydrated plants relative to continuously hydrated controls, although net uptake of N was similar between treatments (Bates 1997). For *Racomitrium lanuginosum*, experimental N supplementation and light reduction suppressed growth, with the negative effects additive (van der Wal *et al.* 2005).

Explanations for the negative fertilizer effect on bryophytes range from direct toxicity to bryophyte tissues to indirect effects on bryophytes mediated by overgrowth and subsequent shading out of bryophytes by vascular plants (van der Wal *et al.* 2005). Direct toxic effects of an increased N load resulted in increased mortality, decreased cover, and a decline in inducibility of NRA (nitrate reductase activity) of 53% relative to control plots (low N dose) in *Racomitrium lanuginosum* (Pearce *et al.* 2003), with the authors postulating that these effects are due to a loss of membrane function. A clear dose-related response was found in the latter study, in which N loads were identical to the field site studied in the present paper. Grasses and graminoids respond quickly (1–3 yrs) to N additions, thus bringing about a negative effect on bryophyte growth due to shading and litter deposition (Chapin *et al.* 1995). In *Sphagnum* systems receiving high N loads, plants exhibited reduced growth (Berendse *et al.* 2001); shifts in plant competition equilibria can be caused by making N available to the rooting systems of vascular plants (Bragazza *et al.* 2004 and references therein) that can also be utilized by the tall moss *Polytrichum strictum* that increases in density and overtops *Sphagnum* (Berendse *et al.* 2001; Mitchell *et al.* 2002).

Other nutrient budgets are affected by high N loads. Above a N deposition load of 1 g m^{-2} yr^{-1}, *Sphagnum* plants change from being N-limited to being K + P co-limited (Bragazza *et al.* 2004). Limpens *et al.* (2003) found that the expansion of *S. fallax* at the high N deposition sites in The Netherlands was limited by P availability. Some studies, however, failed to find a negative effect on growth and productivity in bryophytes when exposed to a high N environment. In Norway, no effect was found on biomass production, relative growth rates, and carbon content per shoot in *Dicranum majus* subjected to light and N treatments (Bakken 1995), with chlorophyll contents and concentrations significantly higher at low light levels subjected to high N treatments. Similarly, in a single-season experiment, N had no effect on the growth and branching of *Calliergonella cuspidata* (Bergamini & Peintinger 2002). Finally, short-term addition of N (as ammonium nitrate) had little effect on productivity, CO_2 exchange, or species composition in a *Sphagnum* mire (Saarnio 1999). In two species of *Sphagnum*, while nutrient exposures had a positive effect on shoot length extension in one species, no effect on biomass accumulation was observed (Li & Glime 1990) with the authors concluding that *Sphagnum* plants are adapted to low nutrient concentrations. In Scotland, two of the three species studied (in *Dicranum* and *Polytrichum*) were unaffected by N fertilization under normal light conditions (van der Wal *et al.* 2005), with light reduction producing reduced growth in *Polytrichum*.

To our knowledge only two studies have assessed either the monsoon effect or the effects of smaller, less frequent hydration events on the biological soil crust. Using crust portions from the Colorado Plateau (USA) and with attention toward cyanobacteria and lichens rather than bryophytes, Belnap *et al.* (2004) applied three treatment levels of precipitation and measured photosynthesis (PS), pigments, and nitrogenase activity. The authors found that a biological soil crust dominated by the lichen *Collema* showed the greatest decline in PS, chlorophyll a, and protective pigments, noting that these crusts dry the fastest, whereas crusts dominated by the cyanobacterium *Microcoleus vaginatus* were least affected. With respect to *Syntrichia caninervis*, a higher than normal frequency of light rain events (<3.5 mm) during the warmer months in the Mojave Desert (USA) resulted in declines in chlorophyll content and asexual vigor (Barker *et al.* 2005). In both of these studies, either the increased monsoon effect or an increased frequency of smaller rain events during warmer months was postulated to cause declines in soil moss and lichen cover and in lichen/cyanobacterial species richness, with monsoon (or light rain) events producing a more rapid drying cycle that results in a carbon deficit in the crust, especially when these events were large and infrequent.

To anticipate the ecological effects of these global change scenarios, an intact Mojave Desert ecosystem was subjected to fertilization and summer irrigation, simulating increased N deposition and monsoon activity. Our hypothesis was that N supplementation and summer monsoon conditions would each be expected to negatively impact *Syntrichia caninervis* growth, sexual reproduction, and asexual regenerational vigor, and in the case of N, higher deposition rates would have more negative consequences than lower deposition rates. A negative impact of monsoon conditions on *S. caninervis* is expected because summer storms result in shorter hydration intervals, higher hydrated temperatures, and faster drying times, a combination that likely produces a carbon deficit and the physiological stress of rapid desiccation (Schonbeck & Bewley 1981; Oliver *et al.* 2000).

Methods

Experimental field site

The Mojave Global Change Facility (MGCF) was established in 2001 on the US Department of Energy's Nevada Test Site in southern Nevada (36°49′ N, 115°55′ W, elev. 970 m), in order to study long-term effects of climate change on desert ecosystems. Mean annual precipitation at the MGCF is 138 ±62 mm, falling mostly during winter months (Hunter 1994), with highly episodic summer precipitation and a low relative frequency of large rainfall events (Huxman *et al.* 2004). This community is dominated by the xerophytic shrubs *Larrea tridentata*, *Ambrosia dumosa*, *Lycium pallidum*, *L. andersonii*, and the C_4 bunchgrass *Pleuraphis rigida* (Jordan *et al.* 1999). This remote area represents an essentially undisturbed desert ecosystem closed to the public and free of any livestock grazing for at least 60 years. The MGCF spans 36 ha, comprising 96 plots, each measuring 14 m × 14 m. For each treatment, an area 16 m × 16 m was subjected to the treatment (or treatment combination), which allows for a 1 m buffer area so that the entire 14 m × 14 m plot could be used for measurements.

Overall, the experiment involved three factors arranged in a factorial design (8 blocks, each with 12 plots) with two monsoon treatments (+ and 0), three N treatments (0, 10, and 40 kg N ha^{-1} yr^{-1}), and two biological soil disturbance regimes (+ and 0). In this chapter, we focus on results from treatments having no soil disturbance, equating to 48 of the 96 plots. From 2001 forward, irrigation was applied in three 25 mm events once every three weeks from early July to mid-August through a sprinkler system tested for even coverage (Fig. 8.1a). Local ground water was delivered evenly during windless nights at rates below the infiltration rate of the soil. The total supplemental 75 mm H_2O represents a 3-fold increase in mean annual summer precipitation, but only a 50% increase in annual precipitation. Supplemental N was added in November of each year as $CaNO_3$ in solution via sprinklers, approximating the range of N deposition in the Las Vegas, NV, USA (10 kg N ha^{-1} yr^{-1}) and Los Angeles, CA, USA (40 kg N ha^{-1} yr^{-1}) areas. This resulted in a total application of no more than 5 mm of water, which was also added to all non-+N plots to ensure equal watering among treatments.

Sampling and life history

In late March of 2005, desiccated patches of *Syntrichia caninervis* were sampled in each of the 48 plots. This species is the dominant bryophyte at low

Fig. 8.1. (A) One of the 14 m × 14 m plots at the study site, including walkways, stepping stones, and a sprinkler system for application of +monsoon and +N treatments; the pole extending from a square plate marks the rear corner of the plot. (B) Patch of *Syntrichia caninervis*, consisting of a series of clumps (length of forceps = 11 mm). (C) Cluster of shoots sampled from a clump of *Syntrichia*. (D) A *Syntrichia* shoot denuded of leaves and illustrating how the distal two growth intervals were measured in length (G0=most recent growth interval, G1=previous year growth interval). (E) Three basal buds emerging from the larger shoot placed into culture from the field. (F) A cyanobacterial species (filaments in the foreground at arrow) descending from a *Syntrichia* shoot placed into culture.

elevations in the Mojave Desert (Bowker *et al.* 2000) and grows in clusters of ramets that are mostly unbranched. These ramets extend in length each year during the cooler, wetter months, and each ramet can be up to 17 years old (Stark *et al.* 2001). Sexual reproduction is exceedingly rare (Bowker *et al.* 2000); patch expansion and colonization are likely controlled by asexual processes. One patch per plot having at least one contiguous clump measuring >10 cm × 10 cm and which was situated under a shrub or perennial grass canopy was targeted (Fig. 8.1b). Patches had been previously randomly assigned for the present study and for additional studies monitoring phenology and disturbance based on the minimum size requirements above. If more than one of these clumps existed in the patch, then the centralmost one in the patch was selected. Using fine forceps, *c.* 20 desiccated ramets (8 were used in this study) were removed from the clump (Fig. 8.1c), placed in a microenvelope and transported to the laboratory. If the patch did not contain a clump with 10 cm × 10 cm dimensions located in the understory, or if the plot did not contain a suitably sized patch (in 8 instances), then we selected an alternate patch in another plot of the same treatment.

Allocation analyses

From each group of 20 shoots, four similarly sized shoots that were unbranched above the ground level were randomly selected for allocation analyses. If an unbranched shoot was not available, then an individual ramet was cut below the ground surface and used. Each shoot was hydrated, cut at the point along the stem where the ground surface was intersected (as indicated by sand particles), cleaned of sand, particles, and old detached leaves by agitating it in a drop of sterile water, blotted dry, and placed in a micropacket. Shoots were dried for 3 days at 30 °C and *c.* 42% relative humidity (RH) in an incubator (the lower temperature was to ensure that the plants would not be killed and thus preserve the demarcation of the green zones). Following drying, each shoot was weighed to the nearest 0.01 mg. To assess the length of the two most recent growth intervals and sex expression, each shoot was hydrated on a microscope slide, denuded of leaves, examined for gametangia (sex expression) produced in the current season (2005, *G0*), the previous season (2004, *G1*), and pre-2004 (aboveground lifetime sex expression). The denuded shoot apex was photographed and image analysis employed (*SPOT*, Diagnostic Instruments, Sterling Heights, MI, USA) to measure the (i) length of shoot and (ii) lengths of the two distal growth intervals (*G0* + *G1*, sensu Stark 2002a; Fig. 8.1d). The base of the *G1* interval was determined using a lateral bud in conjunction with the end of the brown scales on the surface of the stem: the halfway point between these two internal markers was used, with the brown zone located just below the lateral bud. If the lateral bud could not be located, we used the end of the brown zone (as in Stark *et al.* 1998). Because the demarcation between the *G0* and *G1* intervals was not always obvious (that point where the brown leaf bases ended), rather than separating the two intervals, we added the *G0* and *G1* intervals to produce the length of the "green zone," which equated to the length of shoot produced over the past two years. Biomass of the green zone was calculated using length to biomass ratios, and the "hardiness" of each shoot was calculated as mg mm^{-1}.

Regeneration experiment

Four shoots from each patch were selected, as given above, for the allocation analyses. Each desiccated shoot was cut, using a straight edge, into a 2.0–3.0 mm apical segment exclusive of the awn (but including the leaves) in order to remove most below-ground tissues and their associated sand. Initial dry biomass of each shoot was measured to the nearest 0.01 mg. Each shoot was hydrated on a microscope slide, then transferred into sand in a premoistened Petri dish (inner diameter 35 mm) in an erect position, one shoot per dish. The sand substrate was 4–5 mm in depth, and had been collected from the study site, sieved through a 500 μm mesh, dry autoclaved for 60 min at 131 °C, and apportioned into Petri dishes. The resulting 192 Petri dishes were arranged on two shelves in a growth chamber (*Percival* model E30B, Boone, IA, USA) set at a photoperiod of 12 h, simulating early spring field conditions (20 °C lighted, 8 °C darkened). Light intensity in the chamber ranged from 30 to 130 μmol m^{-2} s^{-1} (PAR sensor, LI-COR LI-250, Lincoln, NE, USA), and relative humidity was 60%–70%. In order to equalize light differences between shelves, dishes were rotated between the two shelves in the chamber on a daily basis through day 7, on day 11, and weekly thereafter, randomly repositioning each dish from day 7 forward. Dishes were watered once per week with sterile water to saturate the sand; from day 14 forward, dishes were watered with 30% Hoagland's solution (Hoagland & Arnon 1938). Observations of each dish occurred on days 4, 5, 6, 7, 10, 14, and weekly thereafter through day 56, and included (i) day of first bud production, classifying buds as subtending (occurring just below the apical meristem), basal (projecting from the shoot base at the sand surface; Fig. 8.1e), and protonemal (produced at a distance from the original shoot and arising from protonemata); (ii) day of protonemal appearance; and (iii) day when any cyanobacterial symbionts (genus *Microcoleus*) appeared (Fig. 8.1f). At the conclusion of the experiment (day 56), dishes were un-lidded and the dishes allowed to dry overnight in the growth chamber. Total final biomass was determined by excavating the shoot/protonemal complex, washing sand particles and any contaminants out of the protonemal/shoot complex, oven drying for 3 days at 40 °C, and measuring dry biomass to the nearest microgram.

Statistics

Differences in allocation and regeneration response means among +monsoon and +N treatments were done using two-way ANOVAs. For the allocation study, the dependent variables were length, biomass, and hardiness of the green zone. For the regeneration study, the dependent variables were days to basal shoot production, overall shoot production, net productivity (final biomass – initial biomass), or days to cyanosymbiont emergence. For each plot we used the mean for each of the four sampled shoots in the analysis. Data that were counts (days, numbers of shoots) were square root transformed. The other data were log transformed. For the allocation data, we used logistic regression to test for associations among monsoon level and nitrogen level with the probability of sex expression. For plants that expressed sex we tested whether N and ±monsoon levels affected number of inflorescences, using a two-way ANOVA as above. For the regeneration data, we used logistic regression to test for associations among monsoon level and N level with the probability of cyanosymbionts emerging. We used the GENMOD procedure of SAS (1994) for the logistic regressions.

Results

Allocation analyses

The length of the green zone (the sum of the growth intervals produced in 2004 and 2005) ranged from 0.59 ±0.02 mm (ambient control) to 0.66 ±0.02 mm (+monsoon, high +N), with no significant main effects (monsoon F (1, 42) = 0.19, $p = 0.6671$; N level F (2, 42) = 1.3, $p = 0.2836$) or interaction effects (F (2, 42) = 0.32, $p = 0.7262$) on the length of the green zone (Table 8.1). The biomass of the green zone ranged from only 0.19 ±0.02 mg (+monsoon, −N) to 0.22 ±0.02 mg (−monsoon, low and high +N levels), with no significant main effects (monsoon F (1, 42) = 1.13, $p = 0.2929$; N level F (2, 42) = 0.71, $p = 0.4961$) or interaction effects (F (2, 42) = 0.02, $p = 0.9754$; Table 8.1). Aboveground shoot biomass ranged from 0.34 ±0.02 mg (+monsoon, −N) to 0.47 ±0.03 mg (−monsoon, low +N). Overall, shoots from −monsoon plots had significantly higher biomass than shoots from the +monsoon (F (1, 42) = 4.82, $p = 0.0334$; Table 8.1). There were no significant main effects for +N (F (2, 42) = 0.40, $p = 0.67$) or interaction effects (F (2, 42) = 0.06, $p = 0.9428$) on biomass. Hardiness, here defined as biomass per unit shoot length (mg mm^{-1}), ranged from 0.30 ± 0.01 mg mm^{-1} (+monsoon, high +N) to 0.35 ±0.02 mg mm^{-1} (−monsoon, low +N). A non-significant tendency was present for a negative influence of the monsoon treatment on hardiness across all N levels (Fig. 8.2). However, there were no significant main effects (monsoon F (1, 42) = 2.51, $p = 0.12$; N level F (2, 42) = 0.39, $p = 0.6826$) or interaction effects (F (2, 42) = 0.00, $p = 0.9970$) on hardiness of the green zone.

Although there was no significant association of monsoon level on the probability of sex expression, there was a significant relationship between N level and the probability of sex expression ($\chi^2 = 9.02$, df = 2, $p = 0.011$). The low +N treatment was associated with greater sex expression compared with the two other N levels ($\chi^2 = 8.99$, df = 1, $p = 0.0027$), with overall probabilities of sex expression in the N treatments at 0.39 (ambient control), 0.65 (low +N), and

Table 8.1. *Length of the "green zone" from field-collected shoots, estimated biomass of the green zone from field-collected shoots, and aboveground biomass of shoots of* Syntrichia caninervis *used in the allocational analyses*

Data are means ± one SE. The "green zone" represents the summed length of the two most recent growth intervals. "Mons" refers to the monsoon field treatment; "Ambient Control" refers to the -Mons, -N treatment. Groups indicated by different superscripts are significantly different ($p = 0.03$).

Field Treatment	N	Green zone length (mm)	Green zone biomass (mg)	Aboveground biomass (mg)
Ambient Control	32	0.59 ± 0.02	0.19 ± 0.01	0.41 ± 0.02
-Mons, Low +N	32	0.63 ± 0.02	0.22 ± 0.01	0.47 ± 0.03 [A]
-Mons, High +N	32	0.63 ± 0.02	0.22 ± 0.02	0.46 ± 0.04 [A]
+Mons, -N	32	0.61 ± 0.02	0.19 ± 0.02	0.34 ± 0.02 [B]
+Mons, Low +N	32	0.62 ± 0.02	0.20 ± 0.01	0.38 ± 0.02 [B]
+Mons, High +N	32	0.66 ± 0.02	0.20 ± 0.01	0.37 ± 0.02 [B]

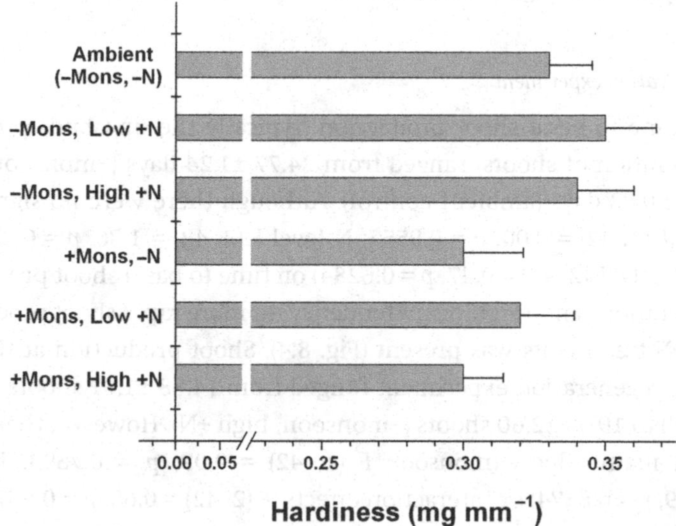

Fig. 8.2. Hardiness of *Syntrichia caninervis* shoots collected from field plots exposed to four years of projected global change conditions of nitrogen supplementation and an enhanced monsoon. "Mons" refers to the monsoon treatment described in methods (mean and SE); "Ambient" refers to the –Mons, -N treatment.

0.43 (high +N). Plants that expressed sex produced fewer inflorescences in +monsoon conditions compared to –monsoon conditions (F (1,42) = 7.91, p = 0.0078; Fig. 8.3). There were no +N effects (F (2, 42) = 0.39, p = 0.679) or interaction effects (F (2, 42) = 0.41, p = 0.669) on inflorescence number.

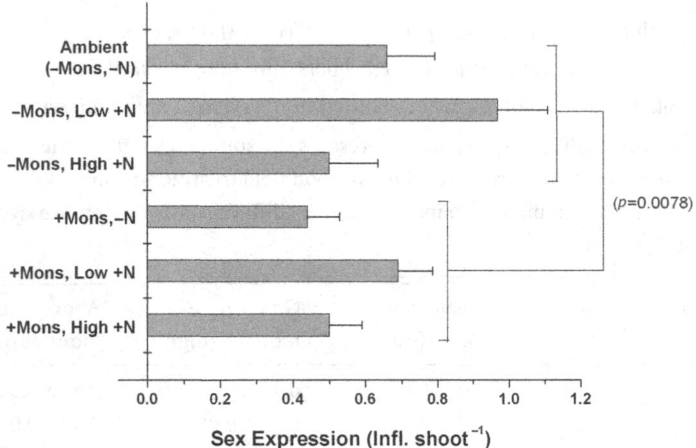

Fig. 8.3. Above-ground sex expression of *Syntrichia caninervis* shoots collected from field plots exposed to four years of projected global change conditions of nitrogen supplementation and enhanced monsoon. "Mons" refers to the monsoon treatment described in methods (mean and SE); "Ambient" refers to the –Mons, -N treatment. Bracketed bars are significantly different.

Regeneration experiment

The time to basal shoot production (typically the first buds produced during regeneration of shoots) ranged from 24.77 ±1.24 days (–monsoon, high +N) to 28.68 ± 0.82 days (ambient control). Although there were no significant main effects (F (1, 42) = 0.00, p = 0.9557; N level F (2, 42) = 1.36, p = 0.2672) or interaction effects (F (2, 42) = 0.47, p = 0.6284) on time to basal shoot production during regeneration, a non-significant tendency for more rapid shoot production in the high +N treatments was present (Fig. 8.4). Shoot production at the conclusion of the regeneration experiment ranged from 14.88 ±1.67 shoots (+monsoon, high +N) to 19.09 ±2.60 shoots (–monsoon, high +N). However, there were no statistical main effects (monsoon F (1, 42) = 0.09, p = 0.7693; N level F (2, 42) = 0.19, p = 0.8274) or interaction effects (F (2, 42) = 0.63, p = 0.5372). Net productivity of the regenerant shoots (final biomass – initial biomass) at the conclusion of the 56-day experiment ranged from 1.33 ±0.13 mg (+monsoon) to 2.02 ±0.31 mg (ambient control), suggesting that at background N levels, the +monsoon treatment had a negative effect on the amount of biomass regenerated (Fig. 8.5). However, there were no significant main effects (monsoon F (1, 42) = 0.94, p = 0.3384; N level F (2, 42) = 0.11, p = 0.8949) or interaction effects (F (2, 42) = 0.77, p = 0.4676) on biomass production during regeneration.

Probability of cyanosymbionts arising from *Syntrichia* shoots ranged from 0.28 (upper and lower 95% confidence limits of 0.42 and 0.26, respectively) for

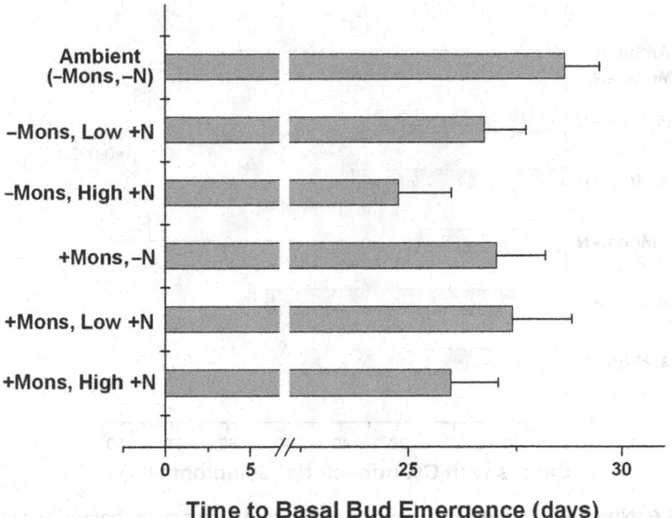

Fig. 8.4. Time to basal shoot production of *Syntrichia caninervis* shoots collected from field plots exposed to four years of projected global change conditions of nitrogen supplementation and enhanced monsoon, and subsequently allowed to regenerate in the laboratory for 56 days. "Mons" refers to the monsoon treatment described in methods (mean and SE); "Ambient" refers to the –Mons, -N treatment.

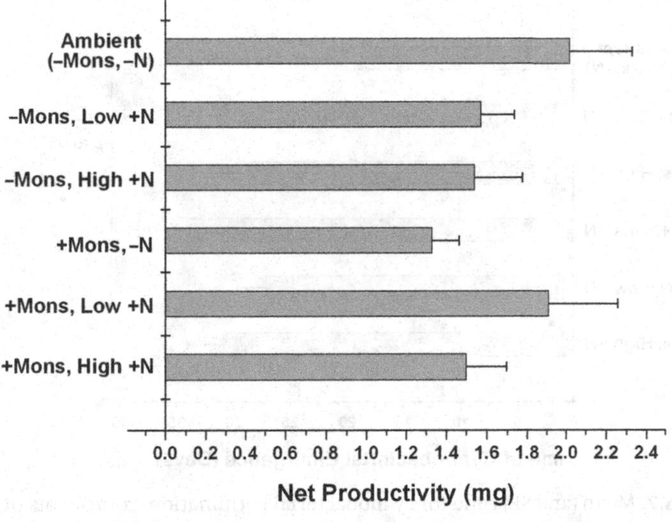

Fig. 8.5. Mean (and SE) biomass produced per shoot of *Syntrichia caninervis* collected from field plots exposed to four years of projected global change conditions of nitrogen supplementation and enhanced monsoon, and subsequently allowed to regenerate in the laboratory for 56 days. "Mons" refers to the monsoon treatment described in methods; "Ambient" refers to the –Mons, -N treatment.

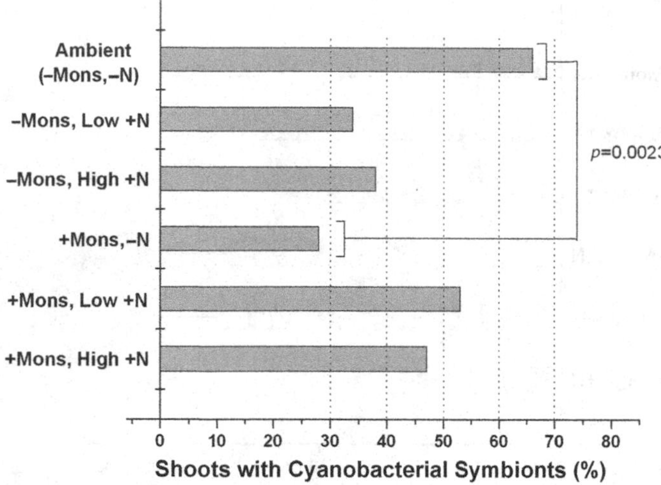

Fig. 8.6. Number of shoots of *Syntrichia caninervis* having cyanobacterial symbionts after collection from field plots exposed to four years of projected global change conditions of nitrogen supplementation and enhanced monsoon, and subsequently allowed to regenerate in the laboratory for 35 days. "Mons" refers to the monsoon treatment described in methods; "Ambient" refers to the –Mons, -N treatment. Bracketed bars are significantly different.

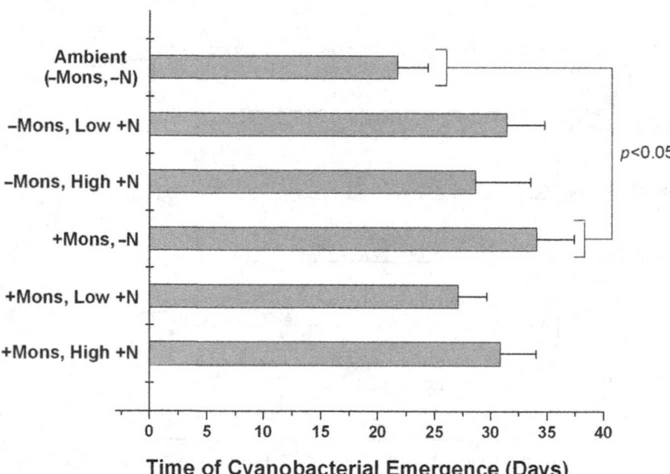

Fig. 8.7. Mean (and SE) time for cyanobacterial germination from shoots of *Syntrichia caninervis* collected from field plots exposed to four years of projected global change conditions of nitrogen supplementation and enhanced monsoon, and subsequently allowed to regenerate in the laboratory for 56 days. "Mons" refers to the monsoon treatment described in methods; "Ambient" refers to the –Mons, -N treatment. Bracketed bars are significantly different.

the +monsoon, −N to 0.66 (upper and lower 95% confidence limits of 0.91 and 0.60, respectively) for −monsoon, −N (ambient control; Fig. 8.6). There were no overall monsoon or N effects (analysis not shown). However, there was a significant interaction effect between monsoon level and N level ($\chi^2 = 11.95$, df = 2, $p = 0.005$). Further analysis across N levels found differences between monsoon conditions within the −N controls ($\chi^2 = 9.26$, df = 1, $p = 0.0023$), with a lower probability of housing a cyanobacterium under monsoon conditions. Days to cyanosymbiont emergence ranged from 21.76 ± 2.63 days (ambient control) to 34.11 ± 3.30 days (+monsoon, −N). There was a monsoon × N level interaction (F (2, 38) = 4.26, $p = 0.0214$; Fig. 8.7). Further analyses showed that at background N levels, cyanosymbionts emerged sooner in the ambient control compared to +monsoon treatment (F (1, 38) = 6.55, $p < 0.05$). There were no other simple effects that were significant (analyses not shown).

Discussion

In the Mojave Desert, second to water, N limits plant productivity (Smith *et al.* 1997). However, bryophytes are buffered from N limitations owing to their receipt of most nutrients from atmospheric deposition rather than from secondary soil sources. When occupying shaded microsites beneath shrub canopies, bryophytes can receive wet deposition from a combination of stem flow and throughfall. Nevertheless, over a period of several years, overgrowth of canopy microsites preferred by the desert moss *Syntrichia caninervis* by cheatgrasses (*Bromus*) could pose a threat by further shading and by physical occupation of the microsite (Bowker *et al.* 2000). In addition to increased N deposition, monsoon (summer) precipitation events are predicted to increase in frequency and severity for the Mojave Desert in the coming decades. These events subject mosses to thermal stresses when both wet and dry, and to a desiccation stress as the plants dry out too quickly to forge a net C gain from the hydration event (Belnap *et al.* 2004).

In the present study, after four years of N supplementation and monsoon treatments in the field, the responses observed in the dominant moss *S. caninervis* indicated that monsoonal conditions will likely produce adverse effects for the growth and reproduction of this moss, whereas increased N deposition effects will be minor. Aboveground biomass of *S. caninervis* shoots was reduced under simulated monsoon conditions in the field, implying that carbon gain is more difficult under enhanced monsoon conditions. Sex expression, the production of gametangia required for fertilization and the production of meiospores in *Syntrichia*, showed an inhibition under +monsoon treatments and a stimulation under low +N conditions. Taken together, these findings indicate that both clonal and sexual reproduction are limited by resources and may be more sensitive to desiccation and thermal stress episodes experienced during enhanced monsoon conditions. Furthermore, when *Syntrichia* shoots taken from the field treatments were allowed to regenerate under growth chamber conditions simulating winter/spring conditions in the desert, the cyanobacterial symbionts of *Syntrichia* were fewer in number and emerged significantly later if they were from +monsoon-treated plots.

Vascular plant responses

Among vascular plants studied to date in this system, the dominant shrub *Larrea tridentata* exhibited increases in leaf-level CO_2 assimilation under +monsoon treatments irrespective of N level. However, during the following spring, leaf N content was lower in plants from +monsoon plots, which brought about lower rates of stomatal conductance and leaf-level CO_2 assimilation (Barker *et al.* 2006). Thus, there was an immediate increase in assimilation and productivity after the monsoon events, but the following spring the effect was lessened. Given their very different responses to hydration events, there is no reason to expect homoiohydric vascular plants and poikilohydric bryophytes to react similarly to changes in monsoon precipitation regimes; in fact, it is not surprising that they react essentially in an opposing fashion.

In more mesic ecosystems, several studies found that N supplementation enhanced the growth of vascular plants and tall weedy mosses, which resulted in "canopy closure" for the bryophytes, with space and light limitations leading to the demise of the bryophyte community (e.g., van der Wal *et al.* 2005 and references therein). The availability of N in soils appears to stimulate seed plant germination and growth and provide a competitive edge to seed plants over mosses (Bragazza *et al.* 2005). Although the present study did not monitor vascular plant effects on space and canopy cover on *Syntrichia*, we noted the proliferation of exotic *Bromus* grasses at the site, and are currently monitoring the long-term capacity of *Syntrichia* patches to recover from mechanical disturbance in the presence of +N and +monsoon treatments. This desert system may require consecutive high-rainfall years for *Syntrichia* to be negatively impacted by overgrowth of vascular species.

Bryophyte responses

In the tundra permafrost, increased summer precipitation (1 mm day^{-1}) and thermal treatments resulted in a stimulation of growth of *Sphagnum* and *Dicranum* mosses (Sonesson *et al.* 2002). However, in deserts, rain events that are short in duration or that occur during the summer have deleterious effects on crustal organisms because these organisms (i) incur a C deficit due to

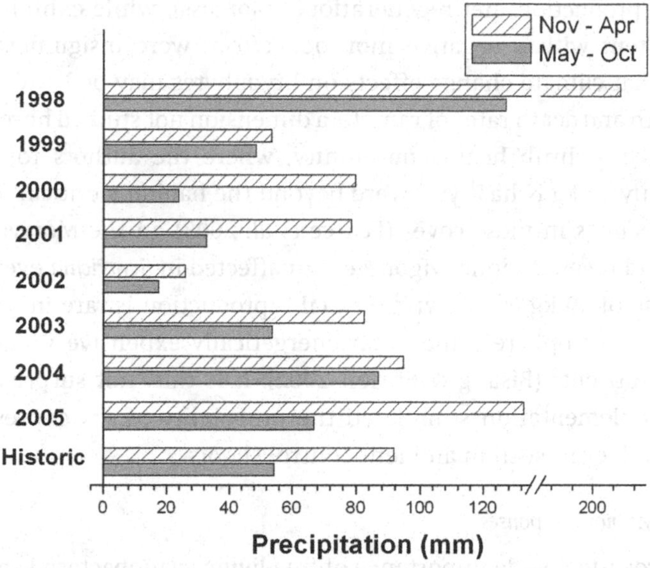

Fig. 8.8. Precipitation at the study site, 1998–2005, including precipitation falling during the growing season (November–April) and precipitation falling during the dormant (for mosses) season (May–October). The historic average levels of precipitation are given for comparison.

insufficient hydration time as the soil surface desiccates quickly (Belnap *et al.* 2004; Barker *et al.* 2005) and (ii) the thermal stress attending summer rains can negatively impact reproduction (Stark 2002b). The +monsoon treatment did not alter recent (past two years) productivity patterns in field-grown *Syntrichia* shoots, but negatively affected aboveground productivity, which may span several more years. The past two years (2004–05) were marked by higher than normal precipitation (especially 2005), whereas the preceeding four years (1999–02) were marked by lower than normal precipitation, including severe drought conditions in 2002 (Fig. 8.8). In addition, the +monsoon treatment negatively affected the ability of the shoots to express sex in the field and to harbor cyanobacterial symbionts. Exposure to summer rainfall appears to impede both sexual and to some extent asexual reproductive capacity in *Syntrichia*, a finding consistent with C deficits manifested as allocational deficiencies. Given the importance of each mode of reproduction to the proliferation of the species (Mishler 1988; Longton 1994), this does not bode well for the future survival of this species under projected climate change conditions. However, a modestly increased N deposition improved the probability of sex expression relative to other field treatments, mitigating the negative effects of the monsoon events. In our regenerational trials, the response variables (time to shoot production, net regenerational biomass), while exhibiting tendencies consistent with a negative monsoon effect, were insignificant. This may indicate that climate change effects on bryophytes may be better seen in analyses of birth and death rates of ramets, a dimension not studied here. Unlike in a *Racomitrium* – shrub heath community, where the authors found that additions of only 10 kg N ha^{-1} yr^{-1} were beyond the habitat's critical load and thus caused declines in moss cover (Pearce *et al.* 2003), in the Mojave Desert productivity and regenerational vigor were unaffected in *Syntrichia* even under the high N dose of 40 kg N ha^{-1} yr^{-1}. Sexual reproduction is rare in dioecious desert species of bryophytes, and is an energetically expensive process that is limited by nutrients (Bisang & Ehrlén 2002). It is thus not surprising that low-dose N supplementation stimulated the probability of sex expression in *Syntrichia* beyond levels seen in ambient control shoots.

Cyanosymbiont responses

The ecosystem-wide importance of free-living cyanobacterial species in the genus *Microcoleus* is well documented. This group of organisms performs a variety of beneficial functions in aridland habitats, including enhancing soil fertility, soil stability, limited N-fixation, soil aeration, nutrient cycling, and water retention among others (Belnap & Lange 2001; Bowker 2007). Species of *Microcoleus* are reported here for the first time as taking refuge in patches of *Syntrichia*, emerging from the moss host during shoot regeneration in culture. Under field ambient and laboratory control conditions, viable *Microcoleus* filaments were present in *c.* 65% of *Syntrichia* shoots (aboveground tissues), and these emerged in about 20–25 days. This finding suggests that, in addition to migrating downward in desert soils when the soil dries (Garcia-Pichel & Pringault 2001), *Microcoleus* may migrate upward into moss shoot microhabitats prior to desiccating. Some cyano-moss associations are known to improve soil fertility following disturbances such as fire (de las Heras & Herranz 1996). Clearly the degree of symbiosis between *Syntrichia* and *Microcoleus* merits further investigation. *Syntrichia* could provide a place of refuge during desiccation periods in exchange for sloughed nutrients and carbon compounds lost as the motile cyanobacteria move through the substrate, and *Microcoleus* could provide *Syntrichia* with carbon compounds and trace nutrients extracted from vacated sheaths. The finding that +monsoon treatments negatively impacted the cyanobacteria harbored in *Syntrichia* patches in both the probability of existing with *Syntrichia* shoots and in the time to descend from a *Syntrichia* shoot indicates that cyanobacteria may be more sensitive than *Syntrichia* (but less sensitive than cyanolichens; Belnap *et al.* 2004) to the two by-products of monsoon conditions: thermal and desiccation stress.

Conclusions

To estimate the responses of desert bryophytes to projected changes in N and summer rainfall frequency in the Mojave Desert (USA), patches of the dominant moss *Syntrichia caninervis* were exposed to field treatments of low and high N supplementation (10 and 40 kg ha^{-1}), added summer rain (+75 mm), and combinations thereof, for a period of four years. Shoots assessed directly from the field exhibited lower biomass from the +monsoon compared to the −monsoon treatments, but exhibited no differences in shoot extension lengths, shoot hardiness (mg mm^{-1}), and recent-season annual biomass accumulation over the last two seasons of growth. The probability of sex expression was suppressed in shoots from background and high +N treatments regardless of the moisture level (+/−monsoon). In shoots that expressed sex, the number of inflorescences was lower in the +monsoon treatment relative to the −monsoon treatment across all N levels. In asexual regeneration trials of these shoots, no differences were seen in the timing, number, or regenerational productivity of shoots among treatments, although a tendency was noted for net regeneration productivity to be lower in the +monsoon treatment compared to the ambient control. Shoots exposed to +monsoon, −N conditions released fewer cyanobacterial symbionts during regeneration compared to −monsoon, −N conditions, and the timing of cyanobacterial release was delayed under +monsoon conditions. Multi-season field biomass, sexual function, and the cyanosymbiosis relationship appear to be adversely affected by increased summer monsoon activity. Increased monsoonal frequency as forecast for the northern Mojave Desert may result in declines in sexual and asexual vigor in desert mosses.

Acknowledgements

We thank Lorenzo Nichols II for assistance in the laboratory, John Brinda, Jayne Belnap, and Brent Mishler for field assistance and/or conceptual contributions, Robin Stark for graphical assistance, Dene Charlet for compiling the precipitation data, and Allen Gibbs for use of his microbalance. Lynn Fenstermaker, Dene Charlet, and Jon Titus installed the initial field site treatments, and Eric Knight and Derek Babcock provided field assistance and site operations. This research was supported by the Office of Science, Biological and Environmental Research Program (BER), US Department of Energy, through the Western Regional Center of the National Institute for Climate Change Research (formerly known as Nat. Inst. for Global Environmental Change under Cooperative Agreement No. DE-FCO2-03ER63613). Financial support does not constitute an endorsement by DOE of the views expressed in this article. DNM was supported by NSF grant IOB 0416407, and SDS was supported by DOE-PER grant DE-FG02-02ER63361. LRS was supported by NSF grant IOB 0416281 and a UNLV faculty sabbatical leave during a portion of this project.

References

Bakken, S. (1995). Effects of nitrogen supply and irradiance on growth and nitrogen status in the moss *Dicranum majus* from differently polluted areas. *Journal of Bryology* **18**: 707–21.

Barker, D. H., Vanier, C., Naumburg, E. *et al.* (2006). Enhanced monsoon precipitation and nitrogen deposition affect leaf traits and photosynthesis differently in spring and summer in the desert shrub *Larrea tridentata. New Phytologist* **169**: 799–808.

Barker, D. H., Stark, L. R., Zimpfer, J. F., McLetchie, N. D. & Smith, S. D. (2005). Evidence of drought-induced stress on biotic crust moss in the Mojave Desert. *Plant, Cell & Environment* **28**: 939–47.

Bates, J. W. (1997). Effects of intermittent desiccation on nutrient economy and growth of two ecologically contrasted mosses. *Annals of Botany* **79**: 299–309.

Bates, J. W. (2000). Mineral nutrition, substratum ecology, and pollution. In *Bryophyte Biology*, ed. A. J. Shaw & B. Goffinet, pp. 248–311. Cambridge: Cambridge University Press.

Belnap, J. & Lange, O. L. (2001). *Biological Soil Crusts: Structure, Function, and Management.* Ecological Studies Vol. 150. Berlin: Springer-Verlag.

Belnap, J., Phillips, S. L. & Miller, M. E. (2004). Response of desert biological soil crusts to alterations in precipitation frequency. *Oecologia* **141**: 306–16.

Berendse, F., van Breemen, N., Rydin, H. *et al.* (2001). Raised atmospheric CO_2 levels and increased N deposition cause shifts in plant species composition and production in *Sphagnum* bogs. *Global Change Biology* **7**: 591–8.

Bergamini, A. & Peintinger, M. (2002). Effects of light and nitrogen on morphological plasticity of the moss *Calliergonella cuspidata. Oikos* **96**: 355–63.

Bisang, I. & Ehrlén, J. (2002). Reproductive effort and cost of sexual reproduction in female *Dicranum polysetum* Sw. *Bryologist* **105**: 384–97.

Bowker, M. A. (2007). Biological soil crust rehabilitation in theory and practice: an underexploited opportunity. *Restoration Ecology* **15**: 13–23.

Bowker, M. A., Stark, L. R., McLetchie, D. N. & Mishler, B. D. (2000). Sex expression, skewed sex ratios, and microhabitat distribution in the dioecious desert moss *Syntrichia caninervis* (Pottiaceae). *American Journal of Botany* **87**: 517–26.

Bragazza, L., Tahvanainen, T., Kutnar, L. *et al.* (2004). Nutritional constraints in ombrotrophic *Sphagnum* plants under increasing atmospheric nitrogen deposition in Europe. *New Phytologist* **163**: 609–16.

Bragazza, L., Limpens, J., Gerdol, R. *et al.* (2005). Nitrogen concentration and $\delta^{15}N$ signature of ombrotrophic *Sphagnum* mosses at different N deposition levels in Europe. *Global Change Biology* **11**: 106–14.

Chapin, F. S. III, Shaver, G. R., Giblin, A. E., Nadelhoffer, K. J. & Laundre, J. A. (1995). Responses of arctic tundra to experimental and observed changes in climate. *Ecology* **76**: 694–711.

Heras, J. de las & Herranz, J. M. (1996). The role of bryophytes in the nitrogen dynamics of soils affected by fire in Mediterranean forests (southeastern Spain). *Ecoscience* **3**: 199–204.

Galloway, J. N. (1998). The global nitrogen cycle: changes and consequences. *Environmental Pollution* **102** (Suppl. 1): 15–24.

Garcia-Pichel, F. & Pringault, O. (2001). Cyanobacteria track water in desert soils. *Nature* **413**: 380–1.

Higgins, R. W. & Shi, W. (2001). Intercomparison of the principal modes of interannual and intraseasonal variability of the North American Monsoon System. *Journal of Climate* **14**: 403–17.

Hoagland, D. & Arnon, D. I. (1938). The water culture method for growing plants without soil. *California Agricultural Experiment Station Circular* **347**: 1–39.

Hunter, R. B. (1994). Status of flora and fauna on the Nevada Test Site, 1994. *National Technical Information Service* **89**: 590–6 (DOE/NV/11432-195 UC-721U.S.). Springfield: Department of Commerce.

Huxman, T. E., Snyder, K. A., Tissue, D. *et al.* (2004). Precipitation pulses and carbon fluxes in semiarid and arid ecosystems. *Oecologia* **141**: 254–68.

Jägerbrand, A. K., Molau, U. & Alatalo, J. M. (2003). Responses of bryophytes to simulated environmental change at Latnjajaure, northern Sweden. *Journal of Bryology* **25**: 163–8.

Jauhiainen, J., Silvola, J. & Vasander, H. (1998). Effects of increased carbon dioxide and nitrogen supply on mosses. In *Bryology for the Twenty-first Century*, ed. J. W. Bates, N. W. Ashton & J. G. Duckett, pp. 343–60. Leeds: British Bryological Society.

Jordan, D. N., Zitzer, S. F., Hendrey, G. R. *et al.* (1999). Biotic, abiotic and performance aspects of the Nevada Desert Free-air CO_2 Enrichment (FACE) Facility. *Global Change Biology* **5**: 659–68.

Li, Y. & Glime, J. M. (1990). Growth and nutrient ecology of two *Sphagnum* species. *Hikobia* **10**: 445–51.

Limpens, J., Tomassen, H. B. M. & Berendse, F. (2003). Expansion of *Sphagnum fallax* in bogs: striking the balance between N and P availability. *Journal of Bryology* **25**: 83–90.

Longton, R. E. (1994). Reproductive biology in bryophytes: the challenge and the opportunities. *Journal of the Hattori Botanical Laboratory* **76**: 159–72.

Mishler, B. D. (1988). Reproductive ecology of bryophytes. In *Plant Reproductive Ecology: Patterns and Strategies*, ed. J. Lovett Doust & L. Lovett Doust, pp. 285–306. New York: Oxford University Press.

Mitchell, E. A. D., Buttler, A., Grosvernier, P. *et al.* (2002). Contrasted effects of increased N and CO_2 supply on two keystone species in peatland restoration and implications for global change. *Journal of Ecology* **90**: 529–33.

Oliver, M. J., Velten, J. & Wood, A. J. (2000). Bryophytes as experimental models for the study of environmental stress tolerance: *Tortula ruralis* and desiccation-tolerance in mosses. *Plant Ecology* **151**: 73–84.

Pearce, I. S. K., Woodin, S. J. & Wal, R. van der (2003). Physiological and growth responses of the montane bryophyte *Racomitrium lanuginosum* to atmospheric nitrogen deposition. *New Phytologist* **160**: 145–55.

Press, M. C., Potter, J. A., Burke, M. J. W., Callaghan, T. V. & Lee, J. A. (1998). Responses of a subarctic dwarf shrub heath community to simulated environmental change. *Journal of Ecology* **86**: 315–27.

SAS (1994). *SAS/STAT User's Guide*, v. 6, 4th edn, vol. I. Cary, NC: SAS Institute.

Saarnio, S. (1999). Carbon gas (CO_2, CH_4) exchange in a boreal oligotrophic mire – effects of raised CO_2 and NH_4NO_3 supply. *University of Joensuu Publications in Sciences* **56**: 1–29.

Schonbeck, M. W. & Bewley, J. D. (1981). Responses of the moss *Tortula ruralis* to desiccation treatments. I. Effects of minimum water content and rates of dehydration and rehydration. *Canadian Journal of Botany* **59**: 2698–706.

Smith, S. D., Monson, R. K. & Anderson, J. E. (1997). *Physiological Ecology of North American Desert Plants*. Berlin: Springer-Verlag.

Sonesson, M., Carlsson, B. A., Callaghan, T. V. *et al.* (2002). Growth of two peat-forming mosses in subarctic mires: species interactions and effects of simulated climate change. *Oikos* **99**: 151–60.

Stark, L. R., McLetchie, D. N. & Mishler, B. D. (2001). Sex expression and sex dimorphism in sporophytic populations of the desert moss *Syntrichia caninervis*. *Plant Ecology* **157**: 183–96.

Stark, L. R. (2002a). New frontiers in bryology: phenology and its repercussions on the reproductive ecology of mosses. *Bryologist* **105**: 204–18.

Stark, L. R. (2002b). Skipped reproductive cycles and extensive sporophyte abortion in the desert moss *Tortula inermis* correspond to unusual rainfall patterns. *Canadian Journal of Botany* **80**: 533–42.

Stark, L. R., Mishler, B. D. & McLetchie, D. N. (1998) Sex expression and growth rates in natural populations of the desert soil crustal moss *Syntrichia caninervis*. *Journal of Arid Environments* **40**: 401–16.

Taylor, K. E. & Penner, J. (1994). Responses of the climate system to atmospheric aerosols and greenhouse gases. *Nature* **369**: 734–7.

van der Wal, R., Pearce, I. S. K. & Brooker, R. W. (2005). Mosses and the struggle for light in a nitrogen-polluted world. *Oecologia* **142**: 159–68.

Van Tooren, B. F., Van Dam, D. & During, H. J. (1990). The relative importance of precipitation and soil as sources of nutrients for *Calliergonella cuspidata* in a chalk grassland. *Functional Ecology* **4**: 101–7.

Vitousek, P. M., Aber, J. D., Howarth, R. W. *et al.* (1997). Human alteration of the global nitrogen cycle: sources and consequences. *Ecological Applications* **7**: 737–50.

Zaady, E., Kuhn, U., Wilske, B., Sandoval-Soto, L. & Kesselmeier, J. (2000). Patterns of CO_2 exchange in biological soil crusts of successional age. *Soil Biology & Biochemistry* **32**: 959–66.

9

Ecology of Bryophytes in Mojave Desert Biological Soil Crusts: Effects of Elevated CO_2 on Sex Expression, Stress Tolerance, and Productivity in the Moss *Syntrichia caninervis* Mitt.

JOHN C. BRINDA, CATHERINE FERNANDO, AND LLOYD R. STARK

Introduction

The arid shrublands and open woodlands of the North American deserts generally support a sparse vascular plant cover. In these and similar environments, the surfaces unoccupied by taller vascular plant species are often colonized by a unique assemblage of organisms collectively referred to as "biological soil crusts" (Belnap & Lange 2003). This assemblage is extremely diverse phylogenetically, including various species of cyanobacteria, algae, fungi, and lichens as well as bryophytes. These disparate species have been thrown together under one umbrella label due to the specialized niche that they inhabit. Their small size and close association with the soil surface make the term "biological soil crust" particularly appropriate.

Biological soil crusts have received a fair amount of attention and study but bryophytes represent one of the less well understood aspects of this community. In arid landscapes like the Mojave Desert, the bryophyte component in soil crusts is generally made up of mosses in the family Pottiaceae (e.g., species in the genera *Syntrichia*, *Pterygoneurum*, *Crossidium*, *Didymodon*, and *Tortula*). In his monographic treatment of the family, Zander (1993) states that the Pottiaceae form a conspicuous part of the vegetation of arid, ruderal, alpine, and Arctic areas. These are also conditions under which biological soil crusts can play a prominent role. As characteristic inhabitants of these extreme environments,

Bryophyte Ecology and Climate Change, eds. Zoltán Tuba, Nancy G. Slack and Lloyd R. Stark. Published by Cambridge University Press.

bryophytes of biological soil crusts exhibit many morphological and physiological adaptations to stress that remain poorly understood (Zander 1993).

Where biological crusts are well developed, turf-forming bryophytes can represent a significant portion of the total ground cover, ranging from a few percent to 30% total cover or more (Thompson *et al.* 2005). Soil crust development can influence susceptibility to erosion, hydrology, nutrient inputs, and the associated vascular plant community structure (Belnap *et al.* 2003). Bryophytes make their unique contributions to this process through the growth of fine rhizoids and protonemata as well as their ability to absorb water and nutrient inputs. In systems with sandy soils, bryophytes have also been shown to influence soil formation as well as its mineral and elemental content (Carter & Arocena 2000). In the Mojave Desert, biological soil crusts are physiologically active mainly in the cooler months, when much of the annual precipitation falls. This is also when peak levels of net ecosystem exchange have been measured despite the dominant shrub species being largely dormant. Thus it is clear that the crust community has a significant effect on the overall carbon balance in this system (Jasoni *et al.* 2005) as has been shown in other arid areas (Zaady *et al.* 2000).

The well-documented rise in global atmospheric CO_2 levels caused by burning of fossil fuels will no doubt affect soil crust bryophytes. Because their gametophytes lack stomata, the internal CO_2 concentration of the main photosynthetic tissues of bryophytes depends solely upon atmospheric concentration and the rate of diffusion into the cell. Thick cell walls and external capillary water in bryophytes often slow diffusion rates (Silvola 1990) but the process is morphologically constrained and essentially passive – unlike that in vascular plants. Therefore the indirect effects of CO_2 on photosynthesis due to increases in water use efficiency cannot be interpreted in the same way in bryophytes.

The direct effects of CO_2 concentration on bryophyte photochemistry have, however, been shown to largely parallel those seen in vascular plants. For example, Dilks (1976) tested the CO_2 compensation point for 27 bryophyte species and found it to be in the normal range for C_3 plants (45–160 ppm [CO_2] at 25 °C and 250 $\mu mol\ m^{-2}\ s^{-1}$). Björkman *et al.* (1968) grew *Marchantia polymorpha* at different CO_2 concentrations for 10 days and found a 42% increase in growth with a doubling of CO_2 from 320 ppm to 640 ppm; Bazzaz *et al.* (1970) showed rates of photosynthesis for *Polytrichum juniperinum* to increase by a factor of 1.4–2.3 when exposed to CO_2 concentrations of 450 ppm compared with 300 ppm. Proctor (1982) showed that substantial increases in CO_2 assimilation could be achieved by bryophytes under elevated CO_2 especially when combined with elevated temperature. Higher temperature optima at elevated CO_2 were confirmed by Silvola (1985) in both forest and peat mosses. This is particularly relevant to desert bryophytes that may be exposed to temperature extremes. Silvola (1985) also varied irradiance levels and higher photosynthetic rates were recorded in *Dicranum majus* at lower irradiance levels and higher CO_2 than at normal levels of each, effectively shifting the light compensation point.

Greater photosynthetic efficiency does not necessarily translate directly into increases in bryophyte growth rates. Studies with boreal and peat mosses have shown that elevated CO_2 can have widely varying effects on plant height and biomass accumulation (Sonesson *et al.* 1996; Berendse *et al.* 2001; Heijmans *et al.* 2001; Mitchell *et al.* 2002; see Chapter 4, this volume) that may be species-specific (van der Heijden *et al.* 2000b). This is similar to the results of many experiments with vascular plants showing species-specific responses to elevated CO_2 (Nowak *et al.* 2004). Poorter (1993) showed that slow-growing vascular plant species had less biomass increase in response to elevated CO_2 than other faster-growing species. According to the scheme described in Grime *et al.* (1990), most of the dominant soil crust bryophyte species in the Mojave Desert would likely be classified as slow-growing stress-tolerators and therefore may not respond with marked growth increases under elevated CO_2.

Aside from growth, bryophytes can alternatively allocate excess photosynthate to stress tolerance and/or reproduction. Takács *et al.* (2004) suggested that elevated CO_2 may allow mosses to cope better with cellular damage caused by heavy metals. Coe *et al.* (2008) showed improved tolerance of stressful temperatures in mosses grown at elevated CO_2, and Tuba *et al.* (1998) showed enhanced photosynthesis during drying in a desiccation-tolerant moss. Restoring cellular integrity following stresses such as these requires energy that is naturally more available to plants at elevated CO_2. Energy released from photosynthate stored as sucrose and/or starch can be used to assemble proteins and perform other tasks even when the photosynthetic machinery is damaged. This is particularly important for desiccation-tolerant bryophytes, which upon rehydration have severely reduced photosynthetic capacity combined with considerable membrane damage (Oliver *et al.* 2005).

Increases in reproductive effort have been found in vascular plants grown at elevated CO_2 (Jablonski *et al.* 2002) resulting in greater seed production (Smith *et al.* 2000) as well an accelerated transition to reproductive maturity (LaDeau & Clark 2001). Enhancement of reproduction in bryophytes under elevated CO_2 has not previously been documented but they are unlikely to differ greatly from vascular plants in this respect. When not immediately used, excess photosynthate can also be stored as sugars and starch in the cell or translocated to other parts of the plant. Csintalan *et al.* (1995, 1997) reported increases in both starch and sugar content in a forest moss (*Polytrichum formosum*) grown at elevated CO_2 for ten months but little difference in an arid land moss (*Syntrichia ruralis*) grown at elevated CO_2 for four months. These stored pools of energy must naturally change over time and their response to stress in particular should be examined more closely.

Limiting factors other than CO_2 availability exist in every natural system and ultimately constrain plant responses to CO_2 within certain bounds (Oechel *et al.* 1994). Long-term experiments on vascular plants have shown that they eventually down-regulate photosynthesis owing to reductions in sink strength and starch accumulation (DeLucia *et al.* 1985) and the same is apparently also true of bryophytes (Jauhiainen *et al.* 1998; Tuba *et al.* 1999; van der Heijden *et al.* 2000a; Ötvös & Tuba 2005). Acclimation to elevated CO_2 may include changes in photosynthetic pigment content (Csintalan *et al.* 2005; Coe *et al.* 2008) and/or Rubisco activity (Wong 1979) leading to a persistent reduction in photosynthetic capacity. In this way resources (most notably nitrogen) can be reallocated to other cellular processes.

Based on these previous studies, bryophytes in desert soil crusts were predicted to exhibit greater carbon gain under elevated CO_2 and thus increased vegetative growth and reproductive allocation. Given that desiccation tolerance requires increased synthesis of an array of repair proteins that draw upon the overall carbon and nitrogen budget of the plants (Oliver *et al.* 2005), elevated CO_2 was predicted to enhance desiccation tolerance. Consequently, elevated CO_2 should result in (i) increased frequency of sex expression (inflorescence production), (ii) greater viability and regeneration of shoots after experimental desiccation stress (as evidenced by protonemata and bud production), (iii) moderate increases in biomass accumulation rates, and (iv) a natural down-regulation of photosynthetic capacity following long-term exposure.

Methods

Field methods

Specimens of the moss *Syntrichia caninervis* Mitt. were collected from the Nevada Desert FACE Facility (NDFF) located on Frenchman Flat, in Nye County, NV, USA (36°49′N, 115°55′W, elev. 970 m) approximately 100 km north of Las Vegas. The acronym FACE is short for Free-Air Carbon dioxide Enrichment and is used to describe experiments where large, open-air vegetation plots are exposed to elevated CO_2 *in situ*. In order to minimize disturbance, the ambient wind currents surrounding the plots are analyzed in real time and used to carry CO_2-enriched air across the plot. At the NDFF the CO_2 delivery system was terminated by upright PVC pipes surrounding the plots (Fig. 9.1) that were individually activated depending upon the current wind direction. The NDFF plots consisted of nine rings; three of the rings used FACE

Fig. 9.1. Photo of NDFF (Nevada Desert Face Facility) site showing elevated scaffolding for access and the FACE CO_2 delivery system (vertical PVC pipes).

technology to raise the CO_2 concentration in the air above the plot to 550 ppm. The remaining six rings were two sets of controls under ambient CO_2 conditions. More details of the operations at NDFF are described by Jordan *et al.* (1999). At the time of sampling, the plots had been treated with elevated CO_2 for approximately eight years.

Because of its location on the Nevada Test Site, a secure US Department of Energy facility, the area surrounding the NDFF had been kept free of grazing and most other disturbances for at least 60 years. In addition, knowledge that undisturbed biological soil crusts are an important part of this ecosystem guided site selection for the NDFF. The health of the soil crust on the plots was maintained by requiring most access to be via an elevated platform hung from scaffolding (Fig. 9.1). Consequently the plots were representative of a relatively intact northern Mojave Desert, *Larrea tridentata* – *Ambrosia dumosa* – *Lycium* spp. vascular plant community and associated biological soil crusts (Jordan *et al.* 1999).

By far the most dominant bryophyte in these soil crusts is *Syntrichia caninervis*, although mosses belonging to other genera (notably *Bryum* and *Pterygoneurum*) occur in smaller numbers. Because the site is relatively flat and exposed, bryophytes are restricted to the highest-quality microsites (mostly defined by shade and/or edaphic factors) and therefore exhibit a patchy distribution across the plots. The rings also varied somewhat in the degree of crust development. Since bryophytes only occur in significant numbers in well-developed crusts, sampling was necessarily restricted to those patches and rings that could absorb the impact. In addition, dramatic variation in the size of *S. caninervis* exists in the field, and patches were selected in order to collect uniformly robust specimens that were large enough to be measured and manipulated in the laboratory. Sixteen patches of *S. caninervis* in four rings were sampled, with the patches evenly split between the control and elevated CO_2 treatments. Approximately 30 ramets were collected from each patch, using forceps and taking care not to destroy the entire patch. Essentially non-destructive sampling was necessary in order to maintain the integrity of the long-term FACE experiment.

Laboratory methods

A total of 216 individual, unbranched *S. caninervis* ramets were separated and cleaned of dirt and debris. Both length and biomass of the stems were measured twice, once for the entire above ground portion and again for the "green zone" only. This green zone was determined as described by Stark *et al.* (1998) and signifies the upper portion of the stem and attached leaves, representing the last few years of growth before the leaves and stem begin to turn brown and senesce (here inclusive of the "yellow green" portion of the shoot). The collected ramets were first trimmed with a razor blade to the above ground portion and length and biomass measurements were taken while still dry. They were then hydrated and denuded of leaves up to the green zone and trimmed with the razor blade once again – this time leaving only the green zone. While hydrated, the stems were examined for evidence of sexual expression in the green zone. Green zone length and biomass measurements were taken after the ramets had dried overnight. Stem lengths were measured with electronic calipers under a dissecting microscope; biomass measurements were taken on an electronic balance to a precision of 0.01 mg. Using shoots trimmed to the green zone alone allowed standardization of the ramets along a biologically relevant axis – namely, age. In addition, the green zone of these shoots was likely made up entirely of biomass that had accumulated under the field treatment at NDFF. This might not have been the case had the shoots instead been cut to the same length.

Following the initial measurements, the relative stress tolerance of the shoots was assessed by exposing them to a series of rapid drying treatments. Three treatment levels were used: a control with no rapid drying, six rapid dry cycles, and twelve rapid dry cycles. These rapid drying treatments simulate the stress effect of rapid desiccation while the plants are physiologically active. Prior to the treatments, the shoots underwent a dehardening treatment in order to bring them out of dormancy. During dehardening they were hydrated and placed on moist sand in well plates. The well plates were placed in a growth chamber and the shoots allowed to recover for three days. The experimental stems were then subjected to a series of rapid desiccation events followed by two-hour hydration periods. The two-hour periods gave the shoots time to expend energy repairing their photosynthetic machinery without allowing time for them to achieve a positive carbon balance. Prior experiments had shown this to be more stressful than rapid desiccation without the intervening hydration period. Rapid drying (less than 15 minutes) was achieved by removing the moss from the wet sand and placing it on dry filter paper in the open air. The shoots were then finished overnight in a desiccation chamber at *c.* 15% relative humidity (RH). The following day they were rehydrated in a drop of sterilized distilled water, once again placed on wet sand and the cycle repeated.

After all the cycles were complete, the shoots were planted upright on moist sand in 216 individual Petri dishes (35 mm inner diameter) and placed back in the growth chamber for a 56 day regeneration assay. The drying treatments were staggered in time such that all the shoots were transferred to Petri dishes and began the final growth period on the same day. The shoots were grown on native sandy soil collected from areas adjacent to NDFF and the sand was sifted (500 μm mesh) and dry autoclaved (60 min at 131 °C) before use. The growth chamber (*Percival* model E30B, Boone, IA, USA) was set to a 12 h photoperiod (20 °C day and 8 °C night) for the entire experiment. Because light intensity and relative humidity vary depending upon position inside the chamber (30–130 $\mu mol\ m^{-2}\ s^{-1}$, 60%–70% RH), the positions of the dishes were randomly rotated daily for the first week and weekly thereafter.

Weekly observations were made of each dish for a total of eight weeks. During these observations the start date of protonemata or regenerant bud production was noted and total buds were counted. Also at this time, sterilized distilled water was added as necessary to keep the sand moist and the shoots fully hydrated between observations. After the full eight weeks were completed a final count of regenerant buds was taken and the dishes were left open in the growth chamber to dry down. After drying, the final area of protonemata on the soil surface was measured using *SPOT* image analysis software (Diagnostic Instruments, Sterling Heights, MI). Protonematal area was used as it has been shown to be a good proxy for total biomass production in similar experiments (e.g., Stark *et al.* 2004).

A follow-up reciprocal transplant experiment was performed using an additional 192 shoots of *S. caninervis* collected and trimmed to the green zone as described above. In this experiment shoots from each of the field treatments were grown at either elevated or ambient CO_2 concentrations in the laboratory. This resulted in four groups of 48 shoots each: (1) both field and laboratory elevated CO_2, (2) field elevated and laboratory ambient CO_2, (3) field ambient and laboratory elevated CO_2, and finally (4) both field and laboratory ambient CO_2. The plants were grown out for 56 days; observations proceeded in a fashion similar to those described above.

Statistical methods

The variability of shoot biomass measures precluded their use to effectively test differences between the CO_2 treatments in the field (due to low power); they were, however, useful as covariates in some of the other analyses. Since length and biomass were naturally correlated and the severed green zone was actually used, green zone biomass was selected as the only covariate in subsequent analyses. A Box–Cox transformation prior to use of green zone biomass was necessary for normality. None of the other response variables tested was normally distributed (even after attempts at transformation); therefore, generalized linear models with the appropriate distributions and survival analyses were employed.

Treatment effects on time to protonemata appearance and protonematal bud emergence were assessed by survival analysis using the Cox proportional hazards model with censoring. Logistic regression was performed to determine whether the field CO_2 treatments affected the likelihood of sexual expression in the green zone. Owing to zero-inflation and overdispersion, final protonematal area and total protonematal bud counts required the use of the Tweedie and negative binomial distributions, respectively. The negative binomial distribution is commonly used for overdispersed discrete count data (Zeileis *et al.* 2008) such as bud counts in this experiment. The Tweedie family of distributions (Jørgensen 1987) includes examples (when the power parameter is between 1 and 2) that are continuous for $y > 0$ with a positive mass at zero. These exponential dispersion models are therefore well suited to the problem of zero-inflated continuous data (final protonematal area in this case).

All statistical analyses were performed using the R software package (v2.8.0, R Development Core Team, 2008). Some R functions were performed using additional packages, including: MASS (Venables & Ripley 2002) for the negative binomial model and Box–Cox transformation, multcomp (Hothorn *et al.* 2008) for multiple comparisons between stress levels, survival (Therneau & Lumley 2008) for survival analyses, and tweedie (Dunn 2007) for Tweedie models.

Results

Shoots collected from the elevated CO_2 rings were nearly twice as likely to have expressed sex at least once in the green zone compared with their counterparts from the ambient rings (45 vs. 23 shoots). A logistic regression for likelihood of green zone sex expression showed both biomass and CO_2 treatment to be significant ($p < 0.001$, Fig. 9.2). Of the shoots that expressed sex in the green zone, those grown in elevated CO_2 were significantly smaller (mean green zone biomass 0.266 mg vs. 0.383 mg, $p < 0.001$, *t*-test, Fig. 9.2). Both the stress and CO_2 treatments had a significant effect on protonematal emergence ($p < 0.001$, Fig. 9.3); however, in this case green zone biomass was not significant ($p = 0.87$). Multiple comparison tests showed that all three stress levels were significantly different from one another (maximum $p = 0.009$). Protonematal bud emergence times were significantly affected by the stress treatments ($p < 0.001$, Fig. 9.4) and green zone biomass ($p = 0.011$) but not by the CO_2 treatments ($p = 0.38$). As with protonematal emergence, all three stress treatments were significantly different from one another (all $p < 0.001$). Stress, CO_2 concentration and green zone biomass all had a significant effect on final

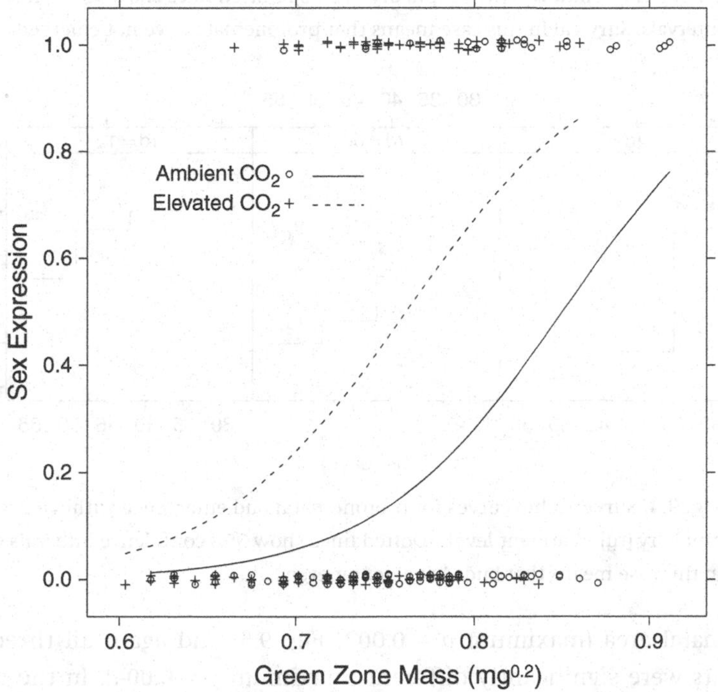

Fig. 9.2. Logistic regression for the likelihood of sex expression in the green zone, given biomass. The predicted response for each of the two CO_2 treatments is shown.

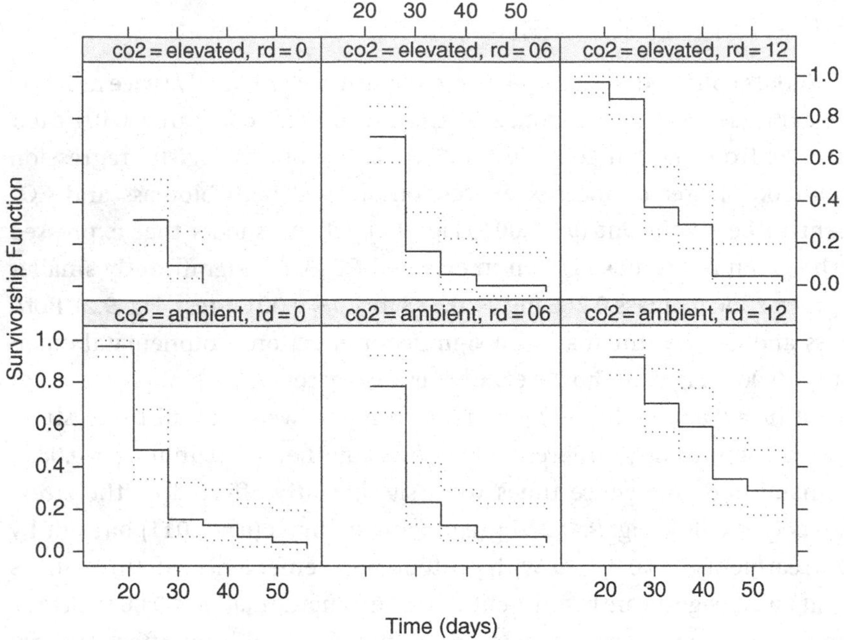

Fig. 9.3. Survivorship curves for protonematal emergence within each of the given treatment levels (rd = rapid-dry events). Dotted lines show 95% confidence intervals. Survival in this case means that protonemata have not emerged.

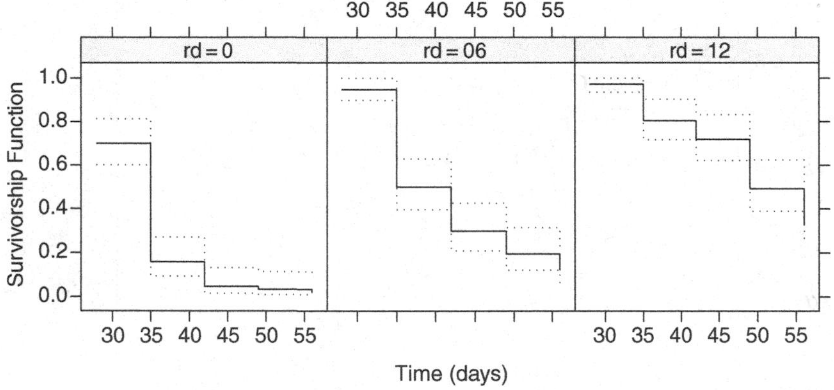

Fig. 9.4. Survivorship curves for protonematal bud emergence within each of the rapid-dry (rd) treatment levels. Dotted lines show 95% confidence intervals. Survival in this case means that buds have not emerged.

protonematal area (maximum $p = 0.003$, Fig. 9.5) and again all three stress treatments were significantly different (maximum $p = 0.004$). In the analysis for protonematal area, two of the two-way interaction effects were significant. These were the interactions between the CO_2 treatments and both the stress

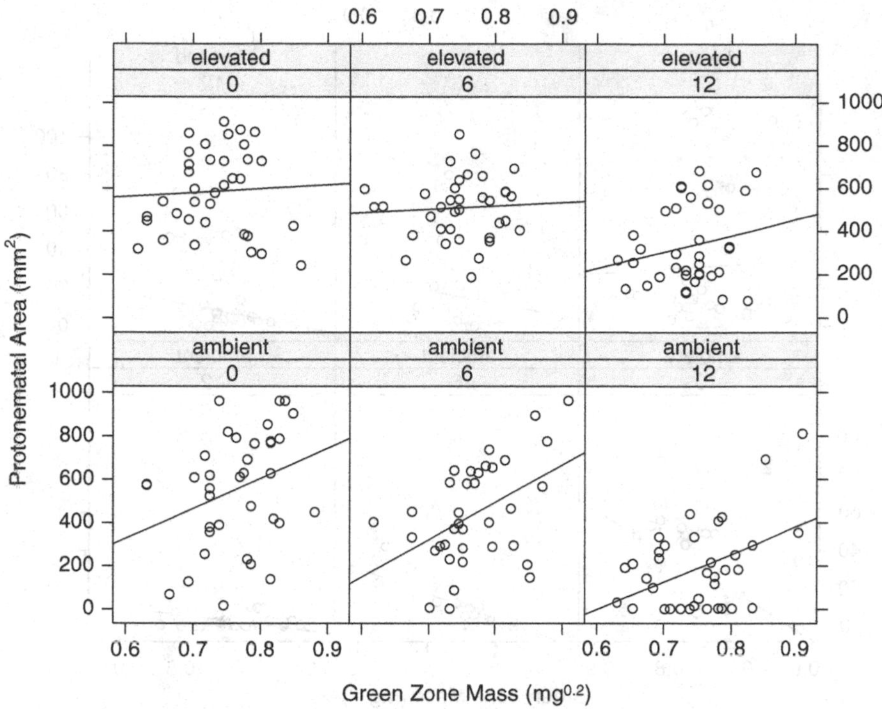

Fig. 9.5. Plots of final protonematal area across CO_2 treatments (ambient, elevated) and rapid-drying treatments (0, 6, 12 cycles); simple regression lines are shown for reference.

treatments ($p = 0.006$) as well as green zone biomass ($p = 0.010$). Finally, stress, CO_2 concentration and green zone biomass also all had a significant effect on final protonematal bud counts ($p < 0.001$, Fig. 9.6) and all three stress treatments were once again significantly different (all $p < 0.001$). In the follow-up reciprocal transplant experiment, green zone biomass ($p < 0.001$), laboratory CO_2 ($p < 0.001$), and field CO_2 ($p = 0.014$) all had a significant effect on final protonematal area (Fig. 9.7). In addition there was a significant interaction effect between field CO_2 and green zone biomass ($p = 0.008$) and a marginally significant interaction between the two CO_2 treatments ($p = 0.054$, Fig. 9.8).

Discussion

Sex expression

Stark *et al.* (1998) showed shoot size and sex expression to be positively correlated and yet despite the fact that they were on average slightly smaller (median green zone biomass 0.23 mg vs. 0.26 mg), shoots from the elevated CO_2

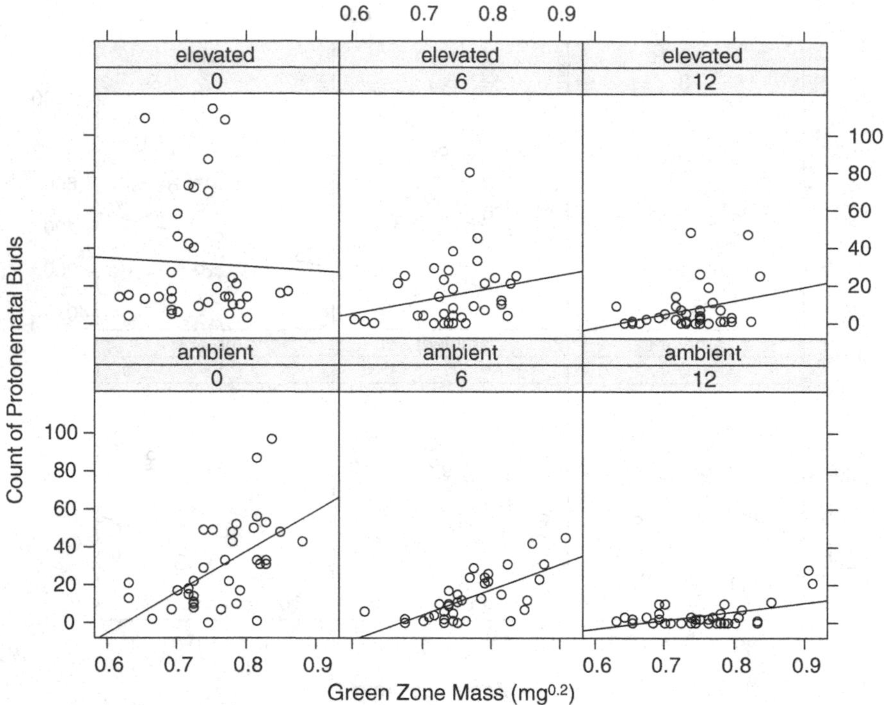

Fig. 9.6. Plots of final protonematal bud counts across CO_2 treatments (ambient, elevated) and rapid drying treatments (0, 6, 12 cycles); simple regression lines are shown for reference.

rings were significantly more likely to have expressed sex. Furthermore, the mean green zone biomass of the expressing shoots grown in elevated CO_2 was less than 70% of that of the expressing shoots grown in ambient air. This suggests that sex expression is at least partly carbon-limited in these plants. It also suggests that under normal conditions these plants depend on carbon that is either stored in and/or translocated from other parts of the shoot for sex expression. Larger shoots store and produce more photosynthate than smaller shoots and sex expression is probably triggered only once some critical capacity is achieved. Elevated CO_2 apparently accelerates this maturation by allowing shoots to achieve this capacity at a smaller size and consequently younger age. This is strikingly similar to findings for loblolly pine (*Pinus taeda*; LaDeau & Clark 2001), where elevated CO_2 also accelerated sexual maturation causing decreases in plant size at maturity.

It has been suggested that bryophytes may not benefit as much as vascular plants under elevated CO_2 owing to limited sink differentiation (Sveinbjörnsson & Oechel 1992). On the contrary, these results show that bryophytes are more likely to initiate sexual reproduction in response to increased CO_2, thereby developing

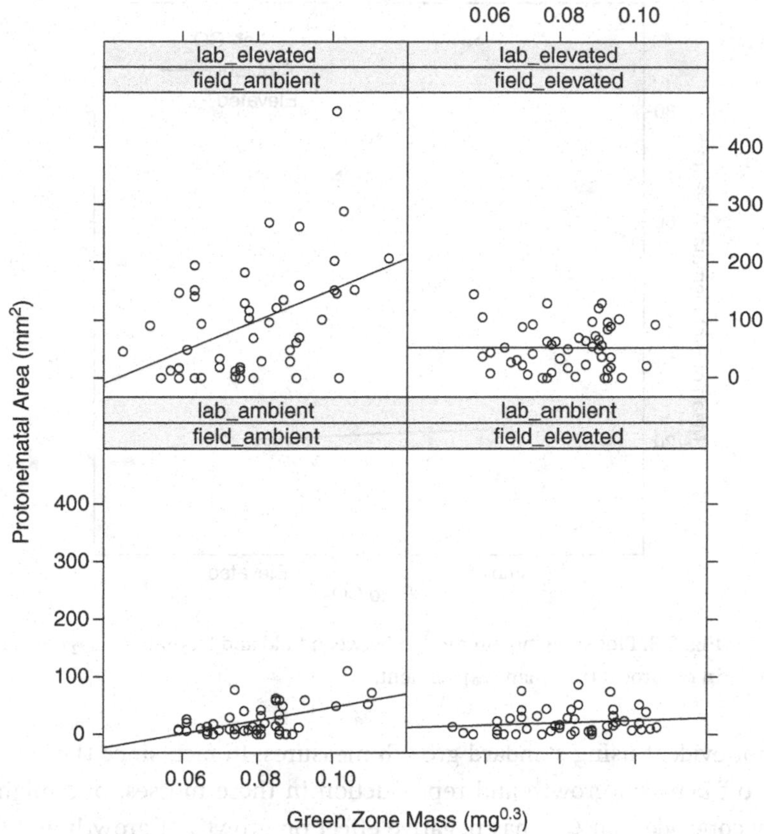

Fig. 9.7. Plots of final protonematal area across transplant treatments; simple regression lines are shown for reference.

additional carbon sinks. Sporophyte production is probably the strongest sink for excess photosynthate available to bryophytes, owing to a morphology specifically adapted for transfer of resources to the next generation (Browning & Gunning 1979). Unfortunately, maturation of sporophytes in the Mojave Desert is a somewhat rare and episodic event not captured in these results. Obviously in areas where sporophytes are more common, these results become even more interesting. For example, Baars and Edwards (2008) found increases in capsule length and spore size in *Leptobryum pyriforme* grown at elevated CO_2. Elevated CO_2 could also affect sporophyte maturation rates, spore number or how quickly the next crop of sporophytes can be produced. Changes in spore size and/or quality may also influence fitness in the next generation.

Many studies use growth rates assessed as either length or biomass increases for measuring bryophyte responses to various conditions. These results show that reproductive status can be useful in discerning population differences that

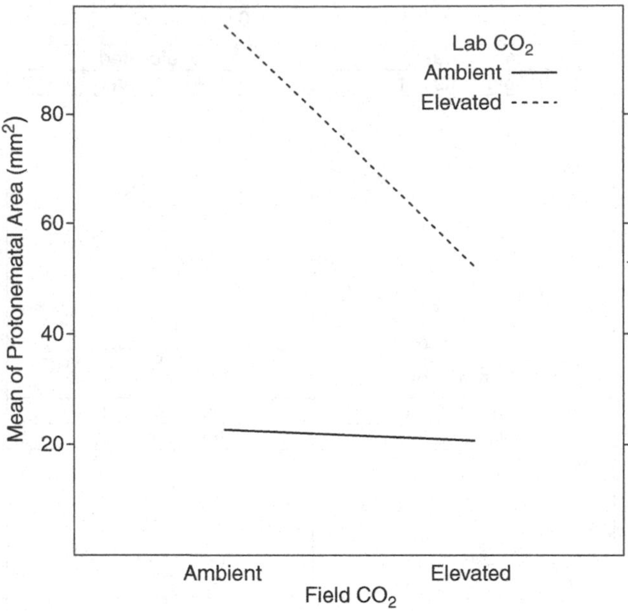

Fig. 9.8. Plot showing interaction between field and laboratory CO_2 concentration in reciprocal transplant experiment.

are not evident using standard growth measures. In fact, since there is likely a trade-off between growth and reproduction in these mosses, one might incorrectly conclude that CO_2 has negative effect on growth, if growth was the only response measured. This is especially true in acrocarpous mosses (e.g., *S. caninervis*) where the apical meristem is consumed in the production of gametangia and growth must resume by branching or subfloral innovation. In these mosses, every expression event not only consumes energy but also consumes an apical meristem, thus delaying height growth in the interim.

Stress tolerance

Besides sexual reproduction, bryophytes can also allocate surplus resources to improve their tolerance to various stresses. The results of this experiment show that elevated CO_2 has a protective effect against the most common stress that arid land mosses encounter – namely desiccation. Upon rehydration these plants have to repair their membranes and photosynthetic machinery before they can begin to grow again (Oliver *et al.* 2005). The weaker the plant is to begin with or the more severe the stress, the longer this repair process will take. If it takes too long, the moss will endure a sustained carbon deficit and eventually die. The resumption of growth (e.g., protonematal emergence) is a good indicator that repair is complete and in healthy, unstressed plants of *S. caninervis* it usually occurs in a few days. All of the plants in this experiment had been exposed to desiccation stress in the field, so protonemata did not begin to emerge until the second to third week of the experiment. As expected, additional laboratory desiccation stress treatments pushed the emergence times back even further; however, plants grown at elevated CO_2 had consistently earlier protonematal emergence times. Faster emergence naturally translated into greater protonematal area and buds produced by the end of the experiment. The enhanced regeneration demonstrated here is clearly important in bryophytes such as *Syntrichia* where asexual reproduction may be the dominant means of propagation due to the rarity of sporophyte production in the harsh desert environment (Newton & Mishler 1994). As with sex expression, these results suggest that the elevated CO_2 grown plants may have more stored energy but it is also likely that they have allocated more resources to repair proteins and other secondary compounds that are used to mitigate desiccation damage and other stresses. The energy expended in building and recycling these compounds is another potential sink for carbon that is probably strengthened in bryophytes as they capture the benefits of elevated CO_2.

Productivity and resource allocation

These results illustrate that allocation of carbon and other resources in bryophytes is not simple, but rather that it is a complicated process involving trade-offs and limiting factors. Teasing apart these influences is not trivial, and it is obvious that more work needs to be done in this area. However, some of the complexities of the carbon economy of these plants are hinted at in the significant interaction effects for protonematal area presented here. Alpert (1989) showed that the xeric-adapted moss *Grimmia laevigata* was able to transport carbon away from its leaves towards other parts of the plant. He hypothesized that this ability, combined with clonal integration, allowed the plant to thrive in areas where desiccation stress might otherwise exclude it. It would appear that similar processes are at work in *S. caninervis*; for example, the effect of shoot biomass on the growth of protonemata is smaller or non-existent in plants grown at elevated CO_2 (Figs. 9.5 and 9.7). This suggests that in plants grown at elevated CO_2 the translocation of carbon from distant parts of the stem to sites of active growth is less important – presumably because the supply of carbon is more plentiful throughout the plant. This also explains why the plants grown at elevated CO_2 begin to show some effects of biomass when additional desiccation stress is added (Fig. 9.5, upper right). Under these experimental conditions carbon is apparently being transferred from the leaves through the stem and out to the actively growing protonemata.

Furthermore, *S. caninervis* ramets in intact soil crust often retain belowground connections that may allow inter-ramet transfer of carbon and other nutrients. This could serve to supplement the carbon balance of connected ramets in less optimal microsites and improve the fitness of the clone as a whole. Evidence for this would demonstrate the existence of yet another strong sink for carbon available to bryophytes, but one that operates on the scale of the entire clone. Of course, community-scale connections also exist via mycorrhizae (Zhang & Guo 2007) and pulse releases of leachate into the soil induced by rehydration damage (Wilson & Coxson 1999); therefore, downstream effects on the soil fauna and vascular plants are also possible.

Another complication in the interpretation of the responses of *S. caninervis* is photosynthetic down-regulation. Shoots of *S. caninervis* showed a dramatic step increase in productivity when moved from ambient conditions in the field to elevated CO_2 in the laboratory growth chamber (Fig. 9.7, upper left, and Fig. 9.8). This increase was not observed in plants that had already experienced elevated CO_2 in the field. The fact that they did not achieve the same productivity as plants from the ambient field treatments is evidence that they had already acclimated to a high CO_2 environment by down-regulating photosynthesis. While this might seem counterintuitive, notice that despite this reduction in maximal photosynthetic capacity, plants grown in and then transplanted to elevated CO_2 conditions still significantly outperform the plants grown at and remaining in ambient conditions (Fig. 9.8). Here the plants demonstrate Liebig's (1840) law of the minimum, which predicts that a plant's response to changes in resource levels is fundamentally constrained by the resource that is least available (see Van der Ploeg *et al.* 1999). Increases in carbon availability have the effect of reshuffling the order of these limiting factors, causing plants to focus less on carbon fixation and more on overcoming other limiting factors. The upshot of this is that when the plants down-regulate photosynthesis it allows them to up-regulate some other process or physiological pathway such as stress tolerance.

Conclusions

In the environments where *S. caninervis* is common there is tremendous evolutionary pressure for stress tolerance even at the expense of slower growth. In fact, height growth beyond that necessary to keep pace with sediment deposition can expose the plant to greater stress by removing it further from the large reservoir of moisture provided by the soil. As a result, *S. caninervis* apparently preferentially allocates resources to stress tolerance and sexual reproduction rather than increased growth. In an environment such as the Mojave Desert where bryophytes are exposed to repeated, chronic stresses (Barker *et al.* 2005), the protective effects of elevated CO_2 are likely to be especially important. However, not all bryophyte species may respond similarly; and when combined with other global changes such as warming and modification of precipitation patterns, elevated CO_2 may cause shifts in the competitive balance among soil crust species. Very little is known concerning species interactions and community assembly in arid land bryophytes, so the overall effect will be difficult to predict. However, the responses of biological soil crusts and their component bryophytes to various aspects of climate change, including elevated CO_2, must be understood in order to predict how arid ecosystems will change in an environment increasingly altered by human influence. It is important that we understand the effects of these changes before they occur so that informed decisions can be made regarding future resource use and management.

Acknowledgements

We thank Jessica Rampy for help with sampling in the field and the UNLV Graduate and Professional Students Association for additional funding assistance. This research was supported by the US Department of Energy's Office of Science (BER) through the Western Regional Center of the National Institute for Climatic Change Research under Cooperative Agreement No. DE-FCO2-03ER63613. Financial support does not constitute an endorsement by DOE of the views expressed in this article.

References

Alpert, P. (1989). Translocation in the nonpolytrichaceous moss *Grimmia laevigata*. *American Journal of Botany* **7**: 1524–9.

Baars, C. & Edwards, D. (2008). Effects of elevated atmospheric CO_2 on spore capsules of the moss *Leptobryum pyriforme*. *Journal of Bryology* **30**: 36–40.

Barker, D. H., Stark, L. R., Zimpfer, J. F., McLetchie, D. N. & Smith, S. D. (2005). Evidence of drought-induced stress on biotic crust moss in the Mojave Desert. *Plant, Cell & Environment* **28**: 939–47.

Bazzaz, F. A., Paolillo, D. J. & Jagels, R. H. (1970). Photosynthesis and respiration of forest and alpine populations of *Polytrichum juniperinum*. *Bryologist* **73**: 579–85.

Belnap, J. & Lange, O. L. (2003). Biological soil crusts: structure, function, and management. *Ecological Studies* **150**: 1–503.

Belnap, J., Prasse, R. & Harper, K. T. (2003). Influence of biological soil crusts on soil environments and vascular plants. *Ecological Studies* **150**: 281–300.

Berendse, F., Van Breemen, N., Rydin, H., *et al.* (2001). Raised atmospheric CO_2 levels and increased N deposition cause shifts in plant species composition and production in *Sphagnum* bogs. *Global Change Biology* **7**: 591–8.

Björkman, O., Gauhl, E., Hiesey, W. M., Nicholson, F. & Nobs, M. A. (1968). Growth of *Mimulus*, *Marchantia*, and *Zea* under different oxygen and carbon dioxide levels. *Carnegie Institution of Washington Yearbook* **67**: 477–8.

Browning, A. J. & Gunning, B. E. S. (1979). Structure and function of transfer cells in the sporophyte haustorium of *Funaria hygrometrica* Hedw: III. Translocation of assimilate into the attached sporophyte and along the seta of attached and excised sporophytes. *Journal of Experimental Botany* **30**: 1265–73.

Carter, D. W. & Arocena, J. M. (2000). Soil formation under two moss species in sandy materials of central British Columbia (Canada). *Geoderma* **98**: 157–76.

Coe, K. K., Belnap, J. & Sparks, J. P. (2008). Physiological ecology of the desert moss *Syntrichia caninervis* after long-term exposure to elevated CO_2: changes in photosynthetic thermotolerance? [abstract]. American Bryological and Lichenological Society, Annual Meeting, Asilomar, California, 2008.

Csintalan, Z., Juhász, A., Benko, Z., Raschi, A. & Tuba, Z. (2005). Photosynthetic responses of forest-floor moss species to elevated CO_2 level by a natural CO_2 vent. *Cereal Research Communications* **33**: 177–80.

Csintalan, Z., Takács, Z., Tuba, Z., *et al.* (1997). Desiccation tolerant grassland cryptogams under elevated CO_2: preliminary findings. *Abstracta Botanica* **21**: 309–15.

Csintalan, Z., Tuba, Z. & Laitat, E. (1995). Slow chlorophyll fluorescence, net CO_2 assimilation and carbohydrate responses in the forest moss *Polytrichum formosum* to elevated CO_2 concentrations. In *Photosynthesis: From Light to Biosphere*, ed. P. Mathis, vol. 5, pp. 925–8. New York, NY: Springer.

DeLucia, E. H., Sasek, T. & Strain, B. (1985). Photosynthetic inhibition after long-term exposure to elevated levels of atmospheric carbon dioxide. *Photosynthesis Research* **7**: 175–84.

Dilks, T. J. K. (1976). Measurement of the carbon dioxide compensation point and the rate of loss of $^{14}CO_2$ in the light and dark in some bryophytes. *Journal of Experimental Botany* **27**: 98–104.

Dunn, P. K. (2007). The tweedie package for R: Tweedie exponential family models. R package version 1.5.1, Toowoomba, Queensland, Australia.

Grime, J. P., Rincon, E. R. & Wickerson, B. E. (1990). Bryophytes and plant strategy theory. *Botanical Journal of the Linnean Society* **104**: 175–86.

Heijmans, M., Berendse, F., Arp, W. J. *et al.* (2001). Effects of elevated carbon dioxide and increased nitrogen deposition on bog vegetation in the Netherlands. *Journal of Ecology* **89**: 268–79.

Hothorn, T., Bretz, F. & Westfall, P. (2008). Simultaneous inference in general parametric models. *Biometrical Journal* **50**: 346–63.

Jablonski, L. M., Wang, X. & Curtis, P. S. (2002). Plant reproduction under elevated CO_2 conditions: a meta-analysis of reports on 79 crop and wild species. *New Phytologist* **156**: 9–26.

Jasoni, R. L., Smith, S. D. & Arnone, J. A. (2005). Net ecosystem CO_2 exchange in Mojave Desert shrublands during the eighth year of exposure to elevated CO_2. *Global Change Biology* **11**: 749–56.

Jauhiainen, J., Silvola, J. & Vasander, H. (1998). The effects of increased nitrogen deposition and CO_2 on *Sphagnum angustifolium* and *S. warnstorfii*. *Annales Botanici Fennici* **35**: 247–56.

Jordan, D. N., Zitzer, S. F., Hendrey, G. R., *et al.* (1999). Biotic, abiotic and performance aspects of the Nevada Desert Free-Air CO_2 Enrichment (FACE) Facility. *Global Change Biology* **5**: 659–68.

Jørgensen, B. (1987). Exponential dispersion models. *Journal of the Royal Statistical Society. Series B (Methodological)* **49**: 127–62.

LaDeau, S. & Clark, J. S. (2001). Rising CO_2 levels and the fecundity of forest trees. *Science* **292**: 95–8.

Liebig, J. (1840). *Die organische Chemie in ihrer Anwendung auf Agricultur und Physiologie.* Braunschweig, Germany: Friedrich Vieweg und Sohn Publ. Co.

Mitchell, E. A. D., Buttler, A., Grosvernier, P., Rydin, H. & Siegenthaler, A. (2002). Contrasted effects of increased N and CO_2 supply on two keystone species in peatland restoration and implications for global change. *Journal of Ecology* **90**: 529–33.

Newton, A. E. & Mishler, B. D. (1994). The evolutionary significance of asexual reproduction in mosses. *Journal of the Hattori Botanical Laboratory* **76**: 127–45.

Nowak, R. S., Ellsworth, D. S. & Smith, S. D. (2004). Functional responses of plants to elevated atmospheric CO_2 – do photosynthetic and productivity data from FACE experiments support early predictions? *New Phytologist* **162**: 253–80.

Oechel, W. C., Cowles, S., Grulke, N., *et al.* (1994). Transient nature of CO_2 fertilization in Arctic tundra. *Nature* **371**: 500–3.

Oliver, M. J., Velten, J. & Mishler, B. D. (2005). Desiccation tolerance in bryophytes: a reflection of the primitive strategy for plant survival in dehydrating habitats? *Integrative and Comparative Biology* **45**: 788–99.

Ötvös, E. & Tuba, Z. (2005). Ecophysiology of mosses under elevated air CO_2 concentration: overview. *Physiology and Molecular Biology of Plants* **11**: 65–70.

Poorter, H. (1993). Interspecific variation in the growth response of plants to an elevated ambient CO_2 concentration. *Vegetatio* **104–105**: 77–97.

Proctor, M. C. F. (1982). Physiological ecology: water relations, light and temperature responses, carbon balance. In *Bryophyte Ecology*, ed. A. J. E. Smith, pp. 333–81. New York, NY: Chapman and Hall.

R Development Core Team. (2008). *R: A Language and Environment for Statistical Computing*. Vienna, Austria. http://www.R-project.org/

Silvola, J. (1985). CO_2 dependence of photosynthesis in certain forest and peat mosses and simulated photosynthesis at various actual and hypothetical CO_2 concentrations. *Lindbergia* **11**: 86–93.

Silvola, J. (1990). Combined effects of varying water content and CO_2 concentration on photosynthesis in *Spagnum fuscum*. *Holarctic Ecology* **13**: 224–8.

Smith, S. D., Huxman, T. E., Zitzer, S. F. *et al.* (2000). Elevated CO_2 increases productivity and invasive species success in an arid ecosystem. *Nature* **408**: 79–82.

Sonesson, M., Callaghan, T. V. & Carlsson, B. (1996). Effects of enhanced ultraviolet radiation and carbon dioxide concentration on the moss *Hylocomium splendens*. *Global Change Biology* **2**: 67–73.

Stark, L. R., Mishler, B. D. & McLetchie, D. N. (1998). Sex expression and growth rates in natural populations of the desert soil crustal moss *Syntrichia caninervis*. *Journal of Arid Environments* **40**: 401–16.

Stark, L. R., Nichols, L. II, McLetchie, D. N., Smith, S. D. & Zundel, C. (2004). Age and sex-specific rates of leaf regeneration in the Mojave Desert moss *Syntrichia caninervis*. *American Journal of Botany* **91**: 1–9.

Sveinbjörnsson, B. & Oechel, W. C. (1992). Controls on growth and productivity of bryophytes: environmental limitations under current and anticipated conditions. In *Bryophytes and Lichens in a Changing Environment*, ed. J. W. Bates & A. M. Farmer, pp. 77–102. Oxford: Oxford University Press.

Takács, Z., Ötvös, E., Lichtenthaler, H. K. & Tuba, Z. (2004). Chlorophyll fluorescence and CO_2 exchange of the heavy metal-treated moss, *Tortula ruralis* under elevated CO_2 concentration. *Physiology and Molecular Biology of Plants* **10**: 291–6.

Therneau, T. & Lumley, T. (2008). survival: Survival analysis, including penalised likelihood. R package version 2.34–1.

Thompson, D. B., Walker, L. R., Landau, F. H. & Stark, L. R. (2005). The influence of elevation, shrub species, and biological soil crust on fertile islands in the Mojave Desert, USA. *Journal of Arid Environments* **61**: 609–29.

Tuba, Z., Csintalan, Z., Szente, K., Nagy, Z. & Grace, J. (1998). Carbon gains by desiccation-tolerant plants at elevated CO_2. *Functional Ecology* **12**: 39–44.

Tuba, Z., Proctor, M. C. F. & Takács, Z. (1999). Desiccation-tolerant plants under elevated air CO_2: A review. *Zeitschrift für Naturforschung – Section C, Journal of Biosciences* **54**: 788–96.

Van Der Heijden, E., Jauhiainen, J., Silvola, J., Vasander, H. & Kuiper, P. J. C. (2000a). Effects of elevated atmospheric CO_2 concentration and increased nitrogen deposition on growth and chemical composition of ombrotrophic *Sphagnum balticum* and oligo-mesotrophic *Sphagnum papillosum*. *Journal of Bryology* **22**: 175–82.

Van Der Heijden, E., Verbeek, S. K. & Kuiper, P. J. C. (2000b). Elevated atmospheric CO_2 and increased nitrogen deposition: effects on C and N metabolism and growth of the peat moss *Sphagnum recurvum* P. Beauv. var. *mucronatum* (Russ.) Warnst. *Global Change Biology* **6**: 201–12.

van der Ploeg, R. R., Bohm, W. & Kirkham, M. B. (1999). On the origin of the theory of mineral nutrition of plants and the law of the minimum. *Soil Science Society of America Journal* **63**: 1055–62.

Venables, W. N. & Ripley, B. D. (2002). *Modern Applied Statistics with S*. New York, NY: Springer.

Wilson, J. A. & Coxson, D. S. (1999). Carbon flux in a subalpine spruce–fir forest: Pulse release from *Hylocomium splendens* feather-moss mats. *Canadian Journal of Botany* **77**: 564–9.

Wong, S. C. (1979). Elevated atmospheric partial pressure of CO_2 and plant growth. *Oecologia* **44**: 68–74.

Zaady, E., Kuhn, U., Wilske, B., Sandoval-Soto, L. & Kesselmeier, J. (2000). Patterns of CO_2 exchange in biological soil crusts of successional age. *Soil Biology and Biochemistry* **32**: 959–66.

Zander, R. H. (1993). Genera of the Pottiaceae: mosses of harsh environments. *Bulletin of the Buffalo Society of Natural Sciences* **32**: 1–378.

Zeileis, A., Kleiber, C. & Jackman, S. (2008). Regression models for count data in R. *Journal of Statistical Software* **27**: 1–25.

Zhang, Y. & Guo, L. (2007). Arbuscular mycorrhizal structure and fungi associated with mosses. *Mycorrhiza* **17**: 319–25.

10

Responses of Epiphytic Bryophyte Communities to Simulated Climate Change in the Tropics

JORGE JÁCOME, S. ROBBERT GRADSTEIN, AND MICHAEL KESSLER

Introduction

Epiphytes are known to respond sensitively to environmental changes. Because of the tight coupling of epiphytes to atmospheric conditions, changes in the chemical and physical conditions of the atmosphere may be expected to have direct effects on epiphytes (Farmer *et al.* 1992; Benzing 1998; Zotz & Bader 2009). In temperate regions, non-vascular epiphytes (bryophytes, lichens) have frequently been used as bioindicators of air quality (Hawksworth & Rose 1970). Owing to the lack of a protective cuticle in many bryophytes and lichens, solutions and gases may enter freely into the living tissues of these plants causing sensitive reactions to changes in the environment. By mapping and monitoring the distribution and abundance of non-vascular epiphytes, changes in environmental conditions can be assessed (Van Dobben & De Bakker 1996; Szczepaniak & Biziuk 2003).

Tropical moist forests, especially mountain forests, are very rich in epiphytes, both vascular and non-vascular. In the Reserva Biológica San Francisco, a small mountain rain forest reserve of approximately 1000 hectares in the Andes of southern Ecuador, about 1200 species of epiphytes have been recorded, with more than half of these bryophytes and lichens (Liede-Schumann & Breckle 2008). About one of every two species of plant in the forests is an epiphyte. The almost constantly saturated air in these mountain forests, due to orographic clouds, mist, and frequent rainfall, allows the epiphytic plants to thrive year-round high up on the trees, favoring high species diversity. Epiphytic biomass also peaks here and plays a significant role in the

Bryophyte Ecology and Climate Change, eds. Zoltán Tuba, Nancy G. Slack and Lloyd R. Stark. Published by Cambridge University Press.

hydrology and nutrient cycling of these forests (Veneklaas *et al.* 1990; Coxson & Nadkarni 1995; Cavelier *et al.* 1996). In a "mossy" mountain forest of Colombia about one-third of the total forest biomass and almost half of the above-ground stocks of nitrogen and phosphorus are contained in the thick epiphytic bryophyte layers (Hofstede *et al.* 1993). The epiphyte mats also facilitate the life of a great variety of animals, fungi and microorganisms (Nadkarni & Longino 1990; Yanoviak *et al.* 2007).

These epiphyte-rich tropical mountain forests are now disappearing rapidly, owing to human activities. Land use changes have been identified as the main current driver of biodiversity changes in the tropics (Sala *et al.* 2000). By the early 1990s it was estimated that *c.* 90% of the mountain forests of South America had been converted into pastures or other forms of land use (Henderson *et al.* 1991), and forest conversion is still continuing at an alarming pace in spite of major efforts to slow down this process (Mosandl & Günter 2008). In spite of the rapid disappearance of tropical mountain forests, the impact of the deforestation on epiphyte diversity remains little documented (Gradstein 2008; Gradstein & Sporn 2009). Bryophyte species richness along land use gradients in the mountains of tropical America and Indonesia declined by 30%–80% towards open, disturbed habitats (Acebey *et al.* 2003; Nöske *et al.* 2008; Aryanti *et al.* 2008; Gradstein & Sporn 2009). Losses varied among different groups of bryophytes and were more severe in mosses than in liverworts. Moreover, the drought-intolerant shade epiphytes of the forest understory were more strongly impacted than the drought-tolerant species of the forest canopy, which in the disturbed habitats moved to lower elevations on the trees (Acebey *et al.* 2003).

Forest canopy closure and microclimate were identified as principal factors driving bryophyte richness, and maintenance of a dense canopy was shown to be of paramount importance for conserving high diversity (Steffan-Dewenter *et al.* 2007; Sporn *et al.* 2009). Indeed, Holz and Gradstein (2005) found no species decline in 15 and 40 yr-old regenerating mountain forests with a closed canopy in Costa Rica. In all cases, however, turnover along the succession gradient was high and up to 40% of the epiphytic bryophyte species were only found in the secondary forest. The very different species compositions of the primary and secondary forests indicated that recovery of the species communities in the regenerating forest is a slow process.

Climate change has been identified as another driver of biodiversity changes (Sala *et al.* 2000) and is gaining in importance worldwide. According to Malcolm *et al.* (2006), climate change may become the principal cause of species extinctions in tropical hotspots. Lovejoy and Hannah (2005) proposed that climate change impacts biological organisms in four different ways: (1) by changes in the local abundance of species, (2) by changes in community structure, (3) by habitat shifts, with species moving towards habitats with cooler microclimates, and (4) by range shifts, with species moving towards higher latitudes or elevations. However, most of the evidence showing climate change effects on biodiversity is derived from modeling or from observations on the behavior of animal taxa (birds, butterflies, amphibia, etc.). Empirical studies on the impact of climate change on plants are still scarce and only very few deal with epiphytes, in spite of their predicted sensitivity to global warming (Zotz & Bader 2008).

The first empirical data on the impact of climate change on non-vascular epiphytes originate in Europe. Based on analyses of historical herbarium records, Frahm and Klaus (1997, 2001) found that 32 subtropical bryophyte species had extended their ranges several hundred kilometers east- and northeastwards into Central Europe during the twentieth century. These range extensions correlated with an increase of the mean winter temperature in the area by *c.* 1.5 degrees Celsius. In The Netherlands, Van Herk *et al.* (2002) observed significant shifts in the geographical ranges of epiphytic lichens based on a monitoring of the lichen flora in permanent plots over nearly 25 years. During the first 15 years, significant floristic changes occurred in relation to changes in air pollution levels. In the 1990s, however, when changes in air pollution levels became insignificant, major range shifts occurred which correlated significantly with a measured rise in air temperature. Thus, they observed a significant increase in the frequency of warm-temperate species and a significant decrease in that of cold-temperate ones. Moreover, several tropical species were newly detected in the country. The recent expansion of tropical and warm-temperate bryophytes and lichens in Europe is striking first evidence of the possible impact of global warming on epiphytes.

In the tropics, an assessment of the possible impact of global warming on epiphytic plants is hampered by lack of long-term distribution records and monitoring. Nevertheless, first evidence is now arising from transplantation experiments. Nadkarni and Solano (2002) studied the response of vascular epiphytes to simulated climate change in a tropical montane cloud forest in Costa Rica. Global climate models predict increasing temperatures and changes in the hydric regimes for these cloud forests, principally implying an upward shift of the cloud condensation surface during the winter season by hundreds of meters, leading to a reduction in the cloud cover and to increased evapotranspiration (Still *et al.* 1999; Foster 2001; Chen *et al.* 2002; Vuille *et al.* 2003). Because of their reduced root systems and inability to take up water from the soil, epiphytes are expected to be particularly negatively affected by the predicted changes in the hydric regime. In the Costa Rican experiment, four common species of epiphyte, a tank bromeliad, a woody shrub, a scandent herb, and a fern, were translocated from cloud forest to drier climatic conditions with less cloud water *c.* 150 m downslope. After one year, the translocated plants had significantly increased leaf mortality and reduced production of new leaves; control plants, however, showed no negative effects (Nadkarni & Solano 2002). The authors also observed the invasion of terrestrial species into the canopy after the death of the transplants. The results indicate that climate change may have rapid negative effects on the productivity and longevity of epiphytes and may lead to compositional changes in the canopy community.

The study of Nadkarni and Solano (2002) demonstrated the individualistic response of epiphyte species to simulated climatic change. Information is still lacking on the effects of climate change on tropical epiphytes at the community level. In temperate regions, studies of climate change effects on plant communities have shown species-specific and functional group responses (Price & Waser 2000; Weltzin *et al.* 2003). For tropical epiphyte communities it would be reasonable to expect species-specific responses, since communities are constituted of species with different microclimatic requirements, stress tolerances, and geographical distributions. However, group responses relative to community structure, composition, and productivity may also be expected.

In a first experimental study at the community level, we have investigated the impact of simulated climate change on epiphytic bryophytes of a tropical mountain forest using the translocation approach employed by Nadkarni and Solano (2002). The objective of our study was to analyze the response of these organisms to simulated warmer and dryer climatic conditions by translocating whole epiphytic bryophyte communities to lower elevations. The results of this case study are presented below.

Materials and methods

Site description

The study site was located between the villages of Chuspipata and Yolosa along the La Paz–Coroico road, Department of La Paz, Bolivia (16°16–17′S, 67°48–53′W), at an elevation of 2500–3000 m. The area is covered by tropical montane cloud forest, precipitation in the area averages 3000–4000 mm per year and average annual temperature ranges from 10 °C to 18 °C (Gerold 2008).

Research approach and measurements

We cut 60 branches one meter in length, each covered by a dense bryophyte mat, from the lower portion of the crown (= Johannson zone 3;

Johansson 1974; Cornelissen & Ter Steege 1989) of ten trees at 3000 m and moved these to other sites, where they were hung on trees. Twenty cut branches were relocated at 3000 m (controls); the remaining ones were moved to 2700 m or 2500 m (20 per elevation). The elevational differences of 300 m and 500 m imply temperature increases of *c.* 1.5 °C and 2.5 °C, respectively, being within the range of predicted temperature increases in the next 50–100 years (Woodward 1992; Pounds *et al.* 1999). At each elevation, care was taken to restore the original growth conditions (irradiation, exposure) of the branch in order to diminish the influence of microclimatic differences that could cause changes in community composition (Gradstein *et al.* 2001). On each branch we recorded the cover of individual bryophyte species at the beginning of the experiment and after one and two years, measuring the maximum and minimum diameter for each species patch. In order to avoid disturbance of the bryophyte mats, only species distinguishable in the field with a 20 × handlens were recorded.

Data analysis

We compared the initial total bryophyte cover on each branch with the cover after one and two years and tested for differences in the bryophyte cover between elevations. Differences in the total bryophyte cover between dates and elevation were statistically tested by ANOVA (Zar 1999). Data on bryophyte cover were log-transformed to adjust them to normal distribution and to increase homogeneity of variance.

We evaluated the effect of the translocation to lower elevations on community structure in two ways, first by observing changes in the community similarity and second by comparing the rank-cover pattern of the community during the two years. To observe changes in the community similarity we calculated Bray–Curtis dissimilarity indices (Magurran 1988) for each branch between the initial community composition and the composition after one and two years, respectively. The dissimilarity values were calculated with the cover data of the ten most important species on each branch, since the Bray–Curtis similarity index tends to overweight the importance of rare species (Legendre & Legendre 1998) and because our data on these rare species, which were often difficult to identify in the field, were less reliable. Because changes in the community could be affected by interspecific interactions, we also calculated the bryophyte saturation of the branches as the total cover of all bryophyte species present on the branch, divided by the branch area (length × diameter). As bryophyte patch overlap occurred, saturation values could exceed the value of one. We tested if there was a difference in the similarity values between elevations after one and two years carrying out an ANCOVA (Zar 1999), after logarithmic or arcsin square root transformation on the data to improve normality and variance homogeneity.

To compare the rank-cover patterns of the community between elevations we selected at each elevation the 12 most important species according to their cover on all branches at a particular elevation. Then we ranked these species by cover and obtained for each rank the cover change after one and two years. Cover change was calculated as the absolute difference of the final and initial relative cover at each elevation. Then, we performed pairwise comparisons between ranks at different elevations, examining for differences in cover change between elevation. For this analysis we used a pairwise *t*-test (Fowler *et al.* 1998).

We also observed the effect of elevation on community dynamics by measuring individual branch cover changes. For this purpose, branch cover of each bryophyte species was proportioned to total branch cover. This adjustment was necessary since in some cases it was difficult to establish clear bryophyte patch limits, leading to over- or underestimation of species cover. Using the proportioned cover data, branch cover changes of each species were expressed as (1) absolute changes (irrespective of increases or decreases) and (2) directional differences between consecutive years. With the absolute value changes it was possible to ascertain differences in the magnitude of the cover changes, whereas with directional value changes general patterns of cover increase or decrease could be determined. Since interspecific interactions could also affect cover changes, we also included bryophyte saturation of the branch as a covariable in the analysis. With these data we carried out a GLM (general lineal model; Crawley 2002), with elevation as main factor, branch as random factor, bryophyte saturation as covariable, and cover change as response variable. Data on cover change were log-transformed to adjust normal distribution and homogeneity of variance.

A further matter to consider was the possible influence of wood degradation on the bryophyte cover. To assess this potential bias we separately analyzed the cover changes in the first and in the second year. In this way we expected that changes caused by wood degradation would obscure the effects of elevation during the first but not during the second year. All statistical tests were carried with Statistica 6 (Statsoft 2003).

To account for differences in cover change by elevation for the most important bryophyte species in the studied communities, we did a descriptive comparison of their cover changes during the two years, since sample size was too low for a species-specific quantitative analysis.

Results

On 60 branches we recorded 46 bryophyte species (32 liverworts, 14 mosses) in 35 genera (Appendix 10.1). In addition, a few lichen species were recorded (data not shown). The dominance of liverworts over mosses in the study site is characteristic of upper montane cloud forests (Gradstein *et al.* 2001). *Plagiochila* was the most species-rich genus with seven different species, and *P. macrostachya* Lindenb. the most common bryophyte species (11 collections). All other genera were represented by only one or two species.

Total bryophyte cover did not show significant differences between epiphyte communities at different elevations during the two years ($F_{(2,56)} = 1.446$; $p = 0.24$). However, epiphyte communities at 3000 m and 2700 m slightly increased in biomass in the second year, whereas at 2500 m biomass remained relatively constant.

After two years, epiphyte communities of branches translocated to 2500 m had changed more strongly than the controls at 3000 m (ANOVA, $F_{(2,51)} = 3.69$, $p = 0.031$). The changes occurred gradually with elevation since dissimilarity values of communities translocated to 2700 m did not differ significantly from the communities at 2500 m and 3000 m after two years (Fig. 10.1). Community saturation affected the similarity changes, with saturated communities being more similar to the original communities after two years than unsaturated communities ($r^2 = 0.063$, $F_{(1,53)} = 3.57$, $p = 0.064$; Fig. 10.2).

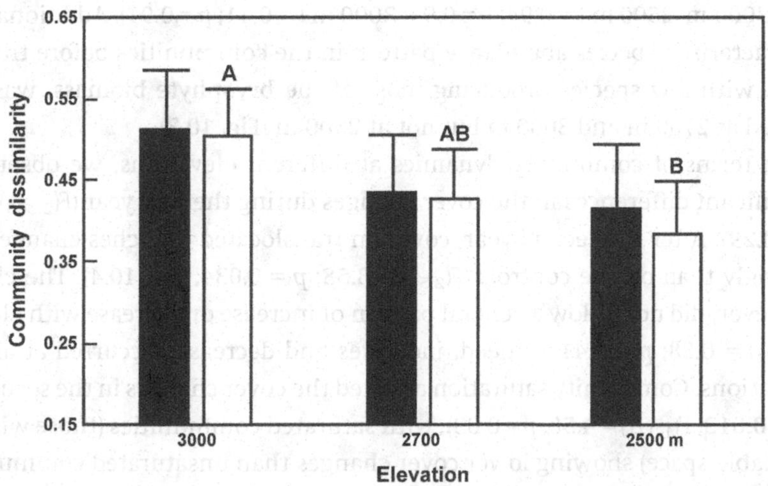

Fig. 10.1. Community dissimilarity of epiphytic bryophyte communities in response to their translocation from 3000 m to 3000 m (control), 2700 m and 2500 m. Dissimilarity values after one year are shown in black, after two years in white (means ± one SD).

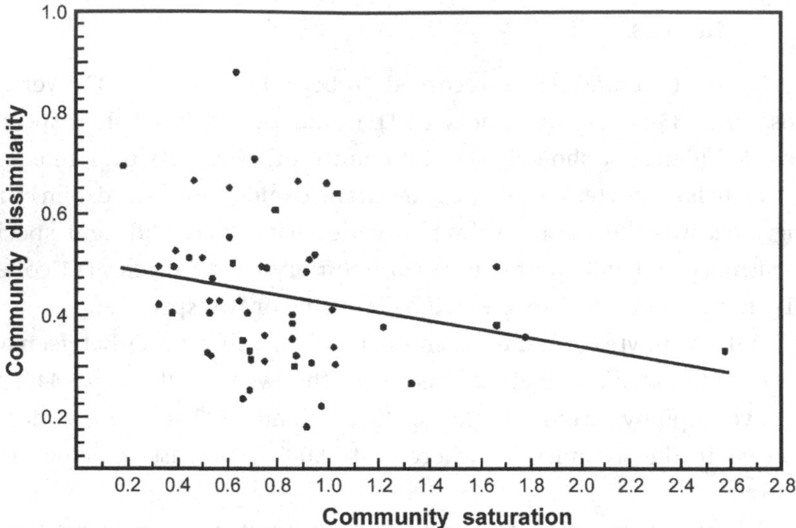

Fig. 10.2. Relationship between community dissimilarity after two years (Bray–Curtis Index) and space saturation in the communities. Each point represents a community; the line is a least-squares linear regression.

According to the rank-cover analysis, communities translocated to 2500 m had a tendency for greater cover change than the controls ($t = 1.84$; $p = 0.094$). These changes in community structure also occurred gradually with elevation, with communities at 2700 m not differing significantly from those at 2500 m and 3000 m (2500 m $t = 0.94$; $p = 0.94$; 3000 m $t = 0.11$; $p = 0.91$). Additionally, the characteristic species abundance pattern in the communities before translocation, with few species producing most of the bryophyte biomass, was maintained at 2700 m and 3000 m but not at 2500 m (Fig. 10.3).

In terms of community dynamics at different elevations, we observed no significant differences in the cover changes during the first year ($F(_{2,325}) = 1.22$; $p = 0.29$). After the second year, cover on translocated branches changed more strongly than on the controls ($F(_{2,54}) = 3.58$; $p = 0.034$; Fig. 10.4). The changes, however, did not follow a general pattern of increase or decrease with elevation ($F(_{2,54}) = 0.88$; $p = 0.41$); indeed, increases and decreases occurred at all three elevations. Community saturation affected the cover changes in the second year ($r^2 = 0.013$, $F(_{1,331}) = 4.58$, $p = 0.03$), with saturated communities (those with little available space) showing lower cover changes than unsaturated communities.

Patterns of cover change with elevation varied among species. Thus, after two years a cover decrease was observed in *Plagiochila bifaria* (Sw.) Lindenb. and *P. gymnocalycina* (Lehm. and Lindenb.) Lindenb. following their translocation to lower elevation. In *Campylopus anderssonii* (Müll. Hal.) Jaeger. and *Plagiochila*

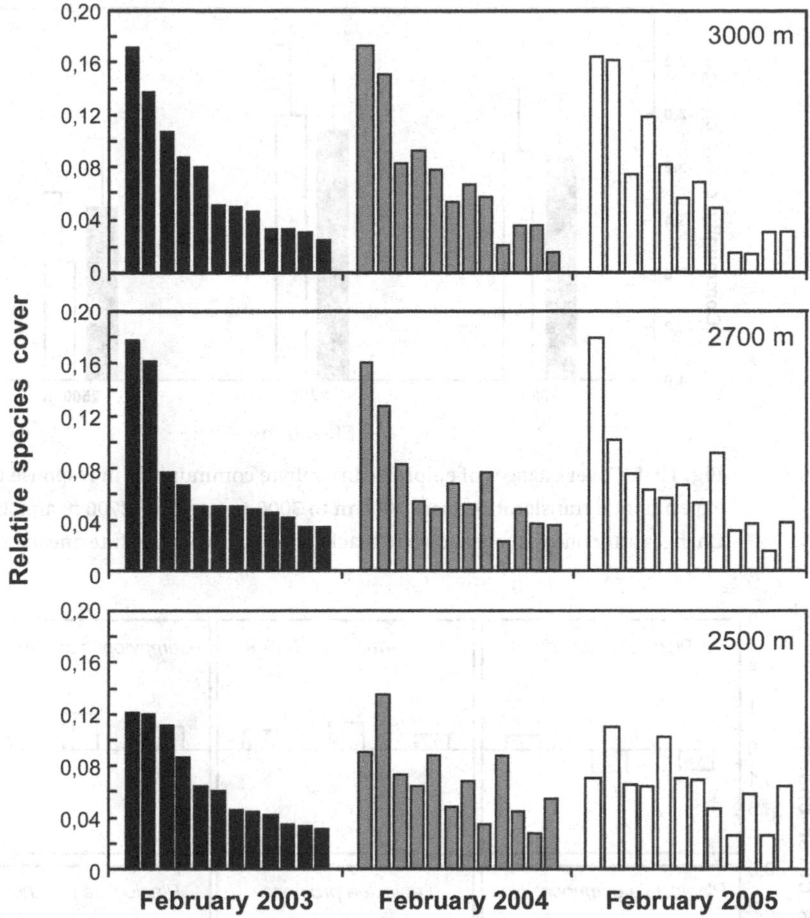

Fig. 10.3. Relative cover changes of the 12 most common epiphytic bryophyte species during the translocation experiment. In black, rank-cover patterns at the beginning of the experiment; in grey, after one year; and in white, after two years.

cf. *stricta* Lindenb., however, we observed a cover increase, and species cover was not affected by the translocation in *Lepicolea pruinosa* (Tayl.) Spruce, *Herbertus acanthelius* Spruce, and *Syzygiella anomala* (Lindenb. and Gott.) Steph. (Fig. 10.5).

Discussion

Our translocation experiment showed that exposure to air temperature increases of 1.5–2.5 °C had a measurable effect on the structure of epiphyte bryophyte communities of tropical montane forest within two years. Our results confirm the sensitivity of non-vascular epiphytes to atmospheric changes and

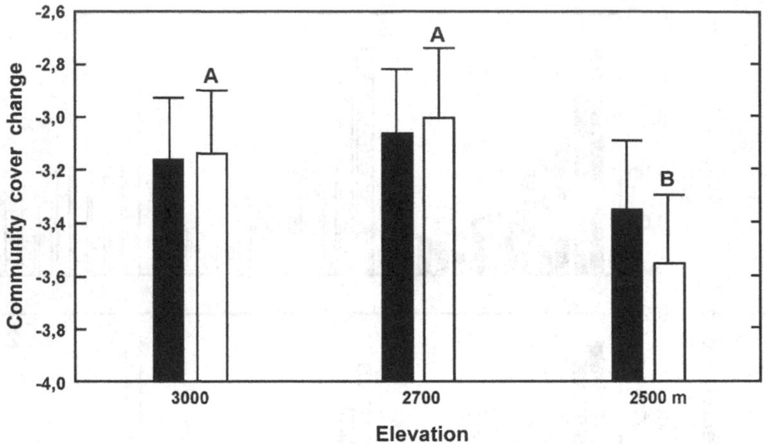

Fig. 10.4. Cover changes of epiphytic bryophyte communities in response to their experimental translocation from 3000 m to 3000 m (control), 2700 m and 2500 m. Changes after one year are shown in black; after two years in white (means ± one SD).

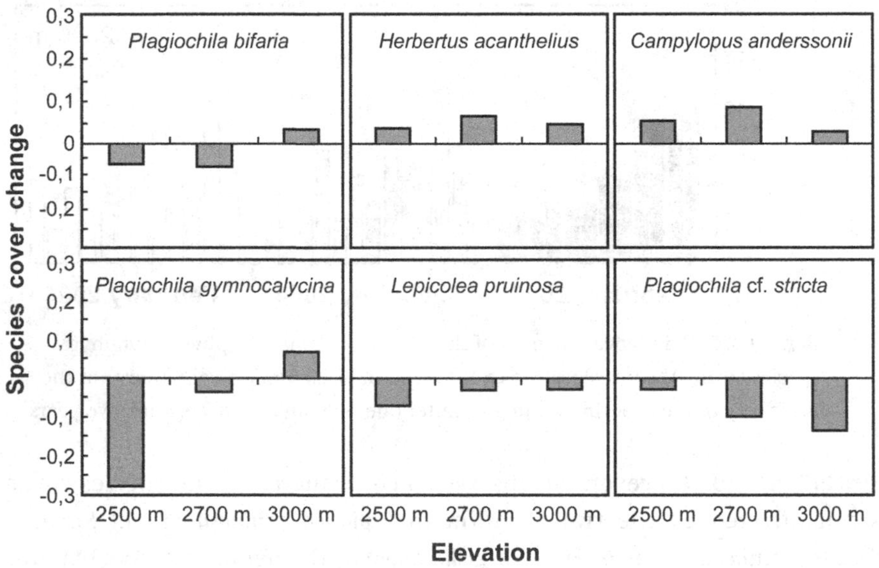

Fig. 10.5. Cover changes of selected epiphytic bryophyte species after two years in response to their experimental translocation from 3000 m to 3000 m (control), 2700 m, and 2500 m.

predict that changes in the climatic regimes of tropical montane forests will rapidly affect the rich non-vascular epiphyte communities. Our data indicate that the response of these epiphyte communities to climate change will not be abrupt but will take place in a gradual manner (Hanson & Weltzin 2000). The gradual change is indicated by the fact that the observed cover changes were species-specific while total epiphyte cover remained unchanged.

Unexpectedly, species cover values in communities translocated to lower elevations became more even, with a decreased dominance of individual species. This structural change indicates a resetting in the community, possibly involving replacement of species with narrow climatic niches by species with broader niches. The latter conclusion is speculative and needs verification. Such a replacement would suggest a return to earlier successional stages since broad niches are characteristic of species occurring in early stages of vegetation development (Bazzaz 1998). We expect that in the long term the translocated bryophyte communities will again become similar to the orginal ones in terms of abundance structure, but with different dominant species.

Our observations are suggestive of increased dynamics of epiphyte communities under changing climatic conditions, with acceleration of changes in species abundances. The observed variation in individual species reponses presumably reflects the heterogeneous composition of the community, being made up of species with different climatic requirements and elevational ranges. Contrary to our expectations, species responses to translocation did not correspond with elevational species ranges as far as known (Appendix 10.1).

Our results suggest that epiphyte communities of tropical montane forests will not migrate as a package but in an individualistic way in response to climate change. Geographical species ranges may not necessarily be affected, at least in the short term, since many factors other than climate (competition, community composition, successional stage, etc.) play a role in determining species distributions (Davis *et al.* 1998; Walther *et al.* 2002; Heegaard & Vandvik 2004; Kladerud & Totland 2005). In fact, the complex reactions of biological species to climate change cast doubt on the validity of recent bioclimatic models predicting the outcome of climate change on vegetation (Pearson & Dawson 2003).

In our study, we observed a correlation of species cover changes not only with elevation but also with initial community saturation. Saturated communities with high cover and without open spaces between bryophyte individuals underwent less cover change following their translocation than unsaturated communities. We assume that this may result from the amount of space or of shelter available for each species within the bryophyte mat. In a study on climate change effects on alpine plant communities, Kladerud and Totland (2005) postulated that vegetation may have a facilitative shelter effect on species, buffering the effect of climatic changes on the community. In this way saturated epiphyte communities could at least temporarily maintain adequate microclimatic conditions (e.g., humidity, nutrient quantities) more effectively than unsaturated ones, and prevent the establishment of invasive species (Buckland *et al.* 1997; Pitelka *et al.* 1997).

Conclusions

Our translocation experiment in the Bolivian Andes suggests that the structure and composition of epiphytic bryophyte communities of tropical mountain forests will change markedly as a result of future temperature increases, but that communities will not collapse. We propose that non-vascular epiphyte communities react rapidly to climatic changes and that individualistic species responses within the community result in a decoupling of ecological interactions, with formation of new floristic relationships and leading to changes in structure and composition of communities (Hughes 2000; Andrew & Hughes 2005). Since the transition towards the new communities is likely to be gradual rather than abrupt, through the die-back of existing species, essential ecosystem functions such as nutrient cycling and water retention mediated by non-vascular epiphyte communities (Nadkarni 1984; Hofstede *et al.* 1993; Coxson & Nadkarni 1995; Benzing 1998) are likely to be maintained. However, care must be taken when predicting long-term responses to climate change from short-term observations, since some responses detected over short time periods may be transitional and may thus poorly reflect long-term community responses (Hollister *et al.* 2005; Jónsdóttir *et al.* 2005).

Acknowledgements

We thank Jorge Uzquiano, Pablo Blatt, and Enrique Domic for field assistance, Katja Poveda for comments on the manuscript and data management, and the National Herbarium of Bolivia, especially Stephan Beck, for facilities and continuous support. For help with voucher identification we are grateful to Katrin Feldberg (*Syzygiella*) and Jochen Heinrichs (*Plagiochila*). Financial support was provided by the German Research Foundation (DFG).

Appendix 10.1

List of bryophyte species recorded in the simulated climate change experiment in the Bolivian Andes. For each species, a collection number (J . . .) and elevational range in the tropical Andes are indicated, based on literature and herbarium specimens. Nomenclature follows Churchill *et al.* (2000) and Gradstein *et al.* (2003), with updates. Vouchers are deposited in the Herbarium of the University of Göttingen (GOET). Asterisks indicate species new to Bolivia.

Mosses (Bryophyta s.str.). *Aptychella proligera* (Broth.) Herz., J1057, 1800–3400 m; *Brachythecium* sp., J1010, J1056; *Bryum* sp., J1008; *Campylopus* cf. *anderssonii* (Müll. Hal.) Jaeger, J1030, 1450–3350 m; *Ctenidium malacodes* Mitt., J1007, 1800–3750 m; *Macromitrium* sp., J1004; *Neckera* sp., J1054; *Porotrichodendron superbum* (Tayl.) Broth., J1058, 2200–3440 m; *Porotrichum longirostre* (Hook.) Mitt., J1006, 500–3200 m; *Racopilum tomentosum* (Hedw.) Brid., J1012, J1013, 500–3100 m; *Rhodobryum* cf. *grandifolium* (Tayl.) Schimp. ex Paris, J1071, 1700–3450 m; *Sematophyllum* sp., J1011; *Squamidium leucotrichum* (Tayl.) Broth., J1055, 1900–3100 m; *Thuidium peruvianum* Mitt., J1009, 1800–3470 m.

Liverworts (Marchantiophyta). *Adelanthus lindenbergianus* (Lehm.) Mitt., J1045, 2000–3400 m; *Anoplolejeunea conferta* (Meissn.) Schiffn., J1060, J1066, 650–3400 m; *Bazzania hookeri* Lindenb., J1026, 1300–3000 m; *Ceratolejeunea grandiloba* Jack & Steph., J1016, J1017, 1400–3000 m; *Cheilolejeunea* cf. *laevicalyx* (Jack & Steph.) Grolle, J1061, J1062, 2600–3620 m; **Cheilolejeunea holostipa* (Spruce) Grolle & Zhu, J1065, 500–3400 m; *Chiloscyphus trapezoideus* (Mont.) Schust. & Engel, J1025, 500–3950 m; **Drepanolejeunea bidens* (Steph.) Evans, J1063, 500–3000 m; *Frullania brasiliensis* Raddi, J1020, J1070, 500–3200 m; *Frullania apiculata* (Reinw. et al.) Nees, J1018, J1019, 500–3000 m; **Harpalejeunea stricta* (Lindenb. & Gott.) Steph., J1067, J1069, 500–3000 m; *Herbertus acanthelius* Spruce, J1023, 2500–4300 m; *Lepicolea pruinosa* (Tayl.) Spruce, J1029, 2000–3800 m; *Lepidozia cupressina* (Sw.) Lindenb., J1003, J1014, 2000–3000 m; *Leptoscyphus amphibolius* (Nees) Grolle, J1002, 1000–3000 m; *Leptoscyphus porphyrius* (Nees) Grolle, J1022, 500–3000 m; *Metzgeria albinea* Spruce, J1032, 1400–3700 m; *Plagiochila bifaria* (Sw.) Lindenb., J1044, 1800–3400 m; *Plagiochila boryana* Gottsche ex Steph., J1053, 2500–4000 m; *Plagiochila fuscolutea* Tayl., J1043, 2200–4000 m; *Plagiochila gymnocalycina* Lindenb., J1034, 500–3000 m; *Plagiochila macrostachya* Lindenb., J1000, J1035, J1036, J1038, J1039, J1040, J1041, J1042, J1046, J1047, J1052, 1000–3200 m; *Plagiochila* cf. *rutilans* Lindenb., J1037, 500–3000 m; *Plagiochila* cf. *stricta* Lindenb., J1049, 1100–3000 m; **Radula tenera* Mitt. ex Steph., J1059, 500–3700 m; *Radula voluta* Tayl., J1028, 1800–3720 m; *Riccardia* sp., J1033; *Scapania portoricensis* Hampe & Gottsche, J1027, 1400–3750 m; *Syzygiella anomala* (Lindenb. & Gott.) Steph., J1015, J1031, 2150–3400 m; *Syzygiella campanulata* Herzog, J1001, 2700–3700 m; *Trichocolea tomentosa* (Sw.) Gottsche, J1021, 500–3700 m; *Tylimanthus laxus* (Lehm. & Lindenb.) Spruce, J1048, 1000–4000 m.

References

Acebey, C., Gradstein, S. R. & Krömer, T. (2003). Species richness and habitat diversification of bryophytes in submontane rain forest and fallows of Bolivia. *Journal of Tropical Ecology* **19**: 9–18.

Andrew, N. R. & Hughes, L. (2005). Diversity and assemblage structure of phytophagous Hemiptera along a latitudinal gradient: predicting the potential impacts of climate change. *Global Ecology and Biogeography* **14**: 249–62.

Aryanti, N. S., Bos, M. M., Kartawiniata, K., *et al.* (2008). Bryophytes on tree trunks in natural forests, selectively logged forests and cacao agroforests in Central Sulawesi, Indonesia. *Biological Conservation* **141**: 2516–27.

Bach, K. (2004). *Vegetationskundliche Untersuchungen zur Höhenzonierung tropischer Bergregenwälder in den Anden Boliviens*. Marburg: Görich und Weiershäuser Verlag.

Bazzaz, F. A. (1998). Tropical forests in a future climate: changes in biological diversity and impact on the global carbon cycle. *Climatic Change* **39**: 317–36.

Benzing, D. H. (1998). Vulnerabilities of tropical forests to climate change: the significance of resident epiphytes. *Climatic Change* **39**: 519–40.

Buckland, S. M., Grime, J. P., Hodgson, J. G. *et al.* (1997). A comparison of plant responses to the extreme drought of 1995 in northern England. *Journal of Ecology* **85**: 875–82.

Cavelier, J., Solis, D. & Jaramillo, M. A. (1996). Fog interception in montane forest across the central cordillera of Panamá. *Journal of Tropical Ecology* **12**: 357–69.

Chen, J., Carlson, B. E. & Del Genio, A. D. (2002). Evidence for strengthening of the tropical general circulation in the 1990s. *Science* **295**: 838–41.

Churchill, A. P., Griffin, D. III & Muñoz, J. (2000). A checklist of the mosses of the tropical Andean countries. *Ruizia* **17**: 1–203.

Cornelissen, J. H. & Ter Steege, H. (1989). Distribution and ecology of epiphytic bryophytes and lichens in dry evergreen forest of Guyana. *Journal of Tropical Ecology* **5**: 29–35.

Coxson, D. & Nadkarni, N. M. (1995). Ecological roles of epiphytes in nutrient cycles of forest ecosystems. In *Forest Canopies*, ed. M. D. Lowman and N. M. Nadkarni, pp. 495–543. San Diego, CA: Academic Press.

Crawley, M. J. (2002). *Statistical Computing*. Chichester: John Wiley and Sons.

Davis, A. J., Jerkinson, L. S., Lawton, J. L. *et al.* (1998). Making mistakes when predicting shifts in species range in response to global warming. *Nature* **391**: 783–6.

Farmer, A. M., Bates, J. W. & Bell, J. N. B. (1992). Ecophysiological effects of acid rain on bryophytes and lichens. In *Bryophytes and Lichens in a Changing Environment*, ed. J. W. Bates & A. M. Farmer, pp. 284–313. Oxford: Clarendon Press.

Foster, P. (2001). The potential negative impacts of global climate change on tropical montane cloud forest. *Earth-Science Reviews* **55**: 73–106.

Fowler, J., Cohen, L. & Jarvis, P. (1998). *Practical Statistics for Field Biology*. Chichester: John Wiley and Sons.

Frahm, J.-P. & Klaus, D. (1997). Moose als Indikatoren von Klimafluktuationen in Mitteleuropa. *Erdkunde* **51**: 181–90.

Frahm, J.-P. & Klaus, D. (2001). Bryophytes as indicators of recent climate fluctuations in Central Europe. *Lindbergia* **26**: 97–104.

Gerold, G. (2008). Soil, climate and vegetation in tropical montane forests – a case study from the Yungas, Bolivia. In *The Tropical Mountain Forest, Patterns and Processes in a Biodiversity Hotspot*, ed. S. R. Gradstein, J. Homeier & D. Gansert, pp. 137–62. Göttingen: Universitätsverlag Göttingen.

Gignac, L. D. (2001). Bryophytes as indicators of climatic change. *Bryologist* **104**: 410–20.

Gradstein, S. R. (2008). Epiphytes of tropical montane forest – impact of deforestation and climate change. In *The Tropical Mountain Forest – Patterns and Processes in a Biodiversity Hotspot*, ed. S. R. Gradstein, J. Homeier & D. Gansert, pp. 179–96. Göttingen: Universitätsverlag Göttingen.

Gradstein, S. R. & Sporn, S. G. (2009). Impact of forest conversion and climate change on bryophytes in the Tropics. *Berichte der Reinhold-Tüxen-Gesellschaft* **21**: 128–41.

Gradstein, S. R., Churchill, S. P. & Salazar-Allen, N. (2001). *Guide to the Bryophytes of Tropical America*. Bronx, New York: The New York Botanical Garden Press.

Gradstein, S. R., Meneses, R. I. & Allain-Arbe, B. (2003). Catalogue of the Hepaticae and Anthocerotae of Bolivia. *Journal of the Hattori Botanical Laboratory* **93**: 1–65.

Hanson, P. J. & Weltzin, J. F. (2000). Drought disturbance from climate change: responses of United States forests. *Science of the Total Environment* **262**: 205–20.

Hawksworth, D. L. & Rose, F. (1970). Qualitative scale for estimating sulphur dioxide air pollution in England and Wales using epiphytic lichens. *Nature* **227**: 145–8.

Heegaard, E. & Vandvik, V. (2004). Climate change affects the outcome of competitive interactions. An application of principal response curves. *Oecologia* **139**: 459–66.

Henderson, A., Churchill, S. P. & Luteyn, J. L. (1991). Neotropical plant diversity. *Nature* **351**: 21–2.

Hofstede, R. G. M., Wolf, J. H. D. & Benzing, D. H. (1993). Epiphytic biomass and nutrient status of a Colombian upper montane rain forest. *Selbyana* **14**: 37–45.

Hollister, R. D., Webber, P. J. & Tweedie, C. E. (2005). The response of Alaskan artic tundra to experimental warming: differences between short- and long-term responses. *Global Change Biology* **11**: 525–36.

Holz, I. & Gradstein, S. R. (2005). Cryptogamic epiphytes in primary and recovering upper montane oak forests of Costa Rica – species richness, community composition and ecology. *Plant Ecology* **178**: 89–109.

Hughes, L. (2000). Biological consequences of global warming: is the signal already apparent? *Trends in Ecology and Evolution* **15**: 56–61.

Johansson, D. (1974). Ecology of vascular epiphytes in West African rain forest. *Acta Phytogeographica Suecica* **59**: 1–136.

Jónsdóttir, I. S., Magnússon, B., Gudmundsson, J. *et al.* (2005). Variable sensitivity of plant communities in Iceland to experimental warming. *Global Change Biology* **11**: 553–63.

Kladerud, K. & Totland, Ø. (2005). The relative importance of neighbours and abiotic environmental conditions for population dynamic parameters of two alpine plant species. *Journal of Ecology* **93**: 493–501.

Legendre, P. & Legendre, L. (1998). *Numerical Ecology*, 2nd English edn. Amsterdam: Elsevier.

Liede-Schumann, S. & Breckle, S.-W. eds. (2008). Provisional checklists of flora and fauna of the San Francisco Valley and its surroundings Reserva Biológica San

Francisco, Province Zamora-Chinchipe, southern Ecuador. *Ecotropical Monographs* **4**: 1–256.

Lovejoy, T. E. & Hannah, L. (2005). *Climate Change and Biodiversity*. New Haven, CT: Yale University Press.

Magurran, A. E. (1988). *Ecological Diversity and its Measurement*. Princeton, NJ: Princeton University Press.

Malcolm, J. R., Liu, C., Neilson, R. P., Hansen, L. & Hannah, L. (2006). Global warming and extinction from biodiversity hotspots. *Conservation Biology* **20**: 438–548.

Mosandl, R. & Günter, S. (2008). Sustainable management of tropical mountain forests in Ecuador. In *The Tropical Mountain Forest – Patterns and Processes in a Biodiversity Hotspot*, ed. S. R. Gradstein, J. Homeier & D. Gansert, pp. 179–96. Göttingen: Universitätsverlag Göttingen.

Nadkarni, N. M. (1984). Epiphyte mats and nutrient capital of a neotropical elfin forest. *Biotropica* **16**: 249–56.

Nadkarni, N. M. & Longino, J. T. (1990). Invertebrates in canopy and ground organic matter in a Neotropical montane forest. *Biotropica* **22**: 358–63.

Nadkarni, N. M. & Solano, R. (2002). Potential effects of climate change on canopy communities in a tropical cloud forest: an experimental approach. *Ecologia* **131**: 580–94.

Nöske, N., Hilt, N., Werner, F. *et al.* (2008). Disturbance effects on the diversity of epiphytes and moths in montane forest of Ecuador. *Basic and Applied Ecology* **9**: 4–12.

Pearson, R. G. & Dawson, T. P. (2003). Predicting the impacts of climate change on the distribution of species: are bioclimate envelope models useful? *Global Ecology & Biogeography* **12**: 361–71.

Pitelka, L. F., Gardner, R. H., Ash, J. *et al.* (1997). Plant migration and climate change. *American Scientist* **85**: 464–73.

Pounds, J. A., Fogden, M. P. L. & Campbell, J. H. (1999). Biological response to climate change on a tropical mountain. *Nature* **398**: 611–15.

Price, M. V. & Waser, N. M. (2000). Responses of subalpine meadow vegetation to four years of experimental warming. *Ecological Applications* **10**: 811–23.

Sala, O. E., Chapin III, F. S., Armesto, J. J. *et al.* (2000). Biodiversity – global biodiversity scenarios for the year 2100. *Science* **287**: 1770–4.

Sporn, S. G., Bos, M. M., Hoffstätter-Müncheberg, M., Kessler, M. & Gradstein, S. R. (2009). Microclimate determines community composition but not richness of epiphytic understory bryophytes of rainforest and cacao agroforest in Indonesia. *Functional Plant Biology* **36**: 171–9.

Statsoft (2003). *Statistica für Windows Software-system für Datenanalyse*, Version 6. www.statsoft.com.

Steffan-Dewenter, I., Kessler, M., Barkman, J. *et al.* (2007). Socioeconomic context and ecological consequences of rainforest conversion and agroforestry intensification. *Proceedings of the National Academy of Sciences of the United States of America* **104**: 4973–8.

Still, C. J., Foster, P. N. & Schneider, S. H. (1999). Simulating the effects of climate change on tropical montane forests. *Nature* **398**: 608–10.

Szczepaniak, K. & Biziuk, M. (2003). Aspects of the biomonitoring studies using mosses and lichens as indicators of metal pollution. *Environmental Research* **93**: 221–30.

Van Dobben, H. F. & De Bakker, A. J. (1996). Re-mapping epiphytic lichen biodiversity in the Netherlands: effects of decreasing SO_2 and increasing NH_3. *Acta Botanica Neerlandica* **45**: 55–71.

Van Herk, C. M., Aptroot, A. & Van Dobben, H. F. (2002). Long-term monitoring in the Netherlands suggests that lichens respond to global warming. *The Lichenologist* **34**: 141–54.

Veneklaas, E. J., Zagt, R. J., Van Leerdam, A., *et al.* (1990). Hydrological properties of the epiphytic mass of a montane tropical rain forest, Colombia. *Vegetatio* **89**: 183–92.

Vuille, M., Bradley, R. S., Werner, M. *et al.* (2003). 20th century climate change in the tropical Andes: observations and model results. *Climatic Change* **59**: 75–99.

Walther, G., Post, E., Convey, P. *et al.* (2002). Ecological responses to recent climate change. *Nature* **416**: 389–95.

Weltzin, J. F., Bridgham, S. D., Pastor, J. *et al.* (2003). Potential effects of warming and drying on peatland plant community composition. *Global Change Biology* **9**: 141–51.

Woodward, F. I. (1992). A review of the effects of climate on vegetation: ranges, competition, and composition. In *Global Warming and Biological Diversity*, ed. R. J. Peters & T. J. Lovejoy, pp. 105–23. New Haven, CT: Yale University Press.

Yanoviak, S. P., Nadkarni, N. M. & Solano, R. (2007). Arthropod assemblages in epiphyte mats of Costa Rican cloud forest. *Biotropica* **36**: 202–10.

Zar, J. H. (1999). *Biostatistical Analysis*, 4th edn. Englewood Cliffs, NJ: Prentice Hall.

Zotz, G. & Bader, M. Y. (2008). Epiphytic plants in a changing world – global change effects on vascular and non-vascular epiphytes. *Progress in Botany* **70**: 147–70.

V ALPINE, ARCTIC, AND ANTARCTIC ECOSYSTEMS

11

Effects of Climate Change on Tundra Bryophytes

ANNIKA K. JÄGERBRAND, ROBERT G. BJÖRK, TERRY CALLAGHAN, AND RODNEY D. SEPPELT

Tundra vegetation in the changing climate

In the 1990s global warming was envisioned scientifically as being highly influential and pronounced at high latitudes (Mitchell *et al.* 1990; Maxwell 1992). Since then, impacts of climate change have been confirmed, especially in the indisputable data of increased air surface temperatures in both the Alaskan Arctic and Europe (Overpeck *et al.* 1997; Keyser *et al.* 2000; Serreze *et al.* 2000; EEA 2004). Ostensibly, climate change is currently affecting life in the world's ecosystems with intensified ramifications of escalating temperatures (IPCC 2007). The Arctic has had a rapid increase in mean temperatures over the past few decades, twice the rate of the rest of the world (ACIA 2005). Its warmest year ever recorded was in 2007 (Richter-Menge *et al.* 2008). Biomes already seem to be changing owing to climate differences, indicated by observations of enhanced plant growth at high northern latitudes (Myneni *et al.* 1997) and mid-latitudes (Nemani *et al.* 2003), landscape-level shifts in species ranges, decline in species populations (McCarthy *et al.* 2001), and changes in species diversity (EEA 2004). Continuing Arctic climate change will therefore have the effect of encouraging forest expansion into tundra biomes, and the tundra vegetation as we know it will greatly change, shifting in its extent, distribution, and species composition. These changes will probably be unprecedented compared with those of past millennia. As climate change is only one of several environmental changes occurring in the Arctic, such as increased UV-B radiation and transboundary contaminants, impacts on vegetation are expected to be compounded.

Tundra vegetation can be described as low shrub vegetation lacking trees in alpine, Arctic, and Antarctic areas (Fig. 11.1). Shrub species have recently

Bryophyte Ecology and Climate Change, eds. Zoltán Tuba, Nancy G. Slack and Lloyd R. Stark. Published by Cambridge University Press.

Fig. 11.1. Photograph from northern Sweden, showing mountains in the sub-Arctic–alpine tundra zone in the background and in the foreground mountain birch forest at lower altitudes. Owing to global climate change, the forest is now expanding into the tundra biome. Photograph by courtesy of Gaku Kudo.

increased in dominance in tundra ecosystems in Arctic Alaska and northern Sweden (Chapin *et al.* 1995; Sturm *et al.* 2001; Jägerbrand *et al.* 2009).

Concurrently, the treeline is migrating towards higher latitudes and elevations in some locations (van Bogaert *et al.* 2008), successively encroaching into tundra vegetation (Sundqvist *et al.* 2008). Even a moderate increase in global mean temperature of 2 °C (above the pre-industrial level) has been predicted to reduce the tundra area by 42%, with an additional 60% of the prostrate dwarf-shrub tundra habitats lost by 2100 (Kaplan & New 2006). Since plants in tundra and high altitudinal areas are adapted to colder environments, for example by having relatively low reproductive output, slow and prolonged growth, and reduced competitive ability, their capacity to adapt to the new climate or to migrate is low (Callaghan *et al.* 2005). Thus, the IPCC has recognized that both polar and alpine ecosystems are highly susceptible and threatened with irreversible damage due to global climate change (McCarthy *et al.* 2001). Tundra is very sensitive to rapid changes, with reductions in numbers of current species, particularly specialist species, although the total number of species may increase as the boreal forest moves northwards. Locally, and particularly where the tundra occupies a narrow coastal strip of land such as in the Western Russian Arctic and Fennoscandia, there may be biome extinction. Consequently, research has been undertaken to examine how tundra ecosystems may respond to future climate and what feedback responses from the Arctic biosphere to the climate system are likely to occur in the future.

The importance of tundra bryophytes

Towards higher latitudes, species richness of both vascular plants and bryophytes decreases, but since the decrease of vascular plants is greater than that of bryophytes, the relative abundance of bryophytes actually increases (Vitt & Pakarinen 1977; Wielgolaski *et al.* 1981). In addition, Arctic bryophytes contribute significantly to global biodiversity: there are approximately 600 species of Arctic moss, representing 4.1% of global diversity, and 250 species of liverwort, representing 2.5% of global diversity (Callaghan *et al.* 2005). Tundra bryophytes have been recognized as very important interactive species contributing to life-sustaining ecological processes and ecosystem functions in, for example:

- Primary production (Oechel & Sveinbjörnsson 1978; Webber 1978; Longton 1982; Glime 2007)
- Biomass (Vitt & Pakarinen 1977; Oechel & Sveinbjörnsson 1978; Wielgolaski *et al.* 1981)
- Cover (Oechel & Sveinbjörnsson 1978; Longton 1988)
- Energy flow and nutrient cycling (Longton 1982, 1984, 1988)
- Species diversity and richness (Longton 1982, 1997; Matveyeva & Chernov 2000)
- Biogeochemistry (Cornelissen *et al.* 2007)
- Feedback responses on climate change (Rosswall *et al.* 1975; Hobbie 1996; Dorrepaal *et al.* 2005)
- Facilitation of other species establishment and growth in some cases (Press *et al.* 1998a) and competitive exclusion in others (Chernov 1985)

Bryophytes can attain high dominance, even coexisting as the major component in habitats of the tundra environment. In such habitats bryophytes play a key role in ecosystem structure and function (Fig. 11.2). Tundra bryophytes in moss layers can exert substantial effects on vascular plants due to effects on temperature regime (van der Wal *et al.* 2001), which influences soil thaw depth (Ng & Miller 1977; Miller *et al.* 1980; Tenhunen *et al.* 1992) and

Fig. 11.2. Bryophytes may attain high dominance in certain habitats: moss-heath vegetation in Thingvellir, sub-Arctic Iceland. Light areas of the ground layer are well-developed moss mats of *Racomitrium lanuginosum*, sometimes reaching a thickness of 30–40 cm.

decomposition (Russell 1990). As mosses are often sensitive to disturbance by grazing animals, there is an important interaction between higher plants, mosses, and nutrient cycling mediated through herbivores (van der Wal & Brooker 2004). A dominant moss layer may not only affect plants but may modify whole ecosystems, as has been shown for *Sphagnum* in Sweden (Svensson 1995).

There are many aspects of the ways in which mosses interact with other mosses, higher plants, and animals that will influence the response of the whole system to climate change. *Sphagnum fuscum* and *Dicranum elongatum* currently co-exist in a sub-Arctic mire (Sonesson *et al.* 2002). During winter the growth of *S. fuscum* was enhanced when *D. elongatum* was a neighbor but *D. elongatum* derived no benefit from growing with *S. fuscum*. Each species responded individualistically when increased temperature and precipitation were simulated. The facilitation of moss growth by *D. elongatum* also occurs in the relationships between a moss and a higher plant. Cushions of bryophytes on high Arctic Svalbard host vascular plants species such as *Poa arctica*. When these higher plant species grow in moss cushions, their growth is significantly increased compared with ambient conditions, and the effect of the facilitation is equivalent to nutrient increase and enhanced temperature (Press *et al.* 1998a). In contrast, Chernov (1985) records the smothering of higher plant seedlings by mosses. In the absence of herbivores, mosses proliferate in the tundra, reduce nutrient cycling, and insulate the soil, thereby often outcompeting vascular plants (van der Wal & Brooker 2004). *Sphagnum* in particular is expected to increase its growth in response to warming and then overgrow shrubs such as *Empetrum*, which have limited capacity to respond to warming (Dorrepaal *et al.* 2006). Consequently, mosses will play a vital role through competition or facilitation in determining species and ecosystem responses to climate change (Brooker *et al.* 2008). Bryophytes have been suggested as predictors of migration of climatically sensitive ecosystems (Gignac 2001). Growth measured by retrospective measurements of the circum-Arctic bryophyte *Hylocomium splendens* was further used to evaluate potential responses to climate change (Callaghan *et al.* 1997). Both shrubs and bryophytes were identified as key indicators of community-level change in tundra vegetation based on field data and simulation models (Epstein *et al.* 2004). As bryophytes play such vital roles in tundra ecosystems, it is important to monitor their different responses to the changing climate.

Field experiments

Simulation of increased temperature

Most recent research on bryophyte responses to climate change in tundra ecosystems is based on field experiments, whereas earlier laboratory experiments focused on the responses of mosses to variations in temperature, moisture, atmospheric CO_2, and UV-B radiation (see Schipperges & Gehrke 1996 for a review). In the field, effects of simulations of global warming on bryophytes have been conducted using *in situ* small greenhouses to simulate increased air temperature. Most of these studies have been performed within the group called ITEX (The International Tundra Experiment), an international network of researchers aiming to use the same experimental design and methods to make results comparable (Henry & Molau 1997). ITEX has developed small greenhouses called open-top chambers (OTCs) (Fig. 11.3a, b) that increase the average temperature by 1.5–3 °C compared with control plots (Marion *et al.* 1997; Molau 2001).

OTCs have many technical advantages. They passively warm the air without maintenance and can be permanently placed in the field, although some researchers remove OTCs during the winter season to avoid damage to them. In addition, microclimatic effects of the OTCs have been validated

Fig. 11.3 (upper) OTC, a hexagonal open-top chamber design that is used in field experiments in tundra areas to raise the temperature by *c.* 1.5–3.0 °C to simulate climate change. Picture by courtesy of Ulf Molau. (lower) OTC placed on heath vegetation in sub-Arctic–alpine tundra in northern Sweden dominated by *Betula nana*. Photograph by courtesy of Ulf Molau.

to correspond with observed regional climatic warming (Hollister & Webber 2000), allowing for natural between-year temperature variations. However, temperature simulations cannot precisely imitate naturally occurring warming (Kennedy 1995). For example, extreme weather events, winter warming (see, for example, Bokhorst *et al.* 2008), and changed snow conditions (but see Dorrepaal *et al.* 2003) are not included in the temperature simulation when using open-top chambers.

Most bryophytes are poikilohydric in that they can desiccate, remain dormant, and then re-initiate photosynthesis after rewetting; their growth is mainly controlled by moisture conditions. OTCs allow precipitation to contact the specimen through its top opening, but relative humidity is generally lower than in untreated areas mainly because increased temperature causes higher evaporation rates and increased desiccation in the chambers

(Hollister & Webber 2000). Although it can be argued that the decreased relative humidity is an experimental artifact, desiccation will also take place under conditions of natural warming. However, when evaluating responses of bryophytes to climate change experiments it is crucial to have accurate information on the moisture conditions of the field site of the natural occurring bryophytes (Sandvik & Heegaard 2003).

Increased nutrient availability

Climate change will cause increased nutrient availability (Nadelhoffer *et al.* 1991, 1992; Bonan & van Cleve 1992; Jonasson *et al.* 1993; Chapin *et al.* 1995; McCarthy *et al.* 2001) in the short term, with indirect successive effects on soil processes (Van Cleve *et al.* 1990; Anderson 1991; Nadelhoffer *et al.* 1991; Bonan & van Cleve 1992; Kane *et al.* 1992). To simulate this, nutrient availability is usually increased experimentally through application of a water-based nutrient solution directly on the vegetation or by granular nutrient supply in the soil at various levels of nutrient additions between 5 and 10 mg N m^{-2}.

Methods for studying plant community responses to climate change

Plant community responses and changes in abundance/cover and height can be analyzed by the point-frame method (Walker 1996; Molau & Alatalo 1998). The point-frame method or the point-intercept method has been widely used within the ITEX group (see Fig. 11.4). For example, its application could include a frame with 100 grid points located 10 cm apart and horizontally positioned above the plots and at each intersect to allow for two or many continuous measurements. This method could be applied to one of the canopy layers and also the bottom layer by analyzing the species intersected underneath (Molau & Alatalo 1998; Press *et al.* 1998b). The total number of intercepted hits is an estimation of species abundance and can also be used to derive biomass. For bryophytes, Jónsdóttir *et al.* (1997) recommended the point-frame method to measure bryophyte community composition either by individual species abundances or through the use of functional groups. There are many methods available for individual measurements of growth, length, or biomass, e.g., the cranked-wire method, the tied-thread method, or the transplant method (Jónsdóttir *et al.* 1997). For some bryophyte species it is also possible to use annual innate markers (Callaghan *et al.* 1978).

Fig. 11.4. When analyzing with a point-frame, measurements are taken at each intersect hitting the canopy layer and the other layers underneath. This plot is fertilized tundra heath vegetation in sub-Arctic–alpine tundra in northern Sweden. Photograph by courtesy of Juha Alatalo.

Responses to warming and increased nutrients

Many studies have been undertaken in tundra ecosystems to investigate responses to simulated environmental change (Arft *et al.* 1999). Unfortunately, most of these studies did not include detailed individual measurements on bryophytes but instead used bryophytes as a functional group, if they were included at all.

Temperature increase

Bryophytes decreased in cover after four years in response to a temperature warming of 1–3 °C when measured in a study that included eleven tundra locations (Walker *et al.* 2006). However, in a similar study that included six tundra sites in northern Sweden and Alaska, no significant effects of warming *per se* on the biomass of bryophytes were detected after 3–20 years of warming (van Wijk *et al.* 2003). Unfortunately it is not known whether there were different responses between those sites included in each of the papers or, for example, if cover is more easily affected by changes in vascular plant abundances compared with changes in biomass. It was also reported that the diversity of bryophytes was not affected after five years of experimental warming (Jägerbrand *et al.* 2006).

At the species level individual bryophyte species showed no significant effects on short-term growth or abundance after three and five years, or increased growth after one to four seasons (Jägerbrand *et al.* 2003; Sandvik & Heegaard 2003; Jägerbrand 2005, 2007). Tundra bryophyte responses to warming might also depend on species-specific ecology. For instance, temperature optima for growth in temperate bryophyte species were significantly different depending on the ecology of the species (Furness & Grime 1982), and similarly, photosynthetic rates produced in Antarctic bryophytes varied widely among species (Davey & Rothery 1997). In the sub-Arctic forest floor environment, growth of *Hylocomium splendens* is limited by temperature, light, and hydration (Sonesson *et al.* 1992). Consequently response to warming will only occur if the co-limiting factors are also ameliorated.

Bryophyte populations in colder areas may have lower temperature optima for photosynthesis in comparison with populations in warmer conditions (Longton 1979; Furness & Grime 1982; Kallio & Saarnio 1986). If bryophytes can acclimate to the new environment, it seems plausible that their growth will be enhanced during a favorable warmer climate. Bryophytes that were transplanted from polar-region areas to warmer conditions showed higher net assimilation rates of photosynthesis across a range of different temperatures, suggesting good acclimatization potential (Kallio & Saarnio 1986). Similarly, populations of *Pleurozium schreberi*, originating from an altitudinal gradient in Hokkaido, Japan, showed different length-growth responses when grown in warm laboratory conditions (20 °C/10 °C; day/night), but the greatest length-growth increase was shown by the population originating from the highest altitude (A.K. Jägerbrand & G. Kudo, unpublished results). Another experiment that transplanted populations of *Hylocomium splendens* from high Arctic Svalbard to sub-Arctic Sweden showed that the predominant monopodial growth form of the high Arctic responded to warming by increasing its ability to grow sympodially, but after 14 years the high Arctic and sub-Arctic populations were still distinct (Ross *et al.* 2001).

Since bryophytes exhibit inter- and intra-species differences in temperature optima for photosynthesis and growth and their acclimatization potential has been poorly studied, it is difficult at present to produce conclusive evidence as to the fate of the tundra bryophytes. Plants specialized for extreme environments may show little acclimation potential to changed temperatures (Atkin *et al.* 2006), and thus it is likely that truly psychrophilic (cold adapted) bryophytes will have little potential for acclimatization. Moreover, bryophytes are C_3 species (Proctor 2000), and in C_3 species the physiological mechanisms behind temperature acclimation and photosynthetic limitation depend on both CO_2 levels and temperatures at suboptimal or supraoptimal levels (Sage & Kubien 2007). In addition, bryophytes are likely to be disadvantaged as more competitive species move northwards or up in elevation. Bryophytes might also be negatively affected by changes in herbivory.

Increased nutrients

Increased nutrient availability in tundra ecosystems usually has very dramatic effects on vegetation; for example, the grass *Calamagrostis lapponica* increased 50 times in biomass after two seasons of nutrient fertilization in a sub-Arctic heath (Parsons *et al.* 1995; see Fig. 11.4 showing fertilized heath vegetation). It is difficult to separate the direct effects of nutrient applications on bryophytes from indirect effects of increased vascular plant abundance in field experiments. Bryophytes are sensitive to nutrient additions applied directly on their leaves, as they lack a protective cuticle, which may lead to direct toxicity (Bates 2000; Solga & Frahm 2006). It has been argued that toxic effects should cause more rapid changes in bryophytes than those observed in field experiments. It is therefore assumed that toxic effects are non-significant and that any changes in bryophyte composition or abundance is a response to changes in the vascular plants (van Wijk *et al.* 2003). When vascular plants increase in dominance as a result of increased nutrients they may affect the vegetation underneath in two ways: through increased shading and through increased deleterious litter. It is difficult for field experiments to separate shade and litter effects, but a few studies have performed light attenuation experiments parallel with field experiments that successfully differentiated effects of shade from other conditions. However, these experiments failed to induce the same responses in bryophytes as the fertilizer experiments (van Wijk *et al.* 2003).

Bryophytes in tundra ecosystems have been shown to decline in abundance (Graglia *et al.* 2001; Jägerbrand *et al.* 2003; Klanderud & Totland 2005), biomass (Press *et al.* 1998b; van Wijk *et al.* 2003), and production (Chapin *et al.* 1995) after treatments of combined increased nutrient availability and temperature. Similarly, bryophyte diversity and species richness decreased during combined treatments of temperature and fertilization (Molau & Alatalo 1998; Klanderud & Totland 2005; Jägerbrand *et al.* 2006). However, responses at the individual level to nutrient additions and temperature increases varied, ranging from decreased short-term growth (Jägerbrand *et al.* 2003) to increased growth (Sandvik & Heegaard 2003).

Nutrient addition altered the growth form of *Hylocomium splendens* from monopodial, typical of cold environments, to sympodial, typical of milder environments (Ross *et al.* 2001). Unfortunately, nutrient addition experiments that focused on bryophytes in tundra regions have involved non-comparable species and do not provide for any general conclusions. Temperate bryophyte species might have different sensitivities to nitrogen pollution than Arctic species (Bates 2000). Different responses by bryophyte species to nitrogen additions have also been confirmed from the high Arctic (Gordon *et al.* 2001) to sub-Arctic areas (Potter *et al.* 1995).

Responses to increased UV-B radiation and other environmental factors

UV-B radiation has increased in recent decades in both polar regions, with the greatest increase seen in the Antarctic. Effects of UV-B radiation on plant species and communities have been determined in two ways: by exclusion of ambient UV-B, particularly in the Antarctic and high Arctic Greenland (Bredahl *et al.* 2004) and by supplementation using UV-B emitting lamps (in the Swedish sub-Arctic and on high Arctic Svalbard; Johanson *et al.* 1995). These field methods produce more realistic radiation regimes than those in controlled growth chambers.

In general, mosses are more sensitive to UV-B radiation than other plant functional types (Gwynn-Jones *et al.* 1999b), but the impacts of enhanced UV-B on mosses are small, owing to their ability to produce UV-B absorbing pigments (Markham *et al.* 1990). Growth-enhancing effects of UV-B have even been recorded (Gehrke *et al.* 1996). During the peak growing season a 20% increase in UV-B over ambient reduced the incremental stem length of *Sphagnum fuscum* in a sub-Arctic mire (Gehrke *et al.* 1996; Sonesson *et al.* 2002) but did not affect its neighbor *Dicranum elongatum* (Sonesson *et al.* 2002). UV-B decreased the growth of *Hylocomium splendens*, but when hydration was increased UV-B stimulated its growth (Gehrke *et al.* 1996); enhanced CO_2 also increased its growth (Gwynn-Jones *et al.* 1996). Increased UV-B also decreased the height growth of *Polytrichum commune* (Gehrke 1999). However, on high Arctic Svalbard, *Aulacomnium turgidum* and *Polytrichum hyperboreum* were unaffected by enhanced UV-B (Gwynn-Jones *et al.* 1999a).

The ecosystem function of mosses can be significantly affected by UV-B, even if the effect is species-specific. On Svalbard in the high Arctic, supplemental UV-B radiation decreased the nitrogen-fixing potential of cyanobacteria associated with the moss *Sanionia uncinata* in a wet year whereas in a dry year, when the cyanobacteria were dormant, there was a smaller effect (Solheim *et al.* 2002). Nitrogen fixation by cyanobacteria hosted by *Hylocomium splendens* was not affected by UV-B. As nitrogen is a key limiting element in tundra ecosystems, any reduction in N-fixation caused by UV-B effects on moss-hosted cyanobacteria can have important ecosystem-wide impacts.

Snowbed communities

Snowbeds develop in areas that accumulate large amounts of winter snow (Gjaerevoll 1956); they are a prominent component throughout the tundra biome. This is particularly the case in alpine areas owing to the rugged topography and wind redistribution of snow. As there are species, and even communities, that are restricted to the snowbed habitat, snowbeds comprise a unique component of the Arctic and alpine biodiversity at multiple scales from species to landscape. In connection with future global warming simulations, the snowbed ecosystems of alpine Europe are described as particularly vulnerable in the IPCC's 2001 assessment report. Kullman (2005) also reported the complete disappearance of a snowbed plant community within only 30 years in the southernmost part of the Swedish Scandes. Snow cover and its duration limit vascular plant distribution and production (Wijk 1986); thus snowbeds have sparse vegetation and snowbed bryophytes have less competition from higher plants than in surrounding plant communities. Indeed, snowbeds contain many bryophyte species and can sustain high diversity; for instance, over 80 species of bryophytes have been found growing in snowbeds in the Latnjajaure catchment (Björk 2007). Of these, 26 species preferentially grow in snow-rich habitats and 13 specifically in snowbeds (Table 11.1).

In some snowbeds more than 40% of the total cover consists of bryophytes. There is an internal snow melt gradient within a snowbed community in that bryophyte cover may range from 3% in the early-melting part to over 80% cover in

Table 11.1. *Total number of bryophytes, boreal-montane bryophytes, Arctic–alpine bryophytes, chionophytic (snow-bound) bryophytes, and snowbed species (bryophytes preferentially growing in snowbed communities) found near Latnjajaure Field Station, northern Sweden*

	Boreal–montane		Arctic–alpine		Chionophytes[a]		Snowbed species		
Total no.	no.	% of total	no.	% of total	no.	% of total	no.	% of total	% of tundra
83	17	20	25	30	26	31	13	16	52

[a] 18 of the 26 chionophytic bryophytes were Arctic–alpine species.
Based on Damsholt (2002) and Hallingbäck (2008).

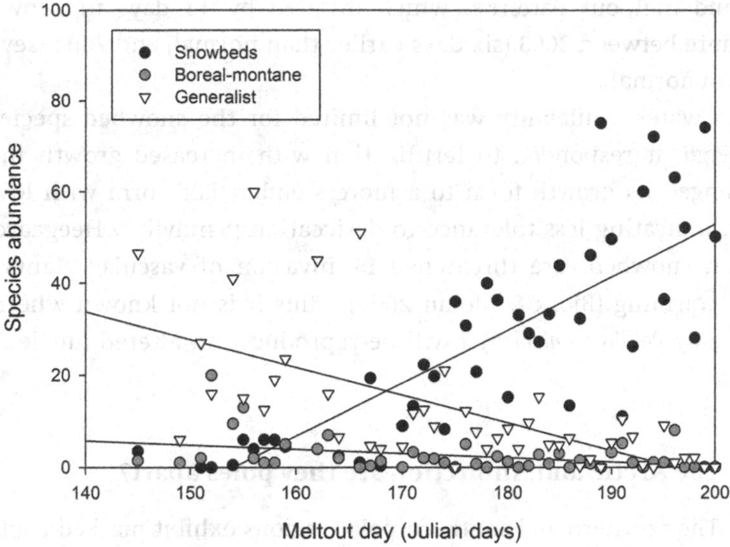

Fig. 11.5. Mean species abundance of snowbed bryophytes (black circles, $r^2 = 0.245$, $p < 0.001$), boreal–montane bryophytes (grey circles, $r^2 = 0.027$, $p = 0.049$), and generalists (white triangles, $r^2 = 0.216$, $p < 0.001$) along the internal snowmelt sequence in moderate melting snowbeds at Latnjajaure, northern Sweden. Mean values are melt out day for all species in a specific category with one value per day for each category. Julian day no. 152 = 1 June and 182 = 1 July. Unpublished data from R. G. Björk.

the late-melting part. In contrast to snowbed bryophytes, both boreal–montane bryophytes and bryophytes that have a widespread distribution (generalists) decrease with longer snow cover (Fig. 11.5). Even though Arctic–alpine bryophytes and chionophytic bryophytes (snowbed specialists excluded) make up a large proportion of the total species pool in snowbeds, they are not the most abundant species.

The large snowpack in snowbeds also has a fertilization effect on the underlying plants (Björk & Molau 2007). Woolgrove and Woodin (1996) have further shown a positive relationship between tissue nitrogen content and the duration of the snowpack in the snowbed bryophyte *Kiaeria starkei*.

In a fertilization experiment in snowbed communities, where a more realistic N deposition level ($1.0\,\mathrm{g\,N\,m^{-2}\,yr^{-1}}$) was used than in many previous studies, Björk (2007) did not find any effect on the bryophytes after three years. This suggests that an increase in N deposition will have a much smaller effect on the bryophytes than changes in the snow regime because snowbed bryophytes are already subjected to high N loads from the snowpack. However, the changes probably need to be larger than the interannual variability in snow cover and melt-out patterns, which differed by 13 days in snowbeds at Latnjajaure between 2003 (six days earlier than normal) and 2004 (seven days later than normal).

When water availability was not limited for the snowbed species *Pohlia wahlenbergii*, it responded to fertilization with increased growth in length and changed its growth form to a more slender 'lax' form with lower leaf density, indicating less tolerance to desiccation (Sandvik & Heegaard 2003). However, snowbeds are threatened by invasion of vascular plants due to climate warming (Björk & Molau 2007). Thus it is not known whether this response by *Pohlia wahlenbergii* will be reproduced in altered nutrient–water regimes.

The Arctic and Antarctic – are they poles apart?

The northern and southern polar regions exhibit marked contrasts in topography, climate, and biota. Significant parts of the North American and Eurasian continents, Greenland, and a number of smaller islands encircle a largely ice-covered Arctic Ocean. The northern lands consist of extensive plains and plateaus with folded mountain ranges reaching an elevation of 2800 m in the Alaskan Brooks Range. Greenland bears an extensive ice cap but elsewhere glaciation is localized.

Plate tectonic movement has relocated the northern continents but they have probably never been as widely separated as the southern hemisphere continental land masses. The Beringian land bridge between northeastern Asia and Alaska united vast continental land masses in the Pleistocene and possibly early Holocene period. The existence of ice-free areas during the Pleistocene glacial maximum around 100 000 years BP and the proximity of ice-free continental land masses to the south have led to a high degree of commonality in the Arctic biota.

The southern polar region provides a marked contrast to the Arctic. The Antarctic continent is a vast expanse of ice-covered land, about 14 million square kilometers in area, centered on the South Pole and separated from the southern continental land masses of South America, South Africa, Australia, and New Zealand by large distances of open ocean. The Antarctic ice cap is up to 4 km thick with surface elevations up to 3800 m. The highest mountain peak is 5140 m. The continental biota is confined largely to the coastal and isolated inland ice-free areas, amounting to only about 1% of the total area of Antarctica. Isolation of the flora and fauna and the extreme climate – cold, dry, windy – has resulted in a depauperate terrestrial biota compared with the Arctic. The flora is almost entirely cryptogam-dominated, with two native vascular plants being confined to the Antarctic Peninsula. However, despite the disparate physical nature of the environment and the biota, are the Arctic and Antarctic regions really poles apart with respect to the biotic adaptations to the environment?

The pronounced seasonality of the polar regions has a marked influence on temperature, incident solar radiation, and moisture availability. Over the spring and summer months terrestrial organisms may be subject to diurnal freeze–thaw events, increasing the levels of physiological stress. The indigenous biota is both physiologically and biochemically adapted to these environmental extremes, but how resilient are these organisms to an accelerating amelioration of global climate?

Throughout much of the winter in both the Arctic and Antarctic the terrestrial flora is insulated from the extreme cold by a thick snow cover. In early spring, as ambient temperatures are rising, there is an accompanying rapid increase in the levels of incident solar radiation. Physiologically, perhaps the most demanding time is early spring, when the temperatures rise, snow cover disappears and free water becomes available. In the Antarctic, and to a lesser extent the Arctic, rehydration and exposure coincide with a rapid increase in the levels of incident UV radiation due to destruction of ozone in the stratosphere. Terrestrial organisms are then subjected to rapidly increasing levels of photosynthetically active radiation (PAR) and short-wave UV, together with considerable diurnal variations in temperature.

Terrestrial plant life has evolved with, and has developed a variety of mechanisms to mitigate, the effects of high solar radiation, especially the production of flavonoid compounds (Rozema *et al.* 2002), which absorb strongly in the UV spectral range and can be used to back-project UV-B levels before the period being studied (Markham *et al.* 1990).

Physiologically, Antarctic cryptogams show little effect of increased ambient CO_2 levels as levels of CO_2 in the dense moss turf and cushions may reach well in excess of 2000 ppm (Tarnawski *et al.* 1992) and 1143 ppm in the understory of sub-Arctic birch forest (Sonesson *et al.* 1992). In response to increased levels of solar radiation, Antarctic mosses show a rapid increase in levels of flavonoids, which provide photoprotection from UV-B (Caldwell *et al.* 2007; Ryan *et al.* 2009).

Green *et al.* (2000, 2005) demonstrated the existence of both sun and shade forms of three common mosses in Antarctica – *Bryum argenteum, B. pseudotriquetrum*, and *Ceratodon purpureus* – and showed that, in response to exposure to increased UV-A levels, the shade forms reached the same levels of photoprotection as the sun forms in as little as six days.

Lichens and mosses possess photosynthetic attributes that are well adapted to the rigors of both the present and the experimentally altered Antarctic environment. Algae and cyanobacteria have similar adaptive attributes (Hawes *et al.* 1992) and some motile cyanobacteria are able to alter their position in an apparent avoidance strategy (Quesada & Vincent 1997).

At least in continental Antarctic localities, water and nutrients may be the primary limiting factors governing plant distribution patterns (Hovenden & Seppelt 1995; Melick & Seppelt 1997; Leishman & Wild 2001; Wasley *et al.* 2006). In the wetter Maritime Antarctic, nutrients do not appear to be limiting (Day *et al.* 2008), owing to enrichment of coastal areas from wind-borne and faunal transport of marine-based nutrients.

It would appear that the greatest impacts of global or regional climate change effects will be seen in the wetter and milder Maritime Antarctic, where regional warming has significantly altered the presence, abundance, and distribution of the two vascular plant species, the grass *Deschampsia antarctica* and the cushion-forming *Colobanthus quitensis* (Smith 1994). In the drier and climatically more severe continental Antarctic, water availability and nutrients are the primary determinants of plant distribution patterns (Melick & Seppelt 1997; Wasley *et al.* 2006).

The vegetation of continental Antarctica is being influenced by, in particular, changes in moisture availability that will have a considerable influence on the distribution and abundance of the dominant lichen and moss vegetation (see Seppelt, Chapter 13, this volume). The terrestrial vegetation is physiologically resilient and even repeated freeze–thaw cycling seems to have little effect on the plants (Melick & Seppelt 1992). However, transient events may have significant follow-up implications for the terrestrial ecosystem. In contrast to Arctic areas, the cryptogamic flora of the Antarctic is likely to be particularly sensitive to future arrivals of invasive species that are increasingly reaching the Antarctic.

Conclusions

The destiny of tundra bryophytes

All factors of climate change will act together in changing ecosystems. For non-Antarctic tundra bryophytes, impacts of a changed vegetation composition are probably most important in the longer term because many species in the tundra biome currently experience little or no competition from surrounding plants, particularly in the polar deserts and semi-deserts. When the tundra becomes shrub-like and forested, ecosystems that are now highly dominated by bryophytes will most likely relocate over a longer timescale; high Arctic ecosystems will colonize forefields of retreating glaciers, whereas tundra communities will spread into the polar deserts. In the IPCC's Third Assessment Report, three ecosystems or groups of ecosystems in Europe's mountain regions were highlighted as particularly sensitive to climate change:

mid-alpine snowbeds, high-alpine fellfield vegetation, and cryosoil communities, e.g., patterned ground and tussock tundra (IPCC 2001, p. 661).

The Arctic bryophyte flora consists of many boreal species (perhaps > 40%) and many species have a wide circumpolar distribution. However, there are also many tundra areas with high bryophyte diversity and with many endemics, for example in Svalbard, Iceland, and Arctic Alaska (Hallingbäck & Hodgetts 2000). Arctic mosses and liverworts make up 6.6% of the world's total biota (Matveyeva & Chernov 2000). Mountains are generally high in biodiversity (Körner 2004), and many alpine ecosystems have high fractions of bryophytes (Geissler 1982). Mountains, however, have not been thoroughly investigated for global bryophyte species richness.

Boreal and temperate bryophyte species with high growth rates, high temperature optima for photosynthesis and growth rates, and tolerance of shaded conditions, might benefit from climate change since both precipitation and temperature will increase in some parts of the tundra. If these bryophyte species are currently growing in Arctic and alpine areas, they might increase in dominance together with shrubs.

Effects of climate change on genetic diversity and the distribution of genotypes in these areas are difficult to evaluate. Very few studies have investigated genetic diversity of tundra bryophytes and possible effects of climate change. The temperate species *Hylocomium splendens* has been shown to have different genetic variants in the alpine and lowland areas in Sweden (Cronberg 2004), but it is not known whether the two variants may have different ecophysiological responses to climate change. The same species has different forms (monopodial and sympodial) associated with habitats in the high altitude/high Arctic and sub-Arctic. When species from the colder environments experience a warmer environment, they shift their growth form to resemble that of the native populations but a genetic separation still remains (Ross *et al.* 2001). Further, mechanical adaptations such as strength of the stem are present in the sub-Arctic populations and this allows plants to grow vertically and to withstand more competition than the weaker horizontally growing plants of the high Arctic and higher altitudes. However, in general it is unknown whether there are geographical differences in bryophyte responses to climate change since no results have been published on *in situ* experiments conducted across several sites that include the same species.

Similarly, there are few studies using global climate change models that focus on bryophytes. There is a great need to include ecologically important factors such as cover, biogeophysical properties and complex responses to temperature and water availability in such models. Furthermore, extreme events and environmental stress will become more frequent in the changing climate; for example, drought, intense precipitation, and winter warming events, but bryophyte responses to such extreme weather events have been insufficiently investigated as yet, either *ex situ* or *in situ*.

To conclude, many tundra habitats in the non-Antarctic regions will decrease in extent or relocate due to climate-change-induced shrub and forest expansion, directly decreasing bryophyte cover, diversity and abundance of species and ecotypes that are adapted to the tundra environment or have low potential for acclimatization to the changing conditions. As the biodiversity of bryophytes in the Arctic is high, climate change will most probably have significant effects on the global diversity of bryophytes.

Acknowledgements

AKJ gratefully acknowledges financial support from the Japan Society for Promotion of Science (PE06524 and P07727).

References

ACIA (2005). *Impacts of a Warming Arctic: Arctic Climate Impact Assessment*. Cambridge: Cambridge University Press.

Anderson, J. M. (1991). The effects of climate change on decomposition processes in grassland and coniferous forests. *Ecological Applications* **1**: 326–47.

Arft, A. M., Walker, M. D., Gurevitch, J. *et al.* (1999). Responses of tundra plants to experimental warming: meta-analysis of the International Tundra Experiment. *Ecological Monographs* **69**: 491–511.

Atkin, O. K., Scheurwater, I. & Pons, T. L. (2006). High thermal acclimation potential of both photosynthesis and respiration in two lowland *Plantago* species in contrast to an alpine congeneric. *Global Change Biology* **12**: 500–15.

Bates, J. W. (2000). Mineral nutrition, substratum ecology, and pollution. In *Bryophyte Biology*, ed. A. J. Shaw & B. Goffinet, pp. 248–311. Cambridge & New York: Cambridge University Press.

Björk, R. G. (2007). *Snowbed Biocomplexity: A Journey from Community to Landscape*. Göteborg: Göteborg University.

Björk, R. G. & Molau, U. (2007). Ecology of alpine snowbeds and the impact of global change. *Arctic, Antarctic and Alpine Research* **39**: 34–43.

Bokhorst, S., Bjerke, J. W., Bowles, F. W. *et al.* (2008). Impacts of extreme winter warming in the sub-Arctic: growing season responses of dwarf-shrub heath land. *Global Change Biology* **14**: 1–10.

Bonan, G. B. & van Cleve, K. (1992). Soil-temperature, nitrogen mineralisation, and carbon source sink relationships in boreal forests. *Canadian Journal of Forest Research* **22**: 629–39.

Bredahl, L., Ro-Poulsen, H. & Mikkelsen, T. N. (2004). Reduction of the ambient UV-B radiation in the high-Arctic increases Fv/Fm in *Salix arctica* and *Vaccinium uliginosum* and reduces stomatal conductance and internal CO_2 concentration in *Salix arctica*. *Arctic, Antarctic and Alpine Research* **36**: 364–9.

Brooker, R. W., Maestre, F. T., Callaway, R. M. *et al.* (2008). Facilitation in plant communities: the past, the present, and the future. *Journal of Ecology* **96**: 18–34.

Caldwell, M. M., Bornman, J. F., Ballare, C. L., Flint, S. D. & Kulandaivelu, G. (2007). Terrestrial ecosystems, increased solar ultraviolet radiation, and interactions with other climate change factors. *Photochemical and Photobiological Sciences* **6**: 252–66.

Callaghan, T. V., Björn, L. O., Chernov, Y. *et al.* (2005). Tundra and Polar Desert Ecosystems. In *ACIA (Arctic Climate Impacts Assessment)*, pp. 243–352. Cambridge: Cambridge University Press.

Callaghan, T. V., Carlsson, B. A., Sonesson, M. & Temesvary, A. (1997). Between-year variation in climate-related growth of circumarctic populations of the moss *Hylocomium splendens*. *Functional Ecology* **11**: 157–65.

Callaghan, T. V., Collins, N. J. & Callaghan, C. H. (1978). Strategies of growth and population dynamics of tundra plants 4. Photosynthesis, growth and reproduction of *Hylocomium splendens* and *Polytrichum commune* in Swedish Lapland. *Oikos* **31**: 73–88.

Chapin, F. S., III, Shaver, G. R., Giblin, A. E., Nadelhoffer, K. J. & Laundre, J. A. (1995). Response of arctic tundra to experimental and observed changes in climate. *Ecology* **76**: 694–711.

Chernov, Y. I. (1985). *The Living Tundra*. Cambridge: Cambridge University Press.

Cornelissen, J. H. C., Lang, S. I., Soudzilovskaia, N. A. & During, H. J. (2007). Comparative cryptogam ecology: a review of bryophyte and lichen traits that drive biogeochemistry. *Annals of Botany (London)* **99**: 987–1001.

Cronberg, N. (2004). Genetic differentiation between populations of the moss *Hylocomium splendens* (Hedw.) Schimp. from low versus high elevation in the Scandinavian mountain range. *Lindbergia* **29**: 64–72.

Damsholt, K. (2002). *Illustrated Flora of Nordic Liverworts and Hornworts*. Lund: Nordic Bryological Society.

Davey, M. C. & Rothery, P. (1997). Interspecific variation in respiratory and photosynthetic parameters in Antarctic bryophytes. *New Phytologist* **137**: 231–40.

Day, T. A., Ruhland, C. T. & Xiong, F. S. (2008). Warming increases aboveground plant biomass and C stocks in vascular-plant-dominated Antarctic tundra. *Global Change Biology* **14**: 1827–43.

Dorrepaal, E., Aerts, R., Cornelissen, J. H. C., Callaghan, T. V. & van Logtestijn, R. S. P. (2003). Summer warming and increased winter snow cover affect *Sphagnum fuscum* growth, structure and production in a sub-arctic bog. *Global Change Biology* **10**: 93–104.

Dorrepaal, E., Aerts, R., Cornelissen, J. H. C., van Logtestijn, R. S. P. & Callaghan, T. V. (2006). *Sphagnum* modifies climate-change impacts on subarctic vascular bog plants. *Functional Ecology* **20**: 31–41.

Dorrepaal, E., Cornelissen, J. H. C., Aerts, R., Wallen, B. & Van Logtestijn, R. S. P. (2005). Are growth forms consistent predictors of leaf litter quality and decomposability across peatlands along a latitudinal gradient? *Journal of Ecology* **93**: 817–28.

EEA (2004). *Impacts of Europe's Changing Climate, an Indicator-Based Assessment*. Copenhagen: European Environmental Agency.

Epstein, H. E., Calef, M. P., Walker, M. D., Chapin, F. S. I. & Starfield, A. M. (2004). Detecting changes in arctic tundra plant communities in response to warming over decadal time scales. *Global Change Biology* **10**: 1325–34.

Furness, S. B. & Grime, J. P. (1982). Growth rate and temperature responses in bryophytes. II. A comparative study of species of contrasted ecology. *Journal of Ecology* **70**: 525–36.

Gehrke, C. (1999). Impacts of enhanced ultraviolet-B radiation on mosses in a subarctic heath ecosystem. *Ecology* **80**: 1844–51.

Gehrke, C., Johanson, U., Gwynn-Jones, D. *et al.* (1996). Effects of enhanced ultraviolet-B radiation on terrestrial subarctic ecosystems and implications for interactions with increased atmospheric CO_2. In *Plant Ecology in the sub-Arctic Swedish Lapland*, ed. P. S. Karlsson & T. V. Callaghan. *Ecological Bulletins* **45**: 192–203.

Geissler, P. (1982). Alpine communities. In *Bryophyte Ecology*, ed. A. J. E. Smith, pp. 167–89. London: Chapman and Hall.

Gignac, L. D. (2001). Bryophytes as indicators of climate change. *Bryologist* **104**: 410–20.

Gjaerevoll, O. (1956). The plant communities of the Scandinavian alpine snowbeds. *Det Kongeliga Norske Videnskabernas Selskabs Skrifter* **1**: 1–405.

Glime, J. M. (2007). *Bryophyte Ecology*. Ebook sponsored by Michigan Technological University and the International Association of Bryologists. http://www.bryoecol.mtu.edu

Gordon, C., Wynn, J. M. & Woodin, S. J. (2001). Impacts of increased nitrogen supply on high Arctic heath: the importance of bryophytes and phosphorus availability. *New Phytologist* **149**: 461–71.

Graglia, E., Jonasson, S., Michelsen, A. *et al.* (2001). Effects of environmental perturbations on abundance of subarctic plants after three, seven and ten years of treatments. *Ecography* **24**: 5–12.

Green, T. G. A., Kulle, D., Pannewitz, S., Sancho, L. G. & Schroeter, B. (2005). UV-A protection in mosses growing in continental Antarctica. *Polar Biology* **28**: 822–7.

Green, T. G. A., Schroeter, B. & Seppelt, R. (2000). Effect of temperature, light and ambient UV on the photosynthesis of the moss *Bryum argenteum* Hedw. in continental Antarctica. In *Antarctic Ecosystems*, ed. W. Darison, C. Howard-Williams & P. Broady, pp. 165–70. Christchurch: New Zealand Natural Sciences.

Gwynn-Jones, D., Johanson, U., Gehrke, C. *et al.* (1996). Effects of enhanced UV-B radiation and elevated concentrations of CO_2 in a sub-arctic heathland. In *Carbon Dioxide, Populations, and Communities*, ed. C. Körner & F. A. Bazzaz, pp. 197–207. San Diego, CA: Academic Press.

Gwynn-Jones, D., Johanson, U., Phoenix, G., *et al.* (1999a). UV-B impacts and interactions with other co-occurring variables of environmental change: an Arctic perspective. In *Stratospheric Ozone Depletion, UV-B Radiation and Terrestrial Ecosystems*, ed. J. Rozema, pp. 187–201. Amsterdam: Backhuys Press.

Gwynn-Jones, D., Lee, J. A., Johanson, U. *et al.* (1999b). The responses of plant functional types to enhanced UV-B. In *Stratospheric Ozone Depletion, UV-B Radiation and Terrestrial Ecosystems*, ed. J. Rozema, pp. 173–86. Amsterdam: Backhuys Press.

Hallingbäck, T. (2008). *Ekologisk Katalog över Mossor* (nätversionen). Uppsala: ArtDatabanken, SLU.

Hallingbäck, T. & Hodgetts, N. (2000). *Mosses, Liverworts and Hornworts. Status Survey and Conservation Action Plan for Bryophytes.* IUCN, Gland, Switzerland, and Cambridge, UK: IUCN/SSC Bryophyte Specialist Group.

Hawes, I., Howard-Williams, C. & Vincent, W. F. (1992). Desiccation and recovery of Antarctic cyanobacterial mats. *Polar Biology* **12**: 587–94.

Henry, G. H. R. & Molau, U. (1997). Tundra plants and climate change: the International Tundra Experiment – Introduction. *Global Change Biology* **3**: 1–9.

Hobbie, S. E. (1996). Temperature and plant species control over litter decomposition in Alaskan tundra. *Ecological Monographs* **66**: 503–22.

Hollister, R. D. & Webber, P. J. (2000). Biotic validation of small open-top chambers in a tundra ecosystem. *Global Change Biology* **6**: 835–42.

Hovenden, M. J. & Seppelt, R. D. (1995). Exposure and nutrients as delimiters of lichen communities in continental Antarctica. *Lichenologist* **27**: 505–16.

IPCC (2001). *Climate Change 2001: The Scientific Basis. Contribution of Working Group I to the Third Assessment Report of the Intergovernmental Panel on Climate Change*, ed. J. T. Houghton, Y. Ding, D. J. Griggs *et al.* Cambridge and New York: Cambridge University Press.

IPCC (2007). *Climate Change 2007. IPCC Fourth Assessment Report (AR4).* http://www.ipcc.ch. Cambridge: Cambridge University Press.

Jägerbrand, A. K. (2005). *Subarctic Bryophyte Ecology: Phenotypic Variation and Responses to Simulated Environmental Change.* Göteborg: Göteborg University.

Jägerbrand, A. K. (2007). Effects of an *in situ* temperature increase on the short-term growth of arctic-alpine bryophytes. *Lindbergia* **32**: 82–7.

Jägerbrand, A. K., Alatalo, J. M., Chrimes, D. & Molau, U. (2009). Plant community responses to 5 years of simulated climate change in meadow and heath ecosystems at a subarctic-alpine site. *Oecologia* **161**: 601–10.

Jägerbrand, A. K., Lindblad, K. E. M., Björk, R. G., Alatalo, J. M. & Molau, U. (2006). Bryophyte and lichen diversity under simulated environmental change compared with observed variation in unmanipulated alpine tundra. *Biodiversity and Conservation* **15**: 4453–75.

Jägerbrand, A. K., Molau, U. & Alatalo, J. M. (2003). Responses of bryophytes to simulated environmental change at Latnjajaure, northern Sweden. *Journal of Bryology* **25**: 163–8.

Johanson, U., Gehrke, C., Bjorn, L. O. & Callaghan, T. V. (1995). The effects of enhanced UV-B radiation on the growth of dwarf shrubs in a subarctic heathland. *Functional Ecology* **9**: 713–19.

Jonasson, S., Havstrom, M., Jensen, M. & Callaghan, T. V. (1993). *In situ* mineralization of nitrogen and phosphorus of arctic soils after perturbations simulating climate change. *Oecologia (Heidelberg)* **95**: 179–86.

Jónsdóttir, I. S., Crittenden, P. & Jägerbrand, A. K. (1997). Measuring growth rate in bryophytes and lichens. *Summary document of 8th Annual ITEX Workshop. Royal Holloway Institute for Environmental Research, 19–22 April*, pp. 10–15.

Kallio, P. & Saarnio, E. (1986). The effect on mosses of transplantation to different altitudes. *Journal of Bryology* **14**: 159–78.

Kane, D. L., Hinzman, L. D., Woo, M. & Everett, K. R. (1992). Arctic hydrology and climate change. In *Arctic Ecosystems in a Changing Climate: an Ecophysiological Perspective*, ed. F. S. I. Chapin, R. L. Jefferies, J. F. Reynolds, G. R. Shaver & J. Svoboda, pp. 35–57. San Diego, CA: Academic Press.

Kaplan, J. O. & New, M. (2006). Arctic climate change with a 2 °C global warming: timing, climate patterns and vegetation change. *Climatic Change* **79**: 213–41.

Kennedy, A. D. (1995). Simulated climate change: are passive greenhouses a valid microcosm for testing biological effects of environmental perturbations? *Global Change Biology* **1**: 29–42.

Keyser, A. R., Kimball, J. S., Nemani, R. R. & Running, S. W. (2000). Simulating the effects of climate change on the carbon balance of North American high-latitude forests. *Global Change Biology* **6**: 185–95.

Klanderud, K. & Totland, Ø. (2005). Simulated climate change altered dominance hierarchies and diversity of an alpine biodiversity hotspot. *Ecology* **86**: 2047–54.

Körner, C. (2004). Mountain biodiversity, its causes and function. *Ambio* **13**: 11–17.

Kullman, L. (2005). Gamla och nya träd på Fulufjället – vegetationshistoria på hög nivå. *Svensk Botanisk Tidskrift* **99**: 315–29.

Leishman, M. R. & Wild, C. (2001). Vegetation abundance and diversity in relation to soil nutrients and soil water content in the Vestfold Hills, East Antarctica. *Antarctic Science* **13**: 126–34.

Longton, R. E. (1979). Climatic adaptation of bryophytes in relation to systematics. In *Bryophyte Systematics*, ed. G. C. S. Clarke & J. G. Duckett, pp. 511–31. Systematics Association Special Volume No. 14. London: Academic Press.

Longton, R. E. (1982). Bryophyte vegetation in polar regions. In *Bryophyte Ecology*, ed. A. J. E. Smith, pp. 123–65. London: Chapman and Hall.

Longton, R. E. (1984). The role of bryophytes in terrestrial ecosystems. *Journal of the Hattori Botanical Laboratory* **55**: 147–63.

Longton, R. E. (1988). *The Biology of Polar Bryophytes and Lichens*. Avon: Cambridge University Press.

Longton, R. E. (1997). The role of bryophytes and lichens in polar ecosystems. In *Ecology of Arctic Environments*, ed. S. J. Woodin & M. Marquiss, pp. 69–96. Oxford: Blackwell Science.

Marion, G. M., Henry, G. H. R., Freckman, D. W. *et al.* (1997). Open-top designs for manipulating field temperature in the High Arctic, Alaskan Arctic and Swedish Subarctic. *Global Change Biology* **3**: 20–32.

Markham, K. R., Franke, A., Given, D. R. & Brownsey, P. (1990). Historical Antarctic ozone trends from herbarium specimen flavonoids. *Bulletin de Liaisan – Group Polyphenols* **15**: 230–5.

Matveyeva, N. & Chernov, Y. (2000). Biodiversity of terrestrial ecosystems. In *The Arctic: Environment, People, Policy*, ed. M. Nuttal & T. V. Callaghan, pp. 233–73. Reading: Harwood Academic Publishers.

Maxwell, B. (1992). Arctic climate: potential for change under global warming. In *Arctic Ecosystems in a Changing Climate. An Ecophysiological Perspective*, ed. F. S. I. Chapin, R. L. Jefferies, J. F. Reynolds, G. R. Shaver & J. Svoboda, pp. 11–34. San Diego, CA: Academic Press.

McCarthy, J. J., Canziani, O. F., Leary, N. A., Dokken, D. J. & White, K. S. (eds) (2001). *Climate Change 2001: Impacts, Adaptation and Vulnerability. Contribution of Working Group II to the Third Assessment Report of the Intergovernmental Panel on Climate Change, IPCC*. Cambridge: Cambridge University Press.

Melick, D. R. & Seppelt, R. D. (1992). Loss of soluble carbohydrates and changes in freezing point of Antarctic bryophytes after leaching and repeated freeze-thaw cycles. *Antarctic Science* **4**: 399–404.

Melick, D. R. & Seppelt, R. D. (1997). Vegetation patterns in relation to climatic and endogenous changes in Wilkes Land, continental Antarctica. *Journal of Ecology* **85**: 43–56.

Miller, P. C., Webber, P. J., Oechel, W. C. & Tieszen, L. L. (1980). Biophysical processes and primary production. In *An Arctic Ecosystem: the Coastal Tundra at Barrow, Alaska*, ed. M. P. Brown Jr, L. L. Tieszen & F. L. Bunnell, pp. 66–101. Stroudsburg: Dowden, Hutchinson and Ross.

Mitchell, J. F. B., Manabe, S., Meleshko, V. & Tokioka, T. (1990). Equilibrium climate change – and its implications for the future. In *Climate Change: the IPCC Scientific Assessment*, ed. J. T. Houghton, G. J. Jenkins & J. J. Ephraums, pp. 131–72. Cambridge: Cambridge University Press.

Molau, U. (2001). Tundra plant responses to experimental and natural temperature changes. *Memoirs of the National Institute for Polar Research, Special issue* **54**: 445–66.

Molau, U. & Alatalo, J. M. (1998). Responses of subarctic-alpine plant communities to simulated environmental change: biodiversity of bryophytes, lichens, and vascular plants. *Ambio* **27**: 322–9.

Myneni, R. B., Keeling, C. D., Tucker, C. J., Asrar, G. & Nemani, R. R. (1997). Increased plant growth in the northern high latitudes from 1981 to 1991. *Nature (London)* **386**: 698–702.

Nadelhoffer, K. J., Giblin, A. E., Shaver, G. R. & Laundre, J. A. (1991). Effects of temperature and substrate quality on element mineralization in six arctic soils. *Ecology* **72**: 242–53.

Nadelhoffer, K. J., Giblin, A. E., Shaver, G. R. & Linkins, A. E. (1992). Microbial processes and plant nutrient availability in arctic soils. In *Arctic Ecosystems in a Changing Climate: an Ecophysiological Perspective*, ed. F. S. I. Chapin, R. L. Jefferies, J. F. Reynolds, G. R. Shaver & J. Svoboda, pp. 281–300. San Diego, CA: Academic Press.

Nemani, R. R., Keeling, C. D., Hashimoto, H. *et al.* (2003). Climate-driven increases in global terrestrial net primary production from 1982 to 1999. *Science (Washington, DC)* **300**: 1560–3.

Ng, E. & Miller, P. C. (1977). Validation of a model of the effect of tundra vegetation on soil temperatures. *Arctic and Alpine Research* **9**: 89–104.

Oechel, W. C. & Sveinbjörnsson, B. (1978). Primary production processes in arctic bryophytes at Barrow, Alaska. In *Vegetation and Production Ecology of an Alaskan*

Arctic Tundra, ed. L. L. Tieszen, pp. 269–98. *Ecological Studies* **29**. New York: Springer-Verlag.

Overpeck, J., Hughen, K. & Hardy, D. (1997). Arctic environmental changes of the last four centuries. *Science (Washington DC)* **278**: 1251–6.

Parsons, A. N., Press, M. C., Wookey, P. A. *et al.* (1995). Growth responses of *Calamagrostis lapponica* to simulated environmental change in the Sub-arctic. *Oikos* **72**: 61–6.

Potter, J. A., Press, M. C., Callaghan, T. V. & Lee, J. A. (1995). Growth responses of *Polytrichum commune* and *Hylocomium splendens* to simulated environmental change in the sub-arctic. *New Phytologist* **131**: 533–41.

Press, M. C., Callaghan, T. V. & Lee, J. A. (1998a). How will European arctic ecosystems respond to projected global environmental change? *Ambio* **27**: 306–11.

Press, M. C., Potter, J. A., Burke, M. J. W., Callaghan, T. V. & Lee, J. A. (1998b). Responses of a subarctic dwarf shrub heath community to simulated environmental change. *Journal of Ecology* **86**: 315–27.

Proctor, M. C. F. (2000). Physiological ecology. In *Bryophyte Biology*, ed. A. J. Shaw & B. Goffinet, pp. 225–47. New York: Cambridge University Press.

Quesada, A. & Vincent, W. F. (1997). Strategies of adaptation by Antarctic cyanobacteria to ultraviolet radiation. *European Journal of Phycology* **32**: 335–42.

Richter-Menge, J., Overland, J., Svoboda, M. *et al.* (2008). *Arctic Report Card 2008*. http://www.arctic.noaa.gov/reportcard.

Ross, S. E., Callaghan, T. V., Sheffield, E. & Sonesson, M. (2001). Variation and control of growth form in the moss *Hylocomium splendens*. *Journal of Bryology* **23**: 283–92.

Rosswall, T., Veum, A. & Kärenlampi, L. (1975). Plant litter decomposition at Fennoscandian tundra sites. In *Fennoscandian Tundra Ecosystems* vol. 1, *Plants and Microorganisms*, ed. F. E. Wiegolaski, pp. 268–77. Berlin: Springer.

Rozema, J., Björn, L. O., Bornman, J. F. *et al.* (2002). The role of UV-B radiation in aquatic and terrestrial ecosystems – an experimental and functional analysis of the evolution of UV-absorbing compounds. *Journal of Photochemistry and Photobiology B. Biology* **66**: 2–12.

Russell, S. (1990). Bryophyte production and decomposition in tundra ecosystems. *Botanical Journal of the Linnean Society* **104**: 3–22.

Ryan, K. G., Burne, A. & Seppelt, R. D. (2009). Historical ozone concentrations and flavonoid levels in herbarium specimens of the Antarctic moss *Bryum argenteum*. *Global Change Biology* **15**: 1694–1702.

Sage, R. F. & Kubien, D. S. (2007). The temperature response of C_3 and C_4 photosynthesis. *Plant, Cell & Environment* **30**: 1086–106.

Sandvik, S. M. & Heegaard, E. (2003). Effects of simulated environmental changes on growth and growth form in a late snowbed population of *Pohlia wahlenbergii* (Web et Mohr) André. *Arctic, Antarctic & Alpine Research* **35**: 341–8.

Schipperges, B. & Gehrke, C. (1996). Photosynthetic characteristics of subarctic mosses and lichens. *Ecological Bulletins* **45**: 121–6.

Serreze, M. C., Walsh, J. E., Chapin, F. S. III *et al.* (2000). Observational evidence of recent change in the northern high-latitude environment. *Climatic Change* **46**: 159–207.

Smith, R. I. L. (1994). Vascular plants as bioindicators of regional warming in Antarctica. *Oecologia* **99**: 322–8.

Solga, A. & Frahm, J. P. (2006). Nitrogen accumulation by six pleurocarpous moss species and their suitability for monitoring nitrogen deposition. *Journal of Bryology* **28**: 46–52.

Solheim, B., Johanson, U., Callaghan, T. V. *et al.* (2002). The nitrogen fixation potential of arctic cryptogam species is influenced by enhanced UV-B radiation. *Oecologia* **133**: 90–3.

Sonesson, M., Carlsson, B. A., Callaghan, T. V. *et al.* (2002). Growth of two peat-forming mosses in subarctic mires: species interactions and effects of simulated climate change. *Oikos* **99**: 151–60.

Sonesson, M., Gehrke, C. & Tjus, M. (1992). Carbon dioxide environment, microclimate and photosynthetic characteristics of the moss *Hylocomium splendens* in a subarctic habitat. *Oecologia (Heidelberg)* **92**: 23–9.

Sturm, M., Racine, C. & Tape, K. (2001). Increasing shrub abundance in the Arctic. *Nature (London)* **411**: 546–7.

Sundqvist, M. K., Björk, R. G. & Molau, U. (2008). Establishment of boreal forest species in alpine dwarf shrub heath in subarctic Sweden. *Plant Ecology & Diversity* **1**: 67–75.

Svensson, B. M. (1995). Competition between *Sphagnum fuscum* and *Drosera rotundifolia*: a case of ecosystems engineering. *Oikos* **74**: 205–12.

Tarnawski, M., Melick, D., Roser, D. *et al.* (1992). *In situ* carbon dioxide levels in cushion and turf forms of *Grimmia antarctici* at Casey Station, East Antarctica. *Journal of Bryology* **17**: 241–9.

Tenhunen, J. D., Lange, O. L., Hahn, S., Siegwolf, R. & Oberbauer, S. F. (1992). The ecosystem role of poikilohydric tundra plants. In *Arctic Ecosystems in a Changing Climate*, ed. F. S. I. Chapin, R. L. Jefferies, J. F. Reynolds, G. R. Shaver & J. Svoboda, pp. 213–56. San Diego, CA: Academic Press.

van Bogaert, R., Walker, D., Jia, G. J. *et al.* (2008). Recent Changes in Vegetation. In *The Arctic Report Card 2008*, ed. J. Richter-Menge, J. Overland, M. Svoboda *et al.* http://www.arctic.noaa.gov/reportcard.

Van Cleve, K., Oechel, W. C. & Hom, J. L. (1990). Response of black spruce (*Picea mariana*) ecosystems to soil temperature modification in interior Alaska. *Canadian Journal of Forest Research* **20**: 1530–5.

van der Wal, R. & Brooker, R. W. (2004). Mosses mediate grazer impacts on grass abundance in arctic ecosystems. *Functional Ecology* **18**: 77–86.

van der Wal, R., van Lieshout, S. & Loonen, M. (2001). Herbivore impact on moss depth, soil temperature and arctic plant growth. *Polar Biology* **24**: 29–32.

van Wijk, M. T., Clemmensen, K. E., Shaver, G. R. *et al.* (2003). Long-term ecosystem level experiments at Toolik Lake, Alaska, and at Abisko, Northern Sweden: generalizations and differences in ecosystem and plant type responses to global change. *Global Change Biology* **10**: 105–23.

Vitt, D. H. & Pakarinen, P. (1977). The bryophyte vegetation production and organic components of Truelove Lowland. In *Truelove Lowland, Canada: A High Arctic Ecosystem*, ed. L. C. Bliss, pp. 225–44. Edmonton: University of Alberta Press.

Walker, M. (1996). Community baseline measurements for ITEX studies. In *ITEX manual*, 2nd edn, ed. U. Molau & P. Mølgaard, pp. 39–41. Copenhagen: International Tundra Experiment.

Walker, M. D., Wahren, H. C., Hollister, R. D. *et al.* (2006). Plant community responses to experimental warming across tundra biome. *Proceedings of the National Academy of Sciences of the United States of America* **103**: 1342–6.

Wasley, J., Robinson, S. A., Lovelock, C. E. & Popp, M. (2006). Some like it wet – biological characteristics and underpinning tolerance of extreme water stress events in Antarctic bryophytes. *Functional Plant Ecology* **33**: 443–55.

Webber, P. J. (1978). Spatial and temporal variation of the vegetation and its production, Barrow, Alaska. In *Vegetation and Production Ecology of an Alaskan Arctic Tundra*, ed. L. L. Tieszen, pp. 37–112. Ecological Studies 29. New York: Springer-Verlag.

Wielgolaski, F. E., Bliss, L. C., Svoboda, J. & Doyle, G. (1981). Primary production of tundra. In *Tundra Ecosystems: a Comparative Analysis*, ed. L. C. Bliss, O. W. Heal & J. J. Moore, pp. 187–226. Cambridge: Cambridge University Press.

Wijk, S. (1986). Performance of *Salix herbacea* in an alpine snow-bed gradient. *Journal of Ecology* **74**: 675–84.

Woolgrove, C. E. & Woodin, S. J. (1996). Current and historical relationships between the tissue nitrogen content of a snowbed bryophyte and nitrogenous air pollution. *Environmental Pollution* **91**: 283–8.

12

Alpine Bryophytes as Indicators for Climate Change: a Case Study from the Austrian Alps

DANIELA HOHENWALLNER, HARALD GUSTAV ZECHMEISTER, DIETMAR MOSER, HARALD PAULI, MICHAEL GOTTFRIED, KARL REITER, AND GEORG GRABHERR

Introduction

The climate of Europe has changed in the past century. An increase in mean annual air temperature of +0.90°C could be observed between 1901 and 2005 (Jones & Moberg 2003). For the period 1977–2000, trends are even higher for Europe's mountain regions (Böhm *et al.* 2001). Beniston (2005) showed that for the alpine region minimum temperatures have increased up to 2 °C during the twentieth century, whereas the snow cover period has been reduced (IPCC 2007). The alpine and nival (uppermost altitudinal zone of the Alps above the closed alpine grassland) zones (e.g., Grabherr 1997) of high mountain ecosystems are considered to be particularly sensitive to warming (Diaz & Bradley 1997; Haeberli & Beniston 1998) as these ecosystems are determined by low temperature conditions. This life zone offers ideal conditions to study climate change effects because (1) direct human impact is very low, (2) its ecological systems are comparatively simple, at least in the upper elevation levels, and (3) its systems are dominated by abiotic, climate-related ecological factors. The importance of biotic factors such as competition decreases with altitude (Körner 1994; Callaway *et al.* 2002). Since high mountain plants have proven to respond sensitively to climate change (Grabherr *et al.* 1994, 2001), great efforts were made to establish the large-scale monitoring network GLORIA (Global Observation Research Initiative in Alpine environments) (Pauli *et al.* 2003). As input to this network, and based on the indicative value of bryophytes for climate change impacts, this group

Bryophyte Ecology and Climate Change, eds. Zoltán Tuba, Nancy G. Slack and Lloyd R. Stark. Published by Cambridge University Press.

of plants was investigated at Mt Schrankogel, a high peak of the central Eastern Austrian Alps.

The early responses of bryophytes to climate warming are due to their unique morphological, physiological, and thus functional characteristics (Bates & Farmer 1992; Grabherr *et al.* 2001; Zechmeister *et al.* 2003), as well as to their nearly worldwide distribution (Longton 1997).

The aim of this study is to detect the influence of snow cover duration on the distribution of bryophytes at the alpine–nival ecotone, and to identify ecological groups of bryophytes with a high indicator value for climate change effects.

Methods

Investigation area

Mount Schrankogel (3497 m a.s.l., Tyrol, Austria, 11°05′E, 47°02′N), one of the highest summits of the Stubaier Alps, has been the object of scientific work since 1994. This research has included the setup of permanent plot transects across the alpine–nival ecotone, studies on species–habitat relationships (Pauli *et al.* 1999, 2007), fine-scale modeling of species and community distribution (Gottfried *et al.* 1998), and predictive climate impact modeling (Gottfried *et al.* 1999, 2002). A temperature measurement net based on miniature data loggers, which has been maintained since 1997, has been of substantial interest for this study.

Mount Schrankogel is composed of siliceous bedrock, mostly gneiss and amphibolites rich in Ca and Mg. Its southern slope system offers ridges, steep slopes, and hollows, leading to a scattered mosaic of snow patches. On stable surfaces, closed alpine swards reach up to about 2850–3000 m. Further upward to approximately 3100–3200 m, the alpine grassland disintegrates into open pioneer swards. This transition zone is defined as the alpine–nival ecotone. Above 3200 m, the pioneer swards are replaced by scattered nival vegetation (3200–3497 m) (Pauli *et al.* 1999).

Sampling design

In total, 26 sampling quadrats (1 m × 1 m) positioned at different altitudes and aspects were selected, according to the availability of GLORIA records of explanatory biotic and abiotic variables. Fourteen such quadrats were established at an altitude of 2625–2635 m on the side moraine of Schwarzenbergferner, four between 3076 m and 3108 m on the south ridge, and eight between 2912 m and 3400 m along the SW slope of the mountain. A wooden grid frame (1 m × 1 m), subdivided into 100 cells of 0.1 m × 0.1 m, was mounted at the quadrats. From these 100 cells, five cells were selected

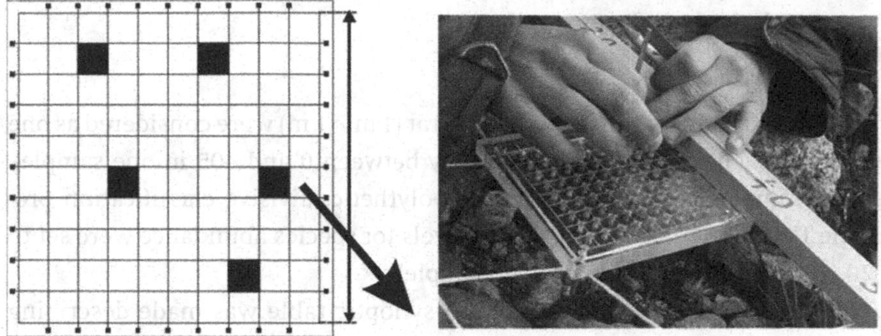

Fig. 12.1. Within a 1 m × 1 m wooden grid frame, five cells (size 0.01 m^2) were chosen randomly. By using a template with 81 holes, bryophytes were sampled.

randomly (Fig. 12.1). On each of these five cells, a perforated transparent fiberglass template (10 cm × 10 cm × 1.5 cm, L × W × D) was attached. This template, with 81 equally sized holes (distance = 0.01 m), was used to pinpoint the exact location of the surface by inserting a metal rod; bryophyte species were recorded on the sampling points (Fig. 12.1). The thickness of the template (0.015 m) guaranteed an exact point recording perpendicular to the grid frameplane. The total number of sampling points was 10 530. This sampling method was non-destructive (except for removing some individuals for species identification) and helped to minimize the disturbance on the long-term monitoring quadrats.

Temperature measurements

Based on previous investigations at Mt Schrankogel, two sets of temperature series were used. (1) From August 1999 to September 2000, miniature data loggers (Tinytag – RS, Gemini Dataloggers, Chichester, England, range –40 °C to +75 °C) measured temperature between 1 and 3 cm above ground in the center of the 1 m × 1 m quadrats on the SW slope and the south ridge (12 quadrats). The temperature sensors were shaded from direct insolation. (2) For the period from August 2000 to September 2001, soil temperature was recorded by using self-contained temperature sensors (StowAway Tidbits TBI32-20+50, Onset Corporation, Bourne, Massachusetts, USA, range –20 °C to +50 °C) buried 5 cm below the surface in the center of the sampling quadrats at the side moraine of the Schwarzenbergferner (14 quadrats). The sampling rate for both types of temperature sensors was set at 1.5 h.

Further, ground coverage (%) of the 1 m × 1 m quadrats for solid rock, scree, bare ground, and vascular plants, as well as altitude (m), were recorded. Nomenclature follows Grims (1999) for mosses, and Grolle and Long (2000) for liverworts.

Statistical analyses

Biotic predictors

The sampling points of each quadrat (1 m × 1 m) were considered as one sample (i.e., each species had a frequency between 0 and 405 in one sample). These 26 samples were ordered by the polythetic, divisive classification programme TWINSPAN (Hill 1979). Cutoff levels for species abundance were set to 10, 20, 40, and 60 (i.e., frequency per sample).

From the TWINSPAN classification, a synoptic table was made describing groups of bryophytes. From these groups only species that showed a distinct consistency concerning their ecological habitat preferences (Grims 1999; Paton 1999; Dierssen 2001; Nebel & Philipi 2001a, b) were selected for further analysis steps. The number of sampling points of each group of bryophytes was summed for each of the 26 quadrats. These values were used as the response variable in a Generalized Linear Model (GLM).

Abiotic predictors

The date of snowmelt was defined by the first incidence of a temperature above +2 °C in spring (Körner *et al.* 2003). To ascertain that no longer period of snow cover occurred after the date of this threshold, all temperature series were checked. Using an ordinal calendar date (value between 1 and 365) of snowmelt allowed the use of both temperature datasets (surface and subsurface), which would not have been possible for absolute temperature values (Fig. 12.2). Furthermore, the parameters scree, solid rock, bare ground, and altitude were used.

Data analysis with Generalized Linear Models (GLM)

Generalized Linear Models (McCullagh & Nelder 1989) were fitted to analyze the relationship between frequencies of the groups of bryophytes obtained from the TWINSPAN classification, and environmental factors, respectively, which were: date of snowmelt, altitude, and ground coverage of vascular plants, of scree, of solid rock, and of bare ground. Assuming a Poisson error distribution of the frequency of species, a logarithmic link function was used to develop the GLMs (Crawley 1993; Austin *et al.* 1996). As the relationships between a certain group of bryophytes and environmental factors were known to be frequently curvilinear (Currie 1991; Austin *et al.* 1996) the significance of quadratic and cubic functions of the predictor variables was tested with a χ^2 statistic. In a first step the univariate relationship between each environmental variable and the species frequencies was tested. Further, in a second step, combinations of the date of snowmelt (for several groups of species) and the altitude variable were tested, respectively, with other environmental variables.

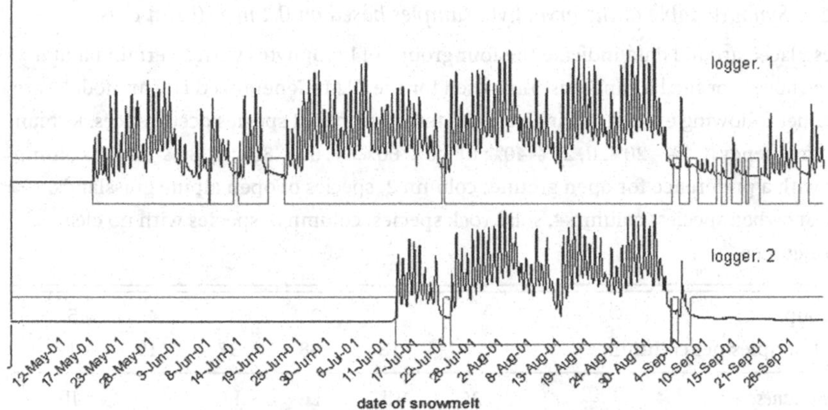

Fig. 12.2. Tidbit data logger temperature series (sections shown from a time series of the ordinal calendar day 122 until day 273) obtained from two sampling quadrats situated at the same altitude (2625 m) but on different aspects (logger 1 SSE, logger 2 NW). The grey line marks the periods of snow cover. Day–night oscillation stopped when the logger was covered by snow and started again with the day of snowmelt (which for logger 1 is day 137 and for logger 2, day 193).

Results

TWINSPAN classification and groups of bryophyte species

The TWINSPAN analysis divided the samples into five groups (Table 12.1). Columns 1, 2, 3, and 4 of the synoptic table clearly corresponded with ecological groups; column 5 did not. The first group consisted of subneutrophytic and basiphytic species growing on open ground mixed with scree in higher altitudes (e.g., *Hypnum revolutum*), whereas the second group comprised bryophytes of the open alpine grasslands like *Polytrichum juniperinum* or *Barbilophozia hatcheri*. Species of the third group were typical snowbed species such as *Polytrichum sexangulare* or *Anthelia julacea*. As a fourth group, cushion bryophytes on solid rock (e.g., *Grimmia incurva*, *Andreaea heinemannii*) were identified. The total number of bryophyte species was 48; single hits were 214 for solid rock species, 339 for open ground species, 642 for snowbed species, and 762 for species of open alpine grasslands.

Temperature measurements

Within all 26 sampling quadrats, the date of snowmelt varied between the ordinal date 121 and 193 (Table 12.2). Quadrats that were dominated by one of the four groups of bryophytes described above could be assigned to the

Table 12.1. *Synoptic table of the bryophyte samples based on 0.1 m × 0.1 m cells*

Rectangles placed around data indicate the four groups of bryophytes with a certain habitat preference chosen for further analysis. Taxa used for the GLM (Generalized Linear Models) are in bold; values following each taxon indicate the total number of species occurrences. Roman numerals, frequency (I, 1% – 20%; II, 20%–40%; III, 40%–60%; IV, 60%–80%; V, 80%–100%); column 1, species with a preference for open ground; column 2, species of open alpine grasslands; column 3, snowbed species; column 4, solid rock species; column 5, species with no clear habitat preference.

Species group:		1	2	3	4	5
Number of samples per column:		18	38	37	19	17
Racomitrium canescens	26	III	I	I	.	III
Bryum sp.	7	II	.	.	.	I
Brachythecium glareosum var. alpinum	7	II	.	.	.	.
Hypnum revolutum	6	II	I	.	.	.
Sanionia uncinata	8	II	I	.	.	.
Bryum stirtonii	1	I	.	.	.	.
Desmatodon latifolius	3	I	I	.	.	.
Didymodon asperifolius	2	I	.	.	.	.
Lophozia excisa	4	I	I	.	.	.
Schistidium papillosum	1	I	.	.	.	.
Syntrichia ruralis	1	I	.	.	.	.
Brachythecium sp.	1	I	.	.	.	.
Bartramia ithyphylla	4	I	.	I	.	.
Grimmia sessinata	6	I	.	I	.	.
Polytrichum alpinum	6	.	I	I	.	.
Dicranoweisia crispula	25	I	I	IV	.	.
Lophozia sp.	1	.	.	I	.	.
Lophozia wenzelii	24	I	I	III	.	I
Pohlia filum	16	.	.	III	.	.
Polytrichum sexangulare	19	.	I	III	.	I
Anthelia julacea	6	.	I	I	.	.
Brachythecium reflexum	2	.	.	I	.	.
Kiaeria starkei	6	.	.	I	.	.
Kurzia trichoclados	2	.	.	I	.	.
Lophozia opacifolia	1	.	.	I	.	.
Marsupella sp.	1	.	.	I	.	.
Moekeria blytii	1	.	.	I	.	.
Ditrichum heteromallum	7	.	I	I	.	.
Lophozia bicrenata	5	.	I	I	.	.
Polytrichum juniperinum	23	.	I	I	.	IV
Dicranum spadiceum	19	.	II	I	.	II

Table 12.1. (*cont.*)

Species group:		1	2	3	4	5
Number of samples per column:		18	38	37	19	17
Bryum capillare	1	.	.	.	.	I
Heterocladium dimorphum	3	.	I	.	.	.
Pohlia elongata var. *greenii*	1	.	I	.	.	.
Dicranum scoparium	7	.	I	I	.	I
Polytrichum piliferum	48	I	V	I	I	III
Barbilophozia hatcheri	15	.	II	.	.	I
Pohlia nutans	9	.	II	I	.	.
Pohlia sp.	20	I	II	I	.	.
Marsupella brevissima	2	.	I	.	.	.
Nardia scalaris	1	.	I	.	.	.
Paraleucobryum enerve	5	.	I	I	.	.
Pohlia cruda	1	.	I	.	.	.
Racomitrium lanuginosum	2	.	I	.	.	.
Andreaea rupestris var. rupestris	11	.	.	.	III	I
Grimmia incurva	17	.	I	.	V	.
Grimmia donniana	3	.	.	.	I	.
Grimmia sp.	2	.	.	.	I	.
Total number of species		18	27	25	5	11

Table 12.2. *Sampling quadrats in rank order according to an ordinal calendar date of snowmelt*

M01–M14: loggers installed at the side moraine of Schwarzenbergferner; S01–S04: loggers of the S ridge; SW01–SW08: loggers of the SW slope. Column 3 indicates the ordinal date of snowmelt. Column 4 shows the proportion of each group of bryophyte species for the species composition in %: SR, solid rock species; AG, alpine grassland species; OG, open ground species; SB, snowbed species.

Quadrat	Altitude (m)	Ordinal date	Ecological bryophyte species group
SW07	3200	121	SR 100%
SW08	3400	123	SR 100%
S02	3089	128	AG 91.9%; OG 8.1%
SW01	2917	131	AG 100%
SW03	2982	131	OG 100%
S04	3103	135	AG 83.6%; SR 16.4%
S03	3108	135	AG 89.7%; SR 5.7%; OG 2.3%; SB 2.3%
M01	2625	136	AG 89.7%; OG 10.3%
SW06	3270	137	SR 100%
M02	2625	137	AG 100%

Table 12.2. (*cont.*)

Quadrat	Altitude (m)	Ordinal date	Ecological bryophyte species group
M03	2625	148	AG 81.4%; SR 18.6%
M04	2625	148	AG 63.6%; SB 19.3%; SR 17.1%
M08	2625	149	AG 89.0%; SB 11.0%
SW05	3100	152	SR 66.0%; OG 32.1%; AG 1.9%
SW04	3000	153	OG 100%
M07	2625	159	OG 47.8%; AG 40.3%; SB 11.9%
M06	2625	174	AG 76.6%; SB 23.4%
M10	2625	178	SB 100%
M09	2625	181	SB 100%
M05	2625	185	SB 100%
M13	2625	190	SB 100%
M14	2625	191	SB 100%
M11	2625	193	SB 100%
M12	2625	193	SB 100%

following dates of snowmelt: (a) snowbed species became snow-free between days 178 and 193, (b) open ground species between days 131 and 153, (c) open alpine grassland species between days 128 and 174, and (d) solid rock species between days 121 and 152.

Environmental factors

The results of the univariate and multivariate regression for the four selected groups of bryophytes are summarized as follows.

Snowbed species: The second-order polynomial of the date of snowmelt ($snowm^2$) accounted for the highest reduction in deviance (89% explained variance). The two-variable models provided little additional explanation (e.g., date of $snowmelt^2$ + $bareground^2$; 93% explained variance).

Open ground species: The date of snowmelt showed the highest non-linear influence on this group, demonstrating a curvilinear response ($snowm^3$ = 63%). Among the two-variable models, the combination of $snowm^3$ with $scree^3$ (84%) and $vasplan^3$ (80%) was the best.

Open alpine grassland species: The results for open alpine grassland species showed a more balanced influence of the environmental variables. Vascular plant cover ($vasplant^3$) accounted for the highest reduction in deviance (34% explained variance); the date of snowmelt accounted for 31%. Together these two variables form the best two-variable model (68%).

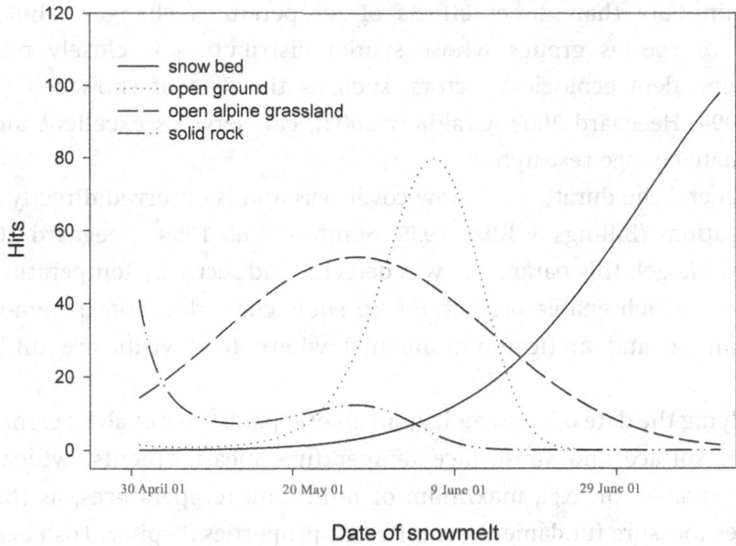

Fig. 12.3. Trendlines derived from the univariate GLM using the date of snowmelt as the predictor variable for the three bryophyte groups: snowbed, open ground, alpine grassland. Altitude was the predictor variable for solid rock species.

Solid rock species: Different from the other ecological groups, the date of snowmelt seemed to be of only secondary importance for explaining solid rock species frequencies. In contrast, the third polynomial of altitude showed the highest influence (alti = 65%) on this group, followed by vascular plant cover ($vasplan^3$ = 58%). Among the two-variable models, the combination of $altitude^3$ and $bareground^3$ accounted for the highest reduction in deviance (74% explained deviance).

Fig. 12.3 visualizes the relationship (trendlines derived for a GLM) between the date of snowmelt and the frequencies of the four groups of bryophytes: from approximately day 170, the abundance of snowbed species increases while the other groups disappear. With an earlier date of snowmelt, shifts in species composition (between open ground and alpine grassland species) can be observed.

Discussion

Models predict that every one degree Celsius temperature rise will push the snowline upwards by about 150 m in altitude (Beniston & Richard 2001). Körner *et al.* (2003) assumed that a change in the duration of snow cover and growing season length would have much more pronounced effects on the mountain flora than direct effects of temperatures changes. Thus, single species or species groups whose spatial distribution is closely related to snow-dependent ecological factors, such as the date of snowmelt (Stanton *et al.* 1994; Heegaard 2002; Grabherr 2003), can serve as excellent indicators for climate change research.

In general the duration of snow cover was and is observed directly by field investigations (Billings & Bliss 1959; Stanton *et al.* 1994; Heegaard 2002). At Mt Schrankogel, this parameter was detected indirectly by temperature measurements, which enable predictions on snow cover duration in remote locations (alpine and arctic environments) where field visits are difficult in winter.

Applying the date of snowmelt as an abiotic parameter enables comparisons between surface and subsurface temperature measurements, which would not be possible for, e.g., maximum or minimum temperatures, as these two variables measure fundamentally different properties (Pepin & Losleben 2002). It is not surprising that the diversity and coverage of snowbed bryophytes is strongly affected by the date of snowmelt (89% explained variance). Information derived from other parameters tested is less important. The close relationship of snowbed species (vascular plants as well as bryophytes) with their microhabitats is well documented (Billings & Bliss 1959; Kudo & Ito 1992; Woolgrove & Woodin 1994; Galen & Stanton 1995). Even within snowbeds, a pronounced snow gradient can be detected in bryophyte species composition and coverage: from areas experiencing long snow lie towards areas with short snow lie (Woolgrove & Woodin 1994). Factors closely related to the snow regime, such as substrate stability and the associated frost heave and moisture regimes (Rothero 1991), soil disturbance, and soil organic content (Stanton *et al.* 1994), also affect the establishment of snowbed communities.

The coverage of alpine grassland bryophytes is influenced by vascular plant cover and the date of snowmelt. Changes in bryophyte species composition seem to be mainly controlled by competition with vascular plants and changes in the snow regime. Therefore, their indicative value is limited.

Solid rock species are characteristic of stable, rocky substrates and are therefore abundant in the alpine–nival ecotone, where ridges bare of closed grassland are a predominant habitat type. This explains the correlation of abundance of "solid rock bryophytes" with altitude. Nevertheless, this does not mean that snow cover duration is longer, because these habitats are often very exposed to wind and thus are snow-free early. A clear prediction on the applicability for climate change research cannot be given because the direct influences of temperature have not yet been analyzed.

Figure 12.3 shows species distribution in relation to the snow-covered period, and thus indicates the potential impacts of a longer growing season on bryophyte diversity patterns. The coverage of snowbed species might decrease rapidly. This is underscored by studies in which species, vascular plants as well as bryophytes, responded rapidly to experimental removal of snow (Walker *et al.* 1993; Galen & Stanton 1995). The frequency of open ground species might increase (if appropriate substrate is available) and could be replaced in a next step by alpine grassland species.

Although most studies predict a decrease of the extent of snow cover, this dataset can also be used to discuss what might happen if snow cover increases. We conclude that in the medium term the cover of snowbed species will increase whereas the species composition of open alpine grassland and open ground species will change.

This study exemplifies that specific bryophyte groups (snowbed species and open ground species) observed at Mt Schrankogel can serve as excellent indicators for climate change research because their distribution is closely related to the date of snowmelt. This is underscored by their ability to occupy habitats where vascular plants are unable to survive (i.e., extreme snowbed situations) (e.g., Kudo & Ito 1992). Although snowbed bryophytes proved to be ideal indicators for climate change in alpine environments, they are spatially limited. Hence, this study points out that other groups of bryophytes (i.e., open ground species) offer additional, valuable information on the effects of modifications in snow cover extent. Further investigation is needed, enlarging the opportunities for ecological climate impact research in the alpine–nival ecotone where bryophytes are still abundant. When evaluating the effects of climate change on different groups of organisms, two approaches are suggested. The first is to manipulate environmental factors (e.g., Callaghan & Jonasson 1995; for a summary, see Zechmeister *et al.* 2003) and the second is to observe the coverage of bryophytes along ecological gradients. The results of this study show that the latter should allow predictions on how groups of bryophytes might respond to warmer climates (Virtanen *et al.* 2003). Consolidated knowledge of the indicator value of certain bryophyte groups can and will serve as a cornerstone for developing sampling methods for long-term monitoring networks.

Conclusions

Based on this study the indicator value of the bryophyte groups investigated for climate change research in the alpine life zone can be evaluated as follows.

- Because the cover of snowbed species is determined by the date of snowmelt (an intrinsic indicator for climate change), this group can serve as an excellent climate change indicator.
- The occurrence of open ground species is strongly affected by the date of snowmelt and depends also on the presence of open patches of soil. Thus their indicator value is restricted and needs further investigation. The cover of alpine grassland species is determined by biotic as well as abiotic factors. Their indicator value is low due to severe competition with vascular plants. Indirect effects, e.g., upward migration of vascular plants (Grabherr *et al.* 1994) can affect the distribution of alpine grassland bryophytes.

References

Austin, M. P., Pausas, J. G. & Nichols, A. O. (1996). Patterns of tree species richness in relation to environment in southeastern New South Wales, Australia. *Australian Journal of Ecology* **21**: 154–64.

Bates, J. W. & Farmer, A. (1992). *Bryophytes and Lichens in a Changing Environment*. Oxford: Clarendon Press.

Beniston, M. (2005). Mountain climates and climate change: an overview of processes focusing on the European Alps. *Pure and Applied Geophysics* **162**: 1587–1606.

Beniston, M. & Richard, S. J. T. (2001). *The Regional Impacts of Climate Change. Cryosphere, Hydrology, Water Resources, and Water Management. Snow and Ice*. Cambridge: Intergovernmental Panel on Climate Change, Working Group I, Cambridge University Press.

Billings, W. D. & Bliss, L. C. (1959). An alpine snowbank environment and its effects on vegetation, plant development, and productivity. *Ecology* **40**: 388–97.

Böhm, R., Auer, I., Brunetti, M. *et al.* (2001). Regional temperature variability in the European Alps: 1760–1998 homogenized instrumental time series. *International Journal of Climatology* **21**: 1779–1801.

Callaghan, T. V. & Jonasson, S. (1995). Implications for changes in arctic plant biodiversity from environmental manipulation experiments. *Ecological Studies* **113**: 151–66.

Callaway, R. M., Brooker, R. W., Choler, P. *et al.* (2002). Positive interactions among alpine plants increase with stress. *Nature* **417**: 844–8.

Crawley, M. J. (1993). *GLIM for Ecologists*. Oxford: Blackwell Science Ltd.

Currie, D. J. (1991). Energy and large-scale patterns of animal- and plant-species richness. *American Naturalist* **137**: 27–49.

Diaz, H. F. & Bradley, R. S. (1997). Temperature variations during the last century at high elevation sites. *Climate Change* **36**: 253–9.

Dierssen, K. (2001). *Distribution, Ecological Amplitude and Phytosociological Characterization of European Bryophytes*. Berlin: Bryophytorium Bibl. 56. J. Cramer.

Galen, C. & Stanton, M. L. (1995). Responses of snowbed plant species to changes in growing-season length. *Ecology* **76**: 1546–57.

Gottfried, M., Pauli, H. & Grabherr, G. (1998). Prediction of vegetation patterns at the limits of plant life: a new view of the alpine-nival ecotone. *Arctic and Alpine Research* **30**: 207–21.

Gottfried, M., Pauli, H., Reiter, K. & Grabherr, G. (1999). A fine-scaled predictive model for changes in species distribution patterns of high mountain plants induced by climate warming. *Diversity and Distributions* **5**: 241–51.

Gottfried, M., Pauli, H., Reiter, K. & Grabherr, G. (2002). Potential effects of climate change on alpine and nival plants in the Alps. In *Mountain Biodiversity – A Global Assessment*, ed. C. Körner and E. M. Spehn, pp. 213–23. London: Parthenon Publishing.

Grabherr, G. (1997). The high-mountain ecosystems of the Alps. In *Polar and Alpine Tundra Ecosystems of the World*, vol. 3, ed. F. E. Wielgolaski, pp. 97–121. Amsterdam: Elsevier.

Grabherr, G. (2003). Overview: Alpine vegetation dynamics and climate change – a synthesis of long-term studies and observations. In *Alpine Biodiversity in Europe, Ecological Studies 167*, ed. L. Nagy, G. Grabherr, C. Körner & D. B. A. Thompson, pp. 399–410. London: Springer.

Grabherr, G., Gottfried, M. & Pauli, H. (1994). Climate effects on mountain plants. *Nature* **369**: 448.

Grabherr, G., Gottfried, M. & Pauli, H. (2001). Long-term monitoring of mountain peaks in the Alps. In *Biomonitoring: General and Applied Aspects on Regional and Global Scales*, ed. C. A. Burga & A. Kratochwil, pp. 153–77. Dordrecht: Kluwer.

Grims, F. (1999). *Die Laubmoose Österreichs, Catalogus Florae Austriae, II. Teil, Bryophyten (Moose), Heft 1. Musci (Laubmoose)*, Biosystematics and Ecology Series 15. Austria: Austrian Academy of Sciences Press.

Grolle, R. & Long, D. G. (2000). An annotated check-list of the Hepaticae and Anthocerotae of Europe and Macronesia. *Journal of Bryology* **22**: 103–40.

Haeberli, W. & Beniston, M. (1998). Climate change and its impacts on glaciers and permafrost in the Alps. *Ambio* **27**: 258–65.

Heegaard, E. (2002). A model of alpine species distribution in relation to snowmelt time and altitude. *Journal of Vegetation Science* **13**: 493–504.

Hill, M. O. (1979). *TWINSPAN – a FORTRAN Program for Arranging Multivariate Data in an Ordered Two-way Table by Classification of Individuals and Attributes*. Ithaca, NY: Cornell University.

IPCC (2007). Fourth Assessment Report. *Working Group II Report "Impacts, Adaptation and Vulnerability"*. Cambridge: Cambridge University Press.

Jones, P. D. & Moberg A. (2003). Hemispheric and large scale surface air temperature variations: an extensive revision and an update to 2001. *Journal of Climatology* **16**: 206–23.

Körner, C. (1994). Impact of atmospheric changes on high mountain vegetation. In *Mountain Environments in Changing Climates*, ed. M. Beniston, pp. 155–66. London: Routlege.

Körner, C., Paulsen, J. & Pelaez-Riedl, S. (2003). A bioclimatic characterisation of Europe's Alpine areas. In *Alpine Biodiversity in Europe, Ecological Studies 167*, ed. L. Nagy, G. Grabherr, C. Körner & D. B. A. Thompson, pp. 2–30. London: Springer.

Kudo, G. & Ito, K. (1992). Plant distribution in relation to the length of the growing season in a snow-bed in the Taisetsu Mountains, northern Japan. *Vegetatio* **98**: 165–74.

Longton, R. E. (1997). Reproductive biology and life-history strategies. *Advances in Bryology* **6**: 65–101.

McCullagh, P. & Nelder, J. A. (1989). *Generalized Linear Models*, 2nd edn. London: Chapman and Hall.

Nebel, M. & Philippi, G. (2001a). *Die Moose Baden-Württembergs*, vol. 1. Stuttgart: Ulmer.

Nebel, M. & Philippi, G. (2001b). *Die Moose Baden-Württembergs*, vol. 2. Stuttgart: Ulmer.

Paton, J. A. (1999). *The Liverwort Flora of the British Isles*. Essex: Harley Books.

Pauli, H., Gottfried, M. & Grabherr, G. (1999). Vascular plant distribution patterns at the low-temperature limits of plant life – the alpine-nival ecotone of Mount Schrankogel (Tyrol, Austria). *Phytocoenologia* **29**: 297–325.

Pauli, H., Gottfried, M., Hohenwallner, D., Reiter, K. & Grabherr, G. (2003). *The GLORIA Field Manual. The Multi-Summit Approach*. Brussels: Ecosystem Research Report of the European Commission.

Pauli, H., Gottfried, M., Reiter, K., Klettner, C. & Grabherr, G. (2007). Signals of range expansions and contractions of vascular plants in the high Alps: observations (1994–2004) at the GLORIA master site Schrankogel, Tyrol, Austria. *Global Change Biology* **13**: 147–56.

Pepin, N. & Losleben, M. (2002). Climate change in the Colorado Rocky Mountains: free air versus surface temperature trends. *International Journal of Climatology* **22**: 311–29.

Rothero, G. (1991). Bryophyte dominated snowbeds in the Scottish Highlands. Unpublished M.Sc. thesis, University of Glasgow.

Stanton, M. L., Rejmánek, M. & Galen, C. (1994). Changes in vegetation and soil fertility along a predictable snowmelt gradient in the Mosquito Range, Colorado, U. S. A. *Arctic and Alpine Research* **4**: 364–74.

Virtanen, R., Eskelinen, A. & Gaare, E. (2003). Long-term changes in alpine plant communities in Norway and Finland. In *Alpine Biodiversity in Europe, Ecological Studies 167*, ed. L. Nagy, G. Grabherr, C. Körner & D. B. A. Thompson, pp. 411–21. London: Springer.

Walker, D. A., Halfpenny, J. C., Walker, M. D. & Wessman, C. A. (1993). Long-term studies of snow-vegetation interaction. *Bioscience* **43**: 287–301.

Woolgrove, C. E. & Woodin, S. J. (1994). Relationships between the duration of snowlie and the distribution of bryophyte communities within snowbeds in Scotland. *Journal of Bryology* **18**: 253–60.

Zechmeister, H. G., Grodzinska, K. & Szarek-Lukaszewska, G. (2003). Bryophytes. In *Bioindicators / Biomonitors (Principles, Assessment, Concepts)*, ed. B. A. Markert, A. M. Breure & H. G. Zechmeister, pp. 329–75. Amsterdam: Elsevier.

13

Bryophytes and Lichens in a Changing Climate: An Antarctic Perspective

RODNEY D. SEPPELT

Introduction

The Antarctic continent occupies about 14.4 million square kilometers, about 99% of which is covered by ice with an average thickness of around 1.8 km. It is the coldest, driest, windiest continent and has the highest mean elevation of all continents. Precipitation, as low as 20 mm on the inland ice plateau and significantly higher in coastal regions (as much as 250 mm annual rainfall equivalent), falls mostly as snow but occasionally as rain, particularly in the climatically milder maritime part of the northern Antarctic Peninsula.

The terrestrial and limnetic plant biota of Antarctica is impoverished and limited to lichens, bryophytes, mostly microscopic algae, cyanobacteria, mostly microscopic fungi, and two small vascular plants found only in the Maritime Antarctic. Invertebrates dominate the terrestrial and limnetic fauna, although large numbers of seabirds breed onshore over the summer months. Despite the severe climate and limited habitat availability, terrestrial plant life flourishes in ice-free areas where moisture is available. The continent of Antarctica with its nearby offshore islands is unique in being the only major land mass with a flora composed almost entirely of cryptogams (Longton 1979; Kappen 1993a; Broady 1996; Green *et al.* 1999; Vincent 2000; Øvstedal & Lewis Smith 2001; Ochyra *et al.* 2008).

The strong negative trend of the global thermal budget from the equator towards the poles is regionally altered by the presence of land masses and by general oceanic circulation. Both polar regions share distinct features that result from their geographic location: very low average temperatures, strong seasonality in incident solar radiation (both photosynthetically active and short wave ultra-violet radiation), and extreme aridity. In Antarctica, the limited

Bryophyte Ecology and Climate Change, eds. Zoltán Tuba, Nancy G. Slack and Lloyd R. Stark. Published by Cambridge University Press.

precipitation accumulates mostly as ice, permanently covering most land areas and a large proportion of the sea surface in winter, significantly affecting the surface albedo. Both polar regions are subject to strong winds, particularly in Antarctica, where winds in excess of 320 km h^{-1} have been recorded and wind speeds greater than 100 km h^{-1} are common. For a comparative review of the abiotic features of the polar regions, see Godard and André (1999).

In an ecological sense, all organisms living in polar and high alpine regions face major constraints that shape their morphology, life history traits and their interactions. While referring specifically to the polar oceans, Hempel (1995) noted:

> Polar seas of both hemispheres are cold, covered by ice, and dark in winter. This statement, which is only partly valid, exhausts the description of common features for Southern and Northern seas.

On land, the situation is even more diverse with the Antarctic continent having less than 1% of the land ice-free and with a not surprisingly depauperate flora and fauna. By contrast, the Arctic has a rich and diverse flora and fauna (Woodin & Marquiss 1997; Hansom & Gordon 1998; Grémillet & Le Maho 2003).

Antarctic organisms have long been isolated from the rest of the world. The ancient continent of Gondwana began to separate from the Laurasian supercontinental land mass about 170 million years ago (mya) and by about 160 mya the African continent separated from Gondwana, followed by the Indian continental mass around 125 mya. Around 65 mya and while still connected to what became Australia, the climate of the Antarctic land mass was tropical to subtropical. Australia separated from Antarctica 35–40 mya as permanent ice began to form on Antarctica following gradual disruption of the equatorial to polar oceanic circulation. About 23 mya, with the opening of the Drake Passage between South America and the Antarctic Peninsula, circumpolar ocean circulation and thus a general cooling of the continent began. The Antarctic continent became virtually covered by permanent ice by about 15 mya. With so little of Antarctica ice-free, the continent is best regarded as a series of ice-free island refugia.

Endemism is low among the bryophytes (Ochyra *et al.* 2008) and considerably higher among the lichens (Øvstedal & Lewis Smith 2001) although with lichens taxonomic difficulties compound the assessment of numbers of endemic species. Taxonomic difficulties also make assessment of the status of algal and fungal endemism somewhat problematic (Vishniac 1995; Ling 1996; Ling & Seppelt 1998a,b; McRae *et al.* 1999; Tosi *et al.* 2002), even utilizing molecular systematic methods (Selbmann *et al.* 2005). The invertebrate fauna is not extensive but contains many apparently endemic taxa (Pugh 1993; Greenslade 1995;

Convey & Block 1996; Andrássy 1998; Convey & McInnes 2005). Recent molecular genetic studies on selected invertebrates support the long-term genetic isolation of individual populations (Stevens & Hogg 2006a, b; Stevens *et al.* 2007), reinforcing the island analogy for ice-free areas.

Water relations and photosynthetic performance

Under the prevailing conditions of extreme aridity, subzero temperatures and large seasonal variations in solar radiation, the terrestrial vegetation of Antarctica is, for the most part, living close to the physiological limits for its survival. The availability of liquid water is a major determinant of the distribution and abundance of the terrestrial and limnetic biota of Antarctica (Hawes *et al.* 1992; Kennedy 1993; Sømme 1995; Block 1996; Leishman & Wild 2001). Significant biological activity only commences when plants become rehydrated with the onset of snow melt and usually begins when they are still covered by snow (Oberbauer & Starr 2002; Schlensog *et al.* 2004; Snell *et al.* 2007). When the insulating layer of snow and ice covering the plants melts, they become fully exposed to solar radiation: both photosynthetically active radiation (PAR: 400–700 nm) and biologically damaging ultraviolet radiation (UV-B 280–315 nm, UV-A 250–400 nm), both of which are absorbed by snow and ice. Just 10 cm of snow absorbs 80% of erythemally weighted UV-B (Cockell *et al.* 2002) but allows up to 60% PAR penetration, sufficient to allow subniveal photosynthesis in Arctic lichens (Kappen *et al.* 1995).

Both mosses and lichens can be considered poikilohydric plants, in which the water status of their tissues is completely dependent on their environment (Walter 1931) so that, in terrestrial habitats, water vapor partial pressure of the plant body is in equilibrium with the atmospheric humidity (Green & Lange 1994). Water content of lichen thalli or moss shoots also has a major influence on gas exchange and chlorophyll *a* fluorescence yield (Kappen & Breuer 1991; Kappen *et al.* 1995; Schlensog *et al.* 2004; Wasley *et al.* 2006). Desiccated lichen thalli and moss shoots under deep snow and ice are probably physiologically inactive (Kappen 1993b) and may only resume activity as they begin to emerge from the snow when thallus and plant temperatures approach the freezing point of water (Pannewitz *et al.* 2003).

Concentrations of photosynthetic pigments increase in poikilohydric plant tissues as they emerge from snow (Kimball *et al.* 1973; Oberbauer & Starr 2002; Snell *et al.* 2007), and similarly UV-protective pigments emerge such as anthocyanins, which are associated both with attenuation of UV-B and with chilling and desiccation tolerance (Chalker-Scott 1999; Oberbauer & Starr 2002; Gould 2004).

Effects of global climate change

Under the influence of global climate change, particularly warmer winters and spring temperatures, emergence of terrestrial plants from beneath the winter snow pack may be expected to commence earlier. Early spring emergence coincides with the onset of depletion of UV-absorbing ozone in the atmosphere, the critical period being from September to December. However, since the production of UV-absorbing pigments responds rapidly to an increase in solar radiation following rewetting, there may be insignificant impacts on the photosynthetic physiology of the terrestrial flora.

Photosynthesis

Snell *et al.* (2007) found with *in situ* observations during snow melt at Rothera Point, western Antarctic Peninsula (67°34′S, 68°W), that the liverwort *Cephaloziella varians* showed no changes in the concentrations of UV-B photoprotective pigments. However, chlorophyll and carotenoid concentrations and maximum photosystem II (PSII) yield (F_v/F_m) were, respectively, 80%, 60%, and 144% higher in plants recently emerged from snow compared with those under 10 cm of snow.

A similar recovery of PSII activity in lichens at Botany Bay, Southern Victoria land (77°S, 162°32′E), was found by Pannewitz *et al.* (2003) and this occurred only when the plants had almost fully emerged from the winter snow pack. Schlensog *et al.* (2004), also at Botany Bay, found that in the mosses *Bryum subrotundifolium* (= *B. argenteum*) and *Hennediella heimii* optimal levels of F_v/F_m after emergence from overwintering in the dehydrated state only occurred after four days following rehydration. In contrast, Schlensog *et al.* (2004) reported that for the lichens *Physcia caesia* and *Umbilicaria aprina*, nearly full recovery was achieved within a few minutes of rehydration. They suggested that the rapid recovery of photosynthesis of these lichens was possibly due to reactivation of conserved photosystems while the slower recovery in mosses may have resulted from the initial repair of the photosystems. Experimental studies by Snell *et al.* (2007) of PSII recovery in *Cephaloziella varians* showed that there was full recovery within about 45 minutes after transferring the plants from beneath snow to a growth cabinet, implying that photosystem recovery or repair may be initiated prior to emergence from the snow.

The level of hydration of these plants prior to emergence from the snow may have a marked influence on the rate of recovery of photosynthetic yield. Other abiotic factors, such as ambient air temperature, the solar radiation spectrum, and season, will also markedly affect recovery of photosynthetic activity.

Nutrient cycling, temperature and freeze–thaw

Because of their dominance in the vegetation in polar regions, cryptogams play a significant role in nutrient cycling (Stoner *et al.* 1982; Lewis Smith 1985; Longton 1988) although in the Antarctic environment the rates of nutrient cycling are low. High concentrations of nutrients are usually found only in the vicinity of seabird nesting sites and seal colonies. Elsewhere nutrient input is largely through precipitation or aerosol transfer in costal regions (Allen *et al.* 1967; Greenfield 1992a, b; Crittenden 1998).

While nutrient requirements are reportedly very low and low levels of soil nutrients non-limiting, particularly in lichens (Greenfield 1992a; Crittenden 1998), a positive correlation between vegetation patterns and nutrient availability has been demonstrated (Gremmen *et al.* 1994; Leishman & Wild 2001). Further, Wasley *et al.* (2006) showed by field manipulation experiments that in both moss and lichen communities, chlorophyll content and electron transport rates respond positively to nutrient enhancement. Nutrient limitations may be more important than once thought (Beyer *et al.* 2000) and may assume significance in the context of future climate change scenarios if the plants are unable to respond to changes in thermal regime, or moisture availability if nutrients become limiting (Hennion *et al.* 2006).

In the maritime Antarctic, studies in fellfield soils showed large increases in soil carbohydrates and also microbial populations, attributed to leaching of soluble carbohydrates and polyols from vegetation during periods of thaw (Wynn-Williams 1980; Tearle 1987). Freezing may increase the rate of leakage from cells, although Hurst *et al.* (1985) found no significant differences in nutrient leaching from mature or senescent leaf material in two subantarctic phanerogams after repeated freeze–thaw. However, it should be emphasized that these studies were undertaken in the comparatively wet and warmer Maritime Antarctic region, utilized dead or senescent plant material, and may not truly reflect what happens with healthy tissues.

Melick and Seppelt (1992) undertook an experimental study examining the loss of carbohydrates and polyols from living bryophyte material in the Windmill Islands region of continental Antarctica, where diurnal freeze–thaw cycling occurs over much of the spring, summer, and autumn growing season. They compared responses of healthy material of the mosses *Schistidium antarctici* (as *Grimmia*), *Ceratodon purpureus*, *Bryum pseudotriquetrum* and the liverwort *Cephaloziella varians* (as *C. exiliflora*) with dead or senescent material of *Schistidium*. The plant material was subjected to leaching in water and up to 16 freeze–thaw cycles. Gas chromatographic studies showed that after 16 days of immersion, loss of carbohydrates (mainly glucose and fructose) from healthy material was relatively low and constituted 10%–29% of the total carbohydrate pool, while from dead *Schistidium* there was a loss of 69%, most of which occurred in the first 24 hours after immersion. Freeze–thaw cycles greatly increased the rate of sucrose leakage with a 2–3-fold increase in total sugar loss from all living samples. After 16 freeze–thaw cycles *B. pseudotriquetrum* lost 65% of the total carbohydrate pool while the other species' losses were below 28%. Carbohydrates are known to confer some freezing protection on plant tissues. But with the Antarctic bryophytes studied, although there were significant inter-species differences in the rates of sugar loss, all showed relatively high tissue freezing points (>–9 °C), suggesting that the mosses do not necessarily avoid freezing but merely tolerate the freeze–thaw process.

In higher plants significant changes in cryoprotection of thylakoid membranes may result from relatively minor changes in sugar concentrations, depending on salt concentrations at the membranes (Santarius & Giersch 1983). The differences observed in sugar loss in Antarctic bryophytes may have a significant impact on plant survival. With climate warming trends resulting from increased freeze–thaw activity it is expected that increased rates of leakage from senescent moss stems may have a major influence on the survivability of the living plants. Carbohydrate and polyol leaching from the standing cryptogam vegetation constitutes a major nutrient input to the ecosystem in areas away from penguin or other seabird colonies and will significantly influence soil microbial activity (Bölter 1990). At least at the microhabitat level, the nutrient balance and biotic structure in moist soils may be considerably altered with increased freeze–thaw events.

Short-wave ultraviolet radiation

Stratospheric ozone (O_3) is an effective filter of short-wave solar ultraviolet-B (UV-B) radiation (Weatherhead & Andersen 2006). Depletion of stratospheric O_3 since the mid-1970s has led to a significant increase in UV-B irradiation in polar regions, particularly over Antarctica. Anthropogenic chlorofluorocarbons (CFCs) accumulate in the stratosphere, where they degrade stratospheric O_3 (McKenzie *et al.* 2007). Prior to 1970, stratospheric O_3 thickness was measured as greater than 300 Dobson units, but from about 1975 a spring – early summer depletion became apparent (Farman *et al.* 1985; McKenzie *et al.* 2007), strongly linked to the polar vortex, which prevents the influx of O_3 from lower latitudes (Roy *et al.* 1990; Rozema *et al.* 1997). For much of the spring and summer, most of Antarctica is subjected to UV-B irradiance levels comparable to those of mid-summer lower latitudes of the northern hemisphere (Frederick & Snell 1988).

Ultraviolet B radiation is damaging to a variety of biologically significant molecules such as nucleic acids, proteins and lipids (Larson & Berenbaum 1988; Paul & Gwynn-Jones 2003). Terrestrial plant life has evolved in the constant presence of UV-B (Greenberg *et al.* 1997) and a variety of mechanisms has evolved to mitigate its effects. Apart from physical attributes (e.g., cuticular structures), secondary plant metabolites have been implicated in protecting plants from the damaging effects of UV-B radiation (Rozema *et al.* 1997; Markham *et al.* 1998; Ryan *et al.* 1998; Newsham *et al.* 2002; Caldwell *et al.* 2007). A rapid increase in levels of flavonoid compounds, which absorb strongly in the UV-B spectrum (280–320 nm), appears to be the most common response to UV-B stress in land plants (Ryan *et al.* 1998; Searles *et al.* 2001; Caldwell *et al.* 2007; Ryan *et al.* 2009). Flavonoids act by dissipating the energy absorbed from UV-B, thus preventing photochemical damage, and they also scavenge free radicals (Husain *et al.* 1987; Smith & Markham 1996; Caldwell *et al.* 2007), making them an important group in photoprotection against the damaging effects of short-wave solar radiation. Over 350 flavonoids have been reported from bryophytes (Mues 2000).

In a study of the photosynthetic physiology of *Bryum subrotundifolium* (= *B. argenteum*), Green *et al.* (2000) suggested that sun and shade forms existed and that the shade form was less protected against the potentially harmful effects of UV-A (250–400 nm). Later, Green *et al.* (2005) expanded their studies to include Antarctic populations of *Ceratodon purpureus*. Both mosses have sun and shade forms that differ markedly in color and in their protection from UV-A, but sun forms have high levels of protection against UV-A. Using a UV-A pulse amplitude modulated fluorometer that allowed the absorption of UV-A prior to arrival at the chloroplast, Green *et al.* (2005) determined photosynthetic yield by chlorophyll fluorescence and suggested that shade forms of *B. subrotundifolium*, initially low in UV-A protection, achieved full sun-form levels in about six days when exposed to ambient sunlight, whereas sun forms showed little change. The shade form also achieved almost the same photosynthetic characteristics as the sun form. It is apparent that Antarctic mosses are, in general, well protected against the impacts of UV radiation and are as adaptable to incident UV as higher plants.

In Antarctic populations of the moss *Ceratodon purpureus* ultrastructural changes occur with increased exposure to solar radiation, with less thylakoid stacking in the chloroplasts in sun-exposed forms, coupled with a preferential increase in violaxanthin (Post 1990). Similar characteristics have been demonstrated with the hepatic *Cephaloziella varians* (Post & Vesk 1992).

Three common and widespread continental Antarctic mosses, the cosmopolitan *Ceratodon purpureus* and *Bryum pseudotriquetrum* and the Antarctic endemic

Schistidium antarctici, have demonstrated, perhaps surprisingly, that the endemic *S. antarctici* had the lowest UV screening potential; however, like *C. purpureus*, most of the UV-screening compounds appear to be cell-wall-bound. Levels of combined intra-cellular and wall-bound UV-screening compounds were similar in *C. purpureus* and *B. pseudotriquetrum*. The cell-wall-bound UV-screening compounds provide in *C. purpureus* and *B. pseudotriquetrum* a more spatially uniform and potentially more effective UV screen than in *S. antarctici*, indicating that the endemic *Schistidium* may be at a disadvantage under continuing springtime ozone depletion (Dunn & Robinson 2006; Clarke & Robinson 2008).

Post and Larkum (1993), in a study comparing Antarctic terrestrial and marine algae, showed that increased exposure to UV reduced chlorophyll and photosynthesis levels. However, particularly in the terrestrial chlorophyte *Prasiola crispa*, there was no change in the ratio of UV absorbing compounds to chlorophyll, indicating that exposure to sunlight with an increased ratio of UV-B to visible light is stressful.

Cyanobacteria are important primary colonists of Antarctic terrestrial environments. Quesada and Vincent (1997) demonstrated differing responses to increased UV radiation in strains of *Phormidium murrayi* and *Oscillatoria priestleyi*, with growth decreasing with increased UV levels but with a 5-fold (UV-A) to 10-fold (UV-B) greater effect with *Oscillatoria* than *Phormidium*. Cellular concentrations of phycobiliproteins and, to a lesser extent chlorophyll a, diminished with increased UV exposure. They also showed the presence of UV-screening compounds in *Oscillatoria priestleyi*, a fast gliding species able to alter its position relative to incident radiation, but not in *Phormidium murrayi*, a non-motile species. The ability of one to escape UV radiation by gliding and the greater ability of the other to tolerate UV-A and UV-B exposure illustrated the differences in UV survival or avoidance strategies even between closely related taxa.

Microscopic fungi are important components of the terrestrial Antarctic ecosystem through their ability to mobilize nutrients from soil organic matter. Temperature, water, and nutrient availability influence the growth of fungi in the natural environment. Exposure to solar radiation, particularly UV, may also limit growth. Fungi on the surface of Antarctic soils are exposed to wide fluctuations in the levels of incident solar radiation due to seasonal variations in snow cover, sun angle, and surface albedo (Wynn-Williams 1996). Hughes *et al.* (2003) demonstrated experimentally that exposure to solar radiation of 287–400 nm (UV-A, UV-B) reduced hyphal extension rates relative to controls kept in the dark, but radiation with a wavelength greater than 400 nm had no effect on hyphal growth. Increases in short-wavelength UV-B occurring in Antarctica during periods of ozone depletion (spring and early summer) are most likely to exacerbate the inhibitory effects of solar radiation on growth of terrestrial fungi. As fungi have a key role to play in soil nutrient cycling, exposure to solar radiation has the potential to affect biogeochemical cycles, with possible consequences for higher-order interactions (Hughes *et al.* 2003).

It is evident that in the Antarctic region, as elsewhere, although there are significant biochemical and biological impacts of increased levels of UV radiation, particularly in spring, indigenous organisms exhibit a variety of stress avoidance mechanisms enabling them to cope with the increased stress levels imposed by a changing world climate.

Transient events and persistent effects

The effects of discrete or transient climatic events (pulses), particularly in dry ecosystems, have been noted for many years (e.g., Stafford Smith & Morton 1990; Gebauer & Ehleringer 2000) and may exert significant controls over the structure and functioning of both natural and managed ecosystems (Jentsch *et al.* 2007). These events are distinguished from longer-term climatic trends, such as regional warming or cooling, on the basis of their frequency or likelihood of occurrence and the magnitude of the consequences relative to the duration of the event (Shar *et al.* 2004; Jentsch *et al.* 2007).

Discrete climatic events may exert a disproportional influence over ecosystems relative to the temporal scales over which they occur (Barrett *et al.* 2008), based on the magnitude of the event (e.g., heavy rain, rapid snowmelt). This is particularly so in polar regions, considering the dual role that temperature and moisture play in determining biological activity and water availability. In the Victoria Land McMurdo Dry Valleys, pulses of liquid water, following surface snow or frozen groundwater melt, rapidly reactivate freeze-dried microbial mats in stream channels (McKnight *et al.* 2007) and contribute to nutrient loading in the adjacent lake ecosystems (Foreman *et al.* 2004). Sudden rewetting of soils and moss beds following the onset of melt flow from these Dry Valley glaciers in early summer, apart from releasing high levels of nutrients, may have significant impacts on the integrity of both soil surficial structure and the moss plants.

A large moss bed along an ephemeral outflow channel from the Canada Glacier, Taylor Valley, is composed largely of *Bryum argenteum* and cyanobacteria, with small amounts of *B. pseudotriquetrum*. *Hennediella heimii* is found at the drier margins. In the winter anhydrobiotic or freeze-dried state, the surface of the moss bed cracks open and the base of the plants becomes only loosely attached to the sandy soil below. On initial rapid rewetting, apical shoots of *B. argenteum* are quickly dislodged and dispersed in vast numbers along the cracks among the moss (personal observations). The cracks rapidly close as the moss and underlying soil becomes saturated, leaving the deciduous shoot tips to grow sandwiched among the existing moss, producing a pronounced reticulate growth appearance to the moss bed. Although no analyses have been undertaken at the Canada Glacier moss bed, based on the findings of Melick and Seppelt (1992), Foreman *et al.* (2004), and McKnight *et al.* (2007), the considerable initial and short-term loss of carbohydrates and polyols and the sudden mobilization of nutrients must be significant and important to the associated microfauna and flora. It is as yet unknown how changes in ambient temperature and pulses in the availability of liquid water may affect not only the moss and algal communities but also the soil microbial and microfaunal communities and linked ecosystem functioning in these extreme polar desert environments.

In the McMurdo Dry Valleys nematodes are the dominant metazoans but their distribution is far from ubiquitous and their distribution varies significantly over environmental gradients determined by the amount of available water, organic carbon, and soluble salts in the surface soil (Freckman & Virginia 1997; Barrett *et al.* 2004; Poage *et al.* 2008). Responses of the common invertebrates to soil physicochemical parameters vary between taxa and more complex communities may show greater resilience to environmental change, particularly following rapid climatic events. Pulses in moisture availability may be important in maintaining soil biodiversity by facilitating the survival, reproduction, and growth of subordinate species. Predictions of the long-term response of Antarctic ecosystems to global climate change that focus on linear trends potentially overlook the importance of extreme or infrequent short-term climatic events on hydrology and linked ecological changes (Barrett *et al.* 2008).

Long-term trends

Ecosystem responses to long-term climate trends and short-term or transient events have been the center of considerable research focus, particularly in dry environments (Peters 2000; Schwinning & Ehleringer 2001; McKay *et al.* 2003; Austin *et al.* 2004; Reynolds *et al.* 2004; Jentsch *et al.* 2007). Compared with vascular plants, mosses generally have little structural complexity and respond to environmental stress at the cellular level (Bates 2000; Christianson 2000).

Mosses may become acclimated to increasing levels of UV (Lappalainen *et al.* 2008), although they may be more susceptible to UV-B when dry and photosynthetically inactive (Takács *et al.* 1999; Proctor 2000). Stratospheric ozone absorbs little UV-A, which biologically appears less harmful than UV-B (Björn 1999), and different organisms react to increased UV levels differently (Rozema *et al.* 2002). As well as being important in providing photoprotection against harmful effects of UV-B and because of their stability, levels of flavonoids have been used in hindsight as a surrogate indicator of former UV levels and, consequently, ozone depletion (Markham *et al.* 1996).

Mosses and lichens in sub-Arctic areas experience carbon dioxide levels (> 1000 $\mu l\ l^{-1}$) well in excess of ambient levels with no apparent ill effects (Callaghan *et al.* 1992). In the Antarctic region, CO_2 concentration within moss turf or cushions depends on both moisture levels and water content (Kappen *et al.* 1989; Tarnawski *et al.* 1992). CO_2 concentrations in turf forms of *Grimmia* (=*Schistidium*) *antarctici* from moist habitats may be up to 10× ambient concentrations, doubling the photosynthetic rate, while in cushions from drier sites CO_2 concentrations were only 30% higher (Tarnawski *et al.* 1992). Seasonal variation in CO_2 concentrations within the moss turf is in part governed by photoinhibition of photosynthesis (Adamson, H. *et al.* 1988; Kappen *et al.* 1989; Wilson 1990; Lovelock *et al.* 1995a, b; Pannewitz *et al.* 2005). CO_2 concentrations of up to 3500 $\mu l\ l^{-1}$ were observed in moist turf in mid-January (summer) and this was attributed to respiratory activity of heterotrophic bacteria, protista, invertebrates, fungi, and bryophyte stems and rhizoids.

Hidden from view: the endolithic habitat

In the extreme arid environments of the McMurdo Dry Valleys of Victoria Land, terrestrial life may assume a decidedly cryptic habit (Friedmann 1977, 1982; Friedmann *et al.* 1980, 1988; Nienow *et al.* 1988; Selbmann *et al.* 2005). These endolithic cryptogam communities consist of algae, cyanobacteria, fungi, and lichens. They are found in sandstone and granitic rocks and the habitat provides shelter from temperature extremes, wind abrasion, and persistent moisture derived from melting snow (Kappen *et al.* 1981). Particularly in the Dry Valleys, mosses are often found living just beneath the surface of the sandy soil. Moisture is derived both from snowmelt and from the active layer of the subsurface permafrost in the soils. Similarly, mosses are often found growing in small cracks or beneath surface flakes of rocks. These habitats may assume much greater importance under the influence of global climate shifts.

Alien invasions and competitive survival

The terrestrial Antarctic ecosystem is simple in structure, particularly in the extreme aridity of the McMurdo Dry Valleys of Victoria Land. The inhospitable environment and long distances from potential diaspore sources serve as effective barriers to potential invasive colonists. However, geothermal sites, as an analog of future warmer and moister conditions, may provide useful insights into the future.

There are six known active volcanic sites in Antarctica. Three are mountains in continental Victoria Land: Mount Erebus, 3794 m at 77°32′S, 167°E; Mount Melbourne, 2730 m at 74°21′S, 164°42′E; and Mount Rittmann, 2500 m at 73°27′S, 165°30′E. All have fumaroles near their summits. The mosses *Campylopus pyriformis* and *Pohlia nutans* have been identified from steam-warmed soils (Broady *et al.* 1987; Bargagli *et al.* 1996; Skotnicki *et al.* 2001). The remaining three volcanic sites are in the Maritime Antarctic region: Bouvetøya, 780 m at 54°26′S, 03°24′E; the South Sandwich Islands, a collection of islands situated in about 56°–59°S, 26°–28°W and varying in height from 190 to 1370 m; and Deception Island, 539 m at 62°57′S, 60°38′W. Deception Island has received most attention because of its documented history of eruptions and accessibility. Geothermal sites on the island have a significant number of bryophyte species not known elsewhere, or that are otherwise rare, in the Antarctic biome. Some 47 (80%) of the island's 59 moss species and all 9 of the hepatics occur in geothermal habitats, with 25 (42%) of the mosses and 4 of the hepatics confined primarily to heated ground (Lewis Smith 2005).

Although there is little evidence of recent colonization or major expansion of the ranges of bryophytes and lichens in the Antarctic biome, the two vascular species, *Deschampsia antarctica* and *Colobanthus quitensis*, have significantly increased in area under increasing regional temperatures (Fowbert & Lewis Smith 1994; Lewis Smith 1994; Gerighausen *et al.* 2003) and there may be significant physiological and reproductive implications (Day *et al.* 1999) providing the vascular plants with a competitive advantage.

Conclusions: what of the future?

Ecophysiological and biochemical studies of the bryophytes and lichens indicate that these taxa are physiologically well adapted to the rigors of the environment. Bryophytes rapidly synthesize photoprotective UV-absorbing compounds in response to increased levels of solar radiation. Such compounds, particularly flavonoids, also provide a useful indicator of former levels of incident UV radiation. Water availability is the dominant limiting factor governing plant distribution patterns. In the wetter and milder Maritime Antarctic region, nutrient availability is not limiting, but in the dry continental Antarctic both nutrient and moisture availability influence plant distribution and abundance.

Warmer conditions over the latter half of the twentieth century resulted in significant ice and snow regression on a number of the subantarctic islands – Signy Island (61°S, 45°W), Heard Island (53°S, 73°E), Îles Kerguelen

(49°S, 69°E) – with subsequent increases in plant cover and diversity (Allison & Keage 1986; Scott 1990; Smith 1990). Increases in ambient temperatures and changes in precipitation have also been reported for Macquarie Island (54°S, 158°E; Adamson, D. *et al.* 1988; Selkirk *et al.* 1990) and Marion Island (47°S, 38°E) where an amelioration of climate has altered ecosystem structure (Smith & Steenkamp 1990). Relatively rapid ecological changes are occurring in the sub-Antarctic region, but the changes are far less understood for the terrestrial ecosystem of continental Antarctica. Owing to the extreme nature of the Antarctic environment, minor fluctuations in microclimate may result in significant changes to the terrestrial environment. In the Windmill Islands region of continental Antarctica (66°S, 110°E), an area that supports the most extensive lichen and moss dominated vegetation of any Antarctic locality, small-scale pattern processes and environmental continua have been correlated with microsite moisture, exposure, and nutrient levels (Hovenden & Seppelt 1995). While plant development is dependent on specific microsite conditions, broad-scale vegetation communities are also identifiable (Melick & Seppelt 1994). Bryophytes appear to be the key ecological markers, with vitality of the moss being indicative of soil moisture. Healthy, lichen-free bryophyte communities make up only a small proportion of the vegetation and are confined almost entirely to the wettest habitats. The vast majority of mosses occur in mixed bryophyte–lichen communities and the predominance of moribund and desiccated moss in these communities is indicative of a formerly far more extensive bryophyte cover. The present restriction of healthy moss turfs to the rare moist microhabitats suggests an ongoing drying of the environment and/or a shift in melt drainage patterns with post glacial uplift in the region over the past 5,000–8,000 years (Goodwin 1993; Melick & Seppelt 1997).

The processes of plant colonization and death of vegetation are far slower in the extreme climate of continental Antarctica than in subpolar regions (Scott 1990; Smith 1990; Havström *et al.* 1995). Havström *et al.* (1995) reported that the sudden onset of snow cover during the last Little Ice Age caused sudden death of dwarf shrubs in the sub-Arctic. By contrast, mosses and lichens in continental Antarctica remain physiologically buffered against sudden environmental fluctuations (Green & Lange 1994; Melick & Seppelt 1994), showing no signs of the seasonal buildup of plant solutes associated with growth flushes and winter hardening in sub-Antarctic cryptogams (Tearle 1987). The poikilohydric frigid Antarctic flora may resist freezing for 5 years or more (Horikawa & Ando 1967; Longton 1988) and it may require a considerably sustained climate shift to radically alter the floristics of continental Antarctic localities.

References

Adamson, D. A., Whetton, P. & Selkirk, P. M. (1988). An analysis of air temperature records for Macquarie Island: decadal warming, ENSO cooling and Southern Hemisphere circulation patterns. *Papers and Proceedings of the Royal Society of Tasmania* **122**: 107–12.

Adamson, H., Wilson, M., Selkirk, P. & Seppelt, R. (1988). Photoinhibition in Antarctic mosses. *Polarforschung* **58**(2/3): 103–11.

Allen, S. E., Grimshaw, H. M. & Holdgate, M. W. (1967). Factors affecting the availability of plant nutrients on an Antarctic island. *Journal of Ecology* **55**: 381–96.

Allison, I. F. & Keage, P. L. (1986). Recent changes in the glaciers of Heard Island. *Polar Record* **23**: 255–71.

Andrássy, I. (1998). Nematodes in the sixth continent. *Journal of Nematode Systematics and Morphology* **1**: 107–86.

Austin, A. T., Yahdjaian, L., Stark, J. M. *et al.* (2004). Water pulses and biogeochemical cycle in arid and semiarid ecosystems. *Oecologia* **141**: 221–35.

Bargagli, R., Broady, P. A. & Walton, D. W. H. (1996). Preliminary investigation of the thermal biosystem of Mount Rittmann fumaroles (northern Victoria Land, Antarctica). *Antarctic Science* **8**: 121–6.

Barrett, J. E., Wall, D. H., Virginia, R. A. *et al.* (2004). Biogeochemical parameters and constraints on the structure of soil biodiversity. *Ecology* **85**: 3105–18.

Barrett, J. E., Virginia, R. A., Wall, D. H. *et al.* (2008). Persistent effects of a discrete warming event on a polar desert ecosystem. *Global Change Biology* **14**: 2249–61.

Bates, J. W. (2000). Mineral nutrition, substratum ecology, and pollution. In *Bryophyte Biology*, ed. A. J. Shaw & B. Goffinet, pp. 248–311. Cambridge: Cambridge University Press.

Beyer, L., Bölter, M. & Seppelt, R. D. (2000). Nutrient and thermal regime, microbial biomass, and vegetation of Antarctic soils in the Windmill Islands region of East Antarctica (Wilkes Land). *Arctic, Antarctic and Alpine Research* **32**: 30–9.

Björn, L.-O. (1999). Ultraviolet-B radiation, the ozone layer and ozone depletion. In *Stratospheric Ozone Depletion – the Effects of Enhanced UV-B Radiation on Terrestrial Ecosystems*, ed. J. Rozema, pp. 21–38. Leiden: Backhuys.

Block, W. R. (1996). Cold or drought – the lesser of two evils for terrestrial arthropods? *European Journal of Entomology* **93**: 325–39.

Bölter, M. (1990). Microbial ecology of soils from Wilkes Land, Antarctica. I. The bacterial population and its activity in relation to dissolved organic matter. *Proceedings of the National Institute of Polar Research: Symposium of Polar Biology* **3**: 104–19.

Broady, P. A. (1996). Diversity, distribution and dispersal of Antarctic terrestrial algae. *Biodiversity and Conservation* **5**: 1307–35.

Broady, P. A., Given, D., Greenfield, L. & Thompson, K. (1987). The biota and environment of fumaroles on Mount Melbourne, northern Victoria Land. *Polar Biology* **7**: 97–113.

Caldwell, M. M., Bornman, J. F., Ballare, C. L., Flint, S. D. & Kulandaivelu, G. (2007). Terrestrial ecosystems, increased solar ultraviolet radiation, and interactions with other climate change factors. *Photochemical and Photobiological Sciences* **6**: 252–66.

Callaghan, T. V., Sonesson, M. & Sømme, L. (1992). Responses of terrestrial plants and invertebrates to environmental change at high latitudes. *Philosophical Transactions of the Royal Society of London* B **338**: 279–88.

Chalker-Scott, L. (1999). Environmental significance of anthocyanins in plant stress responses. *Photochemistry and Photobiology* **70**: 1–9.

Christianson, M. L. (2000). Control of morphogenesis in bryophytes. In *Bryophyte Biology*, ed. A. J. Shaw & B. Goffinet, pp. 199–244. Cambridge: Cambridge University Press.

Clarke, L. J. & Robinson, S. A. (2008). Cell wall-bound ultraviolet-screening compounds explain the high ultraviolet tolerance of the Antarctic moss, *Ceratodon purpureus*. *New Phytologist* **179**: 776–83.

Cockell, C. S., Rettberg, P., Horneck, G. *et al.* (2002). Influence of ice and snow covers on the UV exposure of terrestrial microbial communities: dosimetric studies. *Journal of Photochemistry and Photobiology* B **68**: 23–32.

Convey, P. & Block, W. (1996). Antarctic dipterans: ecology, physiology and distribution. *European Journal of Entomology* **93**: 1–13.

Convey, P. & McInnes, S. J. (2005). Exceptional tardigrade-dominated ecosystems from Ellsworth Land, Antarctica. *Ecology* **86**: 519–27.

Crittenden, P. D. (1998). Nutrient exchange in an Antarctic macrolichen during summer snowfall snow melt events. *New Phytologist* **139**: 697–707.

Day, T. A., Ruhland, C. T., Grobe, C. W. & Xiong, F. (1999). Growth and reproduction of Antarctic vascular plants in response to warming and UV radiation reductions in the field. *Oecologia* **119**: 24–35.

Dunn, J. L. & Robinson, S. A. (2006). Ultraviolet B screening potential is higher in two cosmopolitan moss species than in a co-occurring Antarctic endemic moss: implications of continuing ozone depletion. *Global Change Biology* **12**: 2282–96.

Farman, J. C., Gardiner, B. G. & Shanklin, J. D. (1985). Large losses of total ozone in Antarctica reveal seasonal ClO_x/NO_x interaction. *Nature* **315**: 207–10.

Foreman, C., Wolf, C. E. & Priscu, J. C. (2004). Impact of episodic warming events on the physical, chemical and biological relationships of lakes in the McMurdo Dry Valleys, Antarctica. *Aquatic Geochemistry* **10**: 239–68.

Fowbert, J. A. & Lewis Smith, R. I. (1994). Rapid population increases in native vascular plants in the Argentine Islands, Antarctic Peninsula. *Arctic and Alpine Research* **26**: 290–6.

Freckman, D. W. & Virginia, R. A. (1997). Low-diversity Antarctic soil nematode communities: distribution and response to disturbance. *Ecology* **78**: 363–9.

Frederick, J. E. & Snell, H. E. (1988). Ultraviolet levels during the Antarctic spring. *Science* **241**: 438–40.

Friedmann, E. I. (1977). Microorganisms in Antarctic desert rocks from dry valleys and Dufek Massif. *Antarctic Journal of the United States* **12**: 6–30.

Friedmann, E. I. (1982). Endolithic microorganisms in the Antarctic cold desert. *Science* **215**: 1045–54.

Friedmann, E. I., Garty, J. & Kappen, L. (1980). Fertile stages of cryptoendolithic lichens in the dry valleys of southern Victoria Land. *Antarctic Journal of the United States* **15**: 166.

Friedmann, E. I., Hua, M. & Ocampo-Friedmann, R. (1988). Cryptoendolithic and cyanobacterial communities of the Ross Desert, Antarctica. *Polarforschung* **58** (2–3): 251–9.

Gebauer, R. L. & Ehleringer, J. R. (2000). Water and nitrogen uptake patterns following moisture pulses in a cold desert community. *Ecology* **81**: 1415–24.

Gerighausen, U., Bräutigam, K., Mustafa, O. & Peter, H.-U. (2003). Expansion of vascular plants on an Antarctic island – a consequence of climate change? In *Antarctic Biology in a Global Context*, ed. A. H. L. Huiskes, W. W. C. Gieskes, J. Rozema *et al.*, pp. 79–83. Leiden: Backhaus Publishers.

Godard, A. & André, M.-F. (1999). *Les Milieux Polaires*. Paris: Armand Colin.

Goodwin, I. D. (1993). Holocene deglaciation, sea level change, and the emergence of the Windmill Islands, Budd Coast, Antarctica. *Quaternary Research* **40**: 70–80.

Gould, K. S. (2004). Nature's Swiss army knife: the diverse protective roles of anthocyanins in leaves. *Journal of Biomedicine and Biotechnology* **5**: 314–20.

Green, T. G. A., Kulle, D., Pannewitz, S., Sancho, L. G. & Schroeter, B. (2005). UV-A protection in mosses growing in continental Antarctica. *Polar Biology* **28**: 822–7.

Green, T. G. A. & Lange, O. L. (1994). Photosynthesis in poikilohydric plants: A comparison of lichens and bryophytes. In *Ecophysiology of Photosynthesis*, ed. E.-D. Schulze & M. C. Caldwell. *Ecological Studies* **100**: 319–41.

Green, T. G. A., Schroeter, B. & Sancho, L. G. (1999). Plant life in Antarctica. In *Handbook of Functional Plant Ecology*, ed. F. I. Pugnaire & F. Valladares, pp. 495–543. Basel: Dekker.

Green, T. G. A., Schroeter, B. & Seppelt, R. D. (2000). Effect of temperature, light and ambient UV on the photosynthesis of the moss *Bryum argenteum* Hedw. in continental Antarctica. In *Antarctic Ecosystems: Models for Wider Ecological Understanding*, ed. W. Davison, C. Howard-Williams & P. Broady, pp. 165–70. Christchurch: Caxton Press.

Greenberg, B. M., Wilson, M. I., Huang, X. D. *et al.* (1997). The effects of ultraviolet-B radiation on higher plants. In *Plants for Environmental Studies*, ed. W. Wang, J. W. Gorusuch & J. S. Hughes, pp. 1–36. New York: Lewis Publishers.

Greenfield, L. G. (1992a). Retention of precipitation nitrogen by Antarctic mosses, lichens and fellfield soils. *Antarctic Science* **4**: 205–6.

Greenfield, L. G. (1992b). Precipitation nitrogen at maritime Signy Island and continental Cape Bird, Antarctica. *Polar Biology* **11**: 649–53.

Greenslade, P. (1995). Collembola from the Scotia Arc and Antarctic Peninsula including descriptions of two new species and notes on biogeography. *Polskie Pismo Entomologiczne* **64**: 305–19.

Grémillet, D. & Le Maho, Y. (2003). Arctic and Antarctic ecosystems: poles apart? In *Antarctic Biology in a Global Context*, ed. A. H. L. Huiskes, W. W. C. Gieskes, J. Rozema *et al.*, pp. 169–75. Leiden: Backhuys.

Gremmen, N. J. M., Huiskes, A. H. L. & Francke, J. W. (1994). Epilithic macrolichen vegetation of the Argentine Islands, Antarctic Peninsula. *Antarctic Science* **6**: 463–71.

Hansom, J. G. & Gordon, J. E. (1998). *Antarctic Environments and Resources. A Geographical Perspective*. Harlow: Longman.

Havström, M., Callaghan, T. V., Jonasson, S. & Svoboda, J. (1995). Little Ice Age temperature reduction measured by the reduced growth of an arctic heather. *Functional Ecology* **9**: 650–4.

Hawes, I., Howard-Williams, C. & Vincent, W. F. (1992). Desiccation and recovery of antarctic cyanobacterial mats. *Polar Biology* **12**: 587–94.

Hempel, G. (1995). Epilog. In *Biologie der Polarmeere. Erlebnisse und Ergebnisse*, ed. I. Hempel & G. Hempel, pp. 348–57. Jena: Gustav Fischer.

Hennion, F., Huiskes, A., Robinson, S. & Convey, P. (2006). Physiological traits of organisms in a changing environment. In *Trends in Antarctic Terrestrial and Limnetic Ecosystems: Antarctica as a Global Indicator*, ed. D. M. Bergstrom, P. Convey & A. H. L. Huiskes, pp. 127–57. Dordrecht: Springer.

Horikawa, Y. & Ando, H. (1967). The mosses of Ongul Islands and adjoining coastal areas of the Antarctic continent. *Japanese Antarctic Research Expedition Scientific Reports, Special Issue* **1**: 245–52.

Hovenden, M. J. & Seppelt, R. D. (1995). Exposure and nutrients as delimiters of lichen communities in continental Antarctica. *Lichenologist* **27**: 505–16.

Hughes, K. A., Lawley, B. & Newsham, K. K. (2003). Solar UV-B radiation inhibits the growth of Antarctic terrestrial fungi. *Applied and Environmental Microbiology* **69**: 1488–91.

Hurst, J. L., Pugh, G. J. F. & Walton, D. W. H. (1985). The effects of freeze-thaw cycle and leaching on the loss of soluble carbohydrates from leaf material of two subantarctic plants. *Polar Biology* **4**: 27–31.

Husain, S. R., Cillard, J. & Cillard, P. (1987). Hydroxyl radical scavenging activity of flavonoids. *Phytochemistry* **26**: 2489–91.

Jentsch, A., Kreyling, J. & Beierkuhnlein, C. (2007). A new generation of climate change experiments: events, not trends. *Frontiers in Ecology and the Environment* **5**: 365–74.

Kappen, L. (1993a). Lichens in the Antarctic region. In *Antarctic Microbiology*, ed. E. I. Friedmann, pp. 433–90. Mannheim: Wiley.

Kappen, L. (1993b). Plant activity under snow and ice, with particular reference to lichens. *Arctic* **46**: 297–302.

Kappen, L. & Breuer, M. (1991). Ecological and physiological investigations in continental Antarctic cryptogams. II. Moisture relations and photosynthesis of lichens near Casey Station, Wilkes Land. *Antarctic Science* **3**: 273–8.

Kappen, L., Friedmann, E. I. & Garty, J. (1981). Ecophysiology of lichens in the Dry Valleys of southern Victoria land, Antarctica. I. Microclimate of the cryptoendolithic lichen habitat. *Flora* **171**: 216–35.

Kappen, L., Lewis Smith, R. I. & Meyer, M. (1989). Carbon dioxide exchange of two ecodemes of *Schistidium antarctici* in continental Antarctica. *Polar Biology* **9**: 415–22.

Kappen, L., Sommerkorn, M. & Schroeter, B. (1995). Carbon acquisition and water relations of lichens in polar regions – potentials and limitations. *Lichenologist* **27**: 531–45.

Kennedy, A. D. (1993). Water as a limiting factor in the Antarctic terrestrial environment. *Arctic and Alpine Research* **25**: 308–15.

Kimball, S. L., Bennet, S. D. & Salisbury, F. B. (1973). The growth and development of montane species at near freezing temperatures. *Ecology* **54**: 168–73.

Lappalainen, N. M., Huttunen, S. & Suokanerva, H. (2008). Acclimation of a pleurocarpous moss *Pleurozium schreberi* (Britt.) Mitt. to enhanced ultraviolet radiation *in situ*. *Global Change Biology* **14**: 321–33.

Larson, R. A. & Berenbaum, M. R. (1988). Environmental phototoxicity. Solar ultraviolet radiation affects the toxicity of natural and man-made chemicals. *Environmental Science and Technology* **22**: 354–60.

Leishman, M. R. & Wild, C. (2001). Vegetation abundance and diversity in relation to soil nutrients and soil water content in the Vestfold Hills, East Antarctica. *Antarctic Science* **13**: 126–34.

Lewis Smith, R. I. (1985). Nutrient cycling in relation to biological productivity in Antarctic and sub-Antarctic terrestrial and freshwater ecosystems. In *Antarctic nutrient cycles and food webs. Proceedings of the 4th SCAR Symposium on Antarctic Biology*, ed. W. R. Siegfried, P. R. Condy & R. M. Laws, pp. 138–55. Berlin: Springer-Verlag.

Lewis Smith, R. I. (1994). Vascular plants as bioindicators of regional warming in Antarctica. *Oecologia* **99**: 322–8.

Lewis Smith, R. I. (2005). The thermophilic bryoflora of Deception Island: unique plant communities as a criterion for designating an Antarctic Specially Protected Area. *Antarctic Science* **17**: 17–27.

Ling, H. U. (1996). Snow algae of the Windmill Islands region, Antarctica. *Hydrobiologia* **336**: 99–106.

Ling, H. & Seppelt, R. D. (1993). Snow algae of the Windmill Islands, continental Antarctica. 2. *Chloromonas rubroleosa* sp. nov. (Volvocales, Chlorophyta), an alga of red snow. *European Journal of Phycology* **28**: 77–84.

Ling, H. U. & Seppelt, R. D. (1998a). Non-marine algae and cyanobacteria of the Windmill Islands region, Antarctica, with description of two new species. *Archiv für Hydrobiologie*, Supplement Volume 124, *Algological Studies* **89**: 49–62.

Ling, H. U. & Seppelt, R. D. (1998b). Snow algae of the Windmill Islands, continental Antarctica. 3. *Chloromonas polyptera (*Volvocales, Chlorophyta*) Polar Biology* **20**: 320–4.

Longton, R. E. (1979). Vegetation ecology and classification in the Antarctic Zone. *Canadian Journal of Botany* **57**: 2264–78.

Longton, R. E. (1988). *The Biology of Polar Bryophytes and Lichens*. Cambridge: Cambridge University Press.

Lovelock, C. E., Osmond, C. B. & Seppelt, R. D. (1995a). Photoinhibition in the Antarctic moss *Grimmia antarctici* Card. when exposed to cycles of freezing and thawing. *Plant, Cell & Environment* **18**: 1395–402.

Lovelock, C. E., Jackson, A. E., Melick, D. R. & Seppelt, R. D. (1995b). Reversible photoinhibition in Antarctic moss during freezing and thawing. *Plant Physiology* **109**: 955–61.

Markham, K. R., Franke, A., Given, D. R. & Brownsey, P. (1996). Historical Antarctic ozone level trends from herbarium specimen flavonoids. *Bulletin de Liaison du Groupe Polyphenols* **15**: 230–5.

Markham, K. R., Ryan, K. G., Bloor, S. J. & Mitchell, K. A. (1998). An increase in the luteolin:apigenin ratio in *Marchantia polymorpha* on UV-B enhancement. *Photochemistry* **48**: 791–4.

McKay, C. P., Freidmann, E. I., Gomex-Silva, B., *et al.* (2003). Temperature and moisture conditions for life in the extreme arid region of the Atacama Desert: four years of observations including the El Niño of 1997–1998. *Astrobiology* **3**: 393–406.

McKenzie, R. L., Aucamp, P. J., Bais, A. F., Björn, L. O. & Ilyas, M. (2007). Changes in biologically active ultraviolet radiation reaching the Earth's surface. *Photochemical and Photobiological Sciences* **6**: 218–31.

McKnight, D. M., Tate, C. M., Andrews, E. D. *et al.* (2007). Reactivation of a cryptobiotic stream ecosystem in the McMurdo Dry Valleys, Antarctica: a long-term geomorphological experiment. *Geomorphology* **89**(1/2): 186–204.

McRae, C. F., Hocking, A. D. & Seppelt, R. D. (1999). *Penicillium* species from terrestrial habitats in the Windmill Islands, East Antarctica, including a new species *Penicillium antarcticum*. *Polar Biology* **21**: 97–111.

Melick, D. R. & Seppelt, R. D. (1992). Loss of soluble carbohydrates and changes in freezing point of Antarctic bryophytes after leaching and repeated freeze-thaw cycles. *Antarctic Science* **4**: 399–404.

Melick, D. R. & Seppelt, R. D. (1994). Seasonal investigations of soluble carbohydrates and pigment levels in Antarctic bryophytes and lichens. *Bryologist* **97**: 13–19.

Melick, D. R. & Seppelt, R. D. (1997). Vegetation patterns in relation to climatic and endogenous changes in Wilkes Land, continental Antarctica. *Journal of Ecology* **85**: 43–56.

Mues, R. (2000). Chemical constituents and biochemistry. In *Bryophyte Biology*, ed. A. J. Shaw & B. Goffinet, pp. 150–81. Cambridge: Cambridge University Press.

Newsham, K. K., Hodgson, D. A., Murraya, A. W. A., Peat, H. J. & Lewis Smith, R. I. (2002). Responses of two Antarctic bryophytes to stratospheric ozone depletion. *Global Change Biology* **8**: 972–83.

Nienow, J. A., McKay, C. P. & Friedmann, E. I. (1988). The cryptoendolithic microbial environment in the Ross Desert of Antarctica: light in the photosynthetically active region. *Microbial Ecology* **16**: 271–89.

Oberbauer, S. F. & Starr, G. (2002). The role of anthocyanins for photosynthesis of Alaskan arctic evergreens during snowmelt. *Advances in Botanical Research* **37**: 129–45.

Ochyra, R., Lewis Smith, R. I. & Bednarek-Ochyra, H. (2008). *The Illustrated Moss Flora of Antarctica*. Cambridge: Cambridge University Press.

Øvstedal, D. O. & Lewis Smith, R. I. (2001). *Lichens of Antarctica and South Georgia. A Guide to their Identification and Ecology*. Cambridge, Cambridge University Press.

Pannewitz, S., Green, T. G. A., Maysek, K. *et al.* (2005). Photosynthetic responses of three common mosses from continental Antarctica. *Antarctic Science* **17**: 341–52.

Pannewitz, S., Schlensog, M., Green, T. G. A., Sancho, L. G. & Schroeter, B. (2003). Are lichens active under snow in continental Antarctica? *Oecologia* **135**: 30–8.

Paul, N. D. & Gwynn-Jones, D. (2003). Ecological roles of solar UV radiation: towards an integrated approach. *Trends in Ecology and Evolution* **18**: 48–55.

Peters, D. P. C. (2000). Climatic variation and simulated patterns in seedling establishment of two grasses at a semi-arid grassland ecotone. *Journal of Vegetation Science* **11**: 493–504.

Poage, M. A., Barrett, J. E., Virginia, R. A. & Wall, D. H. (2008). Geochemical control over nematode distribution in soils of the McMurdo Dry Valleys, Antarctica. *Arctic, Antarctic and Alpine Research* **40**: 119–28.

Post, A. (1990). Photoprotective pigment as an adaptive strategy in the Antarctic moss *Ceratodon purpureus*. *Polar Biology* **10**: 241–5.

Post, A. & Larkum, A. W. D. (1993). UV-absorbing pigments, photosynthesis and UV exposure in Antarctica: comparison of terrestrial and marine algae. *Aquatic Botany* **45**: 231–43.

Post, A. & Vesk, M. (1992). Photosynthesis, pigments, and chloroplast ultrastructure of an Antarctic liverwort from sun-exposed and shaded sites. *Canadian Journal of Botany* **70**: 2259–64.

Proctor, M. C. F. (2000). Physiological ecology. In *Bryophyte Biology*, ed. A. J. Shaw & B. Goffinet, pp. 225–47. Cambridge: Cambridge University Press.

Pugh, P. J. A. (1993). A synonymic catalogue of the Acari from Antarctica, the sub-Antarctic Islands and the Southern Ocean. *Journal of Natural History* **27**: 232–41.

Quesada, A. & Vincent, W. F. (1997). Strategies of adaptation by Antarctic cyanobacteria to ultraviolet radiation. *European Journal of Phycology* **32**: 335–42.

Reynolds, J. F., Kemp, P. R., Ogle, K. & Fernandez, R. J. (2004). Modifying the "pulse-reserve" paradigm for deserts in North America: precipitation pulses, soil water, and plant responses. *Oecologia* **141**: 194–210.

Roy, C. R., Gies, H. P. & Elliot, G. (1990). Ozone depletion. *Nature* **347**: 235–6.

Rozema, J., Björn, L. O., Bornman, J. F. *et al.* (2002). The role of UV-B radiation in aquatic and terrestrial ecosystems – an experimental and functional analysis of the evolution of UV-absorbing compounds. *Journal of Photochemistry and Photobiology B: Biology* **66**: 2–12.

Rozema, J., van de Staaij, J., Björn, L. O. & Caldwell, M. (1997). UV-B as an environmental factor in plant life: stress and regulation. *Trends in Ecology and Evolution* **12**: 22–8.

Ryan, K. G., Markham, K. R., Bloor, S. J. *et al.* (1998). UVB radiation induced increase in quercetin:kaempferol ratio in wild-type and transgenic lines of *Petunia*. *Photochemical Photobiology* **68**: 323–30.

Ryan, K. G., Burne, A. & Seppelt, R. D. (2009). Historical ozone concentrations and flavonoid levels in herbarium specimens of the Antarctic moss *Bryum argenteum*. *Global Change Biology* **15**: 1694–1702.

Santarius, K. A. & Giersch, C. (1983). Cryopreservation of spinach chloroplast membranes by low-molecular weight carbohydrates. II. Discrimination between colligative and noncolligative protection. *Cryobiology* **20**: 90–9.

Schlensog, M., Pannewitz, S., Green, T. G. A. & Schroeter, B. (2004). Metabolic recovery of continental Antarctic cryptogams after winter. *Polar Biology* **27**: 399–408.

Schwinning, S. & Ehleringer, J. R. (2001). Water use trade-offs and optimal adaptation to pulse-driven arid ecosystems. *Journal of Ecology* **89**: 464–80.

Scott, J. J. (1990). Changes in vegetation on Heard Island 1947–1987. In *Antarctic Ecosystems: Ecological Change and Conservation*, ed. K. R. Kerry & G. Hempel, pp. 61–76. Berlin: Springer-Verlag.

Searles, P. S., Flint, S. D. & Caldwell, M. M. (2001). A meta-analysis of plant field studies simulating stratospheric ozone depletion. *Oecologia* **127**: 1–10.

Selbmann, L., de Hoog, G. S., Mazzaglia, A., Friedmann, E. I. & Onofri, S. (2005). Fungi at the edge of life: cryptoendolithic black fungi from Antarctic desert. *Studies in Mycology* **51**: 1–32.

Selkirk, P. M., Seppelt, R. D. & Selkirk, D. R. (1990). *Subantarctic Macquarie Island: Environment and Biology*. Cambridge: Cambridge University Press.

Shar, C., Vidale, P. L., Luthi, D. *et al.* (2004). The role of increasing temperature variability in European summer heatwaves. *Nature* **427**: 332–6.

Skotnicki, M. L., Selkirk, P. M., Broady, P., Adam, K. D. & Ninham, J. A. (2001). Dispersal of the moss *Campylopus pyriformis* on geothermal ground near the summits of Mount Erebus and Mount Melbourne, Victoria Land, Antarctica. *Antarctic Science* **13**: 280–5.

Smith, G. J. & Markham, K. R. (1996). The dissipation of excitation energy in methoxyflavones by internal conversion. *Journal of Photochemistry and Photobiology, A* **99**: 97–101.

Smith, R. I. L. (1990). Signy Island as a paradigm of biological and environmental change in Antarctic terrestrial ecosystems. In *Antarctic Ecosystems: Ecological Change and Conservation*, ed. K. R. Kerry & G. Hempel, pp. 32–50. Berlin: Springer-Verlag.

Smith, V. R. & Steenkamp, M. (1990). Climatic change and its ecological implications at a subantarctic island. *Oecologia* **85**: 14–24.

Snell, K. R., Convey, P. & Newsham, K. K. (2007). Metabolic recovery of the Antarctic liverwort *Cephaloziella varians* during snowmelt. *Polar Biology* **30**: 1115–22.

Sømme, L. (1995). *Invertebrates in Hot and Cold Arid Environments*. Berlin: Springer-Verlag.

Stafford Smith, D. M. & Morton, S. R. (1990). A framework for the ecology of arid Australia. *Journal of Arid Environments* **18**: 255–78.

Stevens, M. I., Frati, F., McGaughran, A., Spinsanti, G. & Hogg, I. D. (2007). Phytogeographic structure suggests multiple glacial refugia in northern Victoria Land for the endemic Antarctic springtail *Desoria klovstadi* (Collembola, Isotomidae). *Zoologica Scripta* **36**: 201–12.

Stevens, M. I. & Hogg, I. D. (2006a). Contrasting levels of mitochondrial DNA variability between mites (Penthalodidae) and springtails (Hypogastruridae) from the Trans-Antarctic Mountains suggest long-term effects of glaciation and life history on substitution rates, and speciation processes. *Soil Biology and Biochemistry* **38**: 3171–80.

Stevens, M. I. & Hogg, I. D. (2006b). The molecular ecology of Antarctic terrestrial and limnetic invertebrates and microbes. In *Trends in Antarctic Terrestrial and Limnetic Ecosystems: Antarctica as a Global Indicator*, ed. D. M. Bergstrom, P. Convey & A. D. L. Huiskes, pp. 177–92. Dordrecht: Springer.

Stoner, W. A., Miller, P. & Miller, P. C. (1982). Seasonal dynamics of standing crops of biomass and nutrients in a subarctic tundra vegetation. *Holarctic Ecology* **5**: 172–9.

Takács, Z., Csintalan, Z., Sass, L. *et al.* (1999). UV-B tolerance of bryophyte species with different degrees of desiccation tolerance. *Journal of Photochemistry and Photobiology B: Biology* **48**: 210–15.

Tarnawski, M., Melick, D., Roser, D. *et al.* (1992). *In situ* carbon dioxide levels in cushion and turf forms of *Grimmia antarctici* at Casey Station, East Antarctica. *Journal of Bryology* **17**: 241–9.

Tearle, P. V. (1987). Cryptogamic carbohydrate release and microbial response during spring freeze-thaw cycles in Antarctic fellfield fines. *Soil Biology and Biochemistry* **19**: 381–90.

Tosi, S., Casado, B., Gerdol, R. & Caretta, G. (2002). Fungi isolated from Antarctic mosses. *Polar Biology* **25**: 262–8.

Vincent, W. F. (2000). Evolutionary origins of Antarctic microbiota: invasion, selection and endemism. *Antarctic Science* **12**: 374–85.

Vishniac, H. S. (1995). Biodiversity of yeasts and filamentous microfungi in terrestrial Antarctic ecosystems. *Biodiversity and Conservation* **5**: 1365–78.

Walter, H. (1931). *Die Hydratur der Pflanze und ihre physiologisch-ökologische Bedeutung.* Jena: Fischer.

Wasley, J., Robinson, S. A., Lovelock, C. E. & Popp, M. (2006). Some like it wet – biological characteristics and underpinning tolerance of extreme water stress events in Antarctic bryophytes. *Functional Plant Ecology* **33**: 443–55.

Weatherhead, E. C. & Andersen, S. B. (2006). The search for signs of recovery of the ozone layer. *Nature* **441**: 39–45.

Wilson, M. (1990). Morphology and photosynthetic physiology of *Grimmia antarctici* from wet and dry habitats. *Polar Biology* **10**: 337–41.

Woodin, S. J. & Marquiss, M. (eds.) (1997). *Ecology of Arctic Environments*. Oxford: Blackwell.

Wynn-Williams, D. D. (1980). Seasonal fluctuations in microbial activity in Antarctic moss peat. *Biological Journal of the Linnean Society* **14**: 11–28.

Wynn-Williams, D. D. (1996). Response of pioneer soil microalgal colonists to environmental change in Antarctica. *Microbial Ecology* **31**: 177–88.

VI SPHAGNUM AND PEATLANDS

14

Living on the Edge: The Effects of Drought on Canada's Western Boreal Peatlands

MELANIE A. VILE, KIMBERLI D. SCOTT, ERIN BRAULT, R. KELMAN WIEDER, AND DALE H. VITT

INTRODUCTION

Boreal peatlands of Western Canada

Boreal peatland ecosystems occupy less than 3% of the earth's land surface, yet store between 250 and 455 Pg of C, which is roughly 20%–30% of the world's soil carbon (Gorham 1991; Charman 2002; Joosten & Clarke 2002; Vasander & Kettunen 2006). In continental western Canada, peatlands cover 365,157 km^2 (Vitt *et al.* 2000) and dominate the landscape in northern Alberta, Saskatchewan, and Manitoba (Tarnocai 1984, 1998; Vitt *et al.* 2000; Tarnocai *et al.* 2005). Overall, these western Canadian peatlands have sequestered about 48 Pg of C during the past 10 000 years, with about half of this peat accumulated in the past 4000 years (Vitt *et al.* 2000). Peatlands provide a wide diversity of ecosystem services, not the least of which is the conversion of atmospheric CO_2 into large accumulations of stored organic carbon (C) – a testament to the long-term function of these ecosystems as net C sinks since their widespread initiation after the most recent glacial retreat (Halsey *et al.* 2000). Bryophytes typically are dominant components of the vegetation in northern peatlands and play central roles with regard to nutrient cycling and carbon accumulation.

The initiation, development, succession, and rate of peat accumulation in boreal peatlands are dependent on regional factors such as climate, substrate chemistry, landscape position, and hydrological regime. Water, including its amount, chemistry, fluctuation, and flow, together with moss species

Bryophyte Ecology and Climate Change, eds. Zoltán Tuba, Nancy G. Slack and Lloyd R. Stark. Published by Cambridge University Press.

composition, are the two most important drivers of the biological and physicochemical processes that characterize peatland ecosystems (Vitt 2005). Although peatlands are perceived to be characteristic of wet, oceanic areas, great expanses of peat are found in cool, continental regions of the boreal and sub-Arctic zonobiomes (Vitt 2006) at the dry end of the precipitation spectrum under which peatlands persist (Wieder 2006). In northern Alberta, where most of our research occurs, mean annual precipitation ranges from about 350 to 500 mm (Chetner *et al.* 2003), establishing our sites as end-members for examining the effects of climate change-induced drought and changes in soil moisture on bryophyte communities and peatland ecosystem function.

Bryophyte-dominated ground layers play a critical role in peatland C balance and in preventing surface desiccation. Peatlands often have a nearly continuous moss cover poised at the interface between the pedosphere and the atmosphere. Numerous studies, mostly focusing on *Sphagnum* species, have indicated that growth and/or net primary production are strongly influenced by moisture availability and/or repeated desiccation–wetting cycles at the peat surface (e.g., Titus & Wagner 1984; Wagner & Titus 1984; Gerdol *et al.* 1996; Schipperges & Rydin 1998; McNeil & Waddington 2003; Bortoluzzi *et al.* 2006). These studies collectively suggest that the strength of the peatland C sink may become compromised as moisture availability at the peat surface decreases. While surface peat moisture may vary on a scale of days in response to precipitation and evapotranspiration regimes, changing climatic conditions, including multi-year droughts, are likely to be manifested through changes in water table position and fluctuation, with potential ramifications on C cycling and sequestration (cf. Silvola *et al.* 1996; Alm *et al.* 1999; Tuittila *et al.* 2004).

Peatlands of western Canada can be categorized into three basic types: bogs, poor fens, and rich fens, with bogs receiving precipitation solely from atmospheric inputs, and both rich and poor fens relying on ground water supplies and surface water flows to maintain a relatively stable water table. Bogs tend to have lower water tables than fens, ranging from 20 to 80 cm below the surface. Although both rich and poor fens differ markedly in their chemistry and vegetation, both types have water tables near the surface of the ground layer vegetation. Both fen types have variable surface topography. Much of the fen surface exists as carpets and lawns of vegetation that range in height above the water surface from 1–2 cm (carpets) to 10–15 cm (lawns; Sjörs 1948). Drier, more raised microtopographic features (hummocks) also are abundant within all three peatland types. These local landforms of carpets, lawns, and hummocks have been well studied in terms of their dominant species, especially in western Canada (e.g., Slack *et al.* 1980; Chee & Vitt 1989; Nicholson & Vitt 1990; Vitt *et al.* 1995).

Environmental responses of individual dominant species of these fen types were shown by Gignac *et al.* (1991a) to be very sensitive to water level variation. In 1994, Gignac and Vitt (1994) predicted drastic changes for fens in central Alberta, concluding that with a 2 °C rise in annual temperatures these rich fens would lose substantial percentages of their ground layer species. Across the Canadian Prairie Provinces, mean annual and winter temperatures have increased by 1.3 °C and 3.1 °C, respectively, over the past 58 years, along with decreasing trends in annual, and especially winter, precipitation. Implications of decreasing winter precipitation (i.e., snowfall) and increasing temperatures include less insulation of surface peat, an increase in the depth of peat freezing, and less recharge of peatland water tables during spring snowmelt.

In short, peatlands of continental western Canada are situated in a climatic regime where precipitation is low, while also being at the southernmost edge of their distribution within the boreal forest (Zoltai & Vitt 1990). As climate continues to change, boreal regions of the northern hemisphere, including continental western Canada, are expected to warm. With increasing temperatures, the frequency of surface desiccation of mosses in peatlands is more likely to increase. Long-term predictions about peatland water tables face the same uncertainty as predictions regarding future precipitation regimes, but if both rain and snowfall continue to decline in continental western Canada, peatland water tables will be lowered. These changes pose unknown consequences for bryophyte communities, which are the foundation species of these ecosystems.

Bryophytes: foundation species of boreal peatlands

A ground layer comprising nearly a complete cover of bryophytes is one of the most important distinguishing features of boreal peatland ecosystems, and likely most significant in maintaining a positive soil moisture balance. Bryophytes, including both true mosses and *Sphagnum*, are limited by geochemical and dryness gradients that vary among peatland types (i.e., bog, poor fen, and rich fen). Characteristic indicator species that respond to these moisture gradients have been frequently discussed in the peatland literature (e.g., Chee & Vitt 1989). For example, the Fennoscandian peatland ecologists long ago wrote about the "red and golden alliance" composed of mixed populations of *Sphagnum warnstorfii* [the red] and *Tomenthypnum nitens* [the gold] as a key indicator of rich fens "... de torraste rikkärrmossmattornas röda och gyllene union av röd *Sphagnum Warnstorfianum* och guldgul *Tomenthypnum nitens* ... which translates to ... the red and golden union of red *Sphagnum Warnstorfianum* and golden yellow *Tomenthypnum nitens* of the

Table 14.1. *Foundation species (see text) for continental boreal peatlands in western Canada*

	Bog	Poor fen	Rich fen
Hummock	*Sphagnum fuscum*	*Sphagnum magellanicum*	*Tomenthypnum nitens*
Hollow/lawn/carpet	*Sphagnum angustifolium*	*Sphagnum angustifolium*	*Hamatocaulis vernicosus*
			Scorpidium scorpioides

driest rich fen moss mats . . ." (Du Rietz 1948, translated by Håkan Rydin). More recently, general habitat preferences of many abundant bryophyte species have been summarized by Vitt (www.peatnet.siu.edu/peatguide) and in a number of papers that have developed models of habitat responses for many of these indicator species under different temperature and precipitation regimes (Gignac *et al.* 1991a,b).

In addition to sequestering C, bryophytes are excellent indicators of habitat change, can define gradients within and among northern peatlands, and also perform important community functions. Whitham *et al.* (2006) stated that "species that structure a community by creating locally stable conditions for other species, and by modulating and stabilizing fundamental ecosystem processes are foundation species." Clearly included among the foundation species of northern peatlands are a number of bryophyte species, most of which have habitats that are limited to some extent by water table variability and relative height above the permanently saturated water level. Foundation species for fens and bogs are unique for each of the three peatland types yet provide similar fundamental services to each of these ecosystems. In poor fens and bogs, foundation species are primarily species of the genus *Sphagnum*, whereas in rich fens they are true mosses (Table 14.1). In all cases, these species provide a living, vertically accumulating soil column that creates a varied relief to the peatland, and increases the surface area of the ground layer. These foundation bryophyte species provide four primary services to peatland ecosystems: (1) acidification, (2) water holding capacity and capillarity, (3) formation of the acrotelm, and (4) nutrient limitation. All of these bryophyte-mediated functions are regulated by water availability.

Acidification: Several species of *Sphagnum*, owing to the high cation exchange capacity (CEC) of the their cell walls, are capable of exchanging inorganic ions (e.g., base cations) for H^+, thus producing acidic waters (Clymo

1963; Hemond 1980; Richter & Dainty 1989). The high CEC of mosses is at least partially responsible for the chemistry of the surrounding waters and forms the foundational chemical habitat for co-existing plants. The acidification action of *Sphagnum* not only reduces the pH, but also limits the availability of all other cations including NH_4^+ and K^+.

Water holding capacity and capillarity: all mosses, through a variety of internal and external features, hold relatively large quantities of water, and for some species of *Sphagnum* this is up to 20–25 times their dry mass. Internally this is accomplished through the death of cells and the utilization of these dead cells as storage cavities for water (Malcolm 1996). Included here are the enlarged hyaline cells of *Sphagnum* leaves, and also the hyaline, basal, and alar cells of true mosses. Externally, the large leaf to stem ratio of mosses, abundant stem outgrowths of some mosses, and complicated branching patterns that result in complex canopies all reinforce the external water holding capability of peatland bryophyte species (Vitt & Glime 1984). When combined, these internal and external features of bryophytes allow them to store large amounts of water, as well as to move water upward by capillary action. The water-holding and wicking properties of peatland bryophytes allow moss populations to expand upward and outward and "swamp" the immediate surrounding area, creating a unique environment for co-existing plants.

Acrotelm development: one of the unique features of many peatlands is the development of an aerobic zone (acrotelm) positioned above the deeper, largely anaerobic peat deposit (the catotelm). These acrotelms are developed and maintained by bryophytes. *Tomenthypnum nitens* in rich fens, *Sphagnum magellanicum* in poor fens, and in boreal bogs, *Sphagnum fuscum* all provide the matrix for acrotelmic development. The acrotelm's aerobic waters are maintained within the capillary spaces of the growing bryophyte layer. The resistance of these bryophyte plants to decay under aerobic conditions (Turetsky *et al.* 2008) allows for acrotelmic persistence, and also provides an aerobic zone wherein vascular plant roots reside along with a host of fungal and bacterial populations (Navaratnam 2002).

Nutrient limitation: both true mosses and species of *Sphagnum* strongly compete for incoming nutrients in these characteristically nutrient-poor systems. Recent evidence indicates that some mosses can out-compete other bryophyte species, and perhaps even some vascular plant species for specific macro and micro nutrients (Gotelli *et al.* 2008). Additionally, *Sphagnum* plays a large role in the N cycle in bogs (Wieder 2006). Atmospherically deposited N is readily taken up by bryophytes, and in general, N-sequestration within *Sphagnum* can have a limiting effect on vascular plant growth in peatlands

(Li & Vitt 1997). Slow decomposition rates also play an important role in nutrient availability in both fens and bogs as C/P and C/N ratios affect rates of release of these elements from both vascular and bryophyte litter (Bragazza *et al.* 2007). Bryophytes are adapted to nutrient-poor systems and also aid in creating systems that remain nutrient-depleted.

In summary, bryophytes form the ground layer of peatland ecosystems and are the foundation for structuring the habitats in which other plants also exist. These foundation attributes include controls on chemistry, water levels, and nutrients, as well as the peat substrates themselves. However, the manner in which bryophytes respond to their surroundings is not similar to that of vascular plants, and these differences may result in enhanced sensitivity to climatic changes, especially changes in water levels due to prolonged drought. Differences between bryophyte and vascular plant responses to changes in their local environments involve fundamental aspects of both structure and function.

Bryophytes, although small, *en masse* develop extremely complicated canopy structures. A bryophyte mat or cushion is not a two-dimensional object with only surface features; rather it is series of stems and branches that form a complex interwoven canopy, superficially not unlike that of a vascular-plant-dominated forest or grassland; however, there are several major differences. Within the bryophyte canopy the plant activity levels are determined by the water deficits of the individual leaves. Here a single leaf may be photosynthetically active while its neighbor can be inactive. Secondly, water movement in many bryophyte canopies is largely downward, thus the source of water is largely from precipitation, and nutrient uptake is directly from inorganic sources. Thirdly, bryophyte canopies are home to invertebrates and fungi, and often algae, while vascular plant canopies harbor invertebrates and vertebrates. Fourthly, bryophyte leaves, relative to their stems, are extremely large, and when compared with vascular plant leaves and stems, they never have petioles, but they have the ability to change orientation relative to water content. Finally, light penetration, patterns of nitrogen retention, and C assimilation in *Sphagnum* canopies are all different from those in vascular plant canopies (van der Hoeven *et al.* 1993; Rice *et al.* 2008).

In vascular plants, photosynthesis takes place and C is fixed as long as the component cells retain turgor pressure. When turgor is lost, cells die. Thus, regulation of turgor is of paramount importance to vascular plants and is accomplished by rather complicated structural modifications such as multistratose leaves, a well-developed cuticle, and stomates that help regulate water losses. Water transport systems are extremely well developed and allow for water to be moved upward to replace what is lost through evapotranspiration. Additionally, many vascular plants have deciduous leaves that are not present during times of either drought or cold. Bryophytes, on the other hand, have little ability to regulate turgor. Bryophyte leaves are normally one cell thick and hence have no stomata; the cells have only a thin, poorly developed cuticle, and are never deciduous. Water-conducting systems that transport water from rhizoids to stems to leaves are poorly developed in most peatland bryophytes. The result is that photosynthesis (and hence C sequestration and plant growth) in vascular plants is most often limited by factors other than water deficits, whereas in bryophytes, photosynthesis is always limited first by water availability (Bazzazz *et al.* 1970; Peterson & Mayo 1975). Early studies of photosynthetic rates seemed to show that bryophyte photosynthetic rates were considerably less than those of vascular plants. These studies were based on photosynthetic rates on a plant biomass basis. Thus comparisons were made between entire moss plants and a vascular plant leaf. These comparisons are strongly biased since dead non-photosynthetic stem tissues were included in the moss biomass but excluded for the vascular plants. A study by Martin and Adamson (2001), however, demonstrates that photosynthetic rates of bryophytes are comparable to those of vascular plants when based on chlorophyll amounts. Further, this study underscores the importance of water as a key limiting factor for bryophyte photosynthesis.

In summary, bryophytes fix C and grow when they are wet; however, bryophytes have relatively little ability to control the extent of these periods. Thus, water availability is the key factor in the establishment and maintenance of bryophyte populations, and both climate-change-induced drought frequency and changes in soil moisture have the potential to collectively compromise peatland function.

Drought as a characteristic of the Canadian Prairie Provinces

Drought is a recurring feature in the continental climate of the Prairie Provinces and has received considerable attention in light of its clear relationship to agriculture in the southern parts of Alberta, where crop production dominates land use (Akinremi *et al.* 1996; Schindler & Donahue 2006). Droughts have widespread economic and environmental impacts beyond the agricultural sector that include reduced forest production and increased fire frequency. Costs to other private and public sectors also can be substantial. In the late 1990s, a 50% loss of wetlands, a 16% decline in duck populations, and an increase in waterfowl disease contributed to over $5 million in lost hunting revenue (Wheaton *et al.* 1992). The magnitude of these losses makes recurrent droughts the largest source of risk, uncertainty, and hardship in the western Canadian economy.

Canada is often ranked among the top five countries for per capita water supply; Canada's western Prairie Provinces, however, which cover about 2 million square kilometers, lie in the rain shadow of the Rocky Mountains, and as a result, represent the largest and driest area of southern Canada (Schindler & Donahue 2006). Alberta, the westernmost Prairie Province, has been in a persistent drought for the past 13 years. The period from 1998 to 2004 experienced some of the warmest temperatures and lowest precipitation on record; this period has been considered more severe than "dust bowl" conditions in the 1930s, when 7.3 million hectares of agricultural land were affected and 250 000 people left the Canadian prairies (Wilson *et al.* 2002; Schindler & Donahue 2006). Alberta is a region especially prone to dry growing conditions, and utilizes more irrigation than any other region or province in Canada. About once every four or five years, Alberta experiences conditions that many characterize as a drought (Wilson *et al.* 2002).

Despite considerable effort focussed on the effects of drought on agriculture and economics in the southern parts of the Prairie Provinces, very little attention has been paid to the potential short-term and long-term effects of drought on the peatland ecosystems that dominate the landscape in the northern parts of these provinces. Surface wetness, near-surface wetness, and the position of the water table in peatlands have important ecological, climatic, community, and ecosystem-level consequences for C cycling (Harris *et al.* 2006). We suggest that the key to understanding C cycling in peatlands that are subjected to hydrological changes depends on differentiating effects due to short-term changes in surface wetness versus long-term changes in water table position. Recurring drought cycles that characterize the climate of continental western Canada are hypothesized to have profound effects on peatland surface and near-surface wetness, with somewhat lesser effects on peatland water table position, at least in fens, which are connected to regional surface water and groundwater flow networks. When long-term directional changes in annual precipitation are superimposed on the drought cycles, however, we hypothesized that peatland water tables would begin to drop, with a set of consequences for peatland C cycling that differ from the consequences of changing surface wetness. Here we provide data that lend support to the above-mentioned hypotheses. Specifically, we discuss how drought-induced changes to the bryophyte communities in bogs and fens, along with changes in two major hydrological drivers, namely surface desiccation and water table position, individually and interactively affect C cycling in bogs and fens in northern Alberta, Canada. We highlight results of a field rainfall-exclusion experiment on CO_2 fluxes from a bog, and rates of net ecosystem exchange (NEE) from a drought-impacted fen. Results from the former experiment underscore the impact of excluding rainfall during periods of non-drought and a relatively stable water table, whereas the latter study highlights the impact of lowered water-tables on CO_2 fluxes during periods of drought.

Rainfall exclusion experiment

Methods

The rainfall exclusion experiment was conducted at two *Sphagnum*-dominated bogs near Utikuma Lake (55°48.358′N, 115°10.977′W), in northern Alberta that were part of a 102 year time-since-fire chronosequence where we quantified the trajectory of C balance recovery after fire (Wieder *et al.* 2009). The time-since-fire of the two Utikuma bogs was 58 and 95 years (in 2009). At each site, we had installed 30 aluminum collars (60 cm × 60 cm) for quantification of net CO_2 fluxes using clear polycarbonate static chambers (30 cm tall) and infrared gas analyzers. In 2007, we placed clear, acrylic rain-exclusion roofs (1.2 m × 1.2 m) over 15 plots per bog; the other 15 plots per bog remained unroofed. The roofs were supported by rebar posts pounded into the peatland and were tilted so that the highest to lowest corners were about 30 and 45 cm, respectively, above the peat surface.

Hourly peat moisture and temperature data were collected from June 15 through July 20, 2007, using Decagon Devices ECH$_2$O-TM soil moisture probes placed approximately 5 cm below the peat surface of each plot. The soil moisture probes were calibrated for surface *Sphagnum* peat in the laboratory. A 13.5 cm × 13.5 cm, 10 cm thick intact block of moist peat was placed snugly into a plastic container and a soil moisture probe was placed about 5 cm below the peat surface. Deionized water was added to bring the peat to saturation (100% water content) and was then oven-dried at 60 °C to a constant mass (0% water content). A weather station (WatchDog®) located at each bog measured hourly photosynthetically active radiation (PAR), air temperature (°C), and precipitation (mm).

On eight sampling dates from June through July of 2007, we measured net CO_2 fluxes ($n = 3599$) as a function of *PAR* (100, 75, 50, 25, and 0% full sun) by progressively covering the polycarbonate chambers with different densities of shade cloth. Measurements of CO_2 exchange were taken at intervals of 2 minutes for each *PAR* level for a total of 10 minutes per plot during each sampling day. We characterized understory net ecosystem production (NEP_U) as a function of *PAR* and near-surface air temperature (Wieder *et al.* 2009):

$$NEP_U = \frac{P_{MAX} \times PAR \times \alpha}{PAR \times \alpha + P_{MAX}} - A \times Q_{10}^{T/10}$$

The parameters of the equation are P_{MAX} (maximum gross CO_2 capture at infinite *PAR*; μmol CO_2 m^{-2} s^{-1}), α (photosynthetic quantum efficiency; μmol CO_2 m^{-2} s^{-1} per μmol *PAR* m^{-2} s^{-1}), A (total understory respiration from both vascular plants within the chambers and decomposition in the underlying peat column at 0 °C; μmol CO_2 m^{-2} s^{-1}), and Q_{10} (the temperature-dependence of total understory respiration; dimensionless). We also calculated NEP_{SAT} as the mean NEP_U for $PAR > 1000$ μmol m^{-2} s^{-1} (Bubier *et al.* 2003).

Results and discussion

Although peat moisture probes were all placed at about 5 cm from the peat surface, there was considerable hummock to hummock variation in peat moisture data. Average peat moisture was considerably higher in the burned bog site than in the unburned bog site (Fig. 14.1), even though water table positions were similar at the two sites (Wieder *et al.* 2009), suggesting that peat moisture at the tops of hummocks may not be strongly connected to height above water table. That is of interest since this connection has generally been assumed.

At each of the bogs, surface peat moisture increased in response to rain events, but roofing removed this response (Fig. 14.1). Roofing did not result in a marked drying out of near-surface peat, even during the period of low rainfall in July. At each site, exclusion of rainfall by roofing had minimal effects on the parameters that characterize NEP_U, on NEP_{SAT} (*NEP* at *PAR* values 1000 μmol m^{-2} s^{-1}), or on dark respiration (Table 14.2). Residuals from the fitting of the rectangular parabola relationship between *NEP* and *PAR* were not correlated with relative moisture content of surface peat at either site, in either the roofed or unroofed plots (Table 14.2).

At the two Utikuma bog sites, the bryophyte component is dominated by *Sphagnum fuscum*, with *Pleurozium schreberi*, *Leiomylia anomala*, and *Cladina mitis* present but much less abundant; vascular components of the understory include *Ledum groenlandicum*, *Vaccinium vitis-idaea*, *Oxycoccos microcarpon*, and *Rubus chamaemorus* (Wieder *et al.* 2009). In continental western Canada, post-fire successional development of the moss community coincides with post-fire changes in the source/sink status of bogs (Benscoter & Vitt 2008; Wieder *et al.* 2009).

Numerous studies have focused on the physiological responses of *Sphagnum* species, in particular gross photosynthesis, respiration, and net photosynthesis, to desiccation (e.g., Titus *et al.* 1983; Wagner & Titus 1984; Titus & Wagner 1984; Murray *et al.* 1989; Schipperges & Rydin 1998; McNeil & Waddington 2003). When moss/peat moisture content is manipulated to span wide ranges, often to near complete dryness, responses clearly indicate that desiccation

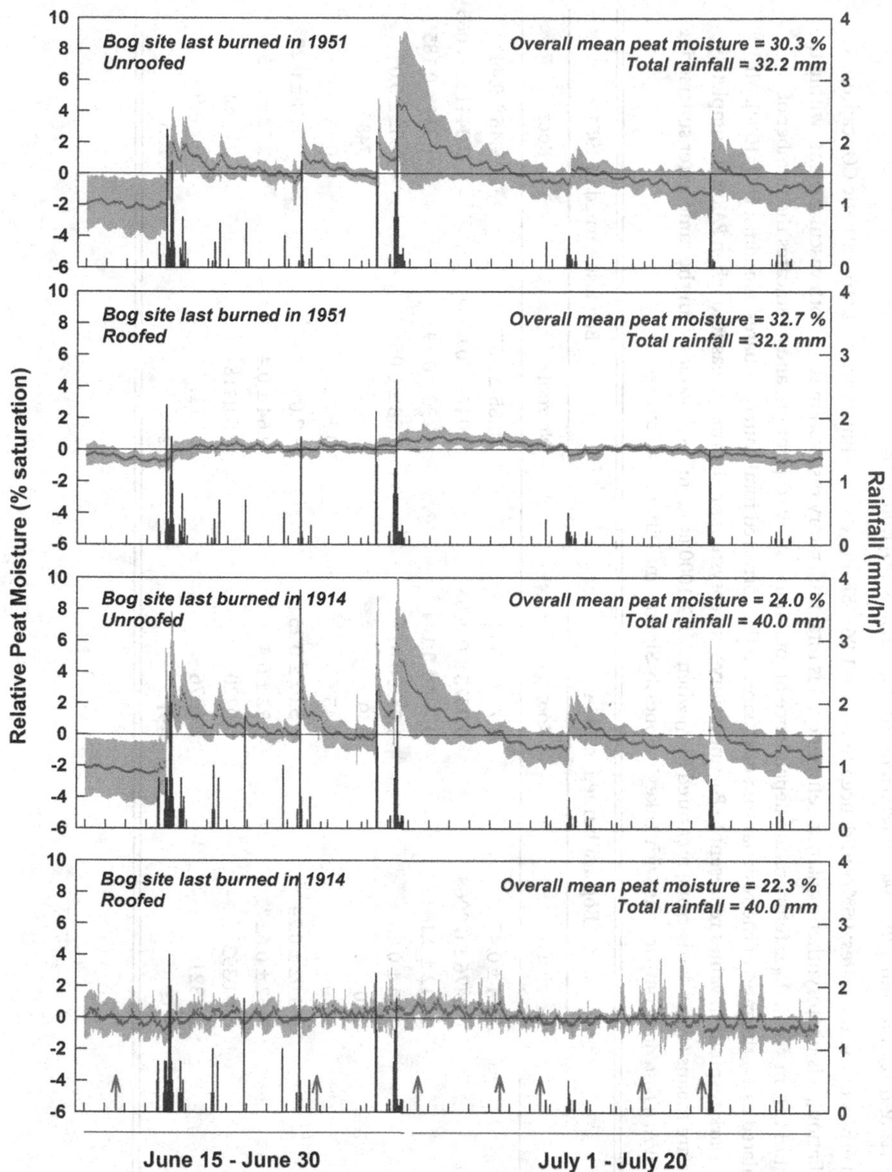

Fig. 14.1. Relative peat moisture (% saturation) plotted continuously as a line with errors, and rainfall (mm) plotted as bars as a function of day beginning on June 15 and ending on July 20, 2007 for both roofed and unroofed plots in each of the bog sites (burned in 1951 and 1914). Arrows indicate CO_2 sampling dates.

compromises a positive C balance of *Sphagnum* species. Our rainfall exclusion study looks not at the effects of desiccation, but at the effects of periodic wetting of the peat surface. Under field conditions without rainfall exclusion, volumetric moisture content of surface mosses, mostly *Sphagnum fuscum*, and

Table 14.2. *Parameter estimates from fitting a rectangular parabola to relationships between* NEP_U *and* PAR

Values are shown ± approximate standard errors of the estimates; 95% confidence intervals = 1.96 × SE. P_{MAX} is maximum gross photosynthetic CO_2 capture at infinite *PAR* (photosynthetically active radiation), α is photosynthetic quantum efficiency, *A* is total understory respiration from both vascular plants within the chambers and decomposition at 0 °C (*A*, μmol CO_2 m^{-2} s^{-1}), Q_{10} is temperature-dependence of total understory respiration, and *n* indicates the number of observations; pseudo-R^2 values were calculated as 1 – the ratio of the residual sum of squares to the corrected total sum of squares (Lindquist *et al.* 1994). All regressions were significant at $p < 0.0001$. Dark respiration values represent NEP_U (understory net ecosystem production) measured when *PAR* was completely blocked from the chambers; NEP_{SAT} values are means (± standard errors) measured NEP_U when $PAR \geq 1000$ μmol m^{-2} s^{-1}; values with the same letter superscript do not differ significantly ($p < 0.0001$; ANOVA, *a posteriori* comparisons with Tukey's Honestly Significant Difference test).

	Bog last burned in 1914		Bog last burned in 1951	
	No roof	Roof	No roof	Roof
P_{MAX} (μmol CO_2 m^{-2} s^{-1})	8.84 ± 0.47	8.72 ± 0.53	10.65 ± 0.57	9.46 ± 0.44
α (μmol CO_2 m^{-2} s^{-1} per μmol *PAR* m^{-2} s^{-1})	0.0476 ± 0.0068	0.0373 ± 0.0058	0.0428 ± 0.0064	0.0441 ± 0.0065
A (μmol CO_2 m^{-2} s^{-1})	1.112 ± 0.164	0.794 ± 0.144	1.290 ± 0.198	1.324 ± 0.185
Q_{10}	1.54 ± 0.06	1.65 ± 0.07	1.58 ± 0.05	1.52 ± 0.05
n	750	749	745	748
Pseudo R^2	0.49	0.45	0.56	0.58
Dark respiration (μmol CO_2 m^{-2} s^{-1})	– 5.02 ± 0.24 [a]	– 5.07 ± 0.25 [a]	– 7.05 ± 0.42[b]	– 6.31 ± 0.23[b]
NEP_{SAT} (μmol CO_2 m^{-2} s^{-1})	3.52 ± 0.62 [a]	3.72 ± 0.49 [a]	2.64 ± 0.41 [a]	2.93 ± 0.35 [a]
Correlation between NEP_U residuals	– 0.0392	0.0220	– 0.0418	0.0762
and relative peat moisture;	0.4321	0.6276	0.5117	0.2449
p value and *n* beneath	404	491	248	235

of surface peat varied by only 8%–10% during the warmest and driest parts of the year (Fig. 14.1). Rainfall exclusion eliminated the wetting of surface peat (Fig. 14.1), but had no clear measurable effects on indicators of bog C cycling (Table 14.2). If ongoing climate change leads to less frequent wetting of the peat surface, but conditions allow the peat surface to maintain fairly constant moisture, implications for bog C balance could be minimal.

Net ecosystem exchange in drought-impacted fens

Methods

We initiated detailed studies of the drought-impacted and dying hummocks in fens of boreal, continental, western Canada. Our main objective was to quantify C losses (measurements of net CO_2 flux) from peatlands where mosses were visually impacted by drought. We established permanent vegetation plots (0.5 m × 0.5 m; 10 plots) at a rich fen near Perryvale, Alberta (54°28′N, 113°17′W) that exhibited severe signs of drought in 2004. We characterized vegetation in each plot during a July (mid-growing season) visit, and revisited annually to document ongoing changes in vegetation. Perryvale lies in a shallow basin and forms the outer treed mat surrounding a small body of water. The peat surface consists of large hummocks separated by wetter lawns. The site is dominated by a mixture of *Larix laricina* and *Picea mariana* that form an open canopy. Shrubs and herbs are abundant, including *Betula glandulifera, Salix pedicellaris, Potentilla palustris, Smilacina trifolia, Carex chordorrhiza, C. interior*, and *Vaccinium macrocarpon*. The ground layer is composed of hummocks of *Tomenthypnum nitens, Aulacomnium palustre, Sphagnum warnstorfii*, and *S. teres*. Lawns were dominated by *Hamatocaulis vernicosus* and *Calliergon giganteum*. In summer of 2004, we found that the hummocks (especially on the south-facing sides) of *Tomenthypnum nitens* were dead. *Hamatocaulis vernicosus* and *Calliergon giganteum* were missing from the lawns, being replaced by species that favor drier habitats (*Aulacomnium palustre, Tomenthypnum nitens*, and *Sphagnum warnstorfii*). From an ecological perspective, it was especially interesting to note that individual hummocks, composed exclusively of *Tomenthypnum nitens*, were completely dry with dead moss on the south-facing portion, but with desiccated live mosses on the north-facing portion.

At each site, we built boardwalks; adjacent to each boardwalk, we installed 10 aluminum collars (Bubier *et al.* 2003; Wieder *et al.* 2009) for CO_2 flux sampling (net ecosystem exchange; NEE). NEE was measured on 6 sampling dates from June through September of 2004 (n = 336) and 12 sampling dates from May through August of 2007 (n = 713). Fluxes of CO_2 were measured exactly as in the rainfall exclusion experiment. The entrapped air was recirculated through an infrared gas analyzer (IRGA; PP Systems, model EGM-4) to measure changes in CO_2 concentrations over 2-min measurement periods under full sunlight conditions and four progressive shading conditions through full darkness (70%, 50%, 30%, and 0% of full sun). CO_2 flux data were combined across all collars and all dates; parameter estimation for the rectangular hyperbola to characterize NEP_U as a function of *PAR* and temperature was accomplished exactly as above, using non-linear regression (PROC NLIN in SAS). The means of the parameters of these equations were compared between 2004 (a dry year) and 2007 (a wetter year relative to 2004) to determine whether there were differences in net ecosystem CO_2 exchange (NEP_U).

Results and discussion

Drought can have serious impacts on peatland C cycling, as evidenced by significantly low or negative NEE in drought-stressed fens relative to healthy fens (Fig. 14.2). In 2004, when precipitation was substantially lower than

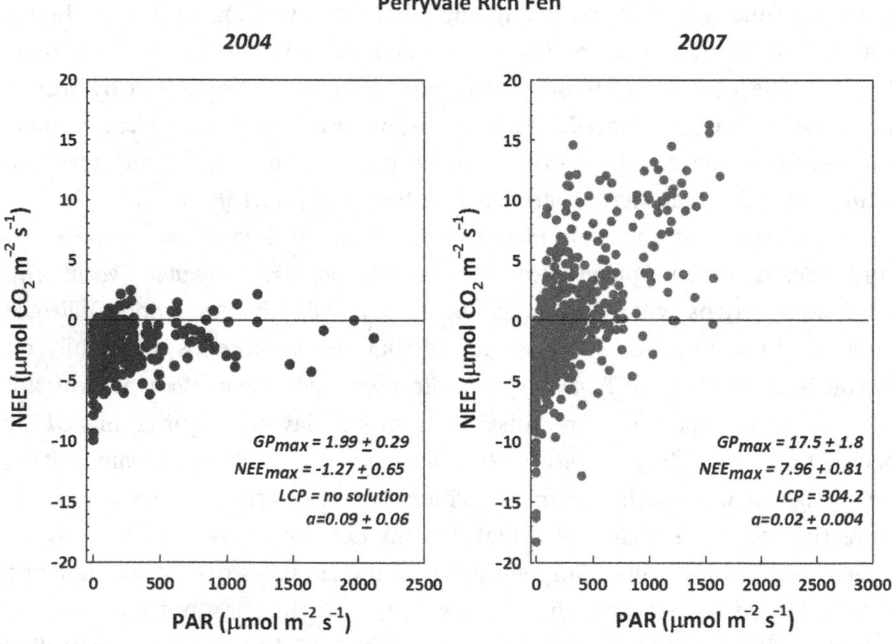

Fig. 14.2. Net ecosystem exchange (μmol CO_2 m^{-2} s^{-1}) as a function of *PAR* (μmol m^{-2} s^{-1}) for the Perryvale fen in 2004 (left panel) and 2007 (right panel). Horizontal lines are plotted to separate positive NEP_U (understory net ecosystem production; points above the line), from NEP_U (points below the line). We adopted the sign convention that positive NEP_U reflects CO_2 uptake by the ecosystem and negative NEP_U represents CO_2 loss to the atmosphere.

predicted relative to historical data (Alberta Environment), NEE rarely exhibited positive values (Fig. 14.2). As a result, the hyperbolic fit of the data yielded no solution for the light compensation point, and both GP_{MAX} and NEP_{MAX} were negative (Fig. 14.2). In 2007, however, when precipitation was normal to above average, NEE in the formerly drought-stressed fen recovered substantially, such that these drought-stressed fens were net sinks for atmospheric C for at least part of the year (Fig. 14.2). In contrast, Bragazza (2008) witnessed no signs of recovery in *Sphagnum* mosses after four years of severe drought in a peatland located in the Italian Alps. Interestingly, Bragazza (2008) also noted that mortality of mosses was restricted to portions of hummocks receiving the highest amount of solar radiation, much as we observed at Perryvale in 2004 during the tail end of the drought period.

The hyperbolic fit of data for 2007 indicated a LCP of 304 $\mu mol\ m^{-2}\ s^{-1}$, which is the amount of *PAR* needed for CO_2 fluxes to switch from net sources to net sinks of atmospheric C (Fig. 14.2). Additionally, α, which is the initial slope of the rectangular hyperbola, was much higher in 2004 than in 2007 (0.09 vs. 0.02, respectively; Fig. 14.2), indicating that mosses were more efficient with their photosynthetic use of C under drought stress. In general, when peatlands are functioning as net sinks for atmospheric C, they tend to have lower LCPs and can have lower or higher values of α depending on the age of the bog or fen (Wieder *et al.* 2009). There is some evidence to suggest that certain moss species have more efficient photosynthetic capacities under stressful conditions, such as when water and light are limiting (Hajek *et al.* 2009). Additionally, Hajek *et al.* (2009) demonstrated that both *Sphagnum* and *Polytrichum* exhibited higher maximum yields and higher photosynthetic capacities under water table draw-down when compared with the same moss species in open mire with no water table stress, suggesting that some species are able to acclimate to changing habit conditions. Under future predicted scenarios of global climate change, adaptation and acclimation may become important mechanisms for maintaining peatland ecosystem function.

Water table heights, measured as depth (in cm) below the peat surface, were quite low at the start of the field season in 2004 and increased over the season to yield substantially higher water tables by September (Fig. 14.3). In 2007, however, water tables were substantially higher in June, when compared with 2004 values, by almost 10 cm, and decreased steadily over the growing season to yield minimum values in August (Fig. 14.3). Values averaged over all dates indicate overall lower water table depths in 2004 than in 2007 (−10.35 ± 0.4 vs. −16.7 ± 0.5; mean ± standard errors for 2004 and 2007, respectively). When averaged over all sampling dates for each year, there was no significant difference in near surface air temperatures between 2004

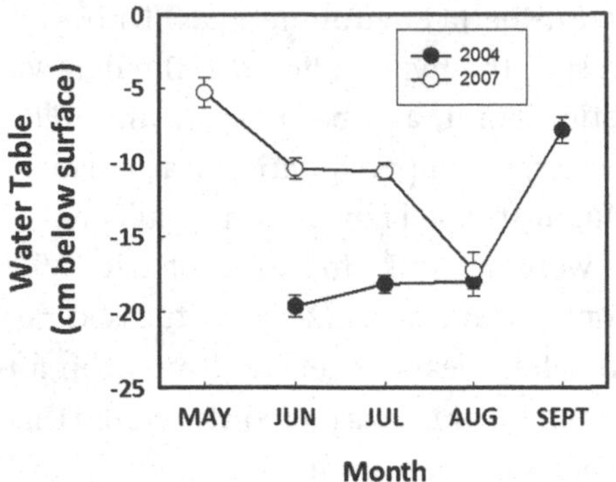

Fig. 14.3. Water table heights (means ± standard errors) plotted in centimeters below the peat surface for Perryvale rich fen in 2004 and 2007 from May through September.

Fig. 14.4. Photograph comparing the same plot at Perryvale rich fen in July 2005 (left) and in June 2008 (right). Increased vegetation cover of both mosses and sedges is evident in the June 2008 image.

and 2007 (27.53 ± 0.45 vs. 29.33 ± 0.27 °C, mean ± standard errors, respectively). In 2005, we began to photograph each plot. In just one year after initiation of the study, the drought-stressed plots exhibited some recovery (Fig. 14.4), and by 2008, the increased cover of mosses and sedges is visibly evident (Fig. 14.4).

Collectively, results from both experiments suggest that drought-induced changes in water table have a greater controlling effect on C fluxes than changes in peat soil moisture. Bryophytes are sensitive to evaporative loss and form a large portion of the C present in fen peat deposits (Yu *et al.* 2002). In rich fens, carpets are dominated by *Drepanocladus aduncus*, *Hamatocaulis lapponicus*, and *Scorpidium revolvens*, and response surfaces (Gignac *et al.* 1991b) indicate that these mosses do not occur higher than 30 cm above the annual average water surface, so it is not surprising that water table draw-down severely impacted moss communities at Perryvale. As Gignac *et al.* (1991b) demonstrated, the microtopographic range of these species is largely limited by height above water level within fens.

Conclusions

Bryophytes form the dominant ground layer of peatland ecosystems, and as such are foundational in structuring their habitats given their ability to engineer the harsh conditions in which they thrive and endure. Key foundational attributes include controls on chemistry, water levels, and nutrients, as well as the peat substrates themselves. Here we have demonstrated via two separate studies that long-term, drought-induced changes in water table have a greater controlling factor on C fluxes than episodic wetting of surface moisture of peat. Bryophytes have relatively little ability to control the extent of wetness periods and longer-term droughts, and thus *water availability* is the key factor in the establishment and maintenance of bryophyte populations. Therefore drought exerts a strong negative pressure on the foundation species of peatlands, and hence, controls on C cycling.

Acknowledgements

This work was funded by the National Science Foundation (DEB-04-36400) to MAV and (DEB-02-12333) to RKW and DHV. We thank Brandon Nichols, Manager of the University of Alberta's Meanook Biological Research Station, for all of his support, logistical help, and assistance. We thank Rose Bloise, Emily Brault, Medora Burke-Scoll, Ingrid Clausen, Phil Jellen, Melissa House, Kathy Kamminga, Dan Lammey, Ashley Nathan, Jim Quinn, Jennie Skanke, Sandi Vitt, James Wood, and Bin Xu for providing valuable assistance in the laboratory and field. We would like to thank Dr. Steven Rice at Union College in the USA, as well as Drs. Håkan Rydin and Hugo Sjörs in the Department of Plant Ecology, Uppsala University, Sweden, for supplying information on both moss canopies and structure–function relationships in bryophytes.

References

Akinremi, O. O., McGinn, S. M. & Barr, A. G. (1996). Evaluation of the Palmer drought index on the Canadian prairies. *Journal of Climate* **9**: 897–905.

Alm, J., Schulman, I., Walden, J. *et al.* (1999). Carbon balance of a boreal bog during a year with an exceptionally dry summer. *Ecology* **80**: 161–77.

Bazzazz, F. A., Paolillo, D. J. & Jagels, R. H. (1970). Photosynthesis and respiration of forest and alpine populations of *Polytrichum juniperinum*. *Bryologist* **73**: 579–85.

Benscoter, B. W. & Vitt, D. H. (2008). Spatial patterns and temporal trajectories of the bog ground layer along a post-fire chronosequence. *Ecosystems* **11**: 1054–64.

Bortoluzzi, E., Epron, D., Siegenthaler, A., Gilbert, D. & Buttler, A. (2006). Carbon balance of a European mountain bog at contrasting stages of regeneration. *New Phytologist* **172**: 708–18.

Bragazza, L. (2008). A climatic threshold triggers the die-off of peat mosses during an extreme heat wave. *Global Change Biology* **14**: 2688–95.

Bragazza, L., Siffi, C., Iacumin, P. & Gerdol, R. (2007). Mass loss and nutrient release during litter decay in peatland: the role of microbial adaptability to litter chemistry. *Soil Biology and Biochemistry* **39**: 257–67.

Bubier, J. L., Bhatia, G., Moore, T. R., Roulet, N. T. & LaFleur, P. M. (2003). Spatial and temporal variability in growing season net ecosystem CO_2 exchange at a large peatland, Ontario, Canada. *Ecosystems* **6**: 353–67.

Charman, D. J. (2002). *Peatlands and Environmental Change*. Chichester: John Wiley & Sons.

Chee, W. L. & Vitt, D. H. (1989). The vegetation, surface-water chemistry and peat chemistry of moderate-rich fens in central Alberta, Canada. *Wetlands* **9**: 227–61.

Chetner, S. & the Agroclimatic Atlas Working Group. (2003). *Agroclimatic Atlas of Alberta, 1971 to 2000*. Edmonton, Alberta: Alberta Agriculture, Food and Rural Development, Agdex 071–1.

Clymo, R. S. (1963). Ion exchange in *Sphagnum* and its relation to bog ecology. *Annals of Botany* **27**: 309–24.

Du Rietz, G. E. (1948). Uppländska myrar. In *Natur i Uppland*, ed. S. Hörstadius & K. Curry-Lindahl, pp. 67–78. Göteborg: Bokförlaget Svensk Natur.

Gerdol, R., Bonora, A., Gualandri, R. & Pancaldi, S. (1996). CO_2 exchange, photosynthetic pigment composition, and cell ultrastructure of *Sphagnum* mosses during dehydration and subsequent rehydration. *Canadian Journal of Botany* **74**: 726–34.

Gignac, L. D. & Vitt, D. H. (1994). Responses of northern peatlands to climate change: effects on bryophytes. *Journal of the Hattori Botanical Laboratory* **75**: 119–32.

Gignac, L. D., Vitt, D. H. & Bayley, S. E. (1991a). Bryophyte response surfaces along ecological and climatic gradients. *Vegetatio* **93**: 29–45.

Gignac, L. D., Vitt, D. H., Zoltai, S. C. & Bayley, S. E. (1991b). Bryophyte response surfaces along climatic, chemical, and physical gradients in peatlands of western Canada. *Nova Hedwigia* **53**: 27–71.

Gorham, E. (1991). Northern peatlands: role in the carbon cycle and probable responses to global warming. *Bryologist* **101**: 572–87.

Gotelli, N. J., Mouser, P. J., Hudman, S. P. *et al.* (2008). Geographic variation in nutrient availability, stoichiometry, and metal concentrations of peatlands and pore-water in ombrotrophic bogs in New England, USA. *Wetlands* **28**: 827–40.

Hajek, T., Tuittila, E. S., Ilomets, M. & Laiho, R. (2009). Light responses of mire mosses – a key to survival after water-level drawdown? *Oikos* **118**: 240–50.

Halsey, L. A., Vitt, D. H. & Gignac, L. D. (2000). *Sphagnum*-dominated peatlands in North America since the last glacial maximum: their occurrence and extent. *Bryologist* **103**: 334–52.

Harris, A., Bryant, R. G. & Baird, A. J. (2006). Mapping the effects of water stress on *Sphagnum*: Preliminary observations using airborne remote sensing. *Remote Sensing of the Environment* **100**: 363–78.

Hemond, H. F. (1980). Biogeochemistry of Thoreau's Bog, Concord, Massachusetts. *Ecological Monographs* **50**: 507–26.

Joosten, H. and Clarke, D. (2002). *Wise Use of Peatlands – Background and Principles Including a Framework for Decision-making.* Finland: International Mire Conservation Group and the International Peat Society.

Li, Y. & Vitt, D. H. (1997). Patterns of retention and utilization of aerially deposited nitrogen in boreal peatlands. *Ecoscience* **4**: 106–16.

Lindquist, J. L., Rhode, D., Puettmann, K. J. & Maxwell, B. D. (1994). The influence of plant population spatial arrangement on individual plant yield. *Ecological Applications* **4**: 518–24.

Malcolm, J. (1996). *Relationships between Sphagnum morphology and absorbency of Sphagnum board.* M.Sc. Thesis, University of Alberta, Edmonton, AB.

Martin, C. F. & Adamson, V. J. (2001). Photosynthetic capacity of mosses relative to vascular plants. *Journal of Bryology* **23**: 319–23.

McNeil, P. & Waddington, J. M. (2003). Moisture controls on *Sphagnum* growth and CO_2 exchange on a cutover bog. *Journal of Applied Ecology* **40**: 354–67.

Murray, K. J., Harley, P. C., Beyers, J., Walz, H. & Tenhunen, J. D. (1989). Water content effects on photosynthetic response of *Sphagnum* mosses from the foothills of the Philip Smith Mountains, Alaska. *Oecologia* **79**: 244–50.

Navaratnam, J. (2002). *A molecular ecological investigation of the archaeal, bacterial and fungal diversity in a Canadian and Siberian peatland complex.* MS Thesis, Villanova University, Villanova, PA.

Nicholson, B. J. & Vitt, D. H. (1990). The paleoecology of a peatland complex in continental western Canada. *Canadian Journal of Botany* **68**: 121–38.

Peterson, W. J. & Mayo, J. M. (1975). Moisture stress and its effect on photosynthesis in *Dicranum polysetum*. *Canadian Journal of Botany* **53**: 2897–900.

Rice, S. K., Aclander, L. & Hanson, D. T. (2008). Do bryophyte shoot systems function like vascular plant leaves or canopies? Functional trait relationships in *Sphagnum* mosses (Sphagnaceae). *American Journal of Botany* **95**: 1366–74.

Richter, C. & Dainty, J. (1989). Ion behavior in plant cell walls. I. Characterization of the *Sphagnum russowii* cell wall ion exchanger. *Canadian Journal of Botany* **67**: 451–9.

Schindler, D. W. & Donahue, W. F. (2006). An impending water crisis in Canada's western prairie provinces. *Proceedings of the National Academy of Sciences of the USA* **103**: 7210–16.

Schipperges, B. & Rydin, H. (1998). Response of photosynthesis of *Sphagnum* species from contrasting microhabitats to tissue water content and repeated desiccation. *New Phytologist* **140**: 677–84.

Silvola, J., Alm, J., Akholm, U., Nykanen, H. & Martikainen, P. J. (1996). CO_2 fluxes from peat in boreal mires under varying temperature and moisture conditions. *Journal of Ecology* **84**: 219–28.

Sjörs, H. 1948. Myrvegetation I Bergslagen. *Acta Phytogeographica Suecica* **21**: 1–229.

Slack, N. G., Vitt, D. H. & Horton, D. G. (1980). Vegetation gradients of minerotrophically rich fens in western Alberta. *Canadian Journal of Botany* **58**: 330–50.

Tarnocai, C. (1984). *Peat Resources of Canada.* Ottawa, Ontario, Canada: Land Resource Research Institute, Research Branch, Agriculture Canada.

Tarnocai, C. (1998). The amount of organic carbon in various soil orders and ecological provinces in Canada. In *Soil Processes and the Carbon Cycle*, ed. R. Lal, J. M. Kimble, R. F. Follett & B. A. Stewart, pp. 81–92. Boca Raton, FL: CRC Press.

Tarnocai, C., Kettles, I. M. & Lacelle, B. (2005). *Peatlands of Canada.* Ottawa: Agriculture and Agri-food Canada, Research Branch.

Titus, J. E., Wagner, D. J. & Stephens, M. D. (1983). Contrasting water relations of photosynthesis for two *Sphagnum* species. *Ecology* **64**: 1109–15.

Titus, J. E. and Wagner, D. E. (1984). Carbon balance for two *Sphagnum* mosses: water balance resolves a physiological paradox. *Ecology* **65**: 1765–74.

Tuittila, E., Vasander, H. & Laine, J. (2004). Sensitivity of C sequestration in reintroduced *Sphagnum* to water-level variation in a cutaway peatland. *Restoration Ecology* **12**: 483–93.

Turetsky, M. R., Crow, S., Evans, R. J., Vitt, D. H. & Wieder, R. K. (2008). Tradeoffs in resource allocation among moss species control decomposition in boreal peatlands. *Journal of Ecology* **96**: 1297–305.

van der Hoeven, E. C., Huynen, C. I. J. & During, H. J. (1993). Vertical profiles of biomass, light intercepting area and light intensity in chalk grassland mosses. *Journal of the Hattori Botanical Laboratory* **74**: 261–70.

Vasander, H. & Kettunen, A. (2006). Carbon in boreal peatlands. In *Boreal Peatland Ecosystems*, ed. R. K. Wieder & D. H. Vitt, pp. 165–94. Ecological studies 188. Berlin: Springer.

Vitt, D. H. (2005). Peatlands: Canada's past and future carbon legacy. In *Climate Change and Carbon in Managed Forests*, ed. J. Bhatti, R. Lal, M. Price & M. J. Apps, pp. 201–16. Boca Raton, FL: CRC Press.

Vitt, D. H. (2006). Functional characteristics and indicators of boreal peatlands. In *Boreal Peatland Ecosystems*, ed. R. K. Wieder & D. H. Vitt, pp. 9–24. Ecological Studies 188. Berlin: Springer.

Vitt, D. H. & Glime, J. M. (1984). The structural adaptations of aquatic Musci. *Lindbergia* **10**: 95–110.

Vitt, D. H., Li, Y. & Belland, R. J. (1995). Patterns of bryophyte diversity in peatlands of continental western Canada. *Bryologist* **98**: 218–27.

Vitt, D. H., Halsey, L. A., Bauer, I. E. & Campbell, C. (2000). Spatial and temporal trends of carbon sequestration in peatlands of continental western Canada through the Holocene. *Canadian Journal of Earth Sciences* **37**: 683–93.

Wagner, D. J. & Titus, J. E. (1984). Comparative desiccation tolerance of two *Sphagnum* mosses. *Oecologia* **62**: 182–7.

Wheaton, E. E., Arthur, M., Chorney, B. *et al.* (1992). The Prairie Drought of 1998. *Climatological Bulletin* **26**: 188–205.

Whitham, T. G., Bailey, J. K., Schweitzer, J. A. *et al.* (2006). A framework for community and ecosystem genetics: from genes to ecosystems. *Nature Reviews Genetics* **7**: 510–23.

Wieder, R. K. (2006). Primary production in boreal peatlands. In *Boreal Peatland Ecosystems*, ed. R. K. Wieder & D. H. Vitt, pp. 145–64. Ecological Studies 188. Berlin: Springer-Verlag.

Wieder, R. K., Scott, K. D. Kamminga, K. *et al.* (2009). Post-fire carbon balance in boreal bogs of Alberta. *Global Change Biology* **15**: 63–81.

Wilson, B. I., Trepanier, I. & Beaulieu, M. (2002). The western Canadian drought of 2001 – how dry was it? *Catalogue no. 21–004-XIE, Vista on the Agri-food Industry and the Farm Community*. Ottawa, Ontario, Canada: Agriculture Division, Statistics Canada.

Yu, Z., Apps, M. J. & Bhatti, J. S. (2002). Implications of floristic and environmental variation for carbon cycle dynamics in boreal forest ecosystems of central Canada. *Journal of Vegetation Science* **13**: 327–40.

Zoltai, S. C. & Vitt, D. H. (1990). Holocene climatic change and the distribution of peatlands in western interior Canada. *Quaternary Research* **33**: 231–40.

15

The Structure and Functional Features of *Sphagnum* Cover of the Northern West Siberian Mires in Connection with Forecasting Global Environmental and Climatic Changes

ALEKSEI V. NAUMOV AND NATALIA P. KOSYKH

Introduction

Changes in structural and functional features of *Sphagnum* cover may be very sensitive indicators of climatic shift in Western Siberia. The spread of raised *Sphagnum* bogs in the West Siberian Plain is limited by low temperatures and the presence of a permafrost earth layer in the north, and by precipitation in the south. It is expected that global warming and increases in ambient CO_2 concentrations may shift bioclimatic zones northward. Comparative ecophysiological analysis of *Sphagnum* indexes for contrasting bioclimatic zones is very important in order to forecast possible changes in northern peatlands and to estimate the tolerance range of *Sphagnum* species.

Western Siberia is located in the central part of the Eurasian continent, covering a vast area from the Urals to the Yenisei River. The extent of the territory is more than 2500 km in the meridional direction; therefore the climate in Western Siberia is very diverse. Within the bounds of the plain territory (West Siberian Plain, WSP) the latitudinal bioclimatic zones (tundra, forest tundra, taiga, forest–steppe, and steppe) are very well distinguished. They replace each other to the south in accordance with temperature and moisture gradients (Richter 1963).

Boggy soils are characteristic of the plain territories. Such types of soils can be explained by the surface slope to the north, high relative humidity, and weak drainage. However, the spread of mires to the north is limited by the presence of

Bryophyte Ecology and Climate Change, eds. Zoltán Tuba, Nancy G. Slack and Lloyd R. Stark. Published by Cambridge University Press.

Table 15.1. *Base wetland types and percentage of the paludal (mire) area*

Bioclimatic zone	Wetland types	Paludal area 10^6 ha	Paludal area % of zone area
Tundra	Polygonal mires	7.1	13
Forest-tundra	Flat- and big-hummock (frozen) palsas	3.8	28
Northern taiga	Flat- and big-hummock (frozen) palsas	17.8	31
Middle taiga	Patterned (ridge–hollow) and domed	19.5	35
Southern taiga	oligotrophic *Sphagnum* bogs	17.7	25
Forest–steppe and steppe	Eutrophic and mesotrophic mires dominated by sedges and rushes	0.7	< 9

Source: Kosykh *et al.* (2008).

a permafrost layer. Its southern extent is located at about 61–62°N. The mires do not violate the general latitudinal and zone pattern and naturally fit into the complex mosaic of forest–mire landscapes, as well as carrying out an important accumulative function. Within the West Siberian Plain several mire zones can be distinguished, which differ in intensity of boggy pedogenesis, typology of mires, typical vegetation, and structure of the *Sphagnum* layer (Kats 1948; Romanova 1977; Liss & Berezina 1981; Liss *et al.* 2001). The total area of mires in Western Siberia, according to the evaluations of different authors, is 58–102 million hectares (Vomperski 1994; Yefremov & Yefremova 2001; Vaganov *et al.* 2005; Velichko *et al.* 2007; Peregon *et al.* 2008). It should be noted, however, that the accuracy of such evaluations greatly depends on the selected cartographic basis and the calculation method. The areas covered by mires, and the characteristic types of mire within different bioclimatic zones, are presented in Table 15.1. The bog-forming process is especially well expressed in the forest zone (northern and middle taiga).

In the forest zone (middle and southern taiga) the mires of atmospheric and mixed atmospheric–ground water sources prevail. Thus, oligotrophic and mesotrophic peat bogs with *Sphagnum* peat occupy more than 65% of boggy soils. For steppe and tundra zones, eutrophic mires (rich fens) dominated by a vegetation structure consisting chiefly of large sedges are characteristic. In transition zones, such as forest–tundra and forest–steppe, the spreading of *Sphagnum* mires is limited by unfavorable environmental conditions.

The vegetation cover of raised *Sphagnum* bogs of the middle and southern taiga is represented by pine – dwarf shrub – *Sphagnum* communities (ryams) in drained landscape areas, and sedge–*Sphagnum* communities in waterlogged hollows. The term "ryam" is a synonym for a raised bog with stunted pine trees. The dense moss cover in ryams is mostly composed of *Sphagnum fuscum* (Schimp.) Klingg. Similar vegetative communities include the patterned ridge–hollow complex bogs. Separate hummocks and intervening local depressions are often covered with *Sphagnum magellanicum* Brid. and *Sphagnum angustifolium* (Russ. *ex* Russ.) C. Jens. True mosses are represented by *Pleurozium schreberi* (Brid.) Mitt. and *Polytrichum strictum* Brid.

Oligotrophic hollows are covered by sedge–*Sphagnum* and cottongrass–*Sphagnum* communities with *Carex limosa* L. and *Eriophorum russeolum* Fries. In the dense *Sphagnum* cover of hollows, *Sphagnum balticum* (Russ.) Russ. ex C. Jens. and *Sphagnum papillosum* Lindb. prevail. On the edges of large boggy areas, poor fens with *Carex rostrata* Stokes, *Sphagnum jensenii* H. Lindb., *S. fallax* (Klinggr.) Klingg., and *S. majus* (Russ.) C. Jens. occur.

To the north of the forest zone on frozen hummocks *S. fuscum* loses its prevailing position, which is then taken by the lichens *Cladonia rangiferina* (L.) Web. and *Cladonia stellaris* (Opiz.) Brodo. The latter two species, together with *Pleurozium schreberi* and *Dicranum* sp., force *S. fuscum* to the edges of hummocks. Relatively favorable conditions for *Sphagnum* moss remain in oligotrophic hollows that thaw down to the mineral bottom, and in places that are lower due to thermokarst subsidence of lakes, in "hasyreis," i.e., drained lakes, in the communities of *Carex rostrata – Eriophorum polystachion – Sphagnum riparium* Aongst., *S. jensenii* and *Carex limosa – Eriophorum russeolum – Sphagnum balticum, S. lindbergii* Schimp. ex Lindb.

Raised *Sphagnum* mires also occur in the forest–steppe zone. They have formed under the influence of pine – dwarf shrubs – *Sphagnum* and birch – dwarf shrubs – *Sphagnum* phytocoenoses. *Sphagnum fuscum* and *S. capillifolium* (Ehrh.) Hedw. are predominant in the moss layer. However, raised *Sphagnum* mires in this region are vulnerable and fragment because the locality is at the limit of the range and physiological tolerance of *Sphagnum*. They do not form any substantial mire areas, but are represented as separate small islands. This is especially true for the mires located near settlements and subject to anthropogenic influence (fires, deforestation, excavation of turf, etc.). Certain types of *Sphagnum* and the degree of *Sphagnum* cover in mire ecosystems are sensitive indicators of ongoing changes happening in the environment. Thus studies of structural and functional characteristics of the *Sphagnum* cover of mires, and the assessment of their ecological condition, represent valuable scientific and practical avenues of research, particularly in relation to recent climatic trends.

Subjects and methods of research

In this study the moss density, linear *Sphagnum* growth, net primary production, moss production efficiency, and total dark respiration were analyzed in relation to geographical latitudes and ecological factors. These included permafrost, temperature, water table level, and hydrological regime.

This work was mostly carried out from 1995 to 2005, and occasionally from 2006 to 2008. The subjects of research were different species of *Sphagnum* moss growing in the West Siberian Plain from tundra to forest–steppe. They were identified according to the keys in Savich-Lyubickaya and Smirnova (1966) and Mul'diyarov (1990).

There are several methods of analysis of the structure and functional state of vegetation. In this work we applied the approach based on distinguishing several ecogroups of *Sphagnum* moss similar in their habit, production potential, and morphophysiological characteristics. Normally such species occupy similar ecological niches and carry out the same functions in the community.

We have distinguished three basic groups of *Sphagnum* with respect to moisture. In doing this we have taken into account microrelief peculiarities of habitats and conditions of mineral nutrition. The first group includes the species that prefer habitats with relatively low levels of bog water, well-drained hills, beds, and hummocks, which sometimes form rising *Sphagnum* "cushions." Normally the representatives of this group relatively easily sustain short desiccation periods and quickly restore their physiological functions after being wetted (*S. angustifolium, S. magellanicum,* and *S. capillifolium*), or form dense sod cover (*S. fuscum*), retaining moisture for a long time, which can also be regarded as an adaptation to desiccation. Compared with other species of this group *S. fuscum* is characterized by its ability to form isolated communities with absolute dominance on hills, beds, and moss "cushions" in forest zone swamps.

Sphagnum species inhabiting oligo- and mesotrophic hollows (*S. balticum, S. fallax, S. majus, S. riparium,* and others), are sharply different in their habit, morphophysiological characteristics, and production potential. Normally the water table level in such habitats during wet periods reaches *Sphagnum* capitula (and sometimes immerses them), and during dry periods it can recede 7–10 cm lower, very rarely 15 cm. These species are typical hygrophytes and we classified them into the second group. It is interesting to note that the representatives of the first and the second groups, which were studied in the southern taiga of Western Siberia (Yefremov & Yefremova 2000), are considerably different in their contents of chlorophylls a and b, and in the correlation between them. The first group had higher values of these pigments.

The third group includes the species occupying the lower parts of slopes and edges of local hills, beds, and hummocks (*Sphagnum compactum* DC. *in* Lat. *et* DC. (= *S. capillifolium*), *S. nemoreum, S. rubellum* Wils., and *S. papillosum*). They often inhabit transitional zones, from elevated swamp areas to waterlogged depressions. However, they do not take the prevailing positions in *Sphagnum* communities.

For the assessment of the functional state of *Sphagnum* moss we used several parameters: the reserve of green phytomass, *G* (g dry mass m^{-2}), the annual linear increase, *L* (mm yr^{-1}), the density of *Sphagnum* sod, *D* (capitula m^{-2}), the annual net production, *ANP* (g $m^{-2}yr^{-1}$), and the intensity of dark respiration, *R* (mg CO_2 g^{-1} h^{-1}). In addition, we determined the dry mass of capitula of each species of *Sphagnum*. We selected all samples and carried out all measurements locally, on homogeneous areas of *Sphagnum* cover.

The linear increase of *Sphagnum* moss was evaluated by using the method of "individual markers." For this we carefully fixed a thin wire ring with a label under the plant's capitulum, as shown in Fig. 15.1. The label was located 5–10 cm away from the plant. When using this method it should be noted that the wire ring should lie freely on the upper branches and should not move down at will. A year later we recorded the increase by taking all *Sphagnum* from 100 cm^2 areas, in the centers of which the marked species were located. The distance between *a* and *b*, between the capitulum and the ring, was measured with a ruler. The new growth portion of the stem was cut off and dried completely; the dry masses of the capitulum and the recent stem growth were determined separately. The data were recorded from 5–10 locations for each habitat.

The moss cover density was evaluated according to the number of capitula of *Sphagnum* on randomly situated surface areas. Under field conditions we made digital photographs of the standard area, 10 cm × 10 cm of homogeneous moss cover. The border areas of the plots examined were marked with a scale frame. The photographs were processed by means of a software image analysis program. For this we marked the *Sphagnum* capitula and counted them automatically (Fig. 15.2). Using this method implies that the annual moss production can be calculated by multiplying the average annual increase of one plant by moss cover density. Thus, the method of digital photography allows us to substantially simplify the calculation of the number of *Sphagnum* plants per unit area.

The intensity of dark respiration of *Sphagnum* moss was measured by two different methods, static and dynamic, using an infrared gas analyzer (IRGA) Infralit-5 (0–0.01 v/v CO_2, Germany). The observations were carried out mostly in the summer. The plant material for evaluations was taken from the areas with homogeneous *Sphagnum* cover. For the dynamic method we cut out 15 cm × 15 cm × 15 cm monoliths from the upper layer and delivered them to the field laboratory. Changes in dark respiration were measured in an open gasometric system with IRGA under the open-air temperature (Voznesenski 1990). For experiments we used only living phytomass, the upper pigmented (autotrophic) and the lower (heterotrophic) *Sphagnum* parts. Before the measurement the moss was carefully separated from the turf and placed in an 0.5 l exposure chamber. Each sample was measured for 10–12 min.

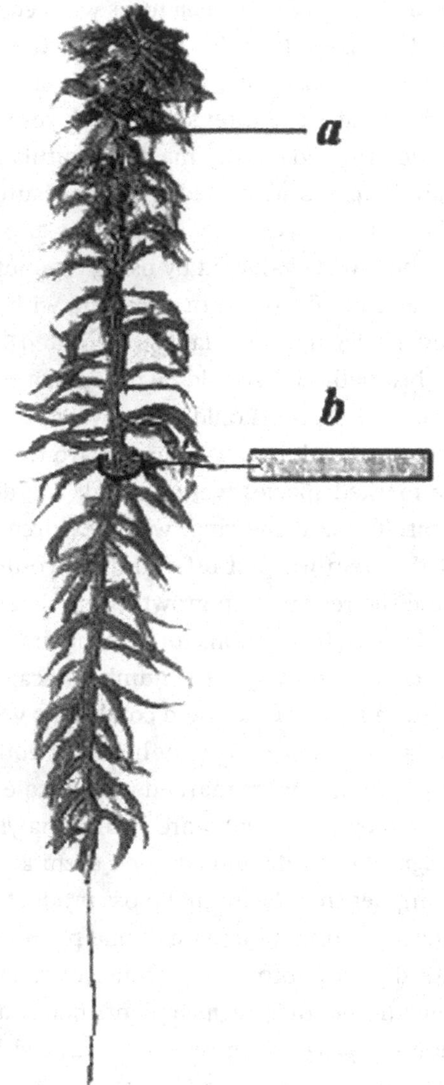

Fig. 15.1. Positions (*a* and *b*) of the individual marks on *Sphagnum* plants at the linear metering; *a*–*b* represents a linear gain in length, rectangle represents the aluminum foil tag.

The static method of exposure in closed chambers was used to measure the respiration of living *Sphagnum* and samples of the upper turf layer directly in field conditions under habitat temperature *in situ*. In the course of experiments the chambers with samples were placed into the moss sod cover. The temperature in the middle part of the moss layer was measured with a mercury thermometer. Air samples from the chambers were taken with syringes every

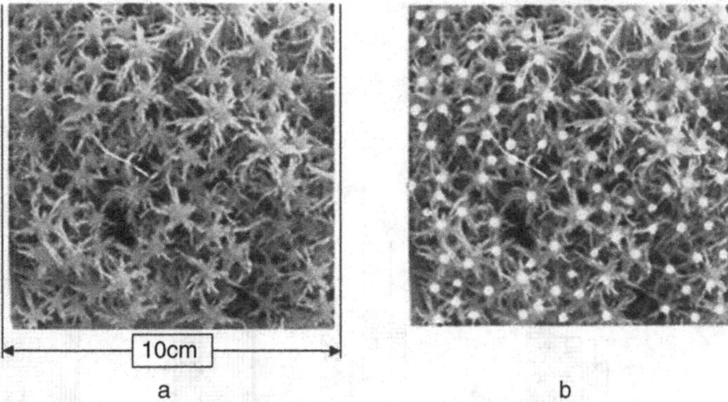

Fig. 15.2. Original (a) and processed (b) pictures of *Sphagnum* cover.

4–5 min and analyzed in a field laboratory by means of IRGA, precalibrated by means of an impulse method with pure CO_2. In the tables the average values of results and their standard deviations, calculated per unit of dry mass, are shown. Both methods showed comparable results.

Results and discussion

Sphagnum species carry out important establishment functions as part of the vegetation of oligotrophic and mesotrophic mires of Western Siberia. They usually dominate the moss layer and determine the qualitative composition and properties of the upper turf. They are especially widespread in the taiga zone. To the north and south of this zone, owing to the change in climatic conditions, the occurrence of *Sphagnum* bogs in the landscape mosaic decreases. How does the change of climatic factors influence morphophysiological characteristics, production potential, and bog building functions of the representative *Sphagnum* species? The answer to this question is of obvious scientific and practical interest because it may throw light on the current and future dynamics of the natural environment of a large (on the global scale) region of the planet.

The diversity of *Sphagnum* moss species in the region examined is relatively high. Out of 43 species registered in the plant guides for the territory of the former USSR in Western Siberia there are 31 described species (Savich-Lyubickaya & Smirnova 1966; Mul'diyarov 1990; Liss *et al.* 2001; Lapshina 2003). By comparison, in the mires of western Canada (Alberta, coniferous forest area) 21 species of *Sphagnum* moss can be found (Vitt & Andrus 1977).

We studied the mass of capitula (heads) of *Sphagnum* moss, growing in different conditions (Fig. 15.3). The analysis of the primary data, including about 650 records, made it possible to assess considerable variation in this parameter

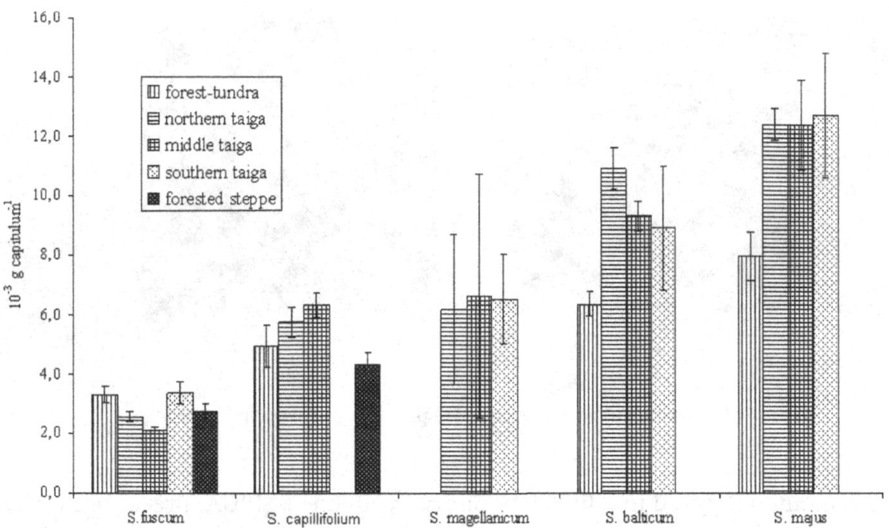

Fig. 15.3. Capitulum biomass of five species of *Sphagnum* in five regions.

within representatives of different groups. Capitulum mass was highest in the typical hygrophytes *S. balticum* and *S. majus*. They had the largest and the heaviest apical segment. *Sphagnum fuscum* had the highest capitulum mass in the moss layer of the pine–dwarf shrub–*Sphagnum* oligotrophic bogs; it is characterized by a shortened stem and a much lighter head.

The impact of the climatic conditions in the corresponding zones influenced the value of capitulum size among the representatives of the first and the second group of *Sphagnum* species in different ways. The minimum size (mass) of heads of *S. fuscum* was found in the middle taiga, and to the north and south of this zone size increased. Regarding *S. balticum* and *S. majus*, no reliable changes in this parameter within the bounds of the taiga zone were found. However, in the forest–tundra both species were characterized by lower values. *Sphagnum magellanicum*, which we referred to the first group, happened to be closer to hygrophytes regarding its parameters. Nevertheless, attention should be paid to wider dispersion of values inside the sampling areas. We suppose that this fact is related to the adaptive capabilities and the ecological flexibility of this species.

One more important parameter for assessing the functional state of different species of *Sphagnum*, is production potential, i.e., the correlation between the annual dry mass increase and the reserve of green phytomass. We demonstrated the existence of the direct dependence of *ANP* on *G* for ecologically different species of *Sphagnum* (Fig. 15.4). It is evident that the angle of the regression line on the graphs characterizes the working efficiency of the assimilation mechanism of plants in certain climatic conditions, and the coefficient of

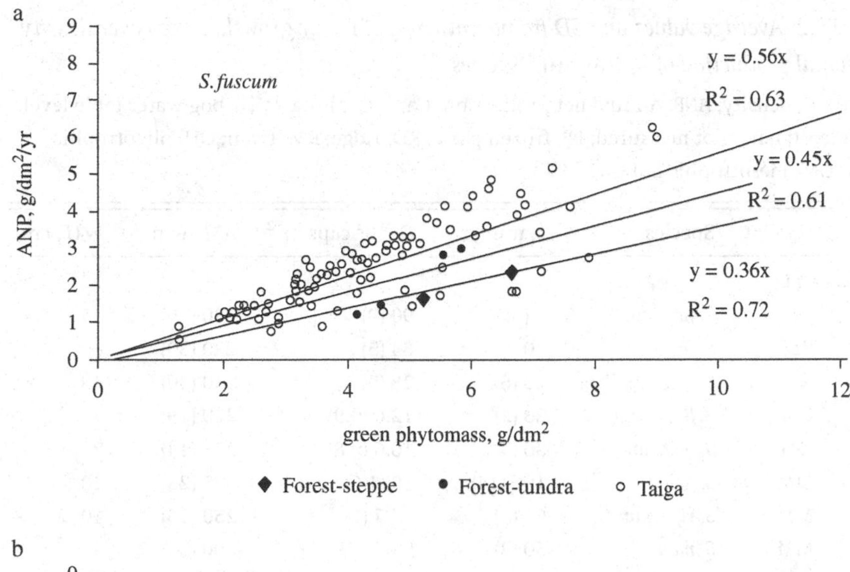

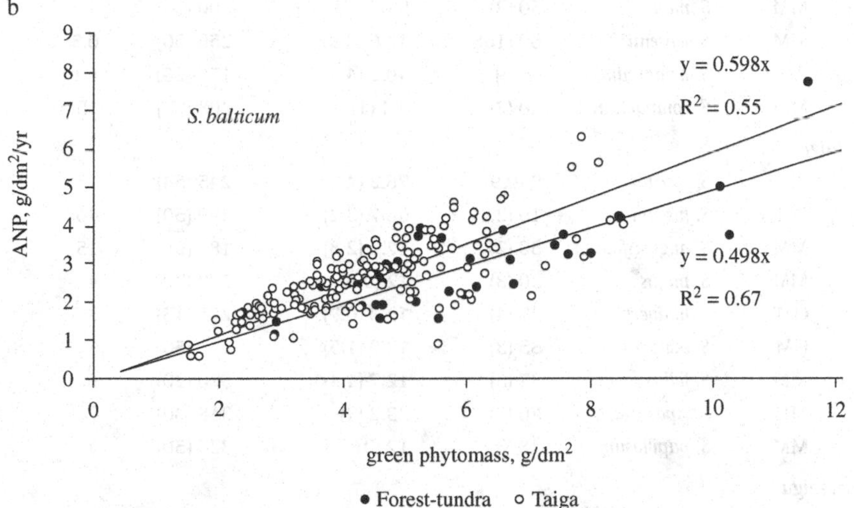

Fig. 15.4. Net production potentials of the photosynthesizing phytomass of *Sphagnum* mosses of dry (a) and wet (b) habitats.

proportionality is a characteristic of the living phytomass renewal. Taiga is apparently the zone of the ecological optimum for *S. fuscum*. In forest–tundra and forest–steppe the efficiency of its production process was noticeably lower. The coefficients *ANP/G*, calculated separately for the southern, middle and northern taiga, nearly coincided at 0.53–0.59. Thus, the renewal of the moss cover of pine – dwarf shrub – *Sphagnum*, dwarf shrub – *Sphagnum* and similar high bogs with prevailing *S. fuscum* takes about two years. The renewal process of hygrophytes inhabiting oligotrophic and mesotrophic sedge–*Sphagnum* bogs in

Table 15.2. *Average values and SD (in parentheses) of linear growth, moss cover density and annual production of* Sphagnum *species*

L, length; *D*, density; *ANP*, Annual net production; Caps, Capitula; WTL, bog water table level; * permafrost; n.m., not measured; FP, frozen palsa; RD, ridge; RM, ryam; OH, oligotrophic hollow; OM, mesotrophic hollow.

Cluster	Habitat	Species	L, mm yr^{-1}	D, 10^3 caps m^{-2}	ANP, g m^{-2}	WTL, cm
Northern taiga						
1	FR	*S. fuscum*	5 (1)	90 (9)	180 (20)	45*
1	RD	*S. fuscum*	20 (5)	64 (5)	220 (30)	60
1	RD	*S. angustifolium*	20 (6)	28 (5)	230 (30)	43
2	OM	*S. lindbergii*	23 (2)	12.6 (0.9)	220 (19)	7
2	OM	*S. balticum*	30 (3)	10.6 (0.8)	298 (14)	7
2	MM	*S. fallax*	106 (14)	19 (1.6)	375 (23)	20
2	MM	*S. riparium*	61 (6)	14.7 (1)	250 (15)	10
2	MM	*S. majus*	60 (3)	9.6 (0.5)	200 (20)	4
2	MM	*S. jensenii*	60 (10)	14.6 (0.9)	250 (50)	0.5
3	RD	*S. nemoreum*	15(5)	40.8 (4)	150 (20)	n.m.
3	MM	*S. compactum*	20 (2)	4.4 (1)	208 (15)	15
Middle taiga						
1	RD	*S. fuscum*	9 (0.9)	76.2 (4.1)	245 (54)	35
1	RM	*S. fuscum*	19 (2.1)	58.7 (3.2)	390 (50)	40
1	MM	*S. angustifolium*	30 (2)	17.2 (2.8)	185 (9)	15
2	MM	*S. majus*	30 (3)	12 (0.5)	290 (20)	4
2	OM	*S. lindbergii*	37 (4)	12.3 (0.7)	279 (15)	8.5
2	OM	*S. balticum*	35 (3)	12.9 (1.5)	255 (50)	8
2	MM	*S. fallax*	87 (5)	12.7 (1.4)	320 (20)	n.m.
2	MM	*S. riparium*	60 (5)	23.2 (3)	318 (30)	7
3	MM	*S. papillosum*	28 (3)	12 (2)	320 (50)	9
Southern taiga						
1	RD	*S. fuscum*	12.1 (1)	54 (1.4)	314 (36)	30
1	RD	*S. angustifolium*	23 (4)	15 (0.3)	150 (40)	40
1	RD	*S. magellanicum*	30 (5)	12 (0.2)	170 (20)	30
2	MM	*S. majus*	70 (10)	10 (1)	250 (20)	12
2	MM	*S. balticum*	100 (20)	12.8 (2.1)	280 (30)	14
2	MM	*S. fallax*	130 (15)	12 (2)	300 (25)	14

forest zone, is somewhat faster ($ANP/G = 0.56$–0.70). Our data provide support for the assumption about the lifetime of *Sphagnum* stems in bogs on the southern taiga (Yefremov & Yefremova 2000).

In Table 15.2 we present the production values of *Sphagnum* moss in different habitat conditions (within the optimum zone, taiga) in more detail. The productivity and the moss cover density are largely determined by the conditions of water–mineral supplies of the habitat. The diversity of species and growth conditions of *Sphagnum* moss in Western Siberia made it possible to assess the production potential and the range of change in annual primary production of the basic turf-formers of high bogs. In spite of the obvious differences of the average habitat elevation of different *Sphagnum* species, the value of primary production showed only 1.5–2× variation (separately for each subzone). This happened in natural conditions owing to the variation of the density of *Sphagnum* cover from 4.4×10^3 stems m^{-2} for *S. compactum* to 90×10^3 stems m^{-2} for *S. fuscum*. For the whole range of habitats, *ANP* of the *Sphagnum* moss studied lay between 150 and 375 g dry mass m^{-2} yr^{-1}.

Low temperatures and the presence of permafrost had great influence on the linear increase and the density of *S. fuscum*. In bogs of the northern taiga it forms more dense sod on frozen peat hills (so-called palsas) in comparison to thawing beds and hummocks of ridge–hollow mires. The linear increase on frozen substrate was 4 times lower, although the annual production was 1.2 times lower. The same was true for *S. fuscum* on narrow beds (edge effect) in comparison with the spacious ryams.

Studies of respiratory gas exchange of *Sphagnum* moss *in situ* are relatively few. It is hard to apply the assessment of respiration made in the conditions of controlled multifactor experiments in microcosms or in phytotrons to the ecological system in general (Kurets *et al.* 1993, 2000). What remains is the problem of determination of the influence of fungi densely inhabiting the *Sphagnum* layer. The range of values of dark respiration is quite large, and it depends on the experimental conditions and the applied method. Nevertheless, the first stage of accumulating the primary information is absolutely necessary in order to improve the method.

Our experiments took place in bogs of the southern and the northern taiga. We conclude that the contrast of conditions revealed some peculiarities and typical reactions of the prevailing species of the moss and moss–lichen layer on high bogs to changes in environmental parameters. As experiments show, in July and August the intensity of respiration of *Sphagnum* moss was low (Table 15.3). This is caused by the fact that the temperature of the air and of the upper peat layers during that period was gradually declining (during the third ten-day period of July and beginning of August). Under such conditions most of the studied species exhibited a similar rate of gas exchange. *Sphagnum majus* was an exception; the temperature drop by about 3° caused a two-fold decrease in its rate of respiration. The second ten-day period of August was hot, and the temperature at a depth of 10–15 cm from the *Sphagnum* head surface was 24–25 °C. Different species reacted to the change in environmental conditions

Table 15.3. *Dark respiration intensity of* Sphagnum *mosses under field conditions in the southern taiga zone*

Values are means ± 1 SE.

Species	Parameter	18 July	7 August	20 August
S. fuscum L.	R, mg CO_2 g^{-1} h^{-1}	0.43 ± 0.10	0.43 ± 0.07	4.03 ± 0.14
	temperature, °C	16.8	13.7	23.9
	n	3	3	2
S. angustifolium (Russ.) C. Jens.	R, mg CO_2 g^{-1} h^{-1}	0.18 ± 0.16	0.17 ± 0.01	0.44 ± 0.42
	temperature, °C	16.8	13.7	23.9
	n	2	2	2
S. magellanicum Brid.	R, mg CO_2 g^{-1} h^{-1}	0.43 ± 0.12	0.38 ± 0.04	0.12 ± 0.05
	temperature, °C	16.8	13.7	24.0
	n	3	3	2
S. majus (Russ.) C. Jens.	R, mg CO_2 g^{-1} h^{-1}	0.48 ± 0.10	0.22 ± 0.05	1.12
	temperature, °C	16.8	13.7	24.0
	n	3	3	1

in a variety of ways. *S. fuscum* and *S. majus* were distinguished by their respiration level. The soil temperature elevation by 10° caused a 6–10-fold respiration increase. The reaction of *S. magellanicum* to the temperature rise was opposite, and this may be connected with greater dehydration of plants during the hot period (McNeil & Waddington 2003). During that period we observed the typical lightening in color of *Sphagnum* heads. We assume that different *Sphagnum* types differ both in the rhythm of the production process and in their physiological reaction to changes in environmental factors. Plants inhabiting areas with variable humidity in transition zones between positive and negative microrelief elements may be subject to higher stress.

The respiration intensity of CO_2 gas exchange of *Sphagnum* in an oligotrophic bog in the northern taiga was considerably lower than in the southern taiga (Table 15.4). These differences are evidently caused by the adaptation of the metabolic system of plants to more severe climate. The rate of carbon dioxide emission by peat samples, taken under different plants, varied from plant to plant. Under the comparatively similar temperature conditions *Sphagnum* moss is able to slow down the peat mineralization by a factor of 3–4, as compared with sedges and lichens. These data confirm the well-known preserving abilities of *Sphagnum* moss. The peculiarities of gas exchange of *Sphagnum* moss were analyzed in more detail in previous publications (Naumov 1997, 2009).

Table 15.4. *Carbon dioxide emission of the components of the moss–lichen layer on a ridge–hollow bog (northern taiga)*

Material	R, mg CO_2 g^{-1} h^{-1}	Temperature, °C	
		in chamber	ambient
Ridge			
S. fuscum	0.10	16.2	17.5
S. magellanicum	0.12	17.8	16.0
S. nemoreum	0.15	16.8	16.0
C. rangiferina (dry)	0.03	18.3	17.4
C. rangiferina (wet)	0.03	15.1	14.0
Peat under the lichen	0.08	12.3	10.0
Peat under *S. fuscum*	0.03	13.7	12.7
Hollow			
S. majus	0.16	17.8	14.7
S. balticum	0.06	15.4	14.7
Peat under the sedge	0.08	13.2	13.6
Peat under (*S. majus* + *S. balticum*)	0.02	12.6	10.0

Conclusions

The accumulation of greenhouse gases in the atmosphere, climatic warming, and changes in the functional state of natural systems are drawing special attention from experts and public institutions. The role of bogs in the carbon balance of the biosphere, the condition of *Sphagnum* cover as an indicator of climatic changes, and the forecast of cryolite zone dynamics are currently under examination by researchers from different countries working within the framework of global environmental problems. The recent climatic warming has caused faster thawing of permafrost, appearance of thermokarst effects, and transformation of the landscape structure and carbon balance in regions of spreading northern bogs of taiga and tundra ecosystems in Alaska and Canada (Billings 1987; Halsey *et al.* 1995; Turetsky 2001; Turetsky *et al.* 2002). The northern parts of Western Siberia exhibit similar phenomena (Kirpotin *et al.* 2007).

According to the assessments of RosHydroMet from 1976 to 2006, the temperature change, corresponding to the linear trend, amounts to 1.0 degree (0.32 °C/10 years) for Western Siberia (Anon 2008). Based on the rules that govern spreading of *Sphagnum* bogs in different climatic zones and known ecological optima for the basic buildup of *Sphagnum* bogs, we assume that further temperature rises will increase the spread of *Sphagnum* and sedge–*Sphagnum* bogs to the

Fig. 15.5. Results of the impact of fire on the upper layer of *Sphagnum* peat in a ryam.

north. The small capacity of peat sediments and proximity of mineral horizons are favorable for formation of mesotrophic communities with participation of the representatives from our second group of *Sphagnum* species (see above).

At the southern border of the range of *Sphagnum* bogs in the forest–steppe zone the environmental situation is also disturbing. Currently, high bogs of the forest–steppe zone under conditions of variable humidity have limited resources for progressive growth. Located at the limit of their natural habitat, they do not form any considerable mire areas, but are represented in landscapes as separate small islands. That is why they are rather vulnerable (Naumov *et al.* 2009). This is especially true for the subjects located near settlements and spots subject to anthropogenic influence (fires, deforestation, peat development, etc.). They are very sensitive to harsh anthropogenic influences and milder climatic changes. The analysis of photos of this area, found in Google Earth, and reconnaissance activities show that the majority of the rare natural areas, even those designated natural monuments, have been influenced by pyrogenic factors (Fig. 15.5). After fires the *Sphagnum* cover is hardly renewed, and the formation of peat stops. The projected moss cover is only 1%–3%. The total number of ryam sets (the average area being 1800–2000 ha) comprises only several dozen, and even in relatively favorable conditions this number consists

Fig. 15.6. Raised virgin bog of the "insular" type in forest–steppe zone, Western Siberia.

of only single units (Fig. 15.6). That is why the priority objective, together with conservation activities, should include continuous complex monitoring aimed at forecasting of natural and climatic changes and preservation of these unique natural ecological complexes.

References

Anon (2008). *Assessment Report on Climate Change and its Consequences in the Russian Federation*, Vol. 1, *Climate Change*. Moscow: RosHydroMet. http://www.meteorf.ru

Billings, W. D. (1987). Carbon balance of Alaskan and taiga ecosystems: past, present and future. *Quaternary Science Reviews* **6**: 165–77.

Camill, P. (1999). Peat accumulation and succession following permafrost thaw in the boreal peatlands of Manitoba, Canada. *Ecoscience* **6**: 592–602.

Halsey, L. A., Vitt, D. H. & Zoltai, S. C. (1995). Disequilibrium response of permafrost in boreal continental western Canada to climate change. *Climatic Change* **30**: 57–73.

Kats, N. Ya. (1948). *Types of Bogs and their Distribution in the Soviet Union and Western Siberia*. Moscow: OGIZ. [In Russian]

Kirpotin, S. N., Naumov, A. V., Vorobiov, S. N. *et al.* (2007). Western-Siberian peatlands: indicators of climate change and their role in global carbon balance.

In *Climate Change and Terrestrial Carbon Sequestration in Central Asia*, ed. R. Lal, M. Suleimenov, B. A. Stewart, D. O. Hansen & P. Draiswamy, pp. 453–72. London: Taylor & Francis.

Kosykh, N. P., Mironycheva-Tokareva, N. P., Peregon, A. M. & Parshina, E. K. (2008). Biological productivity of bogs in the middle taiga subzone of Western Siberia. *Russian Journal of Ecology* **39**: 8–16.

Kurets, V. K., Drozdov, S. N., Talanov, A. V., Popov, E. G. (2000). Light-temperature characteristics of CO_2 gas exchange in some *Sphagnum* species (*SPHAGNACEAE, MUSCI*). *Russian Journal of Botany* **85**: 113–18. [In Russian]

Kurets, V. K., Talanov, A. V., Popov, E. G., Drozdov, S. N. (1993). Light-temperature relationships of visible photosynthesis and dark respiration in some *Sphagnum* species. *Russian Journal of Plant Physiology* **40**: 704–8. [In Russian]

Lapshina, E. D. (2003). *Mire Flora of South-East West Siberia*. Tomsk: Tomsk State University. [In Russian]

Liss, O. L., Abramova, L. I., Avetov, N. A. *et al.* (2001). *The Marsh Systems of West Siberia and their nonconsumptive Importance*. Tula: Grif and Ko. [In Russian]

Liss, O. L. & Berezina, N. A. (1981) *Bogs of the West-Siberian Plain*. Moscow: Moscow State University. [In Russian]

McNeil, P. & Waddington, J. M. (2003). Moisture controls on *Sphagnum* growth and CO_2 exchange on a cutover bog. *Journal of Applied Ecology* **40**: 354–67.

Mul'diyarov, E. Ya. (1990). *A Guide to Cormophyte Mosses for the Tomsk Area*. Tomsk: Tomsk State University. [In Russian]

Naumov, A. V. (1997). Plant respiration and carbon dioxide emission in a mire ecosystem. *Siberian Journal of Ecology* **4**: 385–91. [In Russian]

Naumov, A. V. (2009). *Soil Respiration: Constituents, Ecological Functions, Geographic Patterns*. Novosibirsk: SB RAS Publishing House. [In Russian]

Naumov, A. V., Kosykh, N. P., Parshina, E. K. & Artymuk, S. Yu. (2009). Forest-steppe raised bogs, their condition and monitoring. *Siberian Journal of Ecology* **2**: 251–9.

Peregon, A., Maksyutov, S., Kosykh, N. P. & Mironycheva-Tokareva, N. P. (2008). Map-based inventory of wetland biomass and net primary production in Western Siberia. *Journal of Geophysical Research* **113** (G01007, doi:10.1029/2007JG000441).

Richter, G. D. (ed.) (1963). *Western Siberia*. Moscow: USSR Academy of Science. [In Russian]

Romanova, E. A. (1977). *The Typological Map of Bogs in the Western Siberian Plain, Scale 1:2 500 000*. Moscow: GUGK USSR.

Savich-Lyubickaya, L. I. & Smirnova, Z. N. (1966). *A Guide to the USSR Sphagnum Mosses*. Leningrad: Nauka. [In Russian]

Turetsky, M. R. (2001). Contemporary carbon balance in continental peatlands affected by permafrost melt. In *West Siberian Peatlands and Carbon Cycle: Past and Present* (Proc. Intern. Field Symposium, Noyabrsk, August 18–22, 2001), ed. S. V. Vasiliev, A. A. Titlyanova & A. A. Velichko, pp. 133–5. Novosibirsk: Agenstvo Sibprint.

Turetsky, M. R., Wieder, R. K. & Vitt, D. H. (2002). Boreal peatland C fluxes under varying permafrost regimes. *Soil Biology and Biochemistry* **34**: 907–12.

Vaganov, E. A., Vedrova, E. F., Verkhovets, S. V. *et al.* (2005). Forests and swamps of Siberia in the global carbon cycle. *Siberian Journal of Ecology* **4**: 631–49.

Velichko, A. A., Sheng, Y., Smith, L. C. *et al.* (2007). A high-resolution GIS-based inventory of the West Siberian peat carbon pool. In *West Siberian Peatlands and Carbon Cycle: Past and Present* (Proc. of Second Intern. Field Symposium, Khanty-Mansiysk, August 24–September 2, 2007), ed. S.Ě. Vompersky, p. 10. Tomsk: NTL.

Vitt, D. H. & Andrus, R. E. (1977). The genus *Sphagnum* in Alberta. *Canadian Journal of Botany* **55**: 331–57.

Vomperski, S. E. (1994). The role of bogs in the carbon cycle. In *Readings to Memories of the Academician V.N. Sukachev. XI: Biogeocenotic Particularities of Bogs and Their Rational Use*, ed. I. A. Schilov, pp. 5–37. Moscow: Nauka. [In Russian]

Voznesenski, V. L. (1990). The use of CO_2 gas analyzers in field gas-exchange studies of plants and its constituents. In *Infrared Gas Analyzers in Gas Exchange Study of Plants*, ed. A. A. Nichiporovich, pp. 6–19. Moscow: Nauka. [In Russian]

Yefremov, S. P. & Yefremova, T. T. (2000). Construction and productivity of a *Sphagnum* moss community on West Siberia mires. *Siberian Journal of Ecology* **5**: 615–26. [In Russian]

Yefremov, S. P. & Yefremova, T. T. (2001). Stocks and forms of deposited carbon and nitrogen in bog ecosystems of West Siberia. In *West Siberian Peatlands and Carbon Cycle: Past and Present* (Proc. Intern. Field Symposium, Noyabrsk, August 18–22, 2001), ed. S. V. Vasiliev, A. A. Titlyanova & A. A. Velichko, pp. 148–51. Novosibirsk: Agenstvo Sibprint.

16

The Southernmost *Sphagnum*-dominated Mires on the Plains of Europe: Formation, Secondary Succession, Degradation, and Protection

JÁNOS NAGY

Introduction

There is little doubt that climate change elicits change in plant communities. These changes are conspicuous in those plant associations that lie in the marginal zone of their ranges. The lowland *Sphagnum* dominated mires are frequent in cool and humid climates (e.g., in north and west Europe), but are very rare under continental climatic conditions. Three of the five studied areas are the southernmost occurrences of the *Oxycocco-Sphagnatea* associations on the plains of Europe (Simon 1992a). Their formation and development are caused by edaphic conditions. They were formerly known as "ice age relict" associations. However, current paleobotanical research has documented that these peat moss dominated habitats are much younger (Jakab & Magyari 2000; Magyari 2002). They are extremely sensitive to changes in their environment (which are under marked human influence). These ecosystems are the most sparse and diverse mires of Hungary. All of them are strictly protected. The quality of these peat moss dominated habitats has deteriorated over the past 50 years owing to human and natural influences, and so the Directorate of Hortobágy National Park began their restoration. Restoration measures include the following: blocking of drainage canals; planting of gallery oak forests; initiating artificial water replenishment; and the prohibition of chemical usage on the arable lands around the mires. The aim was to restore these unique habitats so as to provide better conditions for peat mosses to propagate. During the drying period we

Bryophyte Ecology and Climate Change, eds. Zoltán Tuba, Nancy G. Slack and Lloyd R. Stark. Published by Cambridge University Press.

were able to follow the degradation of the peat moss carpet under a willow carr. In the case of restoration processes two of the five flooded mires were over-flooded artificially and one was over-flooded naturally. We were able to follow the secondary succession of the vegetation on these three over-flooded mire basins. This succession has shown a lot of similarities with the primary filling succession and has shown some differences too.

The investigated mires

In the Holocene, alluvial clay and silt layers were deposited onto the gravel sediment of the Pleistocene river basin of the Bereg Plain. The bed of the River Tisza gradually shifted from the eastern part of the plain towards the southwest, leaving behind a labyrinth of oxbow lakes and channels. The five peat moss-dominated mires studied are as follows: Nyíres-tó (48° 11′ 01.25″ N, 22° 30′ 04.94″ E), Báb-tava (48° 11′ 12.90″ N, 22° 28′ 55.64″ E), Navad-patak (48° 10′ 31.26″ N, 22° 30′ 37.19″ E), Zsid-tó (48° 11′ 44.01″ N, 22° 29′ 01.44″ E), and Bence-tó (48° 08′ 45.47″ N, 22° 27′ 00.89″ E). They have been formed by floating mire succession (Nagy *et al.* 1999; Jakab & Magyari 2000) in the above-mentioned abandoned riverbeds in a ring indicated by the villages of Tákos, Hete-Fejércse, Beregdaróc, and Gelénes (Fig. 16.1). These mires belong to the *Samicum* phyto-geographical region on the northeastern part of *Pannonicum*, which lies in the southeastern part of the *Holarcticum* (Soó 1962). In the Köppen (1923) system the climate of the study area is “Cbfx” (between the moderate warm and the moderate cool). The mean annual number of hours of sunshine is *c.* 1950, annual mean temperature is 9.4–9.5 °C, and yearly precipitation is 630–660 mm, of which 370–380 mm falls during the growing season (Marosi & Somogyi 1990). The distribution and amount of precipitation and the groundwater level can

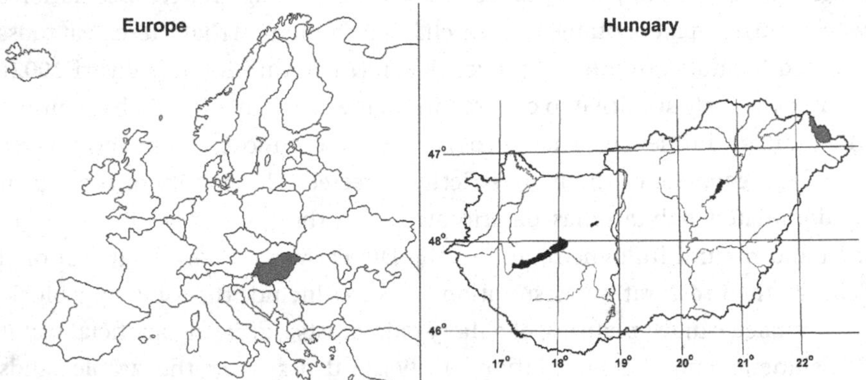

Fig. 16.1. The location of the study area. See text for details.

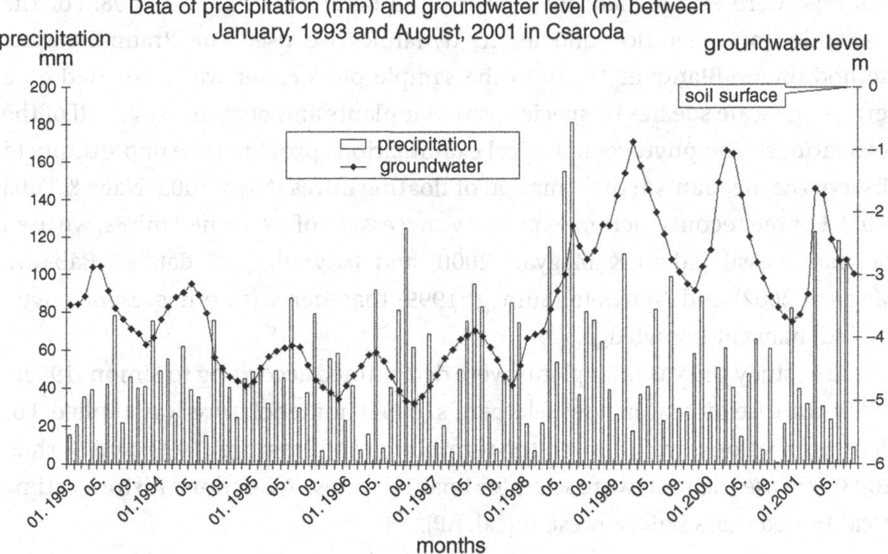

Fig. 16.2. Distribution and amount of precipitation and groundwater level in subsequent years at the weather station of Csaroda from 1 January 1993 to 31 August 2001.

vary greatly in successive years (Fig. 16.2). We have been conducting fieldwork in the area since 1992.

Apart from those at Nyíres-tó, the gallery forests (*Fraxino pannonicae – Ulmetum, Querco robori – Carpinetum*) and the meadows (*Alopecuretum pratensis*) of the other four mires were cut off and were under cultivation. All the mires except Nyíres-tó were damaged by fire. The first two mires still have raised oligotrophic bog associations (mostly *Eriophoro vaginati – Sphagnetum recurvi–magellanici oxycoccetosum*) and fen associations (e.g., *Carici lasiocarpae – Sphagnetum recurvi*) in their centers, and marshes and swamps on the margins. These mires are the southernmost occurrences of *Oxycocco–Sphagnatea* associations on the plains of Europe (Simon 1992a). The Navad-patak mire was similar to the first two but was damaged by fire in 1967, when its *Eriophoro vaginati – Sphagnetum recurvi* association was totally destroyed (Nagy *et al.* 2008). Navad-patak has shown similarities in its vegetation to the last two (Zsid-tó and Bence-tó) mires. These had only marshes, rich fens, and *Sphagnum*-dominated carrs (willow- and alder-dominated associations) in their centers.

Methods

After walking over the ground to become familiar with the area, permanent (more than 1300) and occasional (more than 900) sampling plots (mostly quadrats) were established; these were used between 1992 and 2008. For the description of vegetation and for its dynamics we used the Braun-Blanquet method (Braun-Blanquet 1951). In the sample plots cover was estimated on a percentage scale species by species (vascular plants and peat mosses in all of the associations). The phytocoenological examinations provided the opportunity to observe the mechanism of formation of floating mires (Nagy 2002; Nagy & Tuba 2003). For the reconstruction of primary succession of examined mires, we used the macrofossil (Jakab & Magyari 2000) and palynological data of Báb-tava (Magyari 2002) and Nyires-tó (Sümegi 1999) together with our present phytocoenodynamical knowledge.

In this study the vascular plants were determined according to Simon (1992b) and in the case of peat mosses, Flatberg's (1994) nomenclature was applied. For plant associations, we followed the terminology of Passarge (1999) and Borhidi (2003). We use the Irish word *scraw* for floating mires. We defined historical time as calibrated years before present (cal. BP).

Formation of the southernmost lowland bog communities (Fig. 16.3)

The schematic description of primary succession of the investigated oxbows was based on the analysis as modified by Nagy (2007), an interpretation of the 425 cm deep core of Nyíres-tó (Sümegi 1999) and on the 505 cm deep core of Báb-tava (Jakab & Magyari 2000; Magyari 2002), and is as follows. The Nyíres-tó separated from the main stream about 9500 cal. BP while the separation of Báb-tava occurred about 7800 cal. BP. The connection between the river and the oxbows decreased over the successive centuries up until the river regulations (about 200 years ago). Around the oxbows hardwood (*Fraxino pannonicae – Ulmetum*) and softwood (*Salicetum albae–fragilis*) gallery forests grew. The shores were dominated by tall sedge (*Magnocaricion*) and reed (*Phragmition australis*) associations.

Degradation of *Sphagnum* carpet under the willow carr of Bence-tó

Bence-tó is a C-shaped silted oxbow lake; it is approximately 1.5 km long and 70 m wide. This mire had a *Salici cinereae – Sphagnetum recurvi sphagnetosum squarrosi* association all along the center of its bed until 1998 (Nagy & Réti 2003). Until 1997, open water was observed at the site only sporadically. The draining of the lake accelerated from 1994 and continued until 1996. Throughout the summer of 1997 only shallow water (20–30 cm deep) was found at the deepest

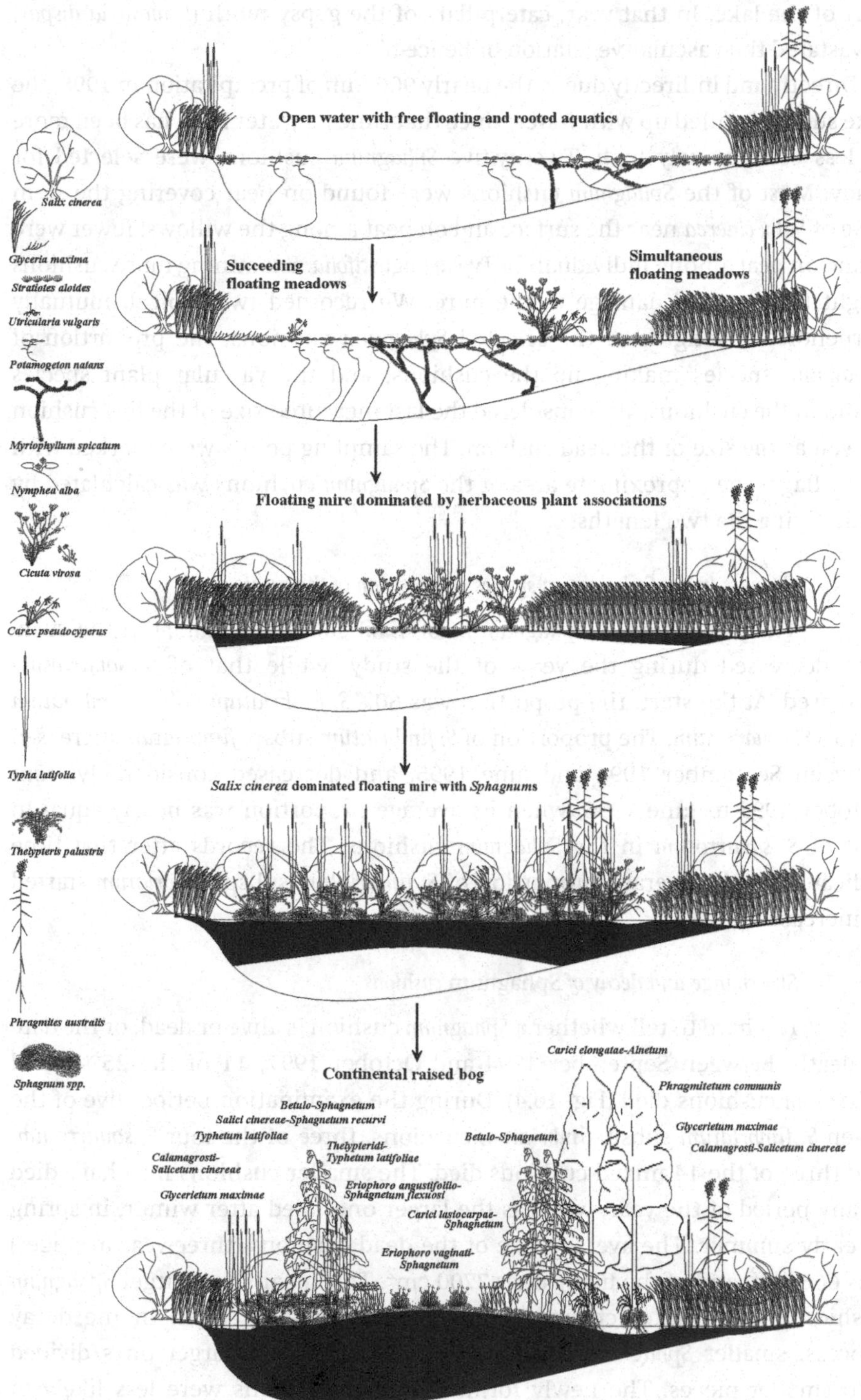

Fig. 16.3. Schematic drawing of primary succession of *Sphagnum*-dominated mires on the Bereg Plain. Drawn by János Nagy (original).

part of the lake. In that year, caterpillars of the gypsy moth (*Lymantria dispar*) devastated the vascular vegetation of Bence-tó.

Directly and indirectly due to the nearly 900 mm of precipitation in 1998, the lake suddenly filled up with water; since that time its water level has been more or less continuously high. Twenty-five *Sphagnum* cushions were selected for study. Most of the *Sphagnum* cushions were found on peat covering the stem base of *Salix cinerea* near the surface and on peat among the willows; fewer were found on peat among individuals of *Typha angustifolia*. Examining more cushions might have caused damage to the mire. We recorded two typical, mutually perpendicular lengths of the selected *Sphagnum* cushions, the proportion of *Sphagnum* species making up the cushions, and the vascular plant species found in the cushions. We considered the last measured size of the live cushion as well as the size of the dead cushion. The sampling points were marked with small flags. The approximate area of the *Sphagnum* cushions was calculated by multiplying the two lengths.

Changes in the Sphagnum *species composition of the cushion*

The proportion of *Sphagnum fimbriatum* subsp. *fimbriatum* in the cushions decreased during the years of the study, while that of *S. squarrosum* increased. At the start, the proportion was 60% *S. fimbriatum* subsp. *fimbriatum* to 40% *S. squarrosum*. The proportion of *S. fimbriatum* subsp. *fimbriatum* increased between September 1994 and June 1995, and decreased considerably from October 1995 to June 1996, when its average proportion was nearly equal to that of *S. squarrosum* in the *Sphagnum* cushions. The records after that time indicate that the average proportion of *S. fimbriatum* subsp. *fimbriatum* started to increase again.

Size change and decay of Sphagnum *cushions*

It is hard to tell whether a *Sphagnum* cushion is alive or dead, or the date of death. Between September 1994 and October 1997, 11 of the 25 studied *Sphagnum* cushions died (Fig. 16.4). During the examination period, five of the seven *S. fimbriatum* subsp. *fimbriatum* cushions, three of the four *S. squarrosum*, and three of the 14 mixed cushions died. The smaller cushions may have died in any period of the year, whereas the larger ones died after winter, in spring or early summer. The average size of the dead cushions (three-year averages) was 660 cm^2, and of the living ones 7700 cm^2. This means that larger *Sphagnum* cushions had a better chance of survival than smaller ones. In the decay process, smaller *Sphagnum* cushions dried out, whereas larger ones divided into smaller pieces. The newly formed smaller cushions were less likely to survive. At first the smaller, apparently live patches were separated by dead

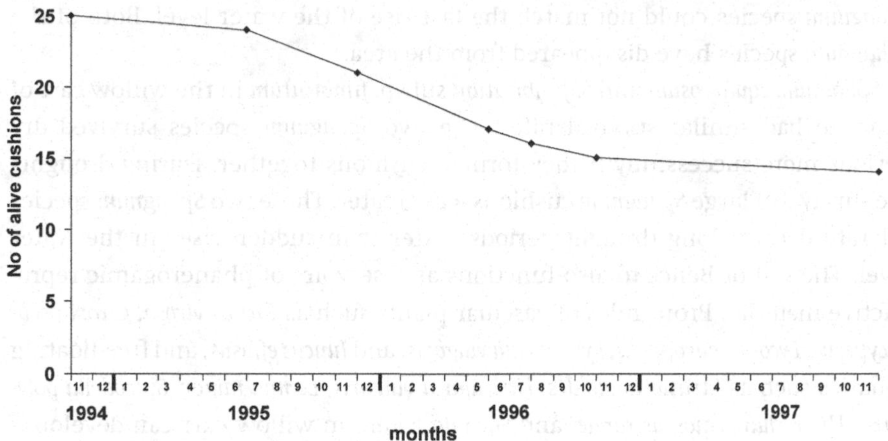

Fig. 16.4. Changes in the number of living cushions at Bence-tó mire during 1994–1997.

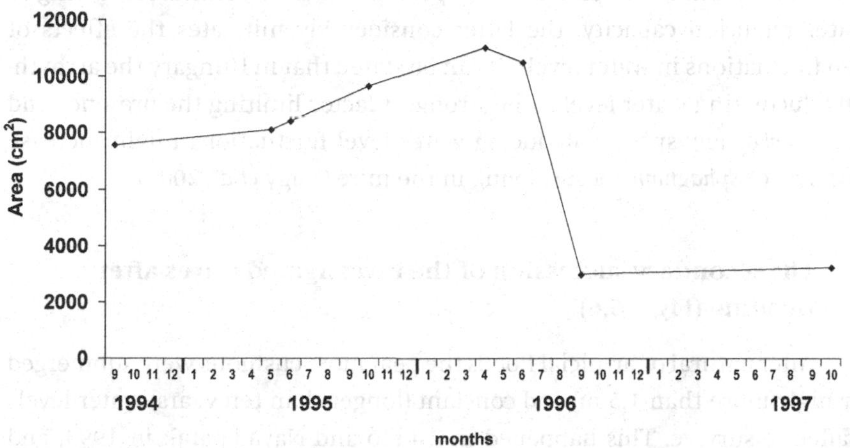

Fig. 16.5. Mean area of *Sphagnum* cushions alive at Bence-tó mire during 1994–1997.

(dry) strips of peat moss. These living patches later became completely separated. *Sphagnum* species were not always able to overgrow *Salix cinerea* fallen leaves on the cushions. Intensive treading by wild boars (*Sus scrofa*) also disrupted the *Sphagnum* carpet. The reason for the decay can be well understood if we observe the size development of *Sphagnum* cushions between September 1994 and October 1997 (Fig. 16.5). The living *Sphagnum* cushions were largest on 24 April 1996, as by that time most of the smaller cushions had died but the larger ones had just started to divide. After that period the average size of the living *Sphagnum* cushions decreased drastically. Since the high water level of 1998 *Sphagnum* species have not occurred on the territory. The rate of growth of

Sphagnum species could not match the fast rise of the water level. Both of the *Sphagnum* species have disappeared from the area.

Sphagnum squarrosum and *S. fimbriatum* subsp. *fimbriatum* in the willow carr of Bence-tó had similar survival rates. The two *Sphagnum* species survived dry periods more successfully if they formed cushions together. During droughts the survival of large *Sphagnum* cushions was greater. These two *Sphagnum* species tolerate decades-long drought periods better than sudden rises in the water level. The soil of Bence-tó also functions as reservoirs of phanerogamic reproductive material. Propagules of vascular plants such as *Cicuta virosa*, *Carex pseudocyperus*, *Lycopus europaeus*, *Lysimachia vulgaris*, and *Juncus effusus*, and free-floating aquatics such as *Stratiotes aloides*, *Utricularia vulgaris*, *Lemna minor*, *Spirodella polyrhiza*, *Hydrocharis morsus-ranae*, and *Salvinia natans* in willow carr can develop if the environmental conditions are suitable. The tolerance of *Sphagnum* species to long drought and sudden flooding is largely dependent on the species composition of the area and the thickness of the peat layer below the mosses. Owing to its water retention capacity, the latter considerably mitigates the effects of sudden fluctuations in water levels. It can be stated that in Hungary the arrhythmically fluctuating water level is the strongest factor limiting the presence and spread of *Sphagnum* species. Reducing water level fluctuations might increase the survival of *Sphagnum* species living in the mire (Nagy *et al.* 2003).

The secondary succession of the investigated mires after flooding (Fig. 16.6)

After natural or artificial floods the peat moss cushions were submerged under high (more than 1.5 m) and constant (longer than ten years) water levels and failed to survive. This happened in Zsid-tó and Navad-patak in 1994, and in Bence-tó in 1998. On these former peat moss mires secondary scraw formation processes can be observed, the first stage of which was called “skirt mire” from the shape. They develop as follows. All the willow species (*Salix cinerea*, *S. pentandra*, *S. fragilis*, *S. alba*, *S. aurita*, and their hybrids) that can be found on the examined areas are able to develop adventitious roots from their shoots near the water surface after flooding. Root formation is independent of the age of the shoots of the willow species.

The dead broken fragments of plants floating in large quantities in the mire water felt (intermingle) with each other, with the long and bushy hair-shaped willow roots of stem origin, and with the plant residues at the lake bottom. Thus a felted carpet forms, which falls as a ‘skirt’ from the water surface to the bottom of the lake. The broken fragments deposited from the water increase mainly at the bottom of the skirt, as the movement of the water erodes more

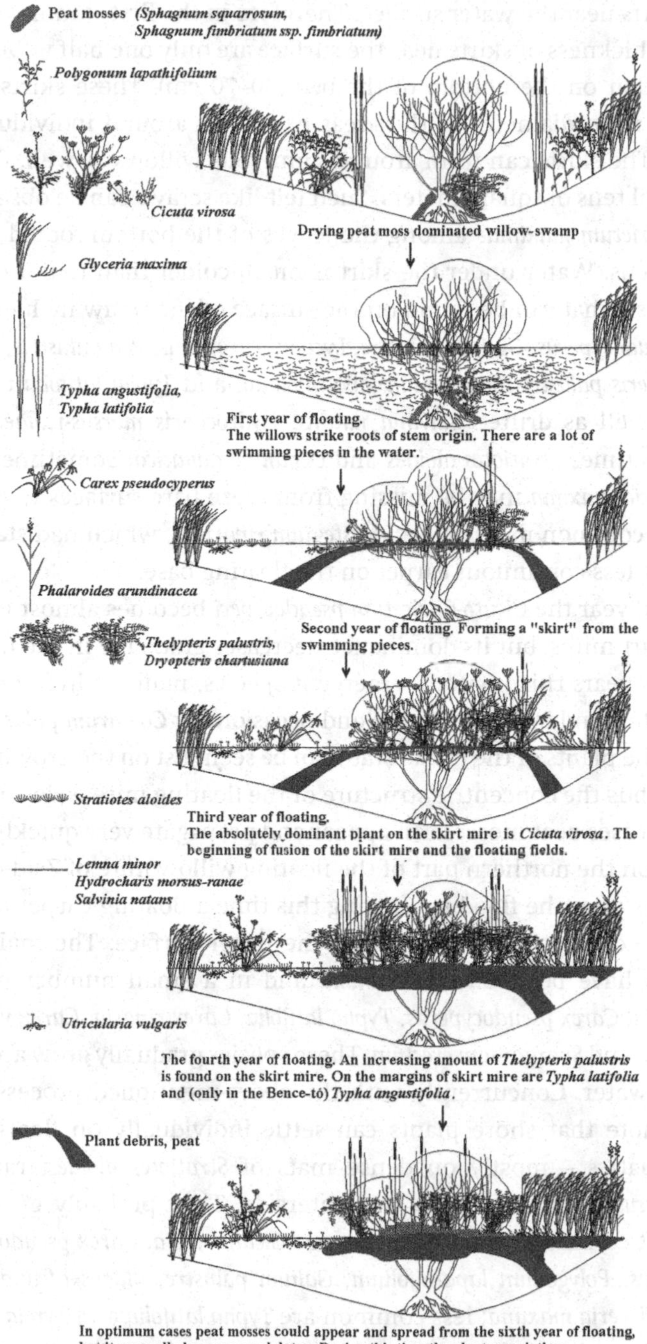

Fig. 16.6. Schematic drawing of secondary succession of flooded *Sphagnum*-dominated willow-swamps on the Bereg Plain. Drawn by János Nagy (original).

strongly the parts near the water surface. Therefore in the first summer of their formation the thickness of skirts near the surface are only one half to one third (20–30 cm) of that on the bottom of the bed (40–70 cm). These skirts are an average of 3–6 m in diameter in the areas examined around individual *Salix cinerea* shrubs. The skirts can form around numerous willows as well, forming scraws of several tens of square meters. Such felt-like scraws can be observed in the dense *Glycerietum maximae* among the leaves of the bottom-rooted *Glyceria maxima* specimens. Water under the skirt is much colder than that around it. The first colonists that can be found on the surface of the scraw include *Cicuta virosa*, *Carex pseudocyperus*, *Galium palustre*, *Lycopus europaeus*, *Poa palustris*, *Glyceria maxima*, *Thelypteris palustris*, *Polygonum lapathifolium*, and *Typha latifolia* or *Typha angustifolia*, as well as drifted *Salvinia natans*, *Hydrocharis morsus-ranae*, *Lemna minor*, and sometimes *Stratiotes aloides* and *Oenanthe aquatica*. Sometimes *Cicuta virosa* and *Glyceria maxima* may be missing from these bare surfaces in the first year. The most common moss was *Amblystegium riparium*, which had started to make a more or less continuous carpet on the floating base.

In the second year the *Cicuto-Caricetum pseudocyperi* becomes almost predominant on the skirt mires, but its dominance decreases gradually in the next few years. For a few years rhizomatous, emergent species, mainly *Glyceria maxima*, *Thelypteris palustris*, and *Lythrum salicaria* and occasionally *Comarum palustre*, will be dominant. The plants of the initial state can be seen just on the growing edge of the scraws, thus the concentric structure of the floating mire is developing.

In optimum cases peat mosses can appear and propagate very quickly, as we have observed on the northern part of the floating willow mire of Zsid-tó since 2003, nine years after the flooding. During this time a floating carpet of shore plants (aquatic community) is formed on the water surface. The main peat-forming plants have been *Glyceria maxima* and in a small number of cases *Typha angustifolia*, *Carex pseudocyperus*, *Typha latifolia*, *Carex riparia*, *Carex vesicaria*, *Carex acutiformis*, and *Sparganium erectum*. These species gradually grow above the surface of the water. Concurrently with the above-mentioned processes it is important to note that shore plants can settle individually on floating leaf rosettes of aquatics – mostly on dense mats of *Stratiotes aloides*, rarely on *Hydrocharis morsus-ranae* – far from the waterside. They probably grow from seeds. The most common of these species are *Cicuta virosa*, *Carex pseudocyperus*, *Lycopus europaeus*, *Polygonum lapathifolium*, *Galium palustre*, *Bidens cernua*, *Bidens tripartita*, and *Glyceria maxima*; less common are *Typha latifolia* and *Typha angustifolia*. The moss layer is dominated by *Amblystegium riparium* in the pioneer state.

The skirt mire formation outlined above can be observed in Hungary in many willow swamps flooded with water, and it is probably widespread where conditions (willows, water flooding, floating plant debris suitable for felting) seem to be suitable. The vegetation changes on these floating mires are very rapid and depend heavily on the rhythm and the rate of water supply (Nagy 2002).

The amount of available nutritive elements decreases from the edge to the center year after year on a scraw, and the appearance of different plant associations follows this progression. After a couple of decades the center of the floating mires may be acidic enough and poor in nutrients for *Sphagna* to appear, as we have seen on the Zsid-tó mire (Nagy *et al.* 2007). *Sphagnum teres* was the most common out of the five pioneer *Sphagnum* species. *Sphagnum squarrosum* has occurred only under the grey willows (*Salix cinerea*) at the floating marginal edges of the *Sphagnum*-dominated willow carr. *Sphagnum angustifolium*, *Sphagnum fimbriatum* subsp. *fimbriatum* and *Sphagnum palustre* have appeared sporadically, mostly on the more open habitats. When the peat mosses appear, they take up cations and release H^+ (Clymo 1963), so the process of acidification and oligotrophy accelerates. All in all the secondary succession of these mires follows this process: *Lemnetea* associations – *Potametea* associations – floating mats of mire plants – fusion of different floating mires – *Sphagnum*-dominated floating carrs. As we have seen above, during primary succession these mires can turn into poor fens and can end, in an optimal case, in continental raised bog associations.

Nature conservation management of the mires

The planting of gallery oak plantation forests began around Báb-tava in the middle of the 1980s and around the Zsid-tó, Navad-patak, and Bence-tó mires between 2000 and 2002, but they have not yet been completed. The trees on the central part of Báb-tava mire have been cut by rangers every year for the past ten years.

The artificial water replenishment was begun in 1986 on Nyíres-tó and Báb-tava, and in 1994 on Navad-patak and Zsid-tó. On the Bence-tó the pump was completed by 1999, but the replenishment was not needed until 2003 owing to the high humidity present between 1997 and 2002 (Nagy *et al.* 2003). The pump at Nyíres-tó (Fig. 16.7) did not work between 1994 and 1995 because of technical factors.

The conditions of the Nyíres-tó and Báb-tava mires quickly showed signs of improvement between 1986 and 1994. During this period the destructive processes (expansion of birch and alder forest, the decline of bog hygrophytes such as *Sphagnum magellanicum*, *S. palustre*, *S. recurvum* s.l., *Drepanocladus exannulatus*, *Eriophorum vaginatum*, *Vaccinium oxycoccos*, *Drosera rotundifolia*, etc.) had stopped and reversed. The amount of hydromesophytes and mesophytes (*Lythrum salicaria*, *Bidens cernua*, *B. tripartita*, *Cirsium arvense*, *Polygonum* spp., *Juncus effusus*, etc.)

Fig. 16.7. The pump at Nyíres-tó in 2008.

also decreased rapidly. In 1987 the inner part started to float again and the vitality of bog hygrophyte populations improved continuously. The state prior to drying out was reached by 1989 (Simon 1992a). The improvement on the Nyíres-tó continued until the breakdown of the pump in 1994. The Nyíres-tó dried up considerably again during the two extremely dry years when there was no flooding. The young trees (mostly *Betula pubescens*, *Alnus glutinosa*, *Populus tremula*), shrubs (*Salix cinerea*, *S. aurita*, *Frangula alnus*) and disturbance-tolerant herbs (e.g., *Bidens cernua*, *B. tripartita*, *Cirsium arvense*, *Juncus effusus*, *Cicuta virosa*, *Epilobium tetragonum*) again crowded the central open places. After the water replenishment the recovery was much slower than before the breakdown. The only fast process could be detected in the wine-red mats of *Sphagnum magellanicum*, which grew from a couple of palm-sizes to some square meters by 1998. This status has not changed since. The undesirable trees, shrubs, and herbs survived for many years. *Drosera rotundifolia* was not able to re-establish; it is now extinct from the mire. *Vaccinium oxycoccos* and *Eriophorum vaginatum* would not spread for three years. The most conspicuous change was the appearance of extremely dense stands of *Juncus effusus*. There were neither *Sphagna* nor *Eriophorum* tufts inside the culms of the rush carpet on tens of square meters until 2002. From this year forward the density and the amount of soft rushes rapidly declined.

The recovery process on the Báb-tava was continuous. The open water surface caused by burning was succeeded by floating mats of *Salix cinerea*, *Typha latifolia*, *Thelypteris palustris*, and *Comarum palustre*, which dominated the edges of the central poor fen (Nagy *et al.* 1999). *Salix cinerea* and young *Betula pubescens* that crowded to the open, central part of this mire have been cleared by felling every year since 1997. During the first two years we detected the absolute dominance of *Thelypteris palustris* and the decline of peat mosses on the cleared places, but after 1999 the competitive ability of peat mosses (*Sphagnum angustifolium*, *S. flexuosum*, *S. palustre*) and *Menyanthes trifoliata* increased quickly. All of this part of the mire was covered by *Sphagna* by 2004, and the population of *Hammarbya paludosa* seems to be stable or slightly increasing. One of the largest problems is the invasion of *Alnus glutinosa* that began after the felling of *Betula pubescens* and *Salix* species; this increased the soluble nitrogen concentration of the water.

The rapid flooding and the settled high water level have caused the temporary extinction of *Sphagna*, and immediately helped the formation of floating mires on the other mires. The processes of floating-mire succession were rapid; the pH of the water decreased year by year, *Thelypteris palustris*-dominated associations appeared, and nine years after the flood we detected a large *Sphagnum* carpet on the peaty floating surface of a mire (only on the Zsid-tó).

The Bence-tó is in the floating mire state. The Navad-patak is in the alder and willow carr state without peat mosses, and the Zsid-tó is in the floating *Sphagnum*-dominated willow carr state. Another problem is that the local hunters failed to remove a sufficient number of wild boar, red deer, and roe deer. These animals directly and indirectly decrease the amount of *Sphagna* as they tread, wallow, graze, and defecate on the compact peat moss surfaces; they also encourage large growths of weeds. This year (2009) the Directorate of Hortobágy National Park will increase fencing to prevent big game entering the area of Nyíres-tó.

Conclusions

As noted above, these lowland *Sphagnum*-dominated mires are very rare under continental climatic conditions. They are extremely sensitive to changes in their environment; both human disturbance and climate change threaten their future existence. Based on our experience the most important practical aspects of the nature conservation management of these mires are the restoration of the surrounding natural vegetation and ensuring the optimal water supply.

Acknowledgements

Thanks to Attila Molnár (Directorate of Hortobágy National Park), Enikő Magyari (Hungarian Natural History Museum, Dept. of Botany), Gábor Figeczky (WWF Hungary), Dániel Cserhalmi, Evelin Péli, and Prof. Zoltán Tuba (Szent István University) for their professional help with these investigations. Special thanks to Réka Borsody and Zoltán Nagy for checking the English version. The research was financed by the Directorate of the Hortobágy National Park, by the Ph. D. School of Biological Sciences (Szent István University), by the KAC K-36-02-00138H, and by the NKFP6-00079/2005 National Project.

References

Braun-Blanquet, J. (1951). *Pflanzensoziologie. (Plant Sociology.)* Wien: Springer-Verlag.

Borhidi, A. (2003). *Magyarország növénytársulásai. (Plant associations of Hungary.)* Budapest: Akadémiai Kiadó.

Clymo, R. S. (1963). Ion exchange in *Sphagnum* and its relation to bog ecology. *Annals of Botany N.S.* **27**: 309–24.

Flatberg, K. I. (1994). *Norwegian Sphagna: A Field Colour Guide. Vitenskapsumseet Rapport Botanisk serie.* 1994(3). University of Trondheim.

Jakab, G. & Magyari, E. (2000). Új távlatok a magyar lápkutatásban: szukcesszió kutatás paleobryológiai és pollenanalitikai módszerekkel. (Progress in the Hungarian mire studies the use of paleobryological and palynological techniques in the reconstruction of hydroseres.) *Kitaibelia* **5** (Suppl. 1): 17–36.

Köppen, W. (1923). *Die Klimate der Erde. (The Climate of the Earth.)* Berlin, Leipzig: De Gruyter.

Magyari, E. (2002). Climatic versus human modification of the Late Quaternary vegetation in Eastern Hungary. Unpublished Ph.D. thesis, University of Debrecen, Faculty of Science and Technology, Department of Mineralogy and Geology.

Marosi, S. & Somogyi, S. (1990). *Magyarország Kistájainak Katasztere. (Cadastre of Micro Regions of Hungary.)* Budapest: MTA Földrajztudományi Kutató Intézet.

Nagy, J. (2002). Szündinamikai folyamatok vizsgálata egy tőzegmoholáp természeti értékeinek megőrzésére. (Research of syn-dynamic processes for conservation of natural values of a *Sphagnum* mire.) Unpublished Ph.D. thesis. Szent István University, Gödöllő.

Nagy, J. (2007). Tavi szukcesszió – úszóláp szukcesszió. (Lake succession – floating mire succession.) In *Botanika (Botany)*, vol.3, ed. Z. Tuba, T. Szerdahelyi, A. Engloner & J. Nagy, pp. 591–3. Budapest: Nemzeti Tankönyvkiadó ZRT.

Nagy, J., Cserhalmi, D. & Gál, B. (2008). The reconstruction of vegetation change in the last 55 years on a mire on the Bereg Plain. *Acta Botanica Hungarica* **50** (Suppl. 1–2): 163–70.

Nagy, J., Figeczky, G., Molnár, M. & Selényi, M. (1999). Adatok a beregi tőzegmohás lápok vegetációjának változásaihoz. (Data on the changes of *Sphagnum* mires of the Bereg Plain.) (In Hungarian with English summary.) *Kitaibelia* **4** (Suppl. 1): 193–5.

Nagy, J., Molnár, A., Cserhalmi, D., Szerdahelyi, T. & Szirmai, O. (2007). The aims and results of the nature-conservation management on the northeast Hungarian mires. *Cereal Research Communications* **35** (Suppl. 2): 813–16.

Nagy, J., Németh, N., Figeczky, G., Naszradi, T. & Lakner, G. (2003). Dynamics of *Sphagnum* cushions in the willow carr of Bence-tó mire and its nature conservation connections. *Polish Botanical Journal (formerly Fragmenta Floristica et Geobotanica Polonica)* **48** (Suppl. 2): 163–9.

Nagy, J. & Réti, K. (2003). The two subassociations of the *Salici cinereae-Sphagnetum recurvi* (Zólyomi 1931) Soó 1954. *Acta Botanica Hungarica* **45** (Suppl. 3–4): 355–64.

Nagy, J. & Tuba, Z. (2003). A preliminary report about a new type of floating mire from Hungary. *Series Historia Naturalis Annales* **13**: 77–82.

Passarge, H. (1999). *Pflanzengesellschaften Nordostdeutschlands II. (Plant Associations of North-East Germany II.)* Berlin: J. Cramer.

Simon, T. (1992a). Vegetation change and the protection of the Csaroda relic mires, Hungary. *Acta Societatis Botanicorum Poloniae* **61** (Suppl. 1): 63–74.

Simon, T. (1992b). *A Magyarországi Edényes Flóra Határozója. (Handbook of the Vascular Flora of Hungary.)* Budapest: Tankönyvkiadó.

Soó, R. (1962). *Növényföldrajz (Plant Geography)*, 5th edn. Budapest: Tankönyvkiadó.

Sümegi, P. (1999). Reconstruction of flora, soil and landscape evolution, and human impact on the Bereg Plain from the late-glacial up to the present, based on palaeoecological analysis. In *The Upper Tisa Valley*, ed. J. Hamar & A. Sārkány-Kiss, Tiscia Monograph Series, pp. 173–204. Szeged: Tisza Club for Environment and Nature & Liga pro Europea.

VII CHANGES IN BRYOPHYTE DISTRIBUTION WITH CLIMATE CHANGE: DATA AND MODELS

17

The Role of Bryophyte Paleoecology in Quaternary Climate Reconstructions

GUSZTÁV JAKAB AND PÁL SÜMEGI

Introduction

The Quaternary covers the past 2.5 million years of Earth history. This unique period is well known for a record of oscillating climatic parameters. If one wishes to understand the trajectory of future climatic changes triggered by human activities, one should also have a clear picture of the climate of the past. Fluctuating climates are reflected in peat bog profiles. Paleoecological studies using plant macrofossils, like bryophyte remains, have an important role in the reconstruction of past hydrological changes in lakes and peat bogs. Plant macrofossil analysis has been used most frequently in the oceanic regions of Europe, where the moisture gradient is reflected clearly in different *Sphagnum* taxa. The method of bog surface wetness predictions has not been adapted to date for the characterization of continental peatbogs. Hungary is located along the southern limit of *Sphagnum*-dominated peat bogs, with peat bogs restricted to the moister regions of the country. Holocene climatic events, such as severe droughts, caused significant changes in mire development and as such are traceable in the paleoenvironmental record of these bogs.

Fossil mosses used as proxies for detecting past climatic changes

Detailed paleoecological investigations of fossil mosses enable us to accurately capture the prevailing conditions in some terrestrial ecosystems, mainly those in littoral parts of various catchment basins. There are two major directions for investigation and interpretation: one is restricted to the

Bryophyte Ecology and Climate Change, eds. Zoltán Tuba, Nancy G. Slack and Lloyd R. Stark. Published by Cambridge University Press.

ecological needs of the individual taxa, whereas the other is based on the ecological requirements of ecological groups and communities in the reconstruction.

The appearance of certain taxa in the fossil communities studied may reflect special environmental conditions. The taxon *Polytrichum norvegicum* quite often appears in the Late Glacial deposits of England and Scotland, implying the presence of scattered late snowbeds in the locality (Dickson 1973). The distributional area of a moss taxon is generally determined by the interplay of one or more climatic variants. For example, the southern lowland distribution line of the species *Polytrichastrum alpinum* follows the 16 °C July isotherm in the region; i.e., it is completely missing from areas characterized by hot summers (Odgaard 1980).

Reconstructions based on the ecological needs or parameters of entire moss communities yield a substantially better and more reliable picture of the prevailing conditions. In this case a cluster of taxa of similar ecological parameters is created on paleoecological diagrams. Such clusters are woodland or aquatic taxa, xerophylous taxa or those preferring a carbonate substrate.

This type of reconstruction is most useful in lacustrine and mire studies to capture the original vegetation. Paleobryological studies can be complemented by other paleoecological, paleobotanical studies to capture the individual stages of a hydroseries of the catchment (Birks & Birks 1980; Birks 1982; Grosse-Brauckmann 1986). Rybníček (1973) arrived at the conclusion that the composition of fossil communities is essentially the same as that of their modern counterparts, based on a comparative study of modern and fossil marshland communities of Central Europe. Since they are excellent syntaxonomical indicators, mosses enable us to reconstruct the entire fossil plant community when macrofossils of flowering plants are also retrieved from the deposits. The continuously growing information on the ecology of mire moss taxa is a great aid in this task (e.g., Janssens 1983a, b; Gignac & Vitt 1990; Vitt & Chee 1990; Rydin 1993; Kooijman 1993; Slack 1994). Peatland mosses also enable us to capture fine-scale hydrological changes of the catchment, which often holds clues to transformations in the climate. During the initial planning of sampling sites it is important to bear in mind that the different mires are not equally suitable for paleoclimatological studies. One of the most essential hallmarks for the accurate assignment of coring points is the hydrology of the peatland. The hydrological parameters are naturally reflected in the morphology of the catchment of the peatland as well as the composition of the vegetation. Several peatland classification systems are known based on the hydrology, morphology or vegetation of the referred areas (e.g., Osvald 1925; Moore & Bellamy 1974; Mitsch & Gosselink 1993; Pakarinen 1995).

Hydrologically, peatlands can be arranged into two major groups. Ombrotrophic peatlands (bogs) rely heavily on precipitation to establish a water balance; thus their evolution and distribution are primarily influenced by the climatic conditions. The water balance is mostly controlled by the mutual interplay of rainfall and evaporation, while the role of the vegetation and the storage capacity of the peat is negligible. Ombrotrophic peatlands are generally restricted to the Atlantic and montane areas of Europe. In addition to local rainfall, water in the second group of minerotrophic peatlands (fens) comes from the surficial watercourses and the groundwater. As such these peatlands are not as dependent on the amount of precipitation as the previous group in maintaining a constant water balance.

These two groups are further divided into various subgroups based on their morphology and vegetation. Several classification systems are known from different parts of Europe and North America. In Atlantic Europe the ombrotrophic peatlands are often subdivided into raised bogs and blanket bogs. Raised bogs, as the name implies, form a positive morphology in a depression. The thinner peat layers of blanket bogs conversely follow the natural morphology of the landscape apart from the steepest slopes.

In continental Europe continental peatlands are known; in the drier summer months an advancement of the arboreal vegetation is characteristic of these peatlands. Floating mats or floating fens are another typical form of continental peatlands, generally developing in the carbonate-rich or alkali-rich shallow littoral parts of lakes and ponds (Balogh 2000a, b). There are two major processes that initiate peatlands. In the first case, a terrestrial area is turned into a peatland; this process is referred to as paludification. In the second case, referred to as terrestrialization, the advancement of a peatland into a lacustrine basin is observable.

The peatlands best suited for paleoclimatic reconstructions are those of ombrotrophic bogs which emerged via paludification. Climatic endowments favoring the evolution of these types of peatlands are mainly restricted to the western parts of Europe under the influence of oceanic climatic conditions. As such, most peatland paleoclimatic records are known from the areas of the UK, Germany, Denmark, and Sweden. Here the moisture gradient is unambiguously reflected in the presence of certain *Sphagnum* species. Barber and Charman (2005) questioned the suitability of strongly continental peatlands for paleoclimatic reconstructions.

Paleoclimatic reconstructions of peatlands fundamentally follow two major approaches. One uses the signal of peat initiation; the other looks for traces of compositional changes within the peat profile to make inferences about possible changes in the climate.

Utilizing peat initiation signals is suitable for capturing past climatic changes at the meso- (regional) and macro-scales (that of the entire continent). The initiation of peat formation at a given site is the outcome of the complex interplay of local vegetation, hydrology, and climatic endowments. Nevertheless, it must be noted that this process may very often be the outcome of some sort of human activity as well, such as deforestation in the area. But at a higher scale, the initiation of peat formation is fundamentally triggered by climatic fluctuations (Halsey *et al.* 1998). The gist of this approach is that the age of the lowermost peat horizons is determined by looking at a large number of samples deriving from a larger area. In certain clearly identifiable periods the number of newly formed peatlands is sufficiently elevated compared with other periods, which ultimately may indicate that these periods were characterized by cooler and wetter climatic conditions. A major drawback of this approach is that there is no evidence for the direct linear relationship between the sudden increase in the number of peatlands and the transformation of the climate to wetter and cooler conditions. Furthermore, as the lowlands and basins of the original landscape are normally covered by peatlands, new ones very rarely develop in areas of unfavorable natural endowments.

This type of approach has a long record and tradition in Canada (Zoltai & Vitt 1990; Halsey *et al.* 1998; Campbell *et al.* 2000; Yu *et al.* 2003). According to the results of these studies, unfavorable drier climatic conditions must have prevented the expansion of peatlands up until 6000 BP. After this period an increase in the precipitation subsequently modified the local hydrological and morphological conditions enabling peatlands to reach their modern state of expansion between 3500 and 2000 years BP. Campbell *et al.* (2000) and Yu *et al.* (2003) managed to identify millennial and centennial-scale changes in the climate as reflected in the initiation of peat formation in Canada. These studies have clearly confirmed that, at the scale of the continent, the development of peatlands is ultimately determined by the natural cycle of climatic changes. This is well correlated with the findings of studies of deep-sea ocean floor and ice cores. Peatland initiation occurred in two stages in West Siberia (Smith *et al.* 2004). The first stage was linked to the warmings after the Late Glacial between 13 000 and 8000 years BP. The second stage took place after 5000 BP. Dry and warm conditions prevailed between these two periods. Similarly, two stages of peat initiation were identified in southern Finland between 8000 and 7300, and between 4300 and 3000 years BP (Korhola 1995).

The second approach is looking for proxies reflecting transformations in the biological and chemical composition of peat sequences as signals of past climatic fluctuations. One of the most frequently used approaches of chemical analysis is the investigation of humification. This approach ultimately relies on the logic that surface humidity ultimately determines the rate of decay of plant matter. When peatlands have dried out, this is reflected in a sudden increase in humic acids within the deposits. These acids are extracted from the deposits using various bases and their concentration is determined in the solution by spectrophotometric approaches.

The approach most widely adopted in the analysis of biological components is the study of plant macrofossils, including those of mosses or testacea. These studies enable us to identify various peatland types and past communities. However, finer-scale short-lived transformations can very rarely be linked to a given plant community. Nevertheless, there is a special feature of peatland plants that can aid interpretations made on the basis of environmental conditions. Certain species are distributed along a gradient reflecting differing water depths. Furthermore, while certain taxa are restricted to waterlogged areas in the peatland, for example *Sphagnum cuspidatum* or *Warnstorfia fluitans*, others show a preference for drier parts, e.g., those of *Sphagnum* sect. *Acutifolia* or *Pohlia nutans*. Among flowering plants, species such as *Scheuchzeria palustris* or *Menyanthes trifoliata* often turn up in the more humid periods. Conversely, occurrences of the taxa *Phragmites communis, Eriophorum vaginatum, Carex elata*, or peatland shrubs increase substantially in drier periods. By the utilization of various multivariate ordination methods (PCA, DCA), the less obviously identifiable moisture gradients can also be assessed in the peatland. In the next section a short overview of the methodology of plant macrofossil analyses is given.

Material and field and laboratory methods of plant macrofossil studies

Although the importance of plant macrofossil studies in paleoecological works was identified and emphasized relatively early (Jessen & Milthers 1928; Jessen 1949), up until the 1970s macrofossil diagrams appear only as a complement of pollen diagrams. Aaby's (1976) impressive study gave an impetus to methodological improvements, which occurred in the 1980s. The most important taxonomic studies are those of Grosse-Brauckmann (1972, 1986), Rybníček (1973), Birks (1980), Birks and Birks (1980), Janssens (1983a, 1987, 1990), Wasylikowa (1996), and Mauquoy and van Geel (2007). A major step forward was the introduction of the so-called

QLCMA (semi-quantitative quadrat and leaf-count macrofossil analysis technique) developed in Southampton (Barber *et al.* 1994), which enabled researchers to achieve an accuracy and resolution in their work known earlier only from palynological studies. Today, macrofossil studies are indispensable components of Quaternary paleoecological investigations (Birks & Birks 2000).

Sampling is made by using a Russian-head corer or Livingstone piston corer, generally used in Quaternary environmental historical studies (Aaby & Digerfeldt 1986). These sampling methods yield undisturbed cores. After transportation to the laboratory, the cores are cut lengthwise for various analyses; the sections for palaeobotanical and geochemical analyses are stored at 4 °C in accordance with international standards. Subsamples are taken in the lab involving a volume of 1–4 cm^3. However, this volume may be larger as well, as much as 8–10 cm^3, when working with samples relatively poor in organic matter. Sampling intervals are determined by the aims of the analysis and the inferred rate of deposition. In the case of plant macrofossil studies this is generally 4 cm, but may also be at the scale of millimeters. Samples are filtered using a sieve of 250 μm mesh. The more consolidated sediments are treated with 10% KOH or NaOH for 5–10 minutes.

In our work a modified version of the QLCMA technique was adopted (Barber *et al.* 1994; Jakab *et al.* 2004). Organic remains from peat and lacustrine sediments rich in organic matter can be divided into two major groups. Some remains can be identified with lower-ranking taxa (specific peat components), whereas others cannot be identified by using this approach (non-specific peat components). Sediment samples can contain significant amounts of non-specific peat components, which reveal much about the hydrologic conditions and chemical composition of the area in which the sediment accumulated. The most important non-specific peat components are the following: undifferentiated monocotyledon remains (Monocot. undiff.), unidentified organic material (UOM), unidentified bryophyte fragments (UBF), unidentified leaf fragments (ULF), charcoal, and wood. In the case of specific peat components, the remains can often be identified to the species level. They are important for reconstructing the sediment depositional environment. The local vegetation often allows identification at the association level. The most important specific peat components are seeds, fruits, sporogonia, mosses, rhizomes and epidermis (e.g., *Carex* species), leaf epidermis, other tissues and organs (hairs, tracheids, etc.), insect remains, and ostracod shells. The identification of herbaceous plant tissues was based on the procedure described by Jakab and Sümegi (2004).

Concentrations are determined by adding a known amount of indicator (0.5 g poppy seed, *c.* 960 pieces) and by counting the poppy seeds and the remains under a stereo microscope in ten 10 mm × 10 mm quadrats in a Petri dish. Similarly to mosses, rhizomes can only be identified with a light microscope. The values for different moss species and UBF are determined using a similar procedure. The concentration can be described by the following equation:

$$\text{macrofossil concentration} = \frac{\text{counted macrofossil (average)} \times 960 \text{ (total poppy seeds)}}{\text{counted poppy seeds (average)} \times \text{sample volume (cm}^3\text{)}}$$

Dominance values are depicted graphically with depth, displaying information on radiocarbon age, the name of local zones, and symbols of sedimentary features in accordance with the system of Troels-Smith (1955), internationally accepted for the description of unconsolidated deposits. The Software packages Psimpoll (Bennett 1992) and Syn-Tax (Podani 1993) are used for plotting the analytical results.

The exact determination of the age of the deposits is indispensable for correct paleoenvironmental reconstructions. Given that more than 50% of the dry peat is composed of elemental carbon, radiocarbon dating is best suited for such purposes. Conventional radiocarbon dating has been a frequently applied technique. The gained raw BP years are calibrated to cal BP or calendar years (AD/BC). The new accelerator mass spectrometry (AMS) approach yields more precise ages and requires smaller input samples too. For example, smaller plant remains such as *Sphagnum* leaves might be sufficient.

Phytogeography of peat bogs in the Carpathian Basin

Hungary is nestled in the heart of the Carpathian Basin, characterized by moderately continental climatic conditions, and enjoys the modifying effect of various climatic influences such as those of the oceanic influences in the west and those of the Mediterranean influences in the south. The basin morphology ultimately determines the distribution of the climatic pattern. Namely, there is a gradual increase in continentality accompanied by a gradual decrease in the rainfall from the margins towards the center. The montane climatic influences prevailing in the surrounding mountains of the Alps in the west and Carpathians in the north and east ultimately determine the climatic conditions of the marginal part of the basin.

The most striking proxy expressing the influences of this basin morphology on the climate is the annual rate of precipitation. This value is often below 500 mm yr^{-1} in the central driest parts of the Alföld (Great Hungarian Plains, hereafter "GHP") and displays a gradually increasing trend to the north and west. The western parts of the country enjoy the highest rates of annual precipitation, sometimes even exceeding 900 mm yr^{-1}. The areas of the GHP are characterized by an average annual rainfall of 500–550 mm; those of the mid-mountains have a rate of 600–800 mm per year (Bacsó 1959).

The general climatic conditions of Hungary are far from ideal for the emergence of *Sphagnum* peatlands. The number of localities harboring *Sphagnum* species hardly reaches 50 in the entire country, and not even scattered occurrences of these are known from the central, driest parts of the GHP. The actual number of *Sphagnum* peatlands is below 20; the majority of these are tiny, covering only a few hectares. Raised bogs are completely absent. The majority of *Sphagnum* peatlands are restricted to the northern areas of the Northern Mid-Mountains and the northern GHP, as well as the eastern parts of the country that enjoy more precipitation due to the oceanic and montane climatic influences (Boros 1968; Szurdoki & Nagy 2002). The southernmost distributions of lowland *Sphagnum* peatlands in the entire continent are found in the area of the northern GHP. In drained *Sphagnum* peatlands, there is an advancement of reed, sedge, and birch into these areas in accordance with the local conditions (Borhidi & Sánta 1999).

Despite all of this, Hungary used to be relatively rich in peatlands and marshy areas. The extent of peatlands exceeded 90 000 hectares in the past, preceding various river regulation and drainage measures. The estimated volume of peat reserves is 973 million tonnes (Dömsödi 1988). However, these peats are not acidic *Sphagnum* peat, but rather basic reed or sedge peat. These peatlands used to cover extensive areas of the country in the neotectonic depressions of the GHP, the abandoned river channels of the river Danube, or the littoral parts of larger lakes such as Lake Balaton or Lake Fertő (Fig. 17.1). In the next section we present the evolutionary history of a Hungarian *Sphagnum* peatland.

Paleoclimate reconstruction from the Nyírjes peat bog

The Nyírjes peat bog of Sirok is found in the northern part of the country at the eastern foothills of the Mátra Mts at an elevation of 250 m (Fig. 17.1). It covers a small area of 9000 m^2. No surficial watercourses feeding or draining the peatland are known. The basin of the peatland is fringed by a woodland of hornbeam and oak. The following plant communities are present, moving from the margins towards the center: *Scirpo-Phragmitetum, Salicetum cinereae-Sphagnetum, Carici lasiocarpae-Sphagnetum.* This peatland harbors the following peat moss taxa: *Sphagnum palustre, S. subsecundum, S. magellanicum, S. recurvum* s. l., *S. fimbriatum, S. squarrosum, S. obtusum,* and *S. angustifolium.* The most common species are *Sphagnum recurvum* s. l. and *S. palustre* (Máthé & Kovács 1958; Szurdoki & Nagy 2002). A detailed palynological study of the peatland was done by Gardner (2002).

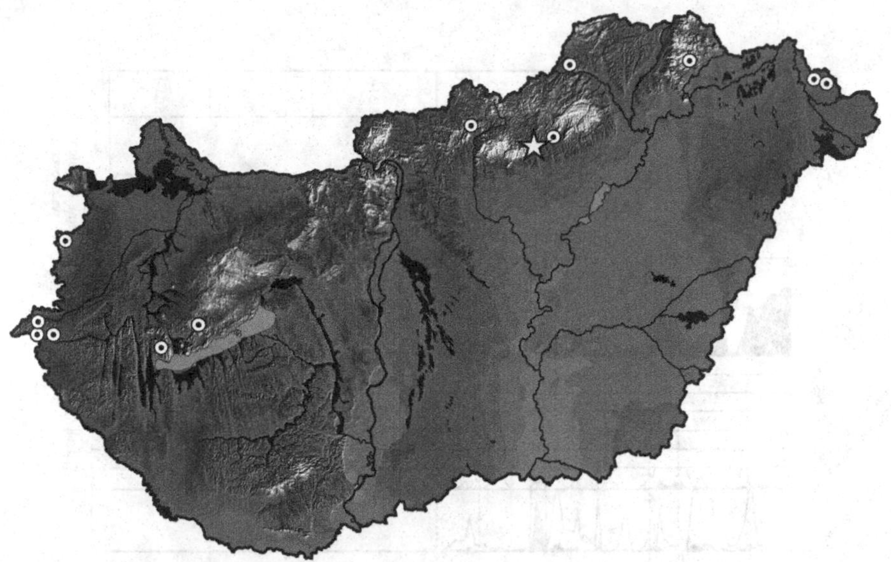

Fig. 17.1. Peatlands (dark areas), *Sphagnum* bogs (circles), and the position of Sirok Nyírjes Peatbog (star) in Hungary. Note that the *Sphagnum* bogs are restricted to the northern and western parts of the country, which experience more precipitation owing to oceanic and montane climatic influences.

Samples taken between depths of 6 and 401 cm were subjected to plant macrofossil analyses. Fig. 17.2 depicts the most important tissue and moss remains of the dominant taxa. In order to reveal the ecological–hydrological gradient of the individual macrofossil zones, a data matrix of the 16 most important peat components was subjected to multivariate statistical analysis (G. Jakab & P. Sümegi, unpublished data). The method of PCA was adopted following Podani (1993) using the software package SYN-TAX 5.0. The PCA values obtained are depicted with depth in Fig. 17.2.

On the basis of the results obtained, the following evolutionary history of the peatland can be derived. The first emergence of aquatic conditions in the depression can be dated to 9500 cal. BP, resulting in the emergence of a relatively deep, oligotrophic lake with scant aquatic vegetation. As shown by the findings of palynological studies, the lake basin was fringed by an open parkland type woodland dominated by *Picea*, *Quercus*, and *Corylus* until about 8950 cal. BP. This was transformed into a woodland dominated by *Tilia* until 8300 cal. BP, which then finally was transformed into a deciduous woodland dominated by *Quercus*, *Tilia*, and *Ulmus* until 6900 cal. BP, with substantial stands of *Corylus*. Despite the clearly observable transformation of the surrounding vegetation, water levels were relatively stable in the basin, apart from minor fluctuations, until 7500 cal. BP. A drop in the water level and peat initiation took

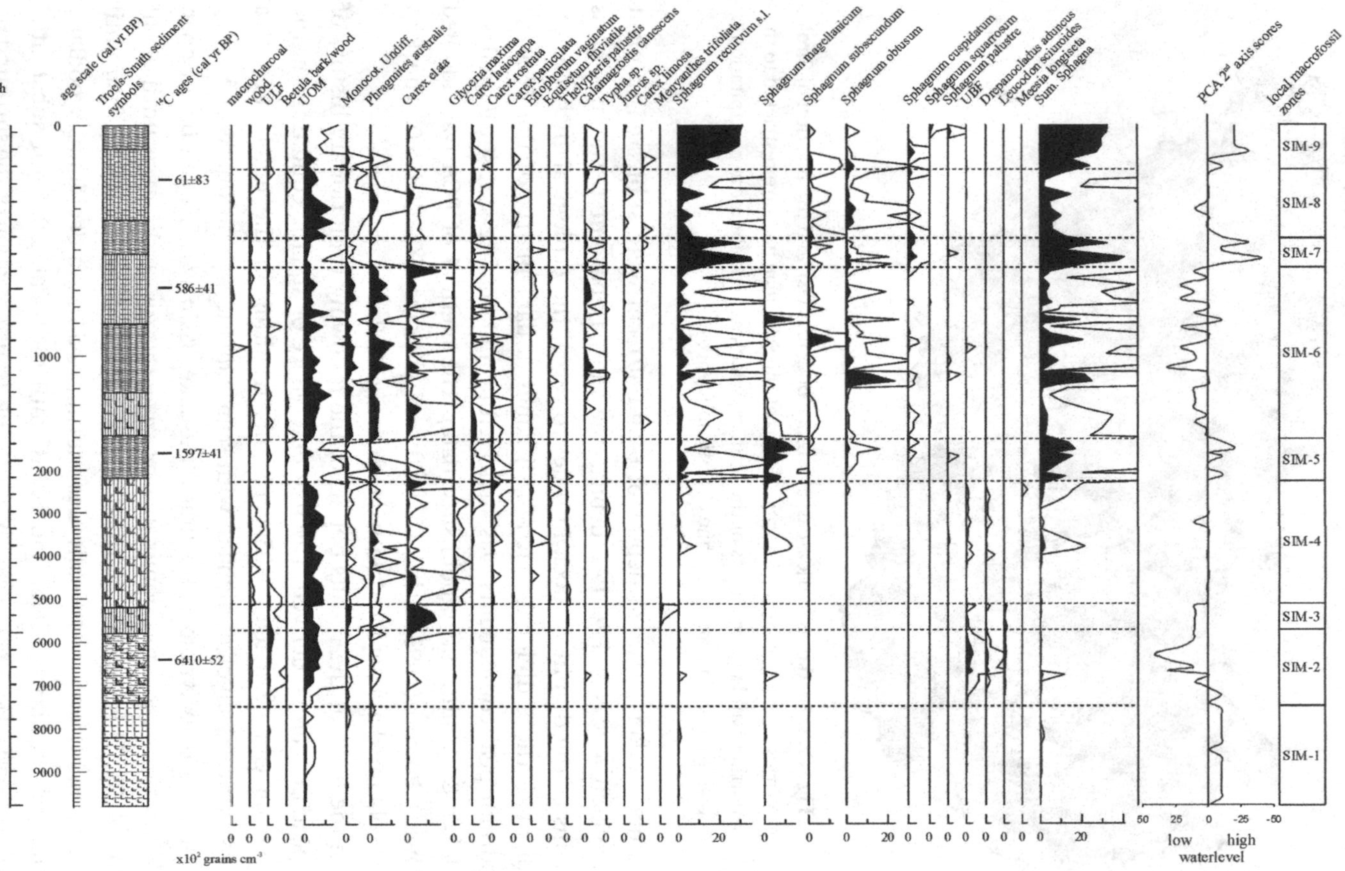

Fig. 17.2. Selected macrofossil diagram of Sirok Nyírjes peat bog.

place almost 1000 years after the development of a closed deciduous woodland. So it is no wonder that there is no direct link between the transformation of the vegetation of the peatland itself and the surrounding terrestrial areas. There is a gradual decrease in the water levels from 7500 cal. BP, reaching an all-time minimum at 6400 cal. BP. Open water areas almost completely disappeared, giving way to the expansion of oak shoots over the major part of the basin. The deepest areas turned into a eutrophic marshland and as such we must assume a gradual increase in the water level from 5800 cal. BP years, yielding a tussock vegetation. During this period the peatland was fringed by a woodland of *Corylus*, *Quercus*, and *Carpinus*.

There was another rise in the water level from 5200 cal. BP, resulting in the expansion of the peatland. This was accompanied by the appearance of floating mats in the expanding shallow eutrophic pond, harboring peat mosses in larger quantities. There was a rapid spread of *Carpinus betulus* in the adjacent closed oak woodlands at the time. Peak distributions of *Fagus sylvatica* and *Carpinus* are found between 3700 and 1750 cal. BP. A similar expansion of *Sphagna* was inferred from 3900 cal. BP onwards in the basin, with the first appearance of real acidophilic *Sphagnum* peatlands dated between 2300 and 1500 cal. BP years.

From 1500 cal. BP there is an alternating succession of *Sphagnum* peatlands with reed and sedge peatland horizons reflecting the alternations of cooler and warmer periods until the present day. Ideal *Sphagnum* peatland conditions were inferred at 1550 cal. AD with such taxa as *Sphagnum cuspidatum*. As shown by the results of Gardner (2002) there is an increase in human influences in the area from 1750 cal. BP, as seen in the drop in the amount of *Fagus* and *Carpinus* accompanied by an advent of *Quercus*.

The past century was also a period of *Sphagnum* peatland expansion. The presence of clayey horizons embedding mollusk shells and carbonate concretions intercalating the peat horizons are the clear signs of soil erosion in the adjacent areas, triggered by deforestation of the nearby slopes. As an outcome of these activities the amount of rainfall reaching the surface substantially increased, resulting in an increase in the water level in the bed of the peatland, triggering the expansion of *Sphagna*. A similar phenomenon was described from several other European sites (Grosse-Brauckmann *et al.* 1973; Rybníček & Rybníčková 1974).

Early to Middle Holocene peatland wetness and lake level records in the Carpathian Basin

The emergence of an oligotrophic lake in the area was dated to 9500 cal. BP with deeper water conditions. Changes in the surficial moisture gradient of peatlands in the Carpathian Basin and those in lake level fluctuations are rather contradictory for this period. High lake level phases are known at 8500 cal. BP for the Szigliget Bay of Lake Balaton (Jakab *et al.* 2005), and Lake Nádas at Nagybárkány (Jakab & Sümegi 2005; Jakab *et al.* 2009). The inferred water levels of Lake Sf Ana in Romania show a highstand at 9500 cal. BP with the emergence of a lowstand at 9000 cal. BP (Magyari *et al.* 2006, 2009). Conversely, studies implemented at various sites of the GHP reconstructed a long-lasting dry and warm period until about 4400 cal. BP (Jakab *et al.* 2004). There seem to be substantial regional differences in the Early and Middle Holocene climate of the Carpathian Basin.

Decreasing water levels inferred at 7500 cal. BP culminated in the driest phase of the peatland, recorded at 6400 cal. BP. This period was the time of the Holocene climatic optimum, when there was a substantial retreat of the Swiss Alp glaciers between 7450 and 6650 cal. BP and between 6200 and 5650 cal. BP years (Joerin *et al.* 2008). Conversely, there is an inferred increase in the water level of Lake Sf Ana in Romania from 7500 cal. BP onwards, interrupted by a short decrease between 5700 and 5500 cal. BP (Magyari *et al.* 2006, 2009). According to Cheddadi *et al.* (1997) and Davis *et al.* (2003) the traditionally postulated Holocene climatic optimum is identifiable only in Northern Europe, and this time period in Southern Europe was characterized by colder conditions, with Central Europe occupying a transitional phase. This assumption is refuted by the findings of paleoecological studies made on lake and marshland basins in the Carpathian Basin.

After the climatic optimum there were two periods when substantial increases in the surface moisture gradient were observable at the study site, at 5800 and 5400 cal. BP years. This change is congruent with the pattern observable in other lacustrine and marshland basins of the Carpathian Basin that also display an increase in the water level. There is a sudden increase in the water level of the Lake Sf Ana from 5500 cal. BP (Magyari *et al.* 2006, 2009). A similar rise in the water level was deduced for Lake Balaton from 5200 cal. BP, resulting in an expansion of the lake's area, exceeding the values of modern day water coverage (Cserny & Nagy-Bodor 2000; Jakab *et al.* 2005). A somewhat delayed similar pattern is observable in the peatlands of the GHP starting at 4400 cal. BP (Jakab *et al.* 2004). This period between 5600 and 5300 cal. BP is referred to as the Middle Holocene Climatic Transition, characterized by a sudden deterioration of the previously warm conditions as a result of the collective transformation of orbital forces, solar activity, and ocean currents (Magny *et al.* 2006; Iizuka *et al.* 2008).

Three short-lived peat formation events were identified at 8200, 6800, and 3800 cal. BP, reflecting cooler conditions. Paleoecological records available from the Carpathian Basin to date yielded no indication of climate change for this period; however, data from Western Europe did. There is a marked cooling related to a global cooling event lasting for merely 200 years known as the "8.2 ky event" (Alley *et al.* 1997; Bond *et al.* 1997; Nesje & Dahl 2001).

At 6000 cal. BP a high lake level phase of Swiss lakes (Magny 1998; Magny & Schoellammer 1999) and changes in the moisture gradient of some British peatlands (Hughes *et al.* 2000) are evidence for the emergence of cooler conditions. Similarly at 3500 cal. BP the higher lake phase of Swiss lakes (Magny 1998; Magny *et al.* 2002), the expansion of Alpine glaciers (Haas *et al.* 1998), and an increase in the moisture gradient of numerous Western European peatlands mark a cooling of the climate (Hughes *et al.* 2000; Barber & Charman 2005).

An increase in the amount of *Sphagna* from 2800 cal. BP in the Nyires peat bog also marks the cooling of the climate and the accompanying rise in the rainfall. This deterioration of the climate starting at 3500 cal BP culminates here in the Carpathian Basin, as shown by numerous records. Water levels were at their highest in the Lake Sf Ana in Romania at this time, and there is information on the development of layering in the waterbody for this period (Magyari *et al.* 2006). Congruent with these data, information from studies of testacea and humic content of peatlands in the Eastern Carpathians shows an increase in the moisture gradient (Schnitchen *et al.* 2003). The resuming peat formation in certain Hungarian peatlands marks the cooling of the climate here (Jakab & Sümegi 2007).

The first real *Sphagnum* peatland developed at Sirok between 2300 and 1500 cal. BP. From here on we have a record of alternating phases of *Sphagnum* peatlands and sedge–reed peatlands. As displayed by the record of vegetation changes, the catchment of the studied peatland was highly prone to climatic fluctuations. Certain periods are characterized by a rapid expansion of *Sphagna*; others are characterized by the expansion of sedge and reed. A sudden expansion of *Sphagna* was recorded at least ten times. Figure 17.3 depicts a comparison of changes inferred for the Nyírjes peat bog with cooler periods determined by Barber *et al.* (1994) and Mauquoy and Barber (1999), emphasizing changes for the past 3000 years. The *Sphagnum* peaks perfectly match the more humid periods identified in the British Isles at 2150, 1750, 1300, 1000, 850, 500, and 200 cal. BP (Barber *et al.* 1994; Mauquoy & Barber 1999; Barber & Charman 2005; Barber 2007), referring to some collective global force as the cause of these changes. Barber and Charman (2005) identified centennial-scale climatic fluctuations in different parts of Western Europe. The length of these cycles varied between 210, 600, 800, and 1100 years in different peatlands. No such cycles have been identified in Central Europe to date.

Historical records as related to climate change

It is worth comparing the paleoecological data from our study site for the past 2000 years with those of written historical records. The interrelatedness of cultural evolution and that of the natural environment is widely

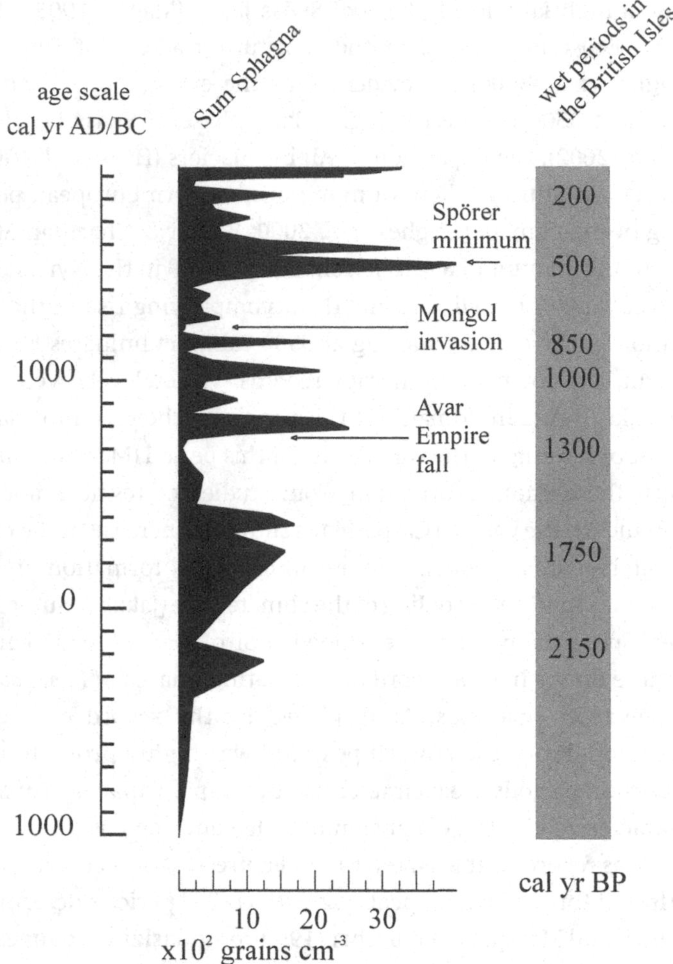

Fig. 17.3. Comparison of bog surface wetness changes of Sirok Nyírjes peat bog and some British peat bogs (Barber *et al.* 1994; Mauquoy & Barber 1999) in the past 3000 yr. The arrows show some medieval events. Note that the spreading periods of *Sphagna* in the Sirok Nyírjes peat bog coincide with the wet periods described from the British Isles.

accepted. Large environmental crises led to the collapse of whole empires, or to famine or wars. Well-known examples are those of the disappearance of the Mayan Empire, the famines of the Little Ice Age, and the abandonment of Viking settlements in Greenland. This interrelatedness is very often indirect, as seen by the ecological carrying capacity of a region and agricultural production influenced by climatic conditions (Berglund 2003).

One major climatic crisis event in the Carpathian Basin is connected to the fall of the Avar Empire in the eighth century AD. Written records blame the famines and wars triggered by the outstanding droughts of the period (Győrffy & Zólyomi 1994). Few environmental historical data are available for this period. As shown by the *Sphagnum* curve of our study site, this period truly was characterized by dry conditions (Fig. 17.3). The same event was also recorded at Nádas Lake in Nagybárkány by Jakab *et al.* (2009) in the area of northern Hungary.

Another major historical crisis was the advancement of Mongolian tribes into the area in 1241–2. Ceratin sources blame this on the severe cold weather; others talk about the severe droughts. Data are highly contradictory. In Europe information is available on the extreme summer droughts for the period in question, whereas in Hungary the extreme cold winter of 1241 is emphasized, when the complete freezing of the Danube enabled the Mongol hordes to cross the river and destroy the settlements of Transdanubia. This seemingly contradictory information can easily be resolved, as shown by Kiss (2000). The freezing of the Danube in the winter of 1241 was not a unique event, and the summer droughts recorded in Europe must have had similar devastating effects in Hungary, amplifying the negative outcome of war and famine. As shown by our paleoecological data for the Nyíres peat bog, Hungary was characterized by extremely warm conditions during this period, resulting in an almost complete desiccation of the *Sphagnum* peatland. Barber *et al.* (2000) also marked this period as the driest one in the past 3000 year history of Europe.

The *Sphagnum* curve of the Nyíres peat bog enables us to identify the period of the Little Ice Age as dated between the middle part of the sixteenth century until the middle part of the nineteenth century (Bradley *et al.* 2003). This was inferred to be the coldest period of the past 2000 years. The coldest period was placed in the terminal part of the sixteenth century (the Spörer minimum), when there was a marked drop in the temperature all over Europe (Pfister 1999; Pfister & Brázdil 1999). This event caused severe problems in food production, triggered by transforming sunspot activities. The later cold periods were less devastating on the economies of Europe and Hungary (Rácz 2001). Environmental records for the Nyíres peat bog are congruent with these events as well. The most diverse *Sphagnum* taxa including the hygrophilous *Sphagnum cuspidatum* lived here at the end of the sixteenth century. Western European peatlands were similarly prone to the fluctuating climate of the Little Ice Age (Mauquoy *et al.* 2002). Wetter conditions were identified from the beginning of the sixteenth and the middle part of the seventeenth centuries.

An increase in the surface wetness of our study area over the past 100 years is attributable not really to climatic changes but rather to human-induced soil erosion and alteration of surficial rainfall conditions via deforestation. This is clearly recorded in the intercalating sand grains and clayey horizons of the peatland studied.

Conclusions

Hungary is characterized by moderately continental climatic conditions. The central part of the country is characterized by an average annual rainfall of 500–550 mm, whereas the mid-mountains have a rainfall of 600–800 mm per annum. These harsh climatic conditions are not favorable for the formation of *Sphagnum* peat bogs. The total number of *Sphagnum* ocurrences hardly exceeds 50, and in the central driest areas of the GHP *Sphagna* are completely absent. Only sporadic *Sphagnum* bogs are known in the country with a total number well below 20. Most of them are extremely small with an area of only a few hectares. True raised bogs are completely missing. The peatlands of Hungary are highly prone to even minor climatic changes because of their marginal biogeographic position.

The small Nyírjes peat bog of Sirok is situated at the eastern foothills of the Mátra Mts. Most of the peat bog is covered with *Scirpo-Phragmitetum* and *Carici lasiocarpae-Sphagnetum* communities. Investigations of this small peat bog provide a late Quaternary record of vegetation development affected by climatic changes. The first emergence of aquatic conditions can be dated to 9500 cal. BP, resulting in the emergence of a relatively deep, oligotrophic lake. A gradual decrease in the water levels is revealed from 7500 cal. BP, reaching a minimum at 6400 cal. BP. Open water areas almost completely disappeared. The expansion of *Sphagna* was inferred from 3900 cal. BP onwards. The first appearance of real acidophilic *Sphagnum* peatlands dated between 2300 and 1500 cal. BP years. From 1500 cal. BP there is an alternating succession of *Sphagnum* peatlands with reed and sedge peatland horizons reflecting the alternations of cooler and warmer periods until the present day. A sudden expansion of *Sphagna* was recorded at least ten times (e.g., at 2150, 1750, 1300, 1000, 850, 500, and 200 cal. BP yr).

It is interesting to ask what climatic factor should be identified as the cause for changes in the moisture gradient of peatlands? Or what magnitude of temperature increases and accompanying precipitation decreases can be inferred for the drier periods? According to Blaauw *et al.* (2004) there is a strong relationship between the moisture gradient of peatlands and solar activity reflected in the correlation of the former parameter with proxy for Δ^{14}C. One may rightly ask what component of the climate controls the moisture gradient of peatlands via fluctuating solar activities? Surficial wetness is controlled by a complex interplay of precipitation and evapotranspiration of the plants, seen in such parameters as annual average rainfall and evaporation influenced by the temperatures of the growing season. There are no surficial water courses feeding the Nyíres peat bog, so the influence of runoff must have been significant to the hydrology of the peatland only during the past 100 years.

As was shown in Western Europe the moisture gradient of peatlands for the past 3000 years was primarily determined by fluctuations in the temperature of the growing season and not really by the amount of rainfall (Barber *et al.* 2000; Barber & Langdon 2001; Barber & Charman 2005; Schoning *et al.* 2005; Barber 2007). According to Charman (2007; Charman *et al.* 2009) in the Atlantic part of Europe summer precipitation and summer temperatures control the moisture gradient of peatlands. Unfortunately macrofossil studies are not acceptable to accurately predict former temperatures or precipitation rates. Only the major trajectories of climate changes can be identified.

In spite of this, the modern distribution of *Sphagnum* peatlands in Hungary enables us to give a rough estimate. *Sphagnum* peatlands appear in areas characterized by a precipitation of 600 mm per annum, assuming modern temperature values. Below this threshold we can find only sporadic occurrences, and there are no *Sphagna* known below the lower limit of 550 mm. Based on our findings for the Nyírjes peat bog we may infer conditions in the lower hilly areas during the drier periods of the past 3000 years to be similar to that of the central parts of the GHP. The complete disappearance of *Sphagna* from the area must be linked to a steady drop in the rainfall, resulting in at least a 50 mm deficit in the local water balance. This can be achieved by an increased evapotranspiration as a result of elevated temperatures of the summer growth season. This value must have exceeded even 100 mm during the Middle Holocene Transition.

Climate scenarios for Hungary predict a 1.5–4.9 °C increase in average summer temperature, and a 25%–35% (*c.* 50 mm) decrease in summer precipitation by the end of the twenty-first century (Szlávik *et al.* 2002; Bartholy *et al.* 2004, 2006). For this reason the frequency of arid (steppe-type) years will increase by 60%! In this case most of the *Sphagnum* peatlands would vanish from Hungary.

Acknowledgements

This research was funded by a Bolyai János Research Scholarship. The authors wish to express their gratitude to the following persons: Zoltán Tuba, Sá ndor Gulyás, Tamás Pócs, Enikő Magyari, Gábor Papp, and András Schmotzer.

References

Aaby, B. (1976). Cyclic climatic variations in climate over the past 5,500 yrs reflected in raised bogs. *Nature* **263**: 281–4.

Aaby, B. & Digerfeldt, G. (1986). Sampling techniques for lakes and bogs. In *Handbook of Holocene Palaeoecology and Palaeohydrology* ed. B. E. Berglund, pp. 181–94. New York: John Wiley and Sons.

Alley, R. B., Mayewski, P. A., Sowers, T. *et al.* (1997). Holocene climatic instability: a prominent, widespread event 8200 yr ago. *Geology* **25**: 483–6.

Bacsó, N. (1959). *Magyarország Éghajlata*. Budapest: Akadémiai Kiadó.

Balogh, M. (2000a). A lápok rendszerezése. In *Tőzegmohás Élőhelyek Magyarországon: Kutatá s, Kezelés, Védelem*, ed. E. Szurdoki, pp. 57–65. Miskolc: CEEWEB Munkacsoport.

Balogh, M. (2000b). Az úszólápi szukcesszió kérdései I. *Kitaibelia* **5**: 9–16.

Barber, K. E. (2007). Peatland records of Holocene climate change. In *Encyclopedia of Quaternary Science*, ed. S. A. Elias, vol. 3, pp. 1883–94. London: Elsevier.

Barber, K. E., Chambers, F. M., Maddy, D. & Brew, J. (1994). A sensitive high resolution record of the Holocene climatic change from a raised bog in northern England. *The Holocene* **4**: 198–205.

Barber, K. E. & Charman, D. (2005). Holocene palaeoclimate records from peatlands. In *Global Change in the Holocene*, ed. A. Mackay, A. Battarbee, R. J. Birks & F. Oldfield, pp. 210–26. London: Hodder Arnold.

Barber, K. E. & Langdon, P. G. (2001). Peat stratigraphy and climate change. In *Handbook of Archaeological Sciences*, ed. D. R. Brothwell & A. M. Pollord, pp. 155–66. Chichester: Wiley.

Barber, K. E., Maddy, D., Rose, N. *et al.* (2000). Replicated proxy-climate signals over the last 2000 yr from two distant UK peat bogs: new evidence for regional palaeoclimate teleconnections. *Quaternary Science Reviews* **19**: 481–7.

Bartholy, J., Pongrácz, R., Matyasovszky, I. & Schlanger, V. (2004). A XX. században bekövetkezett és a XXI. századra várható éghajlati tendenciák Magyarország területére. *AGRO-21 Füzetek* **33**: 1–18.

Bartholy, J., Pongrácz, R., Torma, Cs. & Hunyady, A. (2006). Regional climate projections for the Carpathian Basin. In *Conference Proceedings of the International Conference Climate Change: Impacts and Responses in Central and Eastern European Countries*, ed. I. Lá ng, T. Faragó & Zs. Iványi, pp. 55–62. Budapest: HAS – HMEW – RECCEE.

Bennett, K. D. (1992). PSIMPOLL. A quickBasic program that generates PostScript page description of pollen diagrams. *INQUA Commission for the Study of the Holocene: Working Group on Data Handling Methods, Newsletter* **8**: 11–12.

Berglund, B. E. (2003). Human impact and climate changes – synchronous events and a causal link? *Quaternary International* **105**: 7–12.

Birks, H. H. (1980). Plant macrofossils in Quaternary lake sediments. *Archiv für Hydrobiologie Ergebnisse der Limnologie*, Suppl. **15**: 1–60.

Birks, H. J. B. (1982). Quaternary bryophyte paleo-ecology. In *Bryophyte Ecology*, ed. A. J. E Smith, pp. 437–90. London & New York: Chapman and Hall.

Birks, H. J. B. & Birks, H. H. (1980). *Quarternary Palaeoecology*. Baltimore, MD: University Park Press.

Birks, H. H. & Birks, H. J. B. (2000). Future uses of pollen analysis must include plant macrofossils. *Journal of Biogeography* **27**: 31–5.

Blaauw, M., van Geel, B. & van der Plicht, J. (2004). Solar forcing of climatic change during the mid-Holocene: indicators from raised bogs in The Netherlands. *The Holocene* **14**: 35–44.

Bond, G., Showers, W., Cheseby, M. *et al.* (1997). A pervasive millenial-scale cycle in North Atlantic Holocene and glacial climates. *Science* **278**: 1257–66.

Borhidi, A. & Sánta, A. (eds.)(1999). *Vörös Könyv Magyarország Növénytársulásairól 1–2.* Budapest: Természetbúvár Alapítvány Kiadó.

Boros, Á. (1968). *Bryogeographie und Bryoflora Ungarns.* Budapest: Akadémiai Kiadó.

Bradley, R. S., Hughes, M. K., & Diaz, H. F. (2003). Climate in medieval times. *Science* **302**: 404–5.

Campbell, I. D., Campbell, C., Yu, Z., Vitt, D. H. & Apps, M. J. (2000). Millenial-scale rhythms in peatlands in the western interior of Canada in the global carbon cycle. *Quaternary Research* **54**: 155–8.

Charman, D. J. (2007). Summer water deficit variability controls on peatland water-table changes: implications for Holocene palaeoclimate reconstructions. *The Holocene* **17**: 217–27.

Charman, D. J., Barber, K. E., Blaauw, M. *et al.* (2009). Climate drivers for peatland palaeoclimate records. *Quaternary Science Reviews* **28**(19–20): 1811–19.

Cheddadi, R., Yu, G., Guiot, J., Harrison, S. P. & Prentice, I. C. (1997). The climate of Europe 6000 years ago. *Climate Dynamics* **13**: 1–9.

Cserny, T. & Nagy-Bodor, E. (2000). Limnogeology of Lake Balaton (Hungary). In *Lake Basins though Space and Time: AAPG Studies in Geology*, ed. E. H. Gierlowski-Kordesch & K. R. Kelts, **46**: 605–18.

Davis, B. A. S., Brewer, S., Stevenson, A. C. & Guiot, J. (data contributors) (2003). The temperature of Europe during the Holocene reconstructed from pollen data. *Quaternary Science Reviews* **22**: 1701–16.

Dickson, J. H. (1973). *Bryophytes of the Pleistocene. The British Record and its Chorological and Ecological Implications.* London and New York: Cambridge University Press.

Dömsödi, J. (1988). *Lápképződés, Lápmegsemmisülés.* Budapest: MTA Földrajztudományi Kutató Intézet.

Gardner, A. R. (2002). Neolithic to Copper Age woodland impacts in northeast Hungary? Evidence from the pollen and sediment chemistry records. *The Holocene* **12**: 521–53.

Gignac, D. L. & Vitt, D. H. (1990). Habitat limitations of *Sphagnum* along climatic, chemical and physical gradients in mires of western Canada. *Bryologist* **93**: 7–22.

Grosse-Brauckmann, G. (1972). Über pflanzliche Makrofossilien mitteleuropäischer Torfe. I: Gewebereste krautiger Pfanzen und ihre Merkmale. *Telma* **2**: 19–55.

Grosse-Brauckmann, G. (1986). Analysis of vegetative plant macrofossils. In *Handbook of Holocene Palaeoecology and Palaeohydrology*, ed. B. E. Berglund, pp. 591–618. New York: John Wiley and Sons Ltd.

Grosse-Brauckmann, G., Haussner, W. & Mohr, K. (1973). Über eine kleine Vermoorung im Odenwald, ihre Ablagerungen und ihre Entwicklung der umgebenden Kulturlandschaft. *Zeitschrift für Kulturtechnik und Flurbereinigung* **14**: 132–43.

Győrffy, GY. & Zólyomi, B. (1994). A Kárpát-medence és Etelköz képe egy évezred előtt. In *Honfoglalás és Régészet*, ed. GY. Győrffy & L. Kovács, pp. 13–37. Budapest: Balassi Kiadó.

Haas, J. N., Richoz, I., Tinner, W. & Wick, L. (1998). Synchronous Holocene climatic oscillations recorded on the Swiss Plateau and at the timberline in the Alps. *The Holocene* **8**: 301–4.

Halsey, L. A., Vitt, D. H. & Bauer, I. E. (1998). Peatland initiation during the Holocene in continental western Canada. *Climate Change* **40**: 315–42.

Hughes, P. D. M., Mauquoy, D., Barber, K. E. & Langdon, P. G. (2000). Mire development pathways and palaeoclimatic records from a full Holocene peat archive at Walton Moss, Cumbria, England. *The Holocene* **10**: 465–79.

Iizuka, Y., Hondoh, T. & Fujii, Y. (2008). Antarctic sea ice extent during the Holocene reconstructed from inland ice core evidence. *Journal of Geophysical Research* **113**. 113(D15), Citation D15114.doi: 10.1029/2007JD009326.

Jakab, G. & Sümegi, P. (2004). A lágyszárú növények tőzegben található maradványainak határozója mikroszkópikus bélyegek alapján. *Kitaibelia* **9**: 93–129.

Jakab, G. & Sümegi, P. (2005). The evolution of Nádas-tó at Nagybárkány in the light of the macrofossil finds. In *Environmental History of North-Eastern Hungary*, ed. E. Gál, I. E. Juhász, & P. Sümegi. *Varia Archaeologica Hungarica* **19**: 67–77.

Jakab, G. & Sümegi, P. (2007). The vegetation history of Baláta-tó. In *Environmental History of Transdanubia*, ed. I. E. Juhász, C s. Zatykó & P. Sümegi. *Varia Archaeologica Hungarica* **20**: 251–4.

Jakab, G., Sümegi, P. & Magyari, E. (2004). A new paleobotanical method for the description of Late Quaternary organic sediments (Mire-development pathways and palaeoclimatic records from S Hungary). *Acta Geologica Hungarica* **47**: 1–37.

Jakab, G., Sümegi, P. & Szántó, Zs. (2005). Késő-glaciális és holocén vízszintingadozá sok a Szigligeti-öbölben (Balaton) makrofosszília vizsgálatok eredményei alapján. *Földtani Közlöny* **135**: 405–31.

Jakab, G., Majkut, P., Juhász, I. *et al.* (2009). Palaeoclimatic signals and anthropogenic disturbances from the peatbog at Nagybárkány (N Hungary). *Hydrobiologia* **631**: 87–106.

Janssens, J. A. P. (1983a). A quantitative method for stratigraphic analysis of bryophytes in Holocene peat. *Journal of Ecology* **71**: 189–96.

Janssens, J. A. P. (1983b). Past and extant distribution of *Drepanocladus* in North America, with notes on the differentiation of fossil fragments. *Journal of the Hattori Botanical Laboratory* **54**: 251–98.

Janssens, J. A. P. (1987). Ecology of peatland bryophytes and palaeoenvironmental reconstruction of peatlands using fossil bryophytes. *Manual for Bryological Methods Workshop. Satellite Conference of the XIV. Intenational Botanical Conference*. Mainz: International Association of Bryologists.

Janssens, J. A. P. (1990). Methods in Quarternary Ecology 11. Bryophytes. *Geoscience Canada* **17**: 13–24.

Jessen, K. (1949). Studies in the Late Quaternary deposits and flora history of Ireland. *Proceedings of the Royal Irish Academy* **52**(B): 85–290.

Jessen, K. & Milthers, V. (1928). Stratigraphical and paleontological studies of interglacial fresh-water deposits in Jutland and northwest Germany. *Danmarks Geologiske Undersogelse* Series 11, **48**: 1–378.

Joerin, U. E., Nicolussi, K., Fischer, A., Stocker, T. F. & Schlüchter, C. (2008). Holocene optimum events inferred from subglacial sediments at Tschierva Glacier, Eastern Swiss Alps. *Quaternary Science Reviews* **27**: 337–50.

Kiss, A. (2000). Weather events during the first Tartar invasion in Hungary (1241–42). *Acta Geographica Szegediensis* **37**: 149–56.

Kooijman, A. M. (1993). On the ecological amplitude of four mire bryophytes: a reciprocal transplant experiment. *Lindbergia* **18**: 19–24.

Korhola, A. (1995). Holocene climatic variations in southern Finland reconstructed from peat-initiation data. *The Holocene* **5**: 43–57.

Magny, M. (1998). Reconstruction of Holocene lake-level changes in the Jura (France): methods and results. In *Palaeohydrology as Reflected in Lake-level Changes as Climatic Evidence for Holocene Times*, ed. S. P. Harrison, B. Frenzel, U. Huckried, & M. Weiss. *Paläoklimaforschung* **25**: 67–85.

Magny, M. & Schoellammer, P. (1999). Lake-level fluctuations at Le Locle, Swiss Jura, from the Younger Dryas to the Mid-Holocene: a high-resolution record of climate oscillations during the final deglaciation. *Géographie Physique et Quaternaire* **53**: 183–97.

Magny, M., Miramont, C. & Sivan, O. (2002). Assessment of climate and anthropogenic factors on Holocene Mediterranean vegetation in Europe on the basis of palaeohydrological records. *Palaeogeography, Palaeoclimatology, Palaeoecology* **186**: 47–59.

Magny, M., Leuzinger, U., Bortenschlager, S. & Haas, J. N. (2006). Tripartite climate reversal in Central Europe 5600–5300 years ago. *Quaternary Research* **65**: 3–19.

Magyari, E., Buczkó, K., Jakab, G. *et al.* (2006). Holocene palaeohydrology and environmental history in the South Harghita Mountains, Romania. *Földtani Közlöny* **136**: 249–84.

Magyari, E. K., Buczkó, K., Jakab, G. *et al.* (2009). Palaeolimnology of the last Eastern Carpathian crater lake – a multiproxy study of Holocene hydrological changes. *Hydrobiologia* **631**: 29–63.

Máthé, I. & Kovács, M. (1958). A Mátra tőzegmohás lápja. *Botanikai Közlemények* **47**(3–4): 323–31.

Mauquoy, D. & Barber, K. (1999). A replicated 3000 yr proxy-climate record from Coom Rigg Moss and Felicia Moss, The Border Mires, northern England. *Journal of Quaternary Science* **14**: 263–75.

Mauquoy, D. & van Geel, B. (2007). Mire and peat macros. In *Encyclopedia of Quaternary Science*, ed. S. A. Elias, vol. 3, pp. 2315–36. London: Elsevier.

Mauquoy, D., van Geel, B., Blaauw, M. & van der Plicht, J. (2002). Evidence from northwest European bogs shows 'Little Ice Age' climatic changes driven by variations in solar activity. *The Holocene* **12**: 1–6.

Mitsch, W. J. & Gosselink, J. G. (1993). *Wetlands*. New York: Van Nostrand Reinhold.

Moore, P. D. & Bellamy, D. J. (1974). *Peatlands*. London: Elek Science.
Nesje, A. & Dahl, S. O. (2001). The Greenland 8200 cal. yr BP event detected in loss-on-ignition profiles in Norwegian lacustrine sediment sequences. *Journal of Quaternary Science* **16**: 155–66.
Odgaard, B. V. (1980). Ecology, distribution and late Quaternary history of *Polytrichastrum alpinum* (Hedw.) G. L. Smith in Denmark. *Lindbergia* **6**: 155–8.
Osvald, H. (1925). Die Hochmoortypen Europas. *Veröffentlichungen Geobotanisches Institut Rubel*, Zürich **3**: 707–23.
Pakarinen, P. (1995). Classification of boreal mires in Finland and Scandinavia: a review. *Vegetatio* **118**: 29–38.
Pfister, C. (1999). *Wetternachhersage: 500 Jahre Klimvariationen und Naturkatastrophen (1496–1995)*. Bern: Haupt.
Pfister, C. & Brázdil, R. (1999). Climatic variability in sixteenth-century Europe and its social dimension: a synthesis. *Climatic Change* **43**: 5–53.
Podani, J. (1993). SYN-TAX 5.0: computer programs for multivariate data analysis in ecology and systematics. *Abstracta Botanica* **17**: 289–302.
Rácz, L. (2001). *Magyarország Éghajlattörténete az Újkor Idején*. Szeged: JGYF Kiadó.
Rybníček, K. (1973). A comparison of the present and past mire communities of Central Europe. In *Quaternary Plant Ecology*, ed. H. J. B. Birks & R. G. West, pp. 237–61. Oxford: Blackwell.
Rybníček, K. & Rybníčková, E. (1974). The origin and development of waterlogged meadows in the central part of the Sumava Foothills. *Folia Geobotanica et Phytotaxonomica* **9**: 45–70.
Rydin, H. (1993). Interspecific competition between *Sphagnum* mosses on a raised bog. *Oikos* **66**: 413–23.
Schnitchen, C., Magyari, E., Tóthmérész, B., Grigorszky, I. & Braun, M. (2003). Micropaleontological observations on a *Sphagnum* bog in East Carpathian region – testate amoebae (*Rhizopoda: Testacea*) and their potential use for reconstruction of micro- and macroclimatic changes. *Hydrobiologia* **506–509**: 45–9.
Schoning, K., Charman, D. J. & Wastegard, S. (2005). Reconstructed water tables from two ombrotrophic mires in eastern central Sweden compared with instrumental meteorological data. *The Holocene* **15**: 111–18.
Slack, N. G. (1994). Can one tell the mire type from the bryophytes alone? *Journal of the Hattori Botanical Laboratory* **75**: 149–59.
Smith, L. C., MacDonald, G. M., Velichko, A. A. *et al.* (2004). Siberian peatlands a net carbon sink and global methane source since the early Holocene. *Science* **303**: 353–6.
Szlávik, L., Mika, J. & Bálint, G. (2002). Review of climate change induced modification of hydrological extremes in Hungary. *Proceedings of International Conference on Drought Mitigation and Prevention of Land Desertification*, Bled, Slovenia, 21–25 April 2002.
Szurdoki, E. & Nagy, J. (2002). *Sphagnum* dominated mires and *Sphagnum* occurrences of North-Hungary. *Folia Historico-Naturalia Musei Matraensis* **26**: 67–84.
Troels-Smith, J. (1955). Karakterisering af lose jordater. *Danmarks Geologiske Undersogelse* **4**: 10.

Vitt, D. H. & Chee, W.-L. (1990). The relationships of vegetation to surface water chemistry and peat chemistry in fens of Alberta, Canada. *Vegetatio* **89**: 87–106.

Wasylikowa, K. (1996). Analysis of fossil fruits and seeds. In *Handbook of Holocene Palaeoecology and Palaeohydrology*, ed. B. E. Berglund, pp. 571–90. New York: John Wiley and Sons.

Yu, Z., Campbell, I. D., Campbell, C. *et al.* (2003). Carbon sequestration in western Canadian peat highly sensitive to Holocene wet-dry climate cycles at millennial timescales. *The Holocene* **13**: 801–8.

Zoltai, S. C. & Vitt, D. H. (1990). Holocene climatic change and the distribution of peatlands in western interior Canada. *Quaternary Research* **33**: 231–40.

18

Signs of Climate Change in the Bryoflora of Hungary

TAMÁS PÓCS

Introduction

As the average shift of isotherms in Central Europe has been some 200 km northeastwards during the past 60 years, we might expect changes in the flora of Hungary, especially among the cryptogams, owing to their superior dispersal ability by spores and gemmae. During the past 50 years of global warming in Central Europe the average temperature rose by 0.8 °C, which alone does not mean as much as the increasing extremes both in temperature and in the annual distribution of precipitation. In Hungary 1990 was probably the hottest year of the millennium, followed by 1997, 1995, 1999, and 2000. At the same time the winters have become milder with shorter very cold periods, and we have had prolonged summer droughts. According to the records of the Hungarian Meteorological Service (Szegő 2005; Takács-Sánta 2005) the number of hot days above 25 and 30 °C increased considerably in Hungary. The amount of precipitation, especially during winter, decreased.

It is highly likely that global warming is anthropogenic, due to the greenhouse effect of increasing CO_2 and methane in the atmosphere (Vida 2001). Excessive CO_2 emission began with deforestation in the Bronze Age and contributed to the end of the last glacial period. However, its sudden increase by the industrial revolution and especially during the last century (from 270 ppm to 380 ppm CO_2 in the atmosphere) resulted in the 160 km NW shift of the annual isotherms in the Pannonian basin in Hungary. According to the conservative estimates of the IPCC (Intergovernmental Panel on Climate Change), the CO_2 content will reach 500 ppm by the middle of the twenty-first century, which should cause a 2.1–4.6 °C increase in the average temperature.

Bryophyte Ecology and Climate Change, eds. Zoltán Tuba, Nancy G. Slack and Lloyd R. Stark. Published by Cambridge University Press.

This may cause a further 450 km shift of the isotherms in Hungary, bringing a climate analogous to the Mediterranean Vardar Valley in Macedonia, or even worse, the climate of the steppe and semideserts of Turkmenia, depending on the possible degree of oceanity versus continentality. That in turn depends on many factors, especially on the ever-changing sea and air currents. Global warming is far from even and its effect on sea and air currents can even cause cooling in certain places. Parallel to the warming in some areas, like southwest Germany, an increase of precipitation was observed (Deutscher Wetterdienst 2002), while in the inner part of tropical Africa the dry seasons are prolonged and the rains become erratic. The increasing UV radiation and CO_2 concentration itself can have serious effects on living organisms (Tuba 2005).

Such changes must affect the distribution of land plants and animals (as was the case during former geological and climatic periods). Although the higher plants react relatively slowly to climate changes, the more mobile animals (butterflies, moths, Orthoptera, Diptera) and spore-bearing plants are better adapted to long-range air dispersal; such animals and plants can follow the climate changes by changing their distribution areas more rapidly. Among higher plants a well-known case of such a distributional change is the northward advance of the Atlantic–Submediterranean *Ilex europaea* and the northward movement of the limit of broadleaved deciduous forest belts against coniferous belts (Klötzli & Walther 1999). Most spectacular is the effect of climate change on the distribution of flying insects, e.g., butterflies (Parmesan & Yohe 2003) or on the range and nesting places of migratory birds (Kalela 1949).

As expected, cryptogamic plants are more mobile than the gradually dispersing phanerogams owing to the air dispersal ability of their diaspores, which is inversely proportional to their spore and propagule size and depends much on their survival ability (Zanten & Pócs 1981). If the diaspores can then land on a suitable unoccupied habitat, they can easily colonize it (Frahm 2005). Therefore we can expect their reaction to climate changes to occur more quickly and on a larger scale. That this occurs is evident from study of the bryoflora of Germany. Numerous species have recently appeared or their distributions have broadened, serving as indicators of global warming (Frahm & Klaus 2000; Frahm 2003). *Pterygoneurum lamellatum* (Lindb.) Jur., a xerophytic species considered to be a continental steppe and semidesert element, appeared in Arctic Canada and western Greenland; it was unknown there before 1959 (Mogensen *et al.* 1997). On the other hand, we can expect the disappearance of some boreal species due to global warming and the drying out of wetlands.

New bryophyte Occurrences

A fern species of very oceanic–Mediterranean distribution, *Anogramma leptophylla* (L.) Link., was recorded recently from Hungary (Molnár *et al.* 2008). Its occurrence in Hungary, very far from its known distributional area, colonizing a southern exposed, shaded scree slope with favorable microclimatic conditions, supports the idea of its arrival as a result of climate change. Its anthropogenic introduction in this habitat can probably be excluded.

Among bryophytes, during the investigation of the cryptogamic vegetation of exposed loess cliffs, we have found several species that were unknown 50–60 years ago. Loess cliffs in Hungary bear an open, desert-type vegetation due to their near-vertical surface, which receives a very limited amount of precipitation and a high rate of irradiation during the growing season (Pócs 1999). The flora of loess cliffs in the Pannonian basin surrounded by the Carpathians was studied in detail by T. Pócs and B. O. van Zanten between 1996 and 2001, and their desert-like cryptogamic communities were surveyed by Kürschner & Pócs (2002), who described a new bryophyte community: *Hilpertio velenovskyi – Pterygoneuretum compacti*. As a result, we have found on these cliffs several species in large numbers (some others less so) that were known previously only at a much lower latitude than in Hungary (mostly in the Mediterranean). Some of them had already been discovered during the past half century as rarities elsewhere in central Europe (see Fig. 18.1). Some records from other substrates are also included.

Crossidium laxefilamentosum *Frey & Kürschner*

A species described from the Arabian Peninsula (Frey & Kürschner 1987) was discovered recently in Tunisia, southern Spain, and on the loess cliffs along the Danube River in Romania, Serbia, and in Hungary (Pócs *et al.* 2004). In the Arabian Peninsula, Tunisia, and Spain it occurs in large amounts on flat saltpans, chalk, gypsum, or loess soils. It was recently also found on the large loess areas of north-central China (Kürschner & Wagner 2005). In central Europe it is very rare and occurs in quite scattered locations, exclusively on vertical loess cliffs. The central European collections are very recent; all were made in 2000 or later.

The related *Crossidium crassinerve* (De Not.) Jur., which was published by Galambos and Orbán (1984) as new to Hungary, is more widespread in Europe. The first Hungarian records are from the relatively old (mid-twentieth century) collections of Polgár, Boros and of Móczár, from the loess cliffs at the foothills of the mountains in NW Hungary. Later it was found at several points near the Danube river in southern Hungary, northern Serbia, and Romania (Pócs *et al.* 2004), sometimes together with the previous species.

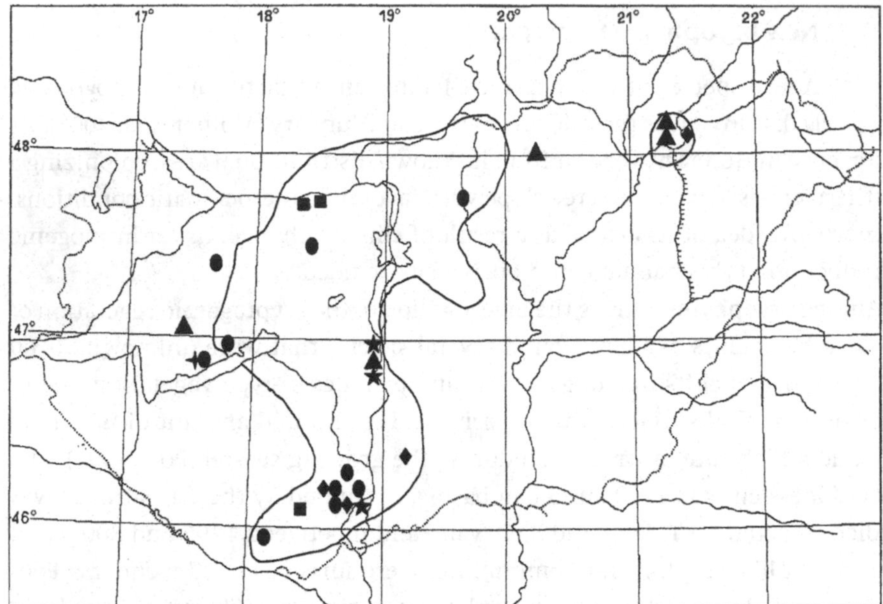

Fig. 18.1. Distribution of thermophilous bryophytes discovered mostly on loess cliffs in Hungary during the past 65 years (after Pócs 2005, modified).

Contour line: *Pterygoneurum compactum* Cano, Guerra and Ros
Ovoid: *Pterygoneurum squamosum* Guerra and Kürschner
Diamond: *Pterygoneurum crossidioides* Frey, Herrnstadt and Kürschner
Triangle: *Tortula brevisssima* Schiffn.
Square: *Dicranella howei* Ren. & Card.
Star: *Crossidium laxefilamentosum* Frey and Kürschner
Cross: *Leptophascum leptophyllum* (Müll. Hal.) Guerra and Cano

Dicranella howei *Ren. et Card.*

This is a common plant in the lowland and lower mountain belt of the Mediterranean region according to Zanten (2005), who discovered it in the loess areas of southern Hungary and in the Gerecse Mountains near the Danube bend. He also remarked that the first records of this moss from southwest Germany are dated from the 1990s. Hence it may be a newcomer in central Europe; nevertheless it would be necessary to study all old herbarium specimens identified as *Dicranella varia*.

Dicranum tauricum *Sapjegin*

This is a southern continental species in Europe, which, according to certain authors seems to have increased its range rapidly within the past few decades. It has many recent records from Hungary (Marstaller 1994, 1995;

Erzberger 1998), partly from disturbed habitats. At the same time the related *Dicranum viride* (Sull. et Lesq.) Lindb. is more montane in distribution and is restricted to less disturbed habitats; it is endangered by forest destruction.

Frullania inflata *Gottsche*

This species is considered to be a thermophilous, sub-Mediterranean, montane element. It is quite widespread in North America but very rare in Europe. It was known until recent times only from four localities: in northern Italy, southern Switzerland, and Moravia, and also in Hungary in 1955 from the shaded basalt rocks of an extinct volcano, Tátika, north of Lake Balaton (Geissler & Bisang 1985). Since then it has become known from Albania, Austria (Frey *et al.* 2006), and from two new localities in Hungary (Papp & Erzberger 2003, 2006), both from volcanic rocks. It is problematic whether the species is an old relic or a more recently arrived newcomer. In any case, its locality on the south-facing basalt rocks at Szarvaskő hill north of Eger is notorious for its thermophilous bryophytes, such as *Mannia fragrans* and *Fabronia ciliaris*. This site has been visited before by many botanists and bryologists. In spite of this, the species was discovered only in 1998 in small patches, together with *Fabronia pusilla* Raddi. Taking into account that the Moravian locality is on artificial substrates (granite walls), it can be supposed that its late arrival, at least in its new localities in Hungary, is due to climate change.

Gymnostomum viridulum *Brid.*

This Atlantic–Mediterranean species was recently found in westernmost Hungary, in a quarry at Cák, on calciferous conglomerate rocks (Papp 2009). It proved to be a new record for the country, together with a revised specimen from limestone rocks at Dunaalmás (northwest Hungary) collected by Á. Boros in 1942. Another specimen was collected by L. Vajda in 1954 in the Sashegy nature reserve in Budapest. Both were misidentified as *Gyroweisia tenuis* (Hedw.) Schimp. and revised recently by B. Papp. The distribution of the species covers the southern and western part of Europe up to SW Germany (Nebel & Philippi 2000) and the Czech Republic, the Mediterranean belt, and the Middle East; it also occurs in California.

Leptophascum leptophyllum *(Müll. Hal.) J. Guerra & M. J. Cano*
Syn.: Chenia leptophylla *(Müll. Hal.) Zander*

A subtropical – warm temperate element, this species was found by Zanten (2000) on the gravelly soil along a tourist path on the southern slope of the extinct Badacsony basalt volcano on the NW side of Lake Balaton. All European records have become known only recently, and from disturbed habitats, and thus are considered to be introduced (Frey *et al.* 2006). In any case, its appearance in Hungary can be connected to global warming.

Pterygoneurum crossidioides *Frey, Herrnstadt & Kürschner*

This species was discovered in the Judean desert near the Dead Sea (Frey *et al.* 1990). Since then it has been found recently at one locality in the Alps and in Hungary in a few places on loess cliffs (Pócs 1999). It can be distinguished from the common *Pterygoneurum ovatum* (Hedw.) Dix. by its fimbriate lamellae ending in branching filaments with smooth cells. The lamellae can reach the leaf base (in *P. ovatum* they taper into the costa above the leaf base).

Pterygoneurum compactum *Cano, Guerra & Ros*

This species was described from southeastern Spain (Cano *et al.* 1994). It proved to be very widespread in Hungary, almost exclusively on loess cliffs (Pócs 1999; Kürschner & Pócs 2002). It has also become known from similar habitats in Serbia and Romania. It differs from the related *Pterygoneurum ovatum* (Hedw.) Dix. in that its lamellae are divided into filaments with papillose cells, and by its much smaller size.

Pterygoneurum squamosum *Segarra & Kürschner*

This is a species described from Alicante Province in Spain (Segarra *et al.* 1998). It is distinguished by the special formation of the two well-developed lamellae on the costa, which have smooth margins and are transversely incised into 3–4 scaly plates, giving the lamellae a squamulose appearance. It occurs mostly in southern Hungary on loess cliffs.

Although the first two *Pterygoneurum* species were acknowledged at the species level in the survey of Spanish representatives of the genus (Guerra *et al.* 1995), the new Iberian moss flora (Guerra *et al.* 2006) treats them as synonyms of *Pterygoneurum ovatum* (Hedw.) Dixon and of *P. lamellatum* (Lindb.) Jur., respectively. I consider them as separate taxa. Searching the twentieth century specimens in the herbaria of BP and EGR, I could not find any of these species collected before 1960; this fact supports the idea that they are newcomers and distinct species. If they were present before, then they should have been collected by such accurate collectors as Ádám Boros and László Vajda, at localities visited by them many times and where these species now occur, often together with other *Pterygoneurum* species. They have collected many *Pterygoneurum ovatum* specimens at these localities, but none of them proved to be the above southern species in collections made before 1960. So it is unlikely that these species were previously overlooked.

Pterygoneurum papillosum *Oesau*

This was described as a new species from Rhinehessia in Germany, where it is quite well distributed (about 70 known localities, almost all from vineyards on calcareous, marly soil) (Oesau 2003). It also differs from the other known *Pterygoneurum* species in the construction of the lamellae. The fact that it was not collected (or was overlooked?) before, and that it grows mostly in human-influenced, disturbed habitats, suggests that this species may also be a newcomer in connection with climate change, although its "original" native locality and habitat are not known.

Tortula brevissima *Schiffn.*

This tiny xerophyllous species was discovered quite recently by Erzberger (1998) at the southern foothills of the Bükk Mountains in northern Hungary, on a small rhyolite tuff rock at the roadside. Later, during the loess studies by Pócs, Zanten and Kürschner, it became known from two loess cliffs, along the River Danube between Kisapostag and Dunaujváros and on the loessy foothills of Nagykopasz at Tokaj in northeast Hungary (Kürschner & Pócs 2002; Pócs 2005). Scattered occurrences of the species have long been known from Germany southwards (Reimers 1941), and the species has a typical Mediterranean distribution area from Iraq to southern Spain.

Changes in bryophyte distribution in relation to climate change

The species occurrences above seem to support the idea of recent or subrecent arrival of these bryophytes, as well as one fern, in connection with global warming. Frahm and Klaus (2000) tried to find a connection between the periodic temperature fluctuations within one century and the appearance and disappearance of Atlantic and Mediterranean species in central Europe. On the other hand, Frey and Kürschner (1983, 1988) consider some of the above species as circum-Tethyan preglacial relicts or xerothermic Pangean elements. Pócs (1999) underscored the special orographic and microclimatic conditions among which desert-like cryptogamic communities can grow on the loess cliffs in the Pannonian basin. It is difficult to decide whether a species is really a newcomer or was overlooked for a long time. As was demonstrated above, herbarium studies can, in some cases, demonstrate their new arrival, and hence their indication of present global warming trends. The fact that almost all of these species belong to the generative colonist life strategy type, producing numerous small spores, providing an easy chance for air dispersal (Frey & Kürschner 1991; Orbán 1992; Kürschner 2002), and colonizing naked, emerging loess cliffs, also support the idea. Pócs (2006) called attention to an "antenna phenomenon" when new colonizers prefer emerging, naked surfaces, which are easily available for air-dispersed spores, and where they can establish themselves without serious competition. This is especially true for the terricolous bryophytes. Further, if they are known only from anthropogenic or human-influenced habitats, their occurrence postdating major climate changes is even more probable.

We need to introduce some words of caution. Franco *et al.* (2006) wrote: "Polewards expansions of species distributions have been attributed to climate warming, but evidence for climate-driven local extinctions at warm (low latitude/elevation) boundaries is equivocal." This is true of the Hungarian bryoflora as well. The bryophyte populations live in biocenoses, which are influenced by the warming climate with increasing extremes. The structure of some communities is highly influenced, or even endangered, by the climate changes; their bryophyte populations can be altered or become extinct. In Hungary this has already been observed in oak (Tóth *et al.* 2008) and in beech (Molnár & Lakatos 2008) forest; their growth is hindered and structure destabilized and opened as a result of climate change. This process is usually indirect: owing to the prolonged drought and warm periods, the resistance of tree species to fungal and insect parasites becomes weak. At the same time wetlands dry out, partly due to human influence and partly due to climate warming. Even north-facing rock habitats harboring montane species desiccate. As a result, their bryoflora is impoverished and their rarer species become extinct. According to the bryophyte chapter of the Hungarian Red Book (Rajczy 1990) the following species became extinct during a few decades up to 1990: *Moerckia hibernica* Gottsche (along the Danube River), *Amblyodon dealbatus* (Sw. ex Hedw.) P. Beauv. (Vértes Mts.), *Sphagnum warnstorfii* Russow (Balaton Upland), and *Ulota hutchinsiae* (J. E. Smith) Hammar. All are species of wet habitats or at least of habitats with high humidity. We can add to this list *Syntrichia norvegica* F. Weber (Bükk Mts) (Tóth 1987) and probably many more oroboreal species. Other species, which were more widespread before, have become very rare now, known only from one or a very few localities, like *Anacamptodon splachnoides* (Brid.) Brid. or *Tomenthypnum nitens* (Hedw.) Loeske. (Papp & Erzberger 2003; Papp 2008). To follow up these changes, a monitoring program has been started (Papp *et al.* 2005, 2007), which will probably shed more light on details of this process.

Conclusions

There are nine bryophyte species in the Hungarian flora that have been discovered during the past half century; according to herbarium records they did not occur in Hungary before 1960. As they occur on sites well collected earlier by experienced bryologists, it is probable that they have arrived recently in connection with global climate warming. This idea is supported by the fact that they all occur in open habitats not yet saturated with species. On the other hand, there are at least six wetland or montane species that have disappeared during the same time period as a result of the desiccation and warming of their habitats, which we presume to be at least in part due to climatic changes.

Acknowledgements

The author acknowledges with thanks the financial support of the Hungarian Research Fund (OTKA T 022575) and of DAAD (Germany) and MöB agencies (Hungary), the cooperation of his colleagues (Dr Bernard Otto van Zanten, Dr Harald Kürschner), to his wife, Sarolta Pócs, for participating in the field and identification work, and finally to Prof. Nancy G. Slack for the critical reading of his manuscript.

References

Cano, M. J., Guerra, J. & Ros, R. M. (1994). *Pterygoneurum compactum* sp. nov. (Musci: Pottiaceae) from Spain. *Bryologist* **97**: 412–15.

Deutscher-Wetterdienst (2002). www.deutscher-wetterdienst.de/research/klis/produkte /monitoring/-t0500/-folgen.html

Erzberger, P. (1998). *Tortula brevissima* Schiffn. Eine für die Flora Ungarns neue Moosart. *Botanikai Közlemények* **85**(1–2): 63–72.

Erzberger, P. (1999). Distribution of *Dicranum viride* and *Dicranum tauricum* in Hungary. *Studia Botanica Hungarica* **29**: 35–47.

Frahm, J.-P. (2003). Weitere Auswirkungen des Klimawandels auf die Moosflora. *Limprichtia* **22**: 147–55.

Frahm, J.-P. (2005). Bryophytes and global change. *Bryological Times* **115**: 8–10.

Frahm, J.-P. & Klaus, D. (2000). Bryophytes as indicators of recent climatic fluctuations in Central Europe. *Lindbergia* **26**: 97–104.

Franco, A. M. A., Hill, J. K., Kitschke, C. *et al.* (2006). Impacts of climate warming and habitat loss on extinctions at species' low-latitude range boundaries. *Global Change Biology* **12**: 1545–53.

Frey, W., Frahm, J.-P., Fischer, E. & Lobin, W. (2006). *The Liverworts, Mosses and Ferns of Europe*. English edition revised and edited by T. L. Blockeel. Essex: Harley Books.

Frey, W. & Kürschner, H. (1983). New records of bryophytes from Transjordan with remarks on phytogeography and endemism in SW Asiatic mosses. *Lindbergia* **9**: 121–32.

Frey, W. & Kürschner, H. (1987). A desert bryophyte synusia from the Jabal Tuwayq mountain systems (Central Saudi Arabia) with the description of two new

Crossidium species (Pottiaceae). (Studies in Arabian bryophytes 8.) *Nova Hedwigia* **45**: 119–36.

Frey, W. & Kürschner, H. (1988). Bryophytes of the Arabian Peninsula and Socotra. Floristics, phytogeography and definition of the Xerothermic Pangean element. (Studies in Arabian bryophytes 12.) *Nova Hedwigia* **46**: 37–120.

Frey, W. & Kürschner, H. (1991). Lebensstrategien von terrestrischen Bryophyten in der Judäischen Wüste. *Botanica Acta* **104**: 172–82.

Frey, W., Herrnstadt, I. & Kürschner, H. (1990). *Pterygoneurum crossidioides* (Pottiaceae, Musci), a new species to the desert flora of the Dead Sea area. *Nova Hedwigia* **50**(1–2): 239–44.

Galambos, I. & Orbán, S. (1984). *Crossidium crassinerve* (De Not.) Jur., new member of the Hungarian bryoflora. *Bryologische Beiträge* **3**: 23–7.

Geissler, P. & Bisang, I. (1985). *Frullania inflata* Gott., ein neues thermophiles Element in der Schweizer Moosflora. *Saussurea* **16**: 95–100.

Guerra, J., Cano, M. J. & Ros, R. M. (1995). El género *Pterygoneurum* Jur. (Pottiaceae, Musci) en la península Ibérica. *Cryptogamie, Bryologie, Lichénologie* **16**: 165–75.

Guerra, J., Cano, M. J. & Ros, R. M. (eds.) (2006). *Flora Briofitica Ibérica III*. Murcia: UMU, SEB.

Kalela, O. (1949). Changes in geographic ranges in the avifauna of northern and central Europe in relation to recent changes in climate. *Bird Banding* **20**: 76–103.

Klötzli, F. & Walther, G.-R. (1999). *Conference on Recent Shifts in Vegetation Boundaries of Deciduous Forests, Especially due to General Global Warming*. Basel: Birkhäuse.

Kürschner, H. (2002). Life strategies of Pannonian loess cliff bryophyte communities. (Studies of the cryptogamic vegetation of loess cliffs, VIII). *Nova Hedwigia* **75**(3–4): 307–18.

Kürschner, H. & Pócs, T. (2002). Bryophyte communities of the loess cliffs of the Pannonian basin and adjacent areas, with the description of *Hilpertio velenovskyi – Pterygoneuretum compacti* ass. nov. (Studies on the cryptogamic vegetation of loess cliffs, VI). *Nova Hedwigia* **75** (1–2): 101–19.

Kürschner, H. & Wagner, D. (2005). Phytosociology and life strategies of a new loess slope bryophyte community from N China (Gansu) including *Crossidium laxefilamentosum* new to China. *Nova Hedwigia* **81**(1–2): 229–46.

Marstaller, R. (1994). Zur Verbreitung bemerkenswerter Moose in der Umgebung von Budapest (Ungarn). *Feddes Repertorium* **105**: 531–47.

Marstaller, R. (1995). Die azidophytische Bryophytenvegetation in einigen Gebirgenm in der Umgebung von Budapest (Ungarn). *Feddes Repertorium* **106**: 247–70.

Mogensen, G. S., Hansen, G. R. & Dalsgaard, M. D. (1997). *Pterygoneurum lamellatum*, a moss new to Greenland (Musci, Pottiaceae). *Bryologist* **100**: 47–9.

Molnár, Cs., Baros, Z., Pintér, I. *et al.* (2008). Remote, inland occurrence of the oceanic *Anogramma leptophylla* (L.) Link. (Pteridaceae: Taenitidoideae) in Hungary. *American Fern Journal* **98**: 128–38.

Molnár, M. & Lakatos, F. (2008). Pusztuló bükköseink. (Our decaying beech forests.). *Természet Világa, 2009 Sept* (abstract, in Hungarian).

Nebel, M. & Philippi, G. (eds) (2000). *Die Moose Baden-Württenbergs* 1. Stuttgart: E. Ulmer.

Oesau, A. (2003). *Pterygoneurum papillosum* (Bryopsida: Pottiaceae), a new moss species from Germany. *Journal of Bryology* **25**: 247–52.

Orbán S. (1992). Life strategies in endangered bryophytes in Hungary. *Biological Conservation* **59**: 109–12.

Papp, B. (2008). Selection of important bryophyte areas in Hungary. *Folia Cryptogamica Estonica* **44**: 101–11.

Papp, B. (2009). *Gymnostomum viridulum* Brid. Bryological Notes. New national and regional bryophyte records, 21. *Journal of Bryology* **31**: 134–5.

Papp, B. & Erzberger, P. (2003). Data about the actual local populations of bryophyte species protected in Hungary. *Studia Botanica Hungarica*. **34**: 33–42.

Papp, B. & Erzberger, P. (2006). Európai Vörös Könyves mohafajok újonnan felfedezett populációi Magyarországon. (Newly discovered Hungarian populations of mosses listed in the European Red Data Book.) *Kitaibelia* **11**: 70.

Papp, B., Ódor, P. & Szurdoki, E. (2005). Methodological overview and a case study of the Hungarian Bryophyte Monitoring Program. *Boletin de la Sociedad Española de Botanica* **26–27**: 23–32.

Papp, B., Ódor, P., & Szurdoki, E. (2007). Bryophyte Biodiversity Monitoring System in Hungary, Eastern Central Europe. *Chenia* **9**: 149–58.

Parmesan, C. & Yohe, (2003). A globally coherent fingerprint of climate change impact across natural systems. *Nature* **421**: 37–42.

Pócs, T. (1999). A löszfalak virágtalan növényzete I. Orografikus sivatag a Kárpát-medencében. (Studies on the cryptogamic vegetation of loess cliffs, I. Orographic desert in the Carpathian Basin). *Kitaibelia* **4**: 143–56.

Pócs, T. (2005). A globális felmelegedés jelei hazánk mohaflórájában. In *A DNS-től a Globális Felmelegedésig. A 70 éves Vida Gábor köszöntése*, ed. F. Jordán, pp. 149–56. Budapest: Scientia.

Pócs, T. (2006). Bryophyte colonization and speciation on oceanic islands, an overview. *Lindbergia* **31**: 54–62.

Pócs, T., Sabovljević, M., Puche, F. *et al.* (2004). *Crossidium laxefilamentosum* Frey and Kürschner (Pottiaceae), new to Europe and to North Africa. (Studies on the vegetation of loess cliffs, VII.) *Journal of Bryology* **26**: 113–24.

Rajczy, M. (1990). Mohák – Bryophyta. In *Vörös Könyv*, ed Z. Rakonczai, pp. 322–5. Budapest: Akadémiai Kiadó.

Reimers, H. (1941). *Tortula brevissima* Schiffn., ein neues vorderasiatisches Wüstensteppenmoos im Zechstein-Kyffhäuser. *Notizblatt des Botanischen Gartens und Museums Zu Berlin-Dahlem* **15**: 402–5.

Segarra, J.-G., Puche, F., Frey, W. & Kürschner, H. (1998). *Pterygoneurum squamosum* (Pottiaceae, Musci), a new moss species from Spain. *Nova Hedwigia* **67**(3–4): 511–15.

Szegő, I. M. (2005). Magyarország és a globális felmelegedés. *National Geographic Magyarország 08.07.*

Takács-Sánta, A. (ed.) (2005). *Éghajlatváltozás a Világban és Magyarországon*. Budapest: Alinea–Védegylet.

Tóth J. A., Papp, M., Krakomperger, Zs. & Kotroczó, Zs. (2008). A klímaváltozás hatása egy cseres-tölgyes erdő struktúrájára (Síkfőkút Project). ('The impact of climate change on the structure of a Turkish oak forest'.) Poster at VAHAVA Conference, Budapest. (In Hungarian.)

Tóth, Z. (1987). A phytogeographic review of *Tortula* Hedw. *Acta Botanica Hungarica* **33**: 249–78.

Tuba, Z. (ed.) (2005). *Ecological Responses and Adaptations of Crops to Rising Atmospheric Carbon Dioxide*. New York: Food Products Press.

Vida, G. (2001). *Helyünk a Bioszférában.* (Our Place in the Biosphere). Budapest: Typotex. (In Hungarian)

Zanten, B. O. (2000). Studies on the cryptogamic vegetation of loess cliffs. IV. *Chenia leptophylla* (C.Müll.) Zander, new to Hungary. *Kitaibelia* **5**: 271–4.

Zanten, B. O. (2005). Studies on the cryptogamic vegetation of loess cliffs. VIII. *Dicranella howei* Ren. et Card., a new addition to the Hungarian bryoflora. *Kitaibelia* **10**: 45–7.

Zanten, B. O. & Pócs, T. (1981). Distribution and dispersal of bryophytes. In *Advances in Bryology*, vol. 1, ed. W. Schultze-Motel, pp. 479–562. Vaduz: J. Cramer.

19

Can the Effects of Climate Change on British Bryophytes be Distinguished from those Resulting from Other Environmental Changes?

JEFFREY W. BATES AND CHRISTOPHER D. PRESTON

Introduction

The old adage that an Englishman's favorite topic of conversation is the weather is surely true, but why bother to consider the flora of Great Britain in an international summary of the effects of climate change on bryophytes? We can offer five main justifications. The first two are the relative thoroughness of Britain's bryological exploration and the long period over which the flora has been repeatedly examined. Third is the high quality of the general recording effort and its documentation in written and computerized records, and in refereed herbarium specimens. Fourth is the exceptional species-richness of the British flora in a regional (European) context.

Our fifth justification concerns the long run of systematic climatic measurements for England which, like the bryophyte records, extends back into the seventeenth century. More recent records are available for a wide range of localities across the whole of the UK. Today, the UK Meteorological Office is one of the world's leading weather forecasters and through its Hadley Centre carries out research into climatic change and publishes regular reports, updating recent weather trends and making available the latest predictions for the future climate of Britain.

Despite the advantages listed above, the task of locating unequivocal examples of recent climate change impacting the British bryophyte flora has not proven to be straightforward. Many areas of Britain are densely populated and almost all its vegetation is managed, often intensively, so that human impacts

Bryophyte Ecology and Climate Change, eds. Zoltán Tuba, Nancy G. Slack and Lloyd R. Stark. Published by Cambridge University Press.

are all-pervasive; there are no wilderness areas. Some of the same factors that have led to the long history of both bryophyte and climate recording, such as the dense population in southern England, are also those that make it difficult to isolate the effects of recent climate changes. There is the danger of confusing the effects of climate change with those of other developments, and also the possibility of interactions that produce changes, which may not be attributable to any one factor alone. Climate change is also a relatively new phenomenon where most work is still in progress; there are few published studies of real or potential impacts on British bryophytes. However, a number of striking changes in ranges of individual bryophytes has been noticed in recent years. This chapter examines the extent to which climate change might be among the causal factors.

Bryological context

History of recording

Although the first localized bryophyte records attributable to current species were made in the mid-seventeenth century, most publications for the next 200 years were devoted to working out the taxonomic composition of the British flora. It was not until the late eighteenth century that bryologists began to devote their energies to recording the distribution of species in any systematic manner. Detailed exploration at the county scale in the second half of the nineteenth century led to published accounts of the bryophytes of numerous counties in Floras devoted entirely to bryophytes, or covering bryophytes alongside vascular plants and sometimes lichens and fungi. The foundation of the Moss Exchange Club in 1896, later the British Bryological Society (BBS), was the key factor in bringing together and integrating the work of individual bryologists. The Club soon produced "census catalogues" for mosses and liverworts (Macvicar 1905; Ingham 1907), standard checklists with the occurrence of species listed in "vice-counties," areas based on the historic administrative counties, subdivided where necessary and with stable boundaries. This systematic record of the occurrence of species in vice-counties is still maintained; updates are published annually and summarized occasionally, most recently by Hill *et al.* (2008).

British bryology was at a low ebb in the 1920s and 1930s, reviving with the influx of academic members after the Second World War. This revival was reflected in the increase in the number of additions to the Census Catalogues in the late 1950s (Watson 1985) and in the launch of the BBS "Mapping Scheme" in 1960. Maps of the occurrence of species in 10 km × 10 km grid squares were published after 30 years of fieldwork (Hill *et al.* 1991, 1992, 1994). These volumes provide a statement of the known distribution of species towards the start of the current warm period.

The systematic collation of records for over a century has documented the changes in our knowledge of the distribution of British species. However, real changes in the distribution of species need to be distinguished from those changes in knowledge resulting from taxonomic changes, improved fieldcraft, and the gradually increased exploration of remote regions, or of accessible areas hitherto dismissed as too dull to merit attention. British bryologists have been and still are too few to achieve a good overall coverage of Britain every generation. Vice-counties tend to be well recorded and then neglected, in cycles that reflect the attraction of under-studied terrain or the random changes in the distribution of bryologists (Preston *et al.* 2009); few areas have a long and continuous sequence of records. Detecting real changes in species distributions requires the critical assessment of the available evidence, using the methods of the historian rather than those of the experimental scientist. Range expansions are more easily detected than contractions in the absence of a systematic repeat national survey.

Phytogeographical groupings in relation to climate

The British bryophytes represent a large proportion (over 60%) of the European total; it is not surprising therefore that they contain species with a wide range of phytogeographical affinities. Hill and Preston (1998) classified species on the basis of their occurrence in major biomes and their eastern limits in relation to the British Isles. Tables 19.1 and 19.2 enumerate the native species in these major biome and eastern limit categories, updating the totals presented by Hill and Preston (1998) on the basis of more recent data from the BRYOATT database (Hill *et al.* 2007). They also summarize the climate of their British ranges. Arctic–montane, Boreo-Arctic Montane and Boreal–montane species have restricted distributions in Britain, occurring in 10 km squares characterized by low winter and summer temperatures. Boreo-temperate, Temperate and Southern-temperate species are wide ranging in the Northern Hemisphere and tend to be frequent in Britain, and Wide-boreal and Wide-temperate species are exceptionally widespread globally and very common in Britain. The plants in the Mediterranean elements are the most southerly members of the British Flora. In Table 19.2, the higher precipitation in the areas occupied by the Hyperoceanic and Oceanic species is apparent, as is the tendency of the more Oceanic groups to occur in areas with milder winters.

Table 19.1. *Native British bryophytes: occurrence in major biomes in relation to climate*

The mean January and July temperatures and annual precipitation are calculated for each species on the basis of the 10 km squares in which they have been recorded. The maximum altitude is that at which the species has been recorded in the field. The values tabulated for January and July temperatures and maximum altitude are the means for all species in the group and the extreme values for group members; only means are given for precipitation.

Biome	No. native spp.	% native flora	Mean no. British 10 km squares since 1950	Mean January temperature (°C)	Mean July temperature (°C)	Precipitation (mm)	Maximum altitude (m)
Arctic–montane	72	7	24	(−2.0) 0.1 (2.8)	(9.9) 11.3 (13.8)	1938	(410) 1115 (1340)
Boreo-Arctic Montane	129	13	162	(−1.4) 1.8 (4.5)	(10.4) 12.7 (15.8)	1602	(10) 937 (1344)
Boreal–montane	205	20	96	(−1.5) 1.9 (5.5)	(10.2) 12.7 (15.8)	1661	(2) 771 (1300)
Wide-boreal	23	2	1183	(2.7) 3.2 (3.6)	(13.4) 14.2 (16.1)	1226	(300) 1063 (1344)
Wide-temperate	17	2	1021	(3.0) 3.5 (4.0)	(13.7) 14.9 (15.6)	1045	(250) 728 (1335)
Boreo-temperate	171	17	719	(1.5)3.2 (5.2)	(12.5) 14.2 (16.1)	1269	(5) 815 (1344)
Temperate	208	20	443	(1.3) 3.4 (5.9)	(11.6) 14.6 (16.3)	1208	(0) 498 (1344)
Southern-temperate	105	10	405	(3.0) 4.1 (6.1)	(13.1) 15.1 (16.7)	1055	(30) 416 (1175)
Mediterranean	94	9	214	(2.7) 4.5 (6.9)	(13.4) 15.4 (16.4)	1031	(30) 290 (800)

Table 19.2. *Native British bryophytes: eastern limit categories in relation to climate*

Species in the Mediterranean elements are excluded from the table, as their eastern limit varies with latitude, but are included in the total used to calculate % native flora. For the derivation of environmental values, see Table 19.1.

Biome	No. native spp.	% native flora	Mean no. British 10 km squares since 1950	Mean January temperature (°C)	Mean July temperature (°C)	Precipitation (mm)	Maximum altitude (m)
Hyperoceanic	47	5	151	(1.6) 3.4 (5.0)	(11.6) 13.4 (15.4)	1761	(100) 561 (1100)
Oceanic	74	7	127	(−0.8) 3.1 (6.2)	(10.2) 13.5 (16.2)	1654	(5) 580 (1300)
Suboceanic	98	10	356	(−1.2) 3.0 (6.7)	(10.3) 13.8 (16.7)	1473	(30) 727 (1340)
European, Eurosiberian, and Eurasian	290	28	413	(−1.5) 2.8 (6.7)	(10.3) 14.0 (16.5)	1310	(2) 660 (1344)
Circumpolar	421	41	409	(−2) 2.3 (5.6)	(9.9) 13.3 (16.3)	1431	(0) 843 (1344)

Reproductive biology and dispersal

The bryophytes are notable for the diversity of their means of reproduction. Species vary greatly in the extent to which they produce sporophytes, from those never recorded as fruiting in Britain to those that fruit abundantly. Freely fruiting species differ in their adaptations to long-distance dispersal. At one extreme are plants with more or less sessile and cleistocarpous capsules and large spores, "shuttle species" (During 1979), many of which appear to be adapted not to dispersal but to persistence in a dormant state in the same area until favorable conditions for growth recur. However, many such species grow in disturbed habitats and in the modern landscape they may be dispersed in soil carried on agricultural machinery, motor vehicles, or accompanying plants sold in the horticultural or silvicultural trades. Species with longer setae and mosses with well-developed peristomes are more obviously adapted to long-distance dispersal. In addition to spores, bryophytes have a wide range of macroscopic asexual propagules (including tubers, gemmae, bulbils, deciduous branches, leaves, and leaf tips). Some species that fruit freely also produce frequent asexual propagules (e.g., *Bryum dichotomum*, *Cololejeunea minutissima*), whereas other species that frequently produce propagules rarely if ever fruit (e.g., *Bryum subelegans*, *Metzgeria temperata*).

Protonemal gemmae are microscopic propagules and their occurrence is therefore less well documented; they presumably provide the means of reproduction of mosses which have been thought until recently to have no propagules (e.g., *Zygodon gracilis*) or where the occurrence of other propagules is rare or doubtful (e.g., *Didymodon nicholsonii*). It is tempting to interpret vegetative reproduction as a means of local dispersal, possibly following the establishment in a new area of plants from long-distance spore dispersal, but recent expansions in the ranges of British species that have only vegetative propagules (such as *Ulota phyllantha*, discussed below) show that they can achieve effective long-distance dispersal and spread.

Britain's climate: recent trends and future predictions

Great Britain lies between latitudes 49° and 61° north, with London situated at 51°30′N. Edmonton, Alberta (53°30′N), one of the most northerly of the major Canadian cities, is only 2° further north than London in latitudinal terms, and Labrador in eastern Canada extends over a similar range of latitudes to Britain. In continental Europe the cities of Moscow (55°45′) and Warsaw (52°13′) lie at comparable latitudes to London. All of these places suffer far colder winters than does London (and the two central European cities have much warmer summers). Famously, the sea freezes along the Labrador coast every winter but it has only been observed to do so very locally in the coldest ("in living memory") winters ever experienced in the UK. These contrasts result from the influence of a major branch of the North Atlantic Drift, or Gulf Stream, which brings warm water diagonally across the Atlantic from the Caribbean to bathe the coasts of Ireland and then Britain, and warm their prevailing westerly air flow. Britain therefore enjoys a temperate, oceanic climate with quite marked seasonal differences in day length (reflecting the high latitude) but comparatively modest extremes in temperature as a result of the oceanic influence. Precipitation occurs mostly as rain and declines steeply from west to east, particularly across England and Wales.

In a detailed summary, Jenkins *et al.* (2007) identified several recent trends which confirm that the climate of the UK has already begun to warm. Central England Temperature (CET), a long-term average based on data from three separate stations, has increased by about 1 °C since 1980. It had remained relatively constant since the early 1930s, but had been gradually rising to this level since 1877. The year 2006 was the warmest since CET records started in 1659. Temperatures in Scotland and Northern Ireland increased by about 0.8 °C over the same period. No significant changes in annual mean precipitation over England and Wales have occurred since records commenced in 1766. However, mean rainfall has decreased in summer (except in northeast England and north Scotland) and increased in winter. Over the past 45 years the contribution of heavy rain events to winter precipitation has increased. Severe windstorms (numbers per decade) have also become more frequent since the 1960s but remain below the level experienced in the 1920s.

The annual mean surface temperature of the seas around the British coast has increased by about 0.7 °C over the past 30 years. Sea level rose by about 1 mm per annum in the twentieth century after allowing for land movement. Sea level rise accelerated in the 1990s and 2000s. The generic temperature increase is reflected in rising values of several indicators including annual average temperatures, seasonal averages (especially winter temperatures), and annual average minimum and maximum temperatures, and by decreasing days of air frost. Potentially of more importance to bryophytes than the precipitation totals are the numbers of rain days (days with ≥ 1 mm) and average relative humidity. A comparison of the 30-year annual means for 1961–1990 and 1971–2000 shows that rain days increased by 1–5 days in western Scotland and decreased by 1–3 days in a broad swath of eastern Britain (Jenkins *et al.* 2007). Annual average relative humidity declined slightly across most of the UK with the most marked changes in southeast England and the Midlands, but also in parts of the Scottish Highlands, the Hebridean Islands, and Shetland.

The currently available predictions for the future climate of the UK (Hulme *et al.* 2002) involved running the Hadley Centre's global climate model with four alternative scenarios of greenhouse gas emissions. The four scenarios (Low Emissions, Medium-Low Emissions, Medium-High Emissions and High Emissions) reflect the considerable uncertainties that currently exist regarding future public desire, about the courage of world politicians, and the inventiveness of scientists and technologists to tackle the difficult issues involved in lowering future emissions. The main predictions under this range of scenarios are as follows. The average temperature of the UK will rise by between 2° and 3.5° by the 2080s, with a rise of 5 °C being possible in parts of the southeast. Warming will be less in the north and west, and possibly greater in summer and autumn than in winter and spring. Unusually hot summers are expected to occur in one out of five years by the 2050s and possibly (High Emissions scenario) in three out of five years by the 2080s. The trend for drier summers and wetter winters is expected to continue with a 50% or larger decrease in summer precipitation by the 2080s and a 30% increase in winter rainfall. Soils over much of England will become arid in summer. Snowfall will decrease everywhere and by 60%–90% over Scotland by the 2080s. The intensities of the heaviest winter precipitation events (rain or snow) are predicted to become 5%–20% greater. Solar radiation is expected to increase in summer with reduced cloudiness, and both relative humidity and frequency of fogs are predicted to decrease under most scenarios. Future changes in sea level are partly dependent on vertical land movements. Sea level in southeast England may rise 26–86 cm above the present level by the 2080s, whereas in western Scotland (where land is rising) it may be slightly below (Low Emissions) or as much as 58 cm above (High Emissions) the current mean tide level. Extreme sea levels on parts of the east coast are predicted, by the 2080s, to become 10–20 times more frequent than at present. Hulme *et al.* (2002) conclude that the Gulf Stream may weaken in the future but they do not anticipate it leading to a cooling of the UK climate over the next century.

Other recent environmental changes in Britain

Great Britain is a small, populous country and was the cradle of the Industrial Revolution. Consequently, its flora has suffered from long exposure to a range of atmospheric, soil, and water-borne pollutants, and its natural habitats from disturbance, fragmentation, and exploitation. An analysis in 2005 showed about 74% of the land surface to be under agriculture (20% crops, 50% grazing, 4% other), 14% urban or unclassified land, and 12% forest or woodland (www.defra.gov.uk). Apart from stretches of inaccessible sea-cliff and some steep mountain fastnesses, very few places appear to have escaped the influence of intentional or unintentional human management.

Although most of Britain and Ireland theoretically comes within the winter-deciduous broad-leaved forest biome, in fact most existing forest cover was either planted or established by secondary succession and only 1.5%–2% of land in the UK and perhaps 1% in Ireland is deemed to be ancient woodland (Peterken 1993). Vast plantations of alien conifers were planted in the twentieth century on upland moorlands that had probably been treeless for centuries but may once have supported broad-leaved woodland. Often, native oak-woods were felled for this purpose, thereby unintentionally exterminating their rich bryophyte floras. For millennia, and until the Great War of 1914–18 decimated the male population, most woodlands in Britain had been intensively managed by regular cutting (coppicing or pollarding) for fuel and a range of timber products, or to employ the bark for tanning leather (Peterken 1993). The woodland bryophyte flora presumably reached equilibrium with this regular opening of the canopy. Modern woodlands, except where managed along traditional lines, are much darker and less disturbed places, and certain bryophytes (e.g., *Leptodontium flexifolium, Pogonatum nanum*) are no longer as common in woodland as older records suggest they once were (Rose 1992). Britain and Ireland have both suffered extensive mining and quarrying activities over several millennia to exploit metals, coal, building stone, lime, sand, and ballast, or to dig peat for fuel. Modern oil and gas exploitation has mainly been offshore, but lately, with some fields becoming exhausted, small enterprises have commenced in formerly unaffected areas on land. Previously undisturbed areas in the Highlands and Islands, and often on fragile peatlands, are increasingly finding favor as sites for wind farms. For bryophytes, it is the building and runoff from the service roads, rather than the whirring turbines, that constitute the main threats to survival.

For most of the twentieth century bryophytes in urban and industrialized parts of Britain had to contend with high emissions of SO_2 released by the burning of coal and oil (Bates 2000, 2002). Populations of many species were seriously depleted by the high atmospheric concentrations of the pollutant in town centers and some were probably also sensitive to acidification of their substratum by the associated “acid rain”. Epiphytes proved to be especially sensitive and were all but exterminated around the main cities and the industrial centers and downwind of power stations and smelters (Gilbert 1968, 1970; Winner 1988). Most *Orthotrichum* and *Ulota* spp., for instance, became quite infrequent near to large cities and a few especially sensitive epiphytes, notably *Antitrichia curtipendula* and *Orthotrichum obtusifolium*, completely vanished from most of the country (Hill *et al.* 1994). Species on exposed rocks and masonry were also affected (Gilbert 1970).

The long history of heavy industrial pollution in Sheffield and Manchester also reduced the occurrence of *Sphagnum* spp. on peat in the adjacent Pennine moorlands (Tallis 1964; Ferguson & Lee 1983; Lee & Studholme 1992; Lee *et al.* 1993). Species on soil and in grasslands were less obviously affected but there is little doubt that some of these also suffered range retractions, e.g., *Hylocomium splendens*, *Rhytidiadelphus triquetrus* (Gilbert 1970; Farmer *et al.* 1992; Bates 1993a) and *Bartramia pomiformis* (Hill *et al.* 1994). Following clean air legislation and socio-economic changes in Britain in the 1950s and 1960s, emissions of SO_2 fell sharply and atmospheric concentrations began to fall progressively from the early 1970s (Adams & Preston 1992; Bates 2002). Whereas SO_2 concentrations in central London were above 300 $\mu g\ m^{-3}$ in the mid-1960s, in the year commencing 1 January 2008, the average daily concentrations varied between 0 and 5 $\mu g\ m^{-3}$ with occasional higher spikes to 15 $\mu g\ m^{-3}$ and only three brief events exceeding 20 $\mu g\ m^{-3}$ (The London Air Quality Network, King's College London, http://www.londonair.org.uk/). The declining SO_2 concentrations have been accompanied by the progressive recolonization of lost territory, most noticeably by epiphytic species (Adams & Preston 1992; Bates *et al.* 2004; Davies *et al.* 2007).

While emissions of SO_2 were declining in many areas, those of nitrogen oxides (NO_x), formed in the combustion chambers of vehicle engines, increased so that these are now among the most important pollutants of town air and on roadsides in Britain (Lee *et al.* 1998; Pearson *et al.* 2000). Another nitrogenous pollutant, NH_3, is derived both from farms that practice intensive animal rearing and also during the catalytic reduction of NO_x in vehicle exhaust, and has increased in importance recently. Both pollutants contribute to eutrophication; the alkaline gas NH_3 is believed to be responsible for neutralizing acid substrata (de Bakker 1989; Van Herk 2001). For some neutrophytic epiphytes the loss of mature elms (*Ulmus* spp.) to Dutch elm disease in the latter half of the twentieth century was a major catastrophe when acidification denied them other potential hosts with neutral bark (Rose 1992).

Thus, between the 1960s and 2000 polluted regions experienced a number of distinct episodes, not directly related to climatic change, which appear to have induced marked changes in the flora, especially among epiphytes. From observations made in southeast England (Jones 1991; Adams & Preston 1992; Bates 1995; Bates *et al.* 1997, 2001) and comparisons with the situation in The Netherlands (Greven 1992; Van Herk 2001) we present the following tentative sequence of events.

- The end of the long era of elevated SO_2 pollution, which had locally eradicated many species, caused others to become restricted to calcareous substrata and among the epiphytes favored only a few extreme acidophiles with efficient SO_2 detoxification mechanisms, e.g., *Dicranoweisia cirrata* and the lichen *Lecanora conizaeoides* (Massara *et al.* 2009).
- A phase when SO_2 was declining sharply but acidification of bark and other substrata was still important and favored some less SO_2-tolerant acidophiles, e.g., *Dicranum montanum*, *D. tauricum*, *Orthodontium lineare*, *Plagiothecium curvifolium*, and *P. laetum*.
- A period of increasing N inputs from NO_x deposition, but in which bark retained its earlier-acquired acidity and favored slightly nitrophilous acidophytes, e.g., *Bryum subelegans*, *Ulota bruchii*, and *Zygodon conoideus*.
- Gradual loss of bark acidity through increasing NH_3 deposition or recruitment of new trees, causing losses of acidophytes and permitting extensive colonization by more neutrophyte and possibly nitrophilous bryophytes, e.g., *Orthotrichum* spp., *Ulota phyllantha*, *Cryphaea heteromalla*, and *Radula complanata*.

The above extensive changes occurred over precisely the period for which we have the best meteorological evidence of climate change in Britain.

Observed and predicted changes in the British bryophyte flora

In the following sections we discuss examples of change, or sometimes the lack of it, in the British bryophyte flora, that may have been driven by recent climate change or were observed under field conditions when artificial simulations of predicted future climates were applied to native vegetation. Where there are no data we offer some tentative predictions about likely effects.

Phenology, growth, and temperature

In a review of the effects of recent climatic warming on British wildlife, Hopkins (2007) cited many studies of individual animal and plant species indicating an advancing of metabolism and life cycle in response to the mild warming that has already occurred. They include flowering and leafing times for various wild species and garden plants. Fitter and Fitter (2002), for instance, found that among 385 species, the average date of first flowering had advanced by 4.5 days in the past decade compared with the previous four decades. Furthermore, 16% flowered earlier in the 1990s, by an average of 15 days, than in previous decades. Regrettably, no comparable data involving repeated observations over a long period appear to exist for British bryophytes.

It is now fairly well established that the vegetative growth of many bryophytes is uncoupled from strict seasonal cues such as day length, but depends primarily on the disposition of "wet days," i.e., periods when the plants are sufficiently hydrated to be metabolically active and experience incremental growth (e.g., Pitkin 1975; Busby *et al.* 1978). In certain situations, temperature has been shown to be the primary predictor of growth, as in *Dawsonia superba* in a New Zealand forest (Green & Clayton-Green 1981), *Leucobryum glaucum* in an English wood (Bates 1989), *Pallavicinia lyellii* on a shaded streambank (Bates 1993b) as well as seven bryophyte species of evergreen forest in the Azores (Gabriel 2000). In many situations where bryophytes grow in Britain, a modest increase in air temperature might be expected to increase growth rate and thereby advance development. This applies as much to the growth of reproductive as to that of vegetative structures.

Except in some of the larger thalloid liverworts (e.g., *Lunularia* and *Marchantia*), there does not appear to be a photoperiod trigger for initiation of sex organs or the development of sporophytes in bryophytes (Hartmann & Jenkins 1984; Knoop 1984). Even so, casual field observations suggest that the timing of sporophyte production in many conspicuous species (e.g., *Mnium hornum*, *Pellia epiphylla* in Britain) is relatively constant from year to year, although perhaps variable from place to place. Miles *et al.* (1989) presented data on gametangial and sporophytic development of five mosses at contrasted localities in Britain obtained over an approximately two-year period. For each species the seasonal timing of maturation of the sporophyte, including spore release, was found to be constant but, intriguingly, and except in *Polytrichum alpestre*, there was much variation in development of the sex organs and the timing of fertilization that preceded it. Potentially, there is much to be learned about bryophyte biology and climatic variation through systematic collection of phenological data of this kind, although with small plants like bryophytes it will necessarily involve more demanding observations than simply noting opening dates of buds or flowers. This is an area where motivated individuals and collaborative schemes organized through the bryological societies could each make valuable contributions to knowledge.

Table 19.3. *Some potential causes of range limitations of bryophytes related to temperature*

Low temperature limitations	High temperature limitations
Chilling injury	Increased evaporation prevents growth in specific habitats
Freezing injury	
Length of season inadequate to complete reproductive cycle or shoot/thallus maturation	Carbon depletion caused by increased respiration rate
Low growth rate undermines competitive ability	Displacement by faster-growing competitors
Increased susceptibility to infection	Increased susceptibility to infection

Temperature responses

Furness and Grime (1982) published growth rate data for 40 common British species showing that temperature responses of bryophytes are relatively broad. The earlier data of Dilks and Proctor (1975) showing temperature responses of net assimilation and respiration, while broadly in agreement, also indicate that some species have lower optima or maxima than the majority of bryophytes (e.g., *Orthothecium rufescens*, *Plagiopus oederi*, and *Hookeria lucens*). A number appear, on the basis of their distributions and ecologies, to be distinctly thermophilous (e.g., *Epipterygium tozeri*, *Habrodon perpusillus*, *Leptodon smithii*, *Scleropodium touretii*, *Scorpiurium circinatum*, and *Tortella nitida*). Equally, many montane bryophytes are clearly limited to cooler habitats, mainly in northern Britain. Perhaps surprisingly, Dilks and Proctor (1975) found that the temperature responses of net assimilation in *Anthelia julacea*, *Andreaea alpina*, *A. nivalis*, and *Racomitrium lanuginosum* did not differ appreciably from those of common lowland species. This was questioned in a later study of the snowbed bryophytes *A. julacea* and *Polytrichum sexangulare* in Sweden by Lösch *et al.* (1983). Very little is known about the precise reasons for the apparent temperature distinctions between bryophytes.

By analogy with what is known about flowering plants it is possible to suggest some temperature-related "filters" (Table 19.3). It should not be forgotten that these may also bear upon phases of the life cycle other than the gametophores and act in concert as well as independently. Temperatures vary widely on a diurnal basis in Britain and it seems unlikely that "chilling injury," essentially the solidification of unsaturated lipids in the plasma membranes below about 10 °C (e.g., Fitter & Hay 2002), would be widespread among native bryophytes. It may, however, explain why species of warmer climes do not colonize temperate countries. Freezing, owing to air and radiation frosts, is frequent in many British habitats. Apart from some thalloid liverworts (*Conocephalum conicum*, *Targionia hypophylla*, and *Pellia epiphylla*), many bryophytes, including the oceanic *Plagiochila spinulosa* and *Myurium hochstetteri*, appear well able to resist freezing without injury (Dilks & Proctor 1975). Survival of freezing appears to be a special case of desiccation tolerance, which is widespread in bryophytes (Dilks & Proctor 1975). Some bryophytes may avoid freezing injury by selecting protected microhabitats or by dying back to underground parts. Snowbed bryophytes are insulated from the lowest temperatures of winter by their snow covering. Reduction of the length of the growth season and the accompanying reduced growth rates are probably a major reason why thermophilous species are unable to expand further north (or to greater altitudes), probably through an inability to reproduce or compete. As chemical reactions like respiration have higher temperature quotients ($Q_{10} \geq 2$) than photochemical processes like photosynthesis ($Q_{10} \sim 1$), increasing respiration rates leading to carbon depletion (Fitter & Hay 2002) are likely consequences of rising temperatures for many bryophytes of cold montane habitats. Most are also probably poor competitors as a result of the burden of cold-resistance physiologies that they carry and liable to be ousted by faster growing species of plants if temperatures increase. Excessive heat and cold may lead to decreased resistance to attacks by pathogens.

What groups would we expect to show increases in range?

The majority of British bryophytes occur in rather northerly phytogeographical groups. Some 61% of the flora are classified in phytogeographical groups with distributions from Arctic-montane to Boreo-temperate (Table 19.1). One would expect the species that benefit from climate change to be those in the more southerly phytogeographical groups, to be plants of habitats that will not themselves be threatened by the changing climate, and to have powers of dispersal which will allow them to colonize new areas. In Table 19.4, ecological data in BRYOATT (Hill *et al.* 2007) have been used to enumerate such species. Native British and Irish species conforming to the following criteria have been selected:

(1) Species in Major Biome Categories: Temperate, Southern-temperate, and Mediterranean–Atlantic.
(2) Species with sporophytes frequent or abundant *or* at least one form of vegetative propagule frequent.

Table 19.4. *The occurrence in Major Biomes and Eastern Limit Categories of native species with southerly distributions in Britain and frequent sexual or vegetative propagules*

					Eastern Limit Category				
Major Biome	No. hornworts	No. liverworts	No. mosses	Total	Hyperoceanic	Oceanic	Suboceanic	European, Eurosiberian and Eurasian	Circumpolar
Temperate	2	14	93	109	3	15	21	58	12
Southern-temperate	0	22	30	52	16	11	3	14	8
Mediterranean-Atlantic	2	13	22	37					
Submediterranean-Subatlantic	0	4	19	23					
TOTAL	4	53	164	221					

(3) Species that are not characteristically found in very infertile or very acidic sites. Ellenberg indicator values represent the habitat preferences of species with respect to these parameters, and plants with Ellenberg $N=1$ (indicators of extremely infertile sites) and $R=1$ (indicators of extreme acidity) have been excluded.
(4) Species recorded from fewer than one thousand 10 km squares in Britain.

Temperate as well as Southern-temperate and Mediterranean–Atlantic species have been included (Criterion 1). Although much of Britain falls within the Temperate biome, the more southerly Temperate species might expand in Britain in response to climate warming. Conversely, the more northerly Temperate species might be adversely affected. Criterion 3 excludes species that are not likely to be favored in the modern environment. Criterion 4 eliminates species that are already frequent, and so have less scope for marked range expansion.

The 221 species enumerated in Table 19.4 represent 22% of the flora of the British Isles. They may be regarded as the pool out of which expanding species are likely to come. However, not all species in the pool are likely to expand – many will be limited by habitat or other factors, and some may be adversely affected.

In addition to the expansion of species northwards in response to increasing temperatures, Crawford (2008) pointed out that one likely result of climate change is the expansion of oceanic species eastwards in response to warmer winters. The extent to which oceanic bryophytes are able to increase will depend on the extent to which the positive effects of warmer winter temperatures are not negated by increasing summer temperatures, especially if accompanied by drought and the loss of habitats with high humidity.

The large group of species inhabiting rocks, screes, and sparsely vegetated habitats, and the arable land group in Table 19.5, include many species of disturbed soil in genera such as *Ephemerum*, *Microbryum*, *Fossombronia*, *Riccia*, and *Weissia*, as well as tuberous species of *Bryum* (many of the latter with Temperate distributions). These tend to be shuttle species, which one would not expect to spread rapidly. The species of surface waters include a similar group of species of seasonally flooded soils, also including *Ephemerum*, *Fossombronia*, and *Riccia* species, as well as the British *Physcomitrium* species. Many of the same genera and species recur in the list of coastal species, which includes species of *Entosthodon*, *Fossombronia*, *Microbryum*, *Riccia*, *Southbya*, and *Tortula*. Many of the grassland species are plants of seasonally droughted habitats including species of *Acaulon*, *Archidium*, *Fossombronia*, *Microbryum*, *Pleuridium*, *Tortula*, and *Weissia*. There is clearly a broad overlap, at least at the generic

Table 19.5. *The occurrence in EUNIS (European Nature Information System) habitats of species with southerly distributions in Britain and frequent sexual or vegetative propagules*

Species may occur in more than one habitat; all plants recorded as occasional (2) or normally occurring (3) in one of the listed EUNIS categories in BRYOATT (Hill *et al.* 2007) are included under the habitat. Key to abbreviations: Hyp: Hyperoceanic; Oc: Oceanic; Suboc: Suboceanic; Eur: European, Eurosiberian and Eurasian; Circ: Circumpolar; MA: Mediterranean–Atlantic; Sm-Sa: Submediterranean–Subatlantic.

		Temperate					Southern-temperate							
Habitat	EUNIS categories	Hyp	Oc	Suboc	Eur	Circ	Hyp	Oc	Suboc	Eur	Circ	MA	Sm-Sa	**Total**
Coast	B		2	8	7	3	12	7	1	8	6	25	12	**91**
Surface waters	C	2	4	6	20	4	4	3		5	2	6	7	**63**
Mires, bogs and fens	D				1		1					1	1	**4**
Grassland	E1-4		2	3	19	4	4			4	3	12	12	**63**
Heathland	F4	1	1	5	9		6	2	1	2		2	2	**31**
Scrub and woodland	F3,9, FA, G1,3		6	13	19	9	14	7	1	3	1	11	8	**92**
Inland sparsely vegetated habitats	G1R, G3R, H	1	4	12	46	6	11	6	2	12	6	28	18	**152**
Agricultural habitats	I1	1	2	3	21	4				2	2	7	7	**49**
Constructed habitats	J		5	8	20	5	2	3	1	8	7	19	13	**91**
Total		**5**	**26**	**58**	**162**	**35**	**54**	**28**	**6**	**44**	**27**	**111**	**80**	

level, between these groups. The plants of scrub and woodland are an ecologically very different group, which include many epiphytes (e.g., species of *Frullania, Orthotrichum*, and *Zygodon*) as well as plants of the woodland floor.

Recent range changes in the British and Irish bryophyte flora

In this section we map the ranges of six species, and discuss the extent to which climate change might be responsible for apparent changes in overall range or in the frequency of the species within that range.

Leptobarbula berica

This is a Mediterranean–Atlantic species. Frahm and Klaus (1997) suggested that its recent occurrence in central Europe is a response to recent climate change, and in particular to milder winters. The British distribution map (Fig. 19.1) suggests that it has increased in frequency in southern England in recent years. However, it is an inconspicuous species, which rarely fruits in Britain and it was only added to the British list in 1985, having been mistaken previously for *Gymnostomum calcareum* and *Gyroweisia tenuis*; there are herbarium specimens dating back to 1948 (Hill *et al.* 1992). There is no reason to believe that the apparent expansion since 1990 results from anything other than the gradual discovery of additional sites as recorders have become familiar with this species. It is slightly surprising that almost all the new localities fall within the boundaries of the pre-1990 distribution.

Cryphaea heteromalla

This conspicuous, easily identified, freely fruiting epiphyte was rare in many areas of central and eastern England during the period of maximum SO_2 pollution and absent from the most polluted areas. It has very clearly increased in frequency and range as air quality has improved (Fig. 19.2). There seems no reason to think that climate change is necessarily involved in its spread, although the species has a Submediterranean–Subatlantic range in Europe and might be expected to benefit from the warmer winters of recent years. Scattered historical records in NE England and SE Scotland, mapped by Hill *et al.* (1994), indicate that even in these northern areas it is recolonizing its previous range rather than expanding into new areas.

Cololejeunea minutissima

This liverwort was classified by Hill and Preston (1998) as having a Hyperoceanic Southern-temperate range in Europe. In Britain it reproduces sexually (sporophytes are frequent) and asexually by numerous foliar

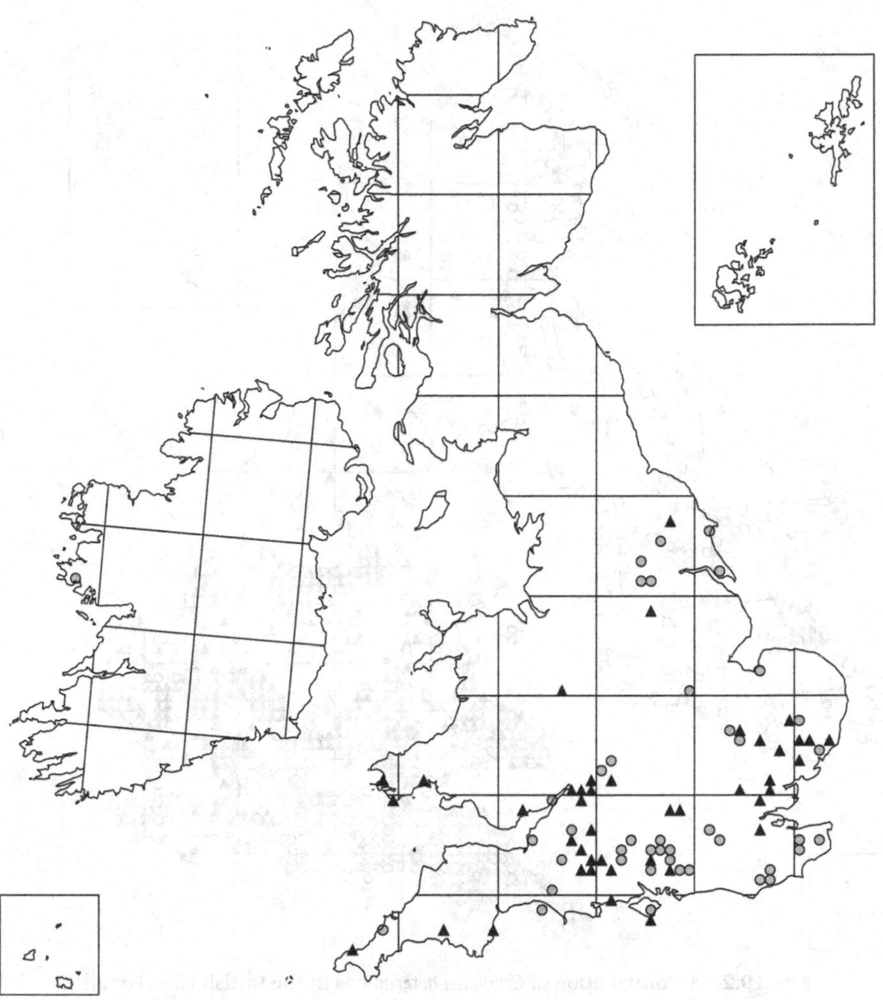

Fig. 19.1. The distribution of *Leptobarbula berica* in the British Isles. Grey circles indicate 10 km squares from which the species had been recorded by 1989; black triangles, squares in which it has been discovered subsequently. Orkney and Shetland (top right) and the Channel islands (bottom left) are plotted as insets. The grid lines are 100 km apart. Mapped by Stephanie Ames, Biological Records Centre, Centre of Ecology and Hydrology, mainly from records collected by members of the British Bryological Society, using Dr Alan Morton's DMAP software.

gemmae. Although the species is tiny, it forms dense patches with numerous perianths on the bark of trees and shrubs and is therefore not as inconspicuous as might be expected from its specific epithet. It has clearly rapidly expanded in range in Britain in recent years (Fig. 19.3), spreading northwards and

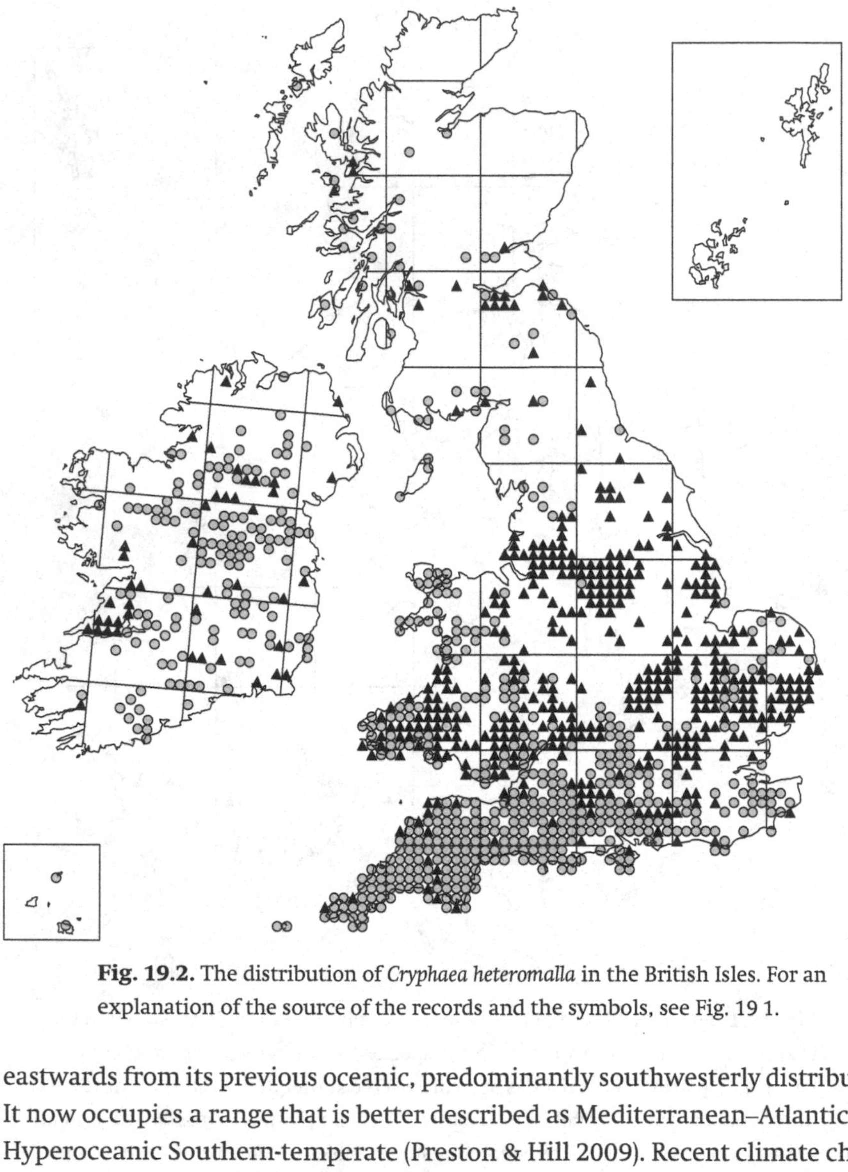

Fig. 19.2. The distribution of *Cryphaea heteromalla* in the British Isles. For an explanation of the source of the records and the symbols, see Fig. 19 1.

eastwards from its previous oceanic, predominantly southwesterly distribution. It now occupies a range that is better described as Mediterranean–Atlantic than Hyperoceanic Southern-temperate (Preston & Hill 2009). Recent climate change appears to be the most likely explanation for this dramatic range expansion. One might, however, ask whether the reduction in SO_2 pollution was a necessary precondition of its expansion.

Colura calyptrifolia

Colura calyptrifolia is another tiny Hyperoceanic Southern-temperate liverwort, which has spread inland from its former strongholds near the west

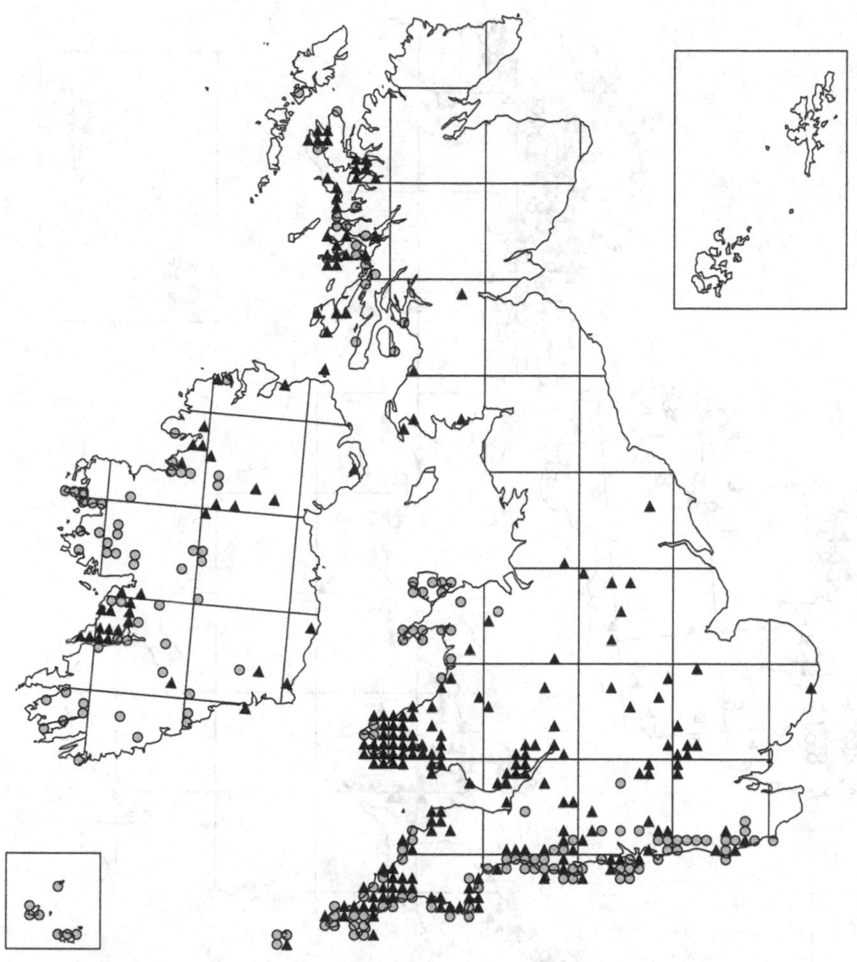

Fig. 19.3. The distribution of *Cololejeunea minutissima* in the British Isles. For an explanation of the source of the records and the symbols, see Fig. 19 1.

coast (Fig. 19.4). Formerly *C. calyptrifolia* had been known as a rare species, principally on rock faces and heather (*Calluna vulgaris*) and gorse (*Ulex europaeus*) stems, and restricted to strongly oceanic parts of SW England, North Wales, and western Scotland. Bosanquet (2004) described its appearance as an epiphyte outside this area in South Wales in 1985 and its gradual spread, usually on boughs of common sallow (*Salix cinerea*), a broad-leaf species associated with conifer plantings or sometimes disused quarries throughout Wales. Similarly, in Scotland it has recently been recorded on a range of deciduous trees and shrubs, and on the bark of planted conifers such as Sitka spruce

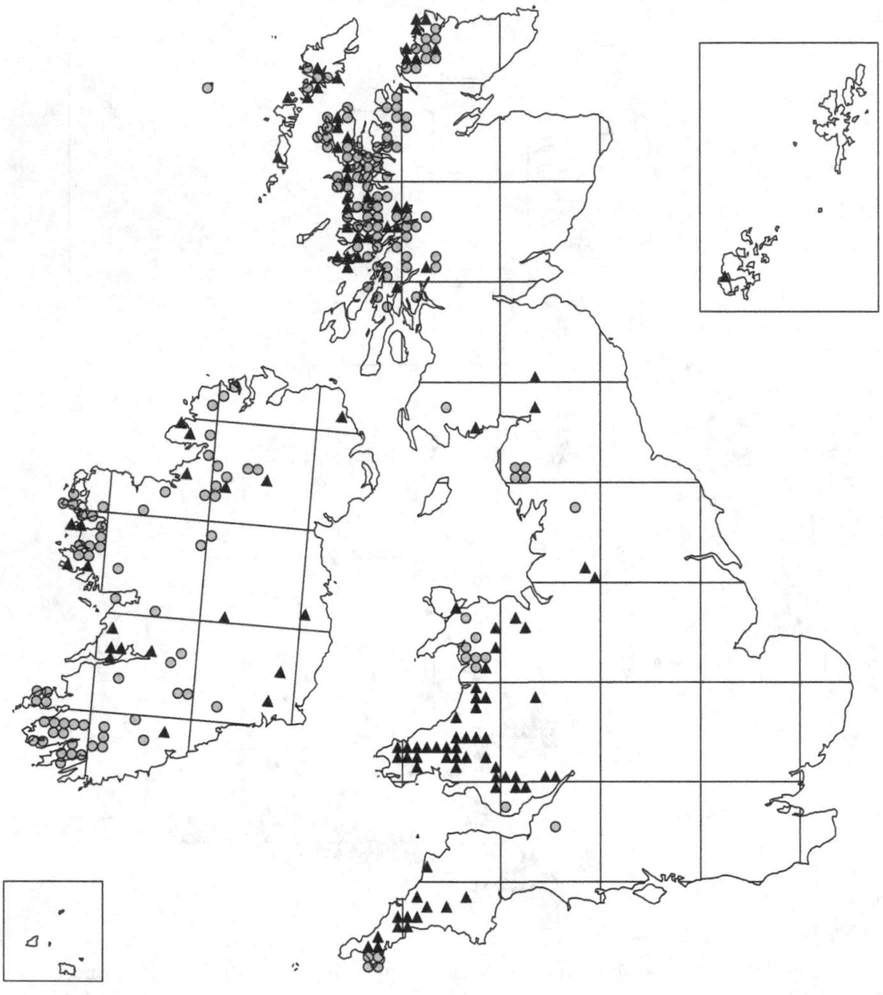

Fig. 19.4. The distribution of *Colura calyptrifolia* in the British Isles. For an explanation of the source of the records and the symbols, see Fig. 19 1.

(*Picea sitchensis*), often in glades in dense conifer plantations or on the damp, north- to east-facing margins of such plantations (Averis 2007). It has even been found in Scotland on a polypropylene rope attached to an old tractor. Bosanquet (2004) offered some possible explanations for this spread, including:

- the canopy of conifers protects *C. calyptrifolia* more effectively than deciduous broad-leaves from injurious frosts;
- the conifers act as more efficient spore traps than deciduous trees;
- the conifers provide constantly humid conditions.

The first and last alternatives represent the most probable explanations. The timing of the expansion fits well with the known increase in average temperatures in the UK. It seems likely that this change has allowed the species to colonize the humid habitats that are now available in modern conifer plantations. An additional or alternative hypothesis for the conifer association is that the recent wave of conifer afforestation has favored a new cohort of *Salix cinerea* individuals as plantation weeds. Older generations of *S. cinerea* are likely to have offered more acidified bark surfaces owing to long range wet-deposited acidity (Bates 2002). The uplands of mid-Wales were formerly significantly affected by acid deposition, leading to a depleted aquatic invertebrate fauna in the upland streams, and effects on their fisheries and aquatic birdlife (e.g., O'Halloran *et al.* 1990). Former high concentrations of gaseous pollutants mainly affected the industrial parts of South Wales, and the "acid rain" effect occurring in the uplands has been measurably reduced by the decline in SO_2 emissions (e.g., Monteith & Evans 2005). Although it remains a possibility, we believe that freezing injury of *C. calyptrifolia* is unlikely to be the reason for its former limitation to the coast (see above). More likely, a small rise in air temperatures now permits this species to reproduce in inland localities, thus facilitating its spread to new habitats. The fact that this is a widespread species in tropical and subtropical countries gives credence to the view that it is relatively thermophilous. The appearance of non-acidified *Salix* bark, perhaps together with raised atmospheric humidity through conifer planting or increased winter precipitation, may be other significant pieces in the jigsaw. Some recent inland localities for *C. calyptrifolia* have also produced *Cololejeunea minutissima*, which suggests that similar factors may be governing the recent spread of both epiphytic liverworts.

Ulota phyllantha

This is another easily identified epiphyte, but one that reproduces in Britain by foliar gemmae (sporophytes are very rare). It is a characteristic species of coastal regions, and it is often abundant on trees (and rocks) exposed to salt-laden winds, conditions which most other epiphytes are unable to withstand. Its expansion inland in recent years, and additional recording in some areas where it was probably overlooked before 1990, are clearly apparent on the map (Fig. 19.5). It does not, however, grow inland in the abundance with which it can occur by the sea. There is no doubt that, unlike *Cryphaea heteromalla*, it has colonized regions in central and eastern England in which it was never previously recorded. This suggests that the recent expansion is not simply a recovery from high levels of SO_2 pollution. The reasons for the spread are unknown. One might imagine that reduction in

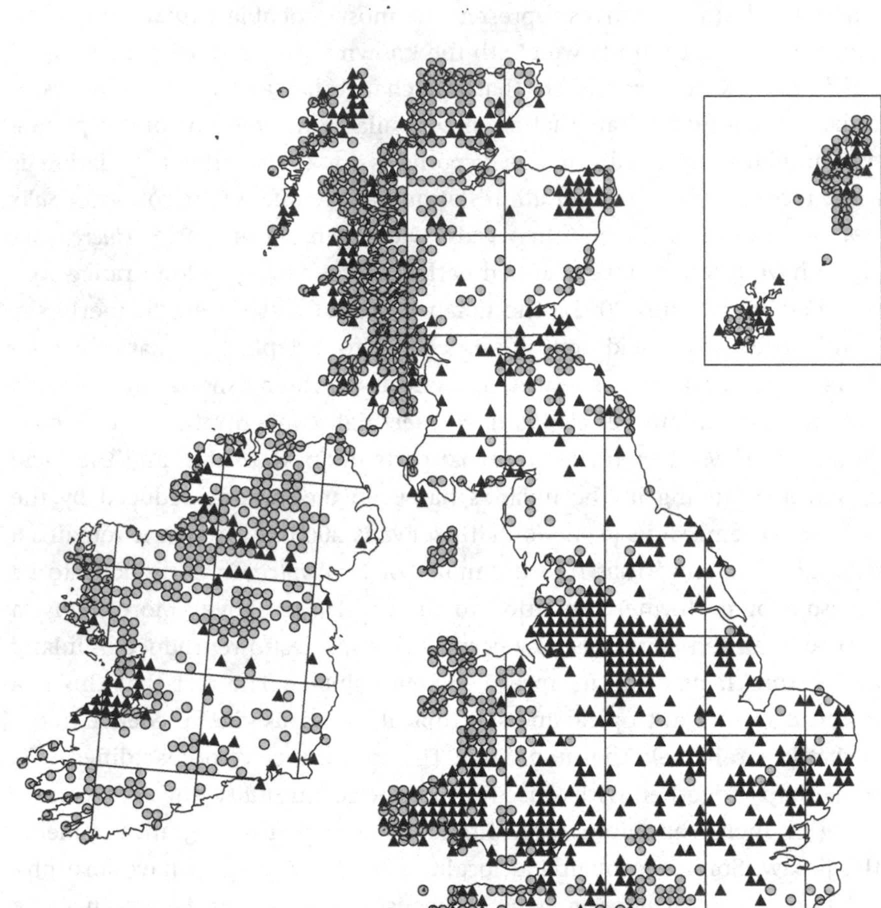

Fig. 19.5. The distribution of *Ulota phyllantha* in the British Isles. For an explanation of the source of the records and the symbols, see Fig. 19 1.

SO_2, increasing N levels from NO_x deposition (or even agricultural fertilizer drift), and climate change are factors that may have played a part, but experimental studies will be needed before we can hope to identify the causes of this very marked range expansion.

Didymodon nicholsonii

This *Didymodon* species was traditionally known "on stones and tree roots on the banks of streams and rivers where subject to periodic flooding" (Hill *et al.* 1992). In recent years it has been recorded with increasing frequency

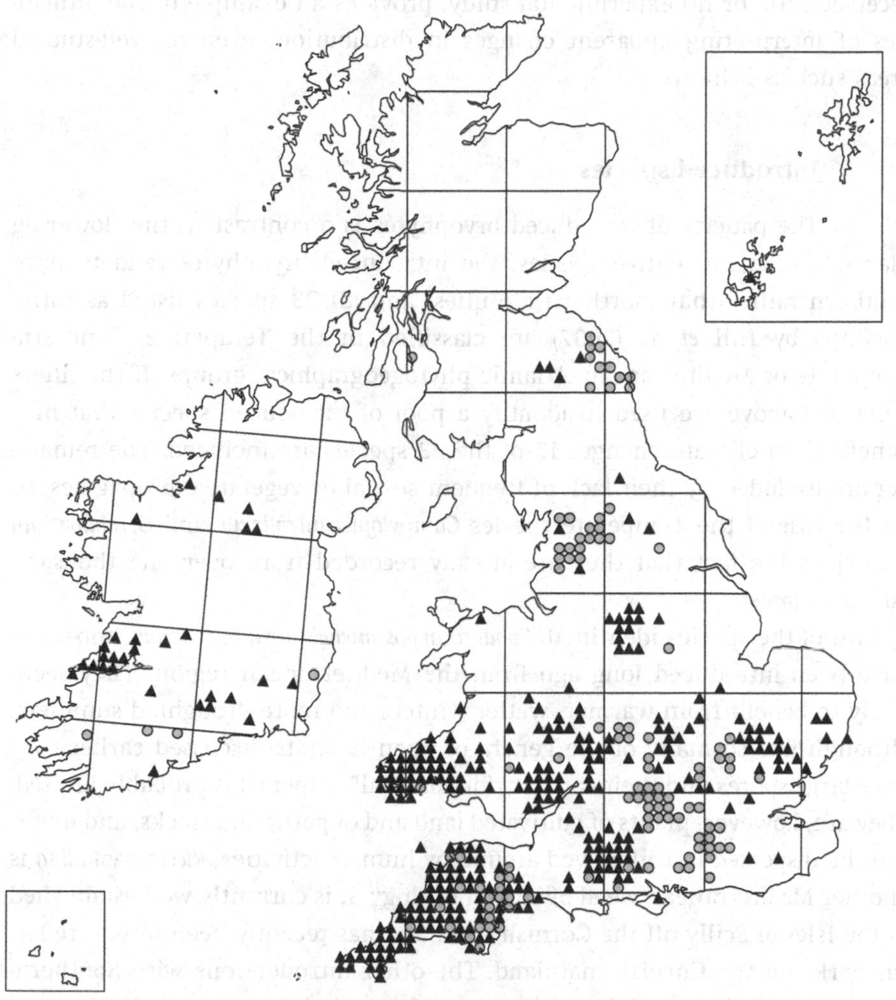

Fig. 19.6. The distribution of *Didymodon nicholsonii* in the British Isles. For an explanation of the source of the records and the symbols, see Fig. 19 1.

on tarmac and other non-riparian substrates, and in S Ireland, SW England and S Wales it is now almost ubiquitous in this habitat (Fig. 19.6). In places it grows with *Syntrichia latifolia*, another riparian species, which has also colonized tarmac. Is this a genuine increase in frequency or has it been overlooked in such habitats by earlier generations of bryologists? We cannot be sure, but it seems very likely that there is a genuine expansion of the species, as it is certainly now frequent in some areas that have been studied previously by bryologists with a good knowledge of *Didymodon*. If the increase is genuine, we can only speculate about the causal factors. This species, somewhat inconspicuous and exhibiting rather individualistic behavior, and which has received little or no experimental study, provides an example of the difficulties of interpreting apparent changes in distribution, even in well-studied areas such as Britain.

Introduced species

The paucity of introduced bryophytes is a contrast to the flowering plants. Unlike the native species, the introduced bryophytes tend to have southern rather than northern affinities, and all 23 species listed as introductions by Hill *et al.* (2007) are classified in the Temperate, Southern-temperate or Mediterranean–Atlantic phytogeographical groups. If the filters described above are used to identify a pool of introduced species that may benefit from climate change, 12 of the 23 species are included. The remainder are excluded by their lack of frequent sexual or vegetative propagules, or (in the case of the Temperate species *Campylopus introflexus* and *Orthodontium lineare*) by the fact that they are already recorded from over one thousand 10 km squares.

Two of the species identified, *Sphaerocarpos michelii* and *S. texanus*, appear to have been introduced long ago from the Mediterranean region. They seem likely to benefit from warmer, wetter winters and more droughted summers, although like so many of the genera of open habitats discussed earlier they have large spores and their capacity for "natural" dispersal is probably limited. They are, however, plants of cultivated land and of paths and tracks, and might thus be dispersed in soil moved around by human activities. *Riccia crystallina* is another Mediterranean weed of similar ecology; it is currently well established in the Isles of Scilly off the Cornish coast and has recently been discovered in car parks on the Cornish mainland. The other introductions with Southern-temperate affinities and frequent propagules include two rare, tuber-bearing weedy species (*Bryum valparaisense* and *Chenia leptophylla*) and the rather more frequent *Didymodon umbrosus* and *Hennediella stanfordensis*. A few introduced bryophytes were first recorded in western gardens and were almost certainly introduced with garden plants from the oceanic climates of the Southern Hemisphere. *Heteroscyphus fissistipus* is a rare established introduction in Ireland but fruits freely and may have the capacity to spread. Further introductions from warmer climates may become established. However, the rate of introduction of bryophytes appears to be low, certainly when compared with the large numbers of flowering plants which have become established following deliberate introduction by the horticultural trade, or accidental importation as contaminants of grain, wool, or other imported products.

Habitat change

Eventually, major changes in a number of British habitats may occur as longer-lived higher plants invade or spread and other species become extinct (Porley & Hodgetts 2005; Hopkins 2007). Progressive infestation of many oceanic oakwoods by the tall alien shrub *Rhododendron ponticum* may possibly have accelerated because of recent climate change. This species was originally planted on large estates as an ornamental and is now causing problems in many important bryophyte sites as its dense cover obliterates the ground flora. British material is believed to have originated from the warmer climate of Spain and its cold tolerance may have been locally accentuated by introgression with another introduced *Rhododendron* species. The European beech (*Fagus sylvatica*) has been shown to be harmed by acute water stress and it has been suggested that it may disappear from parts of southern England suffering arid summers (Peñuelas *et al.* 2008). One probable "loser" under this scenario would be the rare European moss *Zygodon forsteri*, which grows on the edges of water-filled knot-holes and irrigated strips on trunks in a few favored localities with large populations of veteran pollarded beech (Porley & Hodgetts 2005). Major changes in land management, such as cessation of grazing through increased aridity, could have important implications for the bryophyte flora in habitats like chalk and limestone pasture. Increased droughting of semi-natural pastures may also favor annual mosses such *Microbryum curvicolle, M. rectum*, and *Tortula acaulon* by creating gaps, as well as selecting for more desiccation-tolerant mosses (e.g., *Thuidium abietinum*) over their faster-growing competitors (Porley & Hodgetts 2005).

Sea level rise is already occurring as our warming world's glaciers melt. A few important bryophyte localities, particularly for oceanic species (e.g., the only English site for *Cyclodictyon laetevirens* in a Cornish sea cave), will doubtless be lost by eventual inundation. The relatively few bryophytes that require tidal inundation on salt marshes (e.g., *Hennediella heimii*) or regular wetting by salt spray (e.g., *Schistidium maritimum*) are likely to accommodate gradually to changes in the relative positions of land and sea, as they must have done repeatedly in the past. Problems may, however, arise through habitat fragmentation. For instance, in the Prawle area of the South Devon coast, an area of "fossil" geomorphology shaped by periglacial weathering during the Pleistocene, the uncommon oceanic liverwort *Plagiochila bifaria* survives in small relict populations in niches on isolated schist pinnacles. Today *P. bifaria* is sterile in Britain and one imagines that it must have reached these places when continuous forest cover extended to the coast from the hinterland. It is hard to imagine how these populations will avoid the increasing influence of saltwater and eventual annihilation with sea level rise.

Bates and Phoon (2009) described the impact of a severe gale on a population of *S. maritimum* that was being regularly monitored on a rocky shore in southwest England. High spring tides accompanying low atmospheric pressure lifted seawater well into the supralittoral zone and eradicated most of the moss cushions. Subsequent monitoring over several years showed a reasonably rapid recolonization centered on safe sites in the crevices of the schist rocks. These observations suggest that *S. maritimum*, in contrast to *Plagiochila bifaria*, would have little difficulty in keeping pace with predicted rates of sea-level rise by new colonizations.

Community-level changes

Polar tundra and alpine vegetation are considered to be among the most threatened ecosystems by global warming and are likely to reveal its effects sooner than other communities (Harte & Shaw 1995; Arft *et al.* 1999). Much effort by British ecologists has gone into studies of Antarctic vegetation (e.g., Convey & Lewis Smith 2006) and Arctic (e.g., Wookey *et al.* 1993; Potter *et al.* 1995) but comparatively little into UK plant communities. The exceptions considered here all concern upland bryophytes. Average temperatures in the uplands decline by about 0.6 °C per 100 m increase in altitude, the adiabatic lapse rate, principally a consequence of decreasing air density. As Porley and Hodgetts (2005) so eloquently note, many bryophytes of British mountain tops are at the edges of their ranges (e.g., *Anthelia julacea* in the Brecon Beacons of South Wales and *Herbertus stramineus* in Snowdonia, North Wales) and have nowhere to go if the climate warms. In Britain, the plant communities of late-persisting snowbeds in the Scottish Highlands represent the least extensive vegetation type and would appear to be the most vulnerable to the effects of warming. The mountain winds are predominantly westerlies so that snowbeds accumulate in sheltered areas with an easterly or northeasterly aspect that receive very little direct sunlight. They are especially abundant on the Cairngorm plateau, the most extensive area of land in Britain above 1000 m, but also occur on other mountains in the Highlands. As the overall snow cover melts in summer, the thicker snowbeds persist into July in favored spots governed by topography. The extent and persistence of snow cover varies greatly from year to year, but some snow patches may survive until snow falls again. Many mountain plants are chionophilous (snow-loving), favoring snow cover for its insulative properties (Körner 1999), but in the most persistent snowbeds flowering plants give way to bryophytes. A major survey of snowbed sites was carried out 20 years ago by G. P. Rothero; a resurvey is currently underway and should reveal the extent to which these communities have already changed.

Headley and Rumsey (cited in Porley & Hodgetts 2005) made a comparison of the altitudes of extant and extinct colonies of the rare European endemic moss *Zygodon gracilis* in the Yorkshire Dales National Park, England. Extinct colonies occurred at significantly lower altitudes than extant ones, the maximum at present being 520 m. Examination of published records and herbarium specimens led Headley and Rumsey to the conclusion that *Z. gracilis* had been in decline since the early nineteenth century, i.e., from towards the end of the Little Ice Age. The recommendation is made that any future recovery plan for this species should employ sites above 400 m (Porley & Hodgetts 2005).

A relatively long-term study of the effect of various simulations of climate change on an upland limestone grassland community was established at Buxton, Derbyshire (Pennine Hills, central England) in November 1993 (Grime *et al.* 2000). Plots (3 m × 3 m) of ancient sheep pasture were subjected to a factorial arrangement of three moisture and two temperature treatments, each replicated five times. The moisture treatments were: natural precipitation; summer drought (sliding transparent covers were operated by a rain detector); supplemented summer precipitation (20% above the long-term average). The temperature treatments were: ambient temperature; winter heating (by soil heating cables). Even after 13 years, the responses of higher plants to these treatments have been relatively slight, the main effects being an increase in prostrate shrubs (particularly *Thymus polytrichus*) and a decrease in forbs (particularly *Potentilla erecta*) in response to the summer drought treatment (Grime *et al.* 2008). Bryophyte cover of the plots was analyzed critically following seven years of the treatments (Bates *et al.* 2005). Significant treatment effects were detected but, as with the higher plants, they were relatively modest. Percentage cover of *Calliergonella cuspidata* and *Rhytidiadelphus squarrosus*, the two most abundant bryophytes, and total bryophyte cover were all reduced by summer drought, whereas *Fissidens dubius* increased under this treatment. Winter warming benefited *Campyliadelphus chrysophyllus* but was detrimental to *R. squarrosus* and *Lophocolea bidentata*. Remarkably, in places, the moss *Ctenidium molluscum* was discovered growing intimately attached to short lengths of exposed electrical heating cable. Analyses of the bryophyte data, comparing the results from detrended correspondence analysis and canonical correspondence analysis, identified a treatment axis comprising the combined effects of moisture and temperature. However, they also demonstrated that natural factors were more important causes of overall variation in the grassland community than the simulated climatic treatments. Bates *et al.* (2005) speculated that dew may be an important moisture source for bryophytes of upland grasslands and wondered whether natural dewfall at Buxton had reduced the impact of the moisture treatments. The absence of thermophilous bryophytes from the vegetation (notably *Homalothecium lutescens*) may also have reduced the scope for community change under the imposed climate regimes.

Grime *et al.* (2008) drew attention to the surprising resilience of the whole grassland community at Buxton to change induced by their climate treatments. Discussing various possible explanations, they noted that predicted future changes are relatively modest compared with the large diurnal and seasonal fluctuations that the plants endure and have adapted to over centuries. Nevertheless, the possibility remains that genotypic changes have been occurring that are not detectable in species-level analyses. Long-term studies of the effects of simulated climate scenarios on any vegetation type are still few and future work on selected bryophyte-dominated communities is to be encouraged.

Comparison with other groups in Britain

Lichens

Interpreting possible effects of climate change on the British lichen flora is beset with much the same uncertainties as are encountered with the bryophytes. Knowledge of the lichen flora is certainly less complete than that of bryophytes, although better than for most other countries. Little information is available from community level studies in Britain as lichens are more restricted in their occupation of the moist, soil-based communities of higher plants that have attracted most attention. Nevertheless, there are many lichen species that occupy cool, montane habitats or oceanic gullies and woods that are potentially threatened by climate change. The characteristics of many lichens that make them such good bioindicators of ancient woodland, namely an inability to disperse effectively between forest patches, are quite likely to prove their downfall under rapidly changing conditions in the present fragmented landscape of Britain.

In the absence of experimental data or substantial observations of change, Ellis *et al.* (2007) used distributional data based on presences within the 10 km × 10 km national grid to model the "bioclimatic envelope" for each of 26 lichen species chosen subjectively to represent the five main biogeographical groupings of British lichens. These groupings were later confirmed in a cluster analysis of the dataset: northern–montane, northern–boreal, southern–widespread, oceanic–northern, and oceanic–widespread. Sophisticated mathematical models were constructed for each grouping based on the current climate and then predicted distributions for each were computed for the low-emission and high-emission climate scenarios for 2050 (Hulme *et al.* 2002). The main conclusions reached by this approach were: major potential losses of northern–montane and northern–boreal lichens; relatively little change to Britain's important oceanic lichens, although some losses are predicted from localities in southwest England and Wales which might be matched by gains further north; and a major migration northwards of southern–widespread lichens. The authors freely admit that there are a number of important caveats to their findings. These obviously include our ignorance of the abilities of the different species to disperse to new sites as conditions change. Nevertheless, the approach presents a logically constructed set of potential outcomes, which can be used to direct future field monitoring of actual change and help in making difficult decisions regarding allocation of limited resources to conservation. A comparable analysis of the British bryophyte flora might provide similar benefits.

Vascular plants

The strong evidence for changes in the phenology of British flowering plants has already been cited. For vascular plants, unlike bryophytes and lichens, repeat surveys are available which also allow an assessment of both declines and increases of native British plant species. In particular, repeat surveys are available for almost all British 10 km × 10 km squares (Preston *et al.* 2002) and the repeat of a sample of 2 km × 2 km squares thoughout Britain (Braithwaite *et al.* 2006). Analysis of the 10 km square survey, which compared plants recorded in 1930–69 with those recorded in 1987–99, suggested a relative decline in the species in the northern phytogeographical groups, and a relative increase of widespread and southerly species. However, these are attributable almost entirely to the ecological characteristics of the species concerned, rather than the effects of climate change (M. Hill, C. Preston *et al.*, unpublished data). Habitat destruction and eutrophication have had a much greater effect on species in this period than climate change.

Unlike bryophytes and lichens, air quality does not have a marked effect on the distribution of flowering plants, which are less dependent on aerial inputs of nutrients and on conditions at the surface of their substrates. The 2 km survey compared species recorded in 1987–88 with those found in 2003–04. Analyses of these results shows that there were also significant losses of species of infertile habitats, but there were also changes that could only be accounted for by the climatic requirements of species. In several habitats, such as broadleaved woodland, neutral grassland, calcareous grassland, and urban habitats, southerly species had increased relative to more northerly species. In broadleaved woodland and calcareous grassland, warmer January temperatures appear to be the driving factor, whereas warmer summers are probably responsible for the changes in urban habitats.

Conclusions

The ranges of bryophytes in Britain are in a constant state of flux, as species respond at different rates to the varied effects of a dense human population on the biota of a small island. The continuing tradition of bryophyte recording allows changes in species ranges to be detected fairly readily, although expansions are easier to document than declines as detecting the latter relies on repeat surveys, which are not practicable at the national scale. There are problems in attributing changes to climate change as opposed to other causal factors, always assuming that the changes are actually the result of single factors (rather than of interactions between factors). In particular, the recovery of epiphytes and possibly other species from high levels of SO_2 pollution has resulted in major range changes, which may be masking the effects of climate change. These problems of interpretation are exacerbated by the lack of basic physiological information on the responses of most bryophyte species to climatic factors.

The extent to which modeling and other sophisticated analyses of survey results can compensate for deficiencies in basic physiological data remains to be seen. Detecting the effects of climate change on processes such as phenology is much more difficult than detecting potential range changes because of the lack of baseline data for all but a few species. Most British bryophytes have northerly affinities and the Arctic-montane species, in particular, have restricted ranges and may be particularly vulnerable to anticipated changes in climate. Only a minority of British bryophytes have southern ranges and might therefore benefit from a warmer climate. Recent modeling studies for lichens have suggested that there will be little net change in the frequency of typical oceanic species in Britain, as any losses at the southern end of their ranges may be balanced by gains further north. The effects of climate change on the rich oceanic bryophyte flora will depend on the balance between the favorable effects of warmer winters and the adverse effects of drier summers.

References

Adams, K. J. & Preston, C. D. (1992). Evidence for the effects of atmospheric pollution on bryophytes from national and local recording. In *Biological Recording of Changes in British Wildlife*, ed. P. T. Harding, pp. 31–43. ITE Symposium no. 26. London: HMSO.

Arft, A. M., Walker, M. D., Gurevitch, J. *et al.* (1999). Responses of tundra plants to experimental warming: meta-analysis of the international tundra experiment. *Ecological Monographs* **69**: 491–511.

Averis, A. B. G. (2007). Habitats of *Colura calyptrifolia* in north-western Britain. *Field Bryology* **91**: 17–21.

Bates, J. W. (1989). Growth of *Leucobryum glaucum* cushions in a Berkshire oakwood. *Journal of Bryology* **15**: 785–91.

Bates, J. W. (1993a). Regional calcicoly in the moss *Rhytidiadelphus triquetrus*: survival and chemistry of transplants at a formerly SO_2-polluted site with acid soil. *Annals of Botany* **72**: 449–55.

Bates, J. W. (1993b). Comparative growth patterns of the thalloid liverworts *Pallavicinia lyellii* and *Pellia epiphylla* at Silwood Park, Southern England. *Journal of Bryology* **17**: 439–45.

Bates, J. W. (1995). A bryophyte flora of Berkshire. *Journal of Bryology* **18**: 503–620.

Bates, J. W. (2000). Mineral nutrition, substratum ecology, and pollution. In *Bryophyte Biology*, ed. A. J. Shaw & B. Goffinet, pp. 248–311. Cambridge: Cambridge University Press.

Bates, J. W. (2002). Effects on bryophytes and lichens. In *Air Pollution and Plant Life*, ed. J. N. B. Bell & M. Treshow, pp. 309–42. Chichester & New York: John Wiley & Sons.

Bates, J. W., Bell, J. N. B. & Massara, A. C. (2001). Loss of *Lecanora conizaeoides* and other fluctuations of epiphytes on oak in S. E. England over 21 years with declining SO_2 concentrations. *Atmospheric Environment* **35**: 2557–68.

Bates, J. W., Perry, A. R. & Proctor, M. C. F. (1993). The natural history of Slapton Ley National Nature Reserve. XX. The changing bryophyte flora. *Field Studies* **8**: 279–333.

Bates, J. W. & Phoon, X. (2009). Salinity tolerance and survival of the rocky-seashore bryophyte *Schistidium maritimum*. In *Bryology in the New Millenium, 2007*, ed. H. Mohammed, B. H. Bakar, A. N. Boyce & P. Lee, pp. 317–28. Kuala Lumpur: University of Malaya Botanical Garden.

Bates, J. W., Proctor, M. C. F., Preston, C. D., Hodgetts, N. G. & Perry, A. R. (1997). Occurrence of epiphytic bryophytes in a 'tetrad' transect across southern Britain. 1. Geographical trends in abundance and evidence of recent change. *Journal of Bryology* **19**: 685–714.

Bates, J. W., Roy, D. B. & Preston, C. D. (2004). Occurrence of epiphytic bryophytes in a 'tetrad' transect across southern Britain. 2. Analysis and modelling of epiphyte-environmental relationships. *Journal of Bryology* **26**: 181–97.

Bates, J. W., Thompson, K. & Grime, J. P. (2005). Effects of simulated long-term climatic change on the bryophytes of a limestone grassland community. *Global Change Biology* **11**: 757–69.

Bosanquet, S. D. S. (2004). *Colura calyptrifolia* in Wales. *Field Bryology* **82**: 3–5.

Braithwaite, M. E., Ellis, R. W. & Preston, C. D. (2006). *Change in the British Flora 1987–2004*. London: Botanical Society of the British Isles.

Busby, J. R., Bliss, L. C. & Hamilton, C. D. (1978). Microclimatic control of growth rates and habits of the boreal mosses *Tomenthypnum nitens* and *Hylocomium splendens*. *Ecological Monographs* **48**: 95–110.

Convey, P. & Lewis Smith, R. I. (2006). Responses of terrestrial Antarctic ecosystems to climate change. *Plant Ecology* **182**: 1–10.

Crawford, R. M. M. (2008). *Plants at the Margin: Ecological Limits and Climate Change*. Cambridge: Cambridge University Press.

Davies, L., Bates, J. W., Bell, J. N. B., James, P. W. & Purvis, O. W. (2007). Diversity and sensitivity of epiphytes to oxides of nitrogen in London. *Environmental Pollution* **146**: 299–310.

de Bakker, A. J. (1989). Effects of ammonia emission on epiphytic lichen vegetation. *Acta Botanica Neerlandica* **38**: 337–42.

Dilks, T. J. K. & Proctor, M. C. F. (1975). Comparative experiments on temperature responses of bryophytes: assimilation, respiration and freezing damage. *Journal of Bryology* **8**: 317–36.

During, H. J. (1979). Life strategies of bryophytes: a preliminary review. *Lindbergia* **5**: 2–18.

Ellis, C. E., Coppins, B. J., Dawson, T. P. & Seaward, M. R. D. (2007). Response of British lichens to climate change scenarios: trends and uncertainties in the projected impact for contrasting biogeographic groups. *Biological Conservation* **140**: 217–35.

Farmer, A. M., Bates, J. W. & Bell, J. N. B. (1992). Ecophysiological effects of acid rain on bryophytes and lichens. In *Bryophytes and Lichens in a Changing Environment*, ed. J. W. Bates & A. M. Farmer, pp. 284–313. Oxford: Clarendon Press.

Ferguson, P. & Lee, J. A. (1983). Past and present sulphur pollution in the southern Pennines. *Atmospheric Environment* **6**: 1131–7.

Fitter, A. H. & Hay, R. K. M. (2002). *Environmental Physiology of Plants*, 3rd edn. London: Academic Press.

Fitter, A. H. & Fitter, R. S. R. (2002). Rapid change in flowering time in British plants. *Science* **296**: 1689–91.

Frahm, J.-P. & Klaus, D. (1997). Moose als Indikatoren von Klimafluktuationen in Mitteleuropa. *Erdkunde* **51**: 181–90.

Furness, S. B. & Grime, J. P. (1982). Growth rate and temperature responses in bryophytes. II. A comparative study of species of contrasted ecology. *Journal of Ecology* **70**: 525–36.

Gabriel, R. M. de A. (2000). Ecophysiology of Azorean forest bryophytes. Unpublished PhD thesis, Imperial College of Science, Technology & Medicine, London.

Gilbert, O. L. (1968). Bryophytes as indicators of air pollution in the Tyne Valley. *New Phytologist* **67**: 15–30.

Gilbert, O. L. (1970). Further studies on the effect of sulphur dioxide on lichens and bryophytes. *New Phytologist* **69**: 605–27.

Green, T. G. A. & Clayton-Green, K. A. (1981). Studies on *Dawsonia superba* Grev. II. Growth rate. *Journal of Bryology* **11**: 723–31.

Greven, H. C. (1992). *Changes in the Dutch Bryophyte Flora and Air Pollution*. Dissertationes Botanicae Band 194. Berlin & Stuttgart: J. Cramer.

Grime, J. P., Brown, V. K., Thompson, K. *et al.* (2000). The response of two contrasting limestone grasslands to simulated climate change. *Science* **289**: 762–5.

Grime, J. P., Fridley, J. D., Askew, A. P. *et al.* (2008). Long-term resistance to simulated climate change in an infertile grassland. *Proceedings of the National Academy of Sciences, USA* **105**: 10028–32.

Harte, J. & Shaw, R. (1995). Shifting dominance within a montane vegetation community: results of a climate-warming experiment. *Science* **267**: 876–80.

Hartmann, E. & Jenkins, G. I. (1984). Photomorphogenesis of mosses and liverworts. In *The Experimental Biology of Bryophytes*, ed. A. F. Dyer & J. G. Duckett, pp. 203–28. London: Academic Press.

Hill, M. O. & Preston, C. D. (1998). The geographical relationships of British and Irish bryophytes. *Journal of Bryology* **20**: 127–226.

Hill, M. O., Blackstock, T. H., Long, D. G. & Rothero, G. P. (2008). *A Checklist and Census Catalogue of British and Irish Bryophytes: Updated 2008*. Middlewich: British Bryological Society.

Hill, M. O., Preston, C. D., Bosanquet, S. D. S. & Roy, D. B. (2007). *BRYOATT: Attributes of British and Irish Mosses, Liverworts and Hornworts*. Huntingdon: Centre for Ecology and Hydrology.

Hill, M. O., Preston, C. D. & Smith, A. J. E. (1991). *Atlas of the Bryophytes of Britain and Ireland*. Volume 1. *Liverworts (Hepaticae and Anthocerotae)*. Colchester: Harley Books.

Hill, M. O., Preston, C. D. & Smith, A. J. E. (1992). *Atlas of the Bryophytes of Britain and Ireland*. Volume 2. *Mosses (except Diplolepideae)*. Colchester: Harley Books.

Hill, M. O., Preston, C. D. & Smith, A. J. E. (1994). *Atlas of the Bryophytes of Britain and Ireland*. Volume 3. *Mosses (Diplolepideae)*. Colchester: Harley Books.

Hopkins, J. (2007). British wildlife and climate change. 1. Evidence of change. *British Wildlife* **18**: 153–9.

Hulme, M., Jenkins, G. J., Lu, X. *et al.* (2002). *Climate Change Scenarios for the United Kingdom: the UKCIP02 Report*. Norwich: University of East Anglia.

Ingham, W. (ed.) (1907). *A Census Catalogue of British Mosses*. York: Moss Exchange Club.

Jenkins, G. J., Perry, M. C. & Prior, M. J. O. (2007). *The Climate of the United Kingdom and Recent Trends*. Exeter: Meteorological Office Hadley Centre.

Jones, E. W. (1991). The changing bryophyte flora of Oxfordshire. *Journal of Bryology* **16**: 513–49.

Knoop, B. (1984). Development in bryophytes. In *The Experimental Biology of Bryophytes*, ed. A. F. Dyer & J. G. Duckett, pp. 143–76. London: Academic Press.

Körner, C. (1999). *Alpine Plant Life. Functional Plant Ecology of High Mountain Ecosystems*. Berlin & Heidelberg: Springer-Verlag.

Lee, J. A., Caporn, S. J. M., Carroll, J. *et al.* (1998). Effects of ozone and atmospheric nitrogen deposition on bryophytes. In *Bryology for the Twenty-first Century*, ed. J. W. Bates, N. W. Ashton & J. G. Duckett, pp. 331–41. Leeds: Maney & The British Bryological Society.

Lee, J. A., Parsons, A. N. & Baxter, R. (1993). *Sphagnum* species and polluted environments, past and future. *Advances in Bryology* **5**: 297–313.

Lee, J. A. & Studholme, C. J. (1992). Responses of *Sphagnum* species to polluted environments. In *Bryophytes and Lichens in a Changing Environment*, ed. J. W. Bates & A. M. Farmer, pp. 314–32. Oxford: Clarendon Press.

Lösch, R., Kappen, L. & Wolf, A. (1983). Productivity and temperature biology of two snowbed bryophytes. *Polar Biology* **1**: 243–8.

Macvicar, S. M. (1905). *Census Catalogue of British Hepatics*. York: Moss Exchange Club.

Massara, A. C., Bates, J. W. & Bell, J. N. B. (2009). Exploring causes of the decline of the lichen *Lecanora conizaeoides* in Britain: effects of experimental N and S applications. *Lichenologist* **41**: 673–81.

Miles, C. J., Odu, E. A. & Longton, R. E. (1989). Phenological studies on British mosses. *Journal of Bryology* **15**: 607–21.

Monteith, D. T. & Evans, C. D. (2005). The United Kingdom Acid Waters Monitoring Network: a review of the first fifteen years and introduction to the special issue. *Environmental Pollution* **137**: 3–13.

O'Halloran, J., Gribbin, S. D., Tyler, S. J. & Ormrod, S. J. (1990). The ecology of Dipper (*Cinclus cinclus* L.) in relation to stream acidity in upland Wales: time-activity budget and energy expenditure. *Oecologia* **85**: 271–80.

Pearson, J., Wells, D. M., Seller, K. J. *et al.* (2000). Traffic exposure increases natural ^{15}N and heavy metal concentrations in mosses. *New Phytologist*, **147**: 317–26.

Peñuelas, J., Ogaya, R., Hunt, J. M. & Jump, A. S. (2008). Twentieth century changes of tree-ring δ13C at the southern range-edge of *Fagus sylvatica*: increasing water-use efficiency does not avoid the growth decline induced by warming at low altitudes. *Global Change Biology* **14**: 1076–88.

Peterken, G. F. (1993). *Woodland Conservation and Management*, 2nd edn. London: Chapman & Hall.

Pitkin, P. H. (1975). Variability and seasonality of the growth of some corticolous pleurocarpous mosses. *Journal of Bryology* **8**: 337–56.

Porley, R. & Hodgetts, N. (2005). *Mosses and Liverworts*. The New Naturalist Library 97. London: Collins.

Potter, J. A., Press, M. C., Callaghan, T. V. & Lee, J. A. (1995). Growth responses of *Polytrichum commune* and *Hylocomium splendens* to simulated environmental change in the sub-arctic. *New Phytologist* **131**: 533–41.

Preston, C. D. & Hill, M. O. (2009). Bryophyte records. *Nature in Cambridgeshire* **51**: 92–5.

Preston, C. D., Hill, M. O., Bosanquet, S. D. S. & Ames, S. L. (2009). Progress towards a new Atlas of Bryophytes. *Field Bryology* **98**: 14–20.

Preston, C. D., Pearman, D. A. & Dines, T. D. (eds.) (2002). *New Atlas of the British and Irish Flora*. Oxford: Oxford University Press.

Rose, F. (1992). Temperate forest management: its effects on bryophyte and lichen floras and habitats. In *Bryophytes and Lichens in a Changing Environment*, ed. J. W. Bates & A. M. Farmer, pp. 211–33. Oxford: Clarendon Press,

Tallis, J. H. (1964). Studies on the Southern Pennine peats. II. The behaviour of *Sphagnum*. *Journal of Ecology* **52**: 345–53.

Van Herk, C. M. (2001). Bark pH and susceptibility to toxic air pollutants as independent causes of changes in epiphytic lichen composition in space and time. *Lichenologist* **33**: 415–41.

Watson, E. V. (1985). The recording activities of the BBS (1923–83) and their impact on advancing knowledge. In *British Bryological Society Diamond Jubilee*, ed. R. E. Longton & A. R. Perry, pp. 17–29. Cardiff: British Bryological Society.

Winner, W. E. (1988). Responses of bryophytes to air pollution. In *Lichens, Bryophytes, and Air Quality*, ed. T. H. Nash III & V. Wirth, pp. 141–73. Berlin: J. Cramer.

Wookey, P. A., Parsons, A. N., Welker, J. M. *et al.* (1993). Comparative responses of phenology and reproductive development to simulated climate change in sub-arctic and high arctic plants. *Oikos* **67**: 490–502.

20

Climate Change and Protected Areas: How well do British Rare Bryophytes Fare?

BARBARA J. ANDERSON AND RALF OHLEMÜLLER

Introduction

Climate change is affecting biodiversity (Warren *et al.* 2001; Hickling *et al.* 2005; Root *et al.* 2005; Parmesan 2006; Rosenzweig *et al.* 2008), and research into appropriate mitigation and adaptation strategies is now recognized as being of highest priority (Mitchell *et al.* 2007; Hoegh-Guldberg *et al.* 2008). As with other taxa, climate change is likely to affect the physiology, population dynamics, and spatial distributions of bryophytes. Climate-induced range shifts have already been reported for many central European bryophyte species (e.g., Frahm & Klaus 1997). Climate is the ultimate driver of species distributions at large spatial (e.g., country) scales. Although some traits of bryophytes might make them less vulnerable to changes in temperature, many species are likely to be substantially affected by changes in humidity-related parameters (Gignac 2001; Bates *et al.* 2005).

Protected area networks are often established based on well-studied flagship species, and although this is mainly driven by increased computer power and available ecological data, there has been a recent shift towards multi-taxon reserve design (Early & Thomas 2007; Kremen *et al.* 2008; Franco *et al.* 2009). In addition, it is increasingly recognized that current conservation networks might not provide suitable protection for species in light of future climate change (e.g., Dockerty *et al.* 2003; Araújo *et al.* 2004; Pyke *et al.* 2005). An area under protective legislation today might be climatically suitable for an endangered species but the climatic conditions of that area might dramatically change in the future, making it unsuitable for this species. It is essential to at least assess and at best adjust current reserve design practices to incorporate future shifts in climate space (Araújo *et al.* 2004).

Bryophyte Ecology and Climate Change, eds. Zoltán Tuba, Nancy G. Slack and Lloyd R. Stark. Published by Cambridge University Press.

Here we focus on rare species from three overlapping conservation lists: (i) Biodiversity Action Plan (BAP) species (http://www.ukbap.org.uk/), (ii) Red List, (iii) Nationally Scarce List (http://www.britishbryologicalsociety.org.uk/). Both the BAP and Red List species enjoy some statutory protection and species may be listed because of population declines as well as rarity (IUCN 2001). The Nationally Scarce list includes all species with records from fewer than one hundred 10 km squares over Great Britain. We investigated the current conservation provision for species on each of the three lists separately.

Current conservation provision for rare bryophytes in England

We use the Sites of Special Scientific Interest (SSSI) network to investigate the current conservation for rare bryophytes within England based on our current knowledge of bryophyte distributions. We compare and contrast the English situation with that of Great Britain as a whole (England, Wales, and Scotland) and with Wales and Scotland separately. We investigate differences among three phyla: mosses, liverworts, and hornworts. Finally we investigated the spatial distribution of the conservation provision in England for bryophytes to assess whether there were regional differences in the conservation provision. Future work can build on these analyses to determine where the most important areas are for bryophyte conservation within England, and where the current SSSI network might be best augmented for bryophyte conservation especially in light of climate change.

Current and future capacity of protected areas in England to provide suitable climate space for bryophytes

Current climate conditions can explain large proportions of the variation in the geographical distributions of many species (e.g., Thuiller *et al.* 2004). Studies have shown that the current protected area network in several regions of the world will have to be expanded to compensate for likely future range shifts of species from climate change (e.g., Hannah *et al.* 2007). We investigate the overlap of modeled and projected future climatically suitable regions for bryophytes with the current protected area network.

Methods

Distribution data

All records lodged with the national database, the Biological Records Centre (BRC), were considered (see p. 11, Hill *et al.* 1991). Over the 456 species (12 taxa were omitted owing to complications of nomenclature) considered, there were 53 281 records including records from the Threatened Bryophytes Database (TBD). Approximately two thirds of these records were duplicate records for that location and species at the 100 km^2 resolution (different records, different dates, or different spatial locations at a finer resolution). These duplicate records were discarded. At the finer (hectare) resolution less than half the records were unique (3134 in England). We have equated no record with an absence. This means many species will be under-recorded, especially at finer (hectare) resolution, resulting in false absences. However, as we are dealing exclusively with rare species and their protection we felt it was prudent to make no assumptions about possible presences that are undocumented. The climate envelope models considered here are all presence/absence models (Thuiller *et al.* 2008) based on 100 km^2 (hectad) data that should reduce the false absence rate.

Current conservation provision for rare bryophytes in England

Based on existing distribution data at the finest resolution possible (hectare) we calculated the number and proportion of unique records for the bryophytes of England which fall within the Sites of Special Scientific Interest (SSSI) network for the focus species. Record locations were converted to OS grid references and centered in the middle of the hectare square. Rounding error on a hectare record can be up to 70 m (assuming no recording error). Therefore a 100 m buffer was applied to each point. They were then overlaid with the current SSSI network (ESRI vector layer Natural England 2007; http://www.magic.gov.uk/datadoc/metadata.asp?dataset=149). For each species the number and percentage of hectare records (including the 100 m buffer) that fell within an SSSI were calculated.

Climate modeling

We used a series of ecological niche models otherwise known as statistical or pattern-based models (Guisan & Zimmermann 2000). These models define a relationship between the response variable (in this case presence or absence of the bryophyte species within a 100 km^2 cell) and various predictor variables (in this case large-scale climatic variables that characterize the environment). Where the predictor variables are not directly or indirectly related to the distribution of the species the model will not fit well. Where other factors (e.g., land management, habitat, dispersal limitation, or even microclimate) have a larger influence on the distribution of a species than the predictor variables entered into the model (in this case large-scale climatic variables) the model is unlikely to fit well. Where the predictor variables are only indirectly related to the distribution of the species (e.g., when predictor variables are correlated with another unknown "hidden" variable, which is related to the distribution of the species) the models may fit well but will give unreliable predictions for geographical locations and times where this hidden correlation does not hold. This is a well-known limitation of these types of models. This is the same maxim that applies to all correlations, i.e., "Correlation does NOT prove causation."

There are many different types of statistical models (see e.g., Elith *et al.* 2006; Thuiller *et al.* 2008). The assumptions of these models vary considerably. Which model is "best" depends on many things including the species and the unknown underlying relationship between the species distribution and the predictor variables. All modeling was performed in R (version 2.6.1; R Development Core Team 2007). Nine different statistical methods were used: ANN, CTA, GAM, GBM, GLM, MARS, MDA, RF, SRE (implemented in BIOMOD; see Thuiller *et al.* 2008, pp. 43–54 and citations therein) to model the current distribution of the species in order to ensure that conclusions were robust to differences in the modeling method chosen (Elith *et al.* 2006). In order to achieve a more independent evaluation of models the data were split into a building (70% of the data) and a testing (30% of the data) set; care was taken to ensure that the prevalence (i.e., the proportion of presences) in the building and testing sets was constant. The goodness of fit of the models was assessed by using the AUC statistic (the area under the receiver operator characteristic curve) based on the testing data.

All the data were combined and predictions made using the best model for: (i) the reference period (1961–1990); and (ii) three time periods (1991–2020, 2021–50, 2051–80) for the HadCM3 global climate models (gcm) using the A2 storyline (this describes a heterogeneous future world focused on self-reliance, preservation of local identities, and slower and more regional economic development) (IPCC 2007). Thresholds for conversion (cut) of probability of occurrence values into binary presence/absence value based on the receiver–operator characteristic curve (ROC) were applied following the BinRoc procedure in Thuiller *et al.* (2008).

NAn indicates that there is no analogous climate in that time slice (30 year mean – average of a 100 km^2 cell) under that GCM and SRES climate projection (in this case HadCM3, A2), with "analogous" meaning analogous to the modeled climate space (based on the 1961–1990 climate and the GB distribution).

Hectad (100 km²) distribution data

All records for the focus species were converted to 100 km² OS grid references and duplicate records discarded. Small numbers of presence records per species cause many statistical modeling methods to become unstable, therefore only species with 50 or more unique 100 km² records (43 spp.) in England were modeled by using the nine statistical BIOMOD methods. Predictions of suitable climate space based on a small portion of the distribution of a species may seriously underestimate the species' climatic niche and generally a more complete climate envelope is preferred (Pearson *et al.* 2007). However, distribution data are not available for the entire range or even the European range of bryophytes.

Climate variables

The climate in each 100 km² grid cell was characterized by six climate variables: growing degree days at 5 °C (GDD5), mean temperature coldest month (MTCO), mean temperature warmest month (MTWA), ratio of actual to potential evapotranspiration (APET), summer precipitation (SPPT), and winter precipitation (WPPT). We used the average conditions of the last recognized normal period (1961–90) as our reference period assuming that most bryophyte records were obtained since 1961.

Proportion of protected area in 100 km²

At the 100 km² resolution all protected areas with some degree of statutory protection for biodiversity (i.e., SSSI, SPA, SAC, RAMSAR, NNR, but not National Parks and Areas of Outstanding Natural Beauty) were amalgamated (Gaston *et al.* 2006). The proportion of land in each 100 km² cell that is protected was then calculated (Eigenbrod *et al.* 2009).

Overlap with protected areas

Based on the best bioclimate model we determined for each of the 43 species, respectively, which grid cells in the reference period (1961–1990) had climatic conditions considered suitable or analogous to those of the grid cells in which the species is found. We then calculated for each species the average proportion of protected area in the grid cell(s) identified as suitable. A high average proportion indicates that areas which are climatically suitable for the species have a high proportion of protection.

Results and discussion

Current conservation provision for rare bryophytes in England

Although the median number of hectare records per species was low (Table 20.1), within England half of the species had at least 62% of their hectare records within 100 m of a SSSI boundary, which is considerably less than in Scotland, Wales, and GB as a whole. The median for the three conservation lists was similar, with the Nationally Scarce list slightly lower. The percentage of mosses found within 100 m of a SSSI was slightly lower than that of liverworts.

The SSSIs and the hectare bryophyte records within England are not evenly distributed (Fig. 20.1). Although there are concentrations of both, they do not completely overlap (Fig. 20.1b). For instance, the concentration of SSSIs in North Yorkshire (the Yorkshire Moors) is reasonably high; however, there are few hectare records there (Fig. 20.1). In West Yorkshire the congruence of records and SSSIs is even lower with a large concentration of SSSIs within the Peak District and Yorkshire Dales where there are few records, and few SSSIs to the east of this where there are many records. These discrepancies indicate both where more sampling (including the recording of absences) is required and also regions where bryophytes may be less than adequately protected by the current SSSI network.

Table 20.1. *Median and maximum number of hectare records by category and median number and percentage of hectare records within 100 m of an SSSI (Sites of Special Scientific Interest) boundary*

Median omits zeros, i.e., species for which there are no hectare records were excluded.
GB. Great Britain; BAP, Biodiversity Action Plan; NS, species on the Nationally Scarce List.

Country	No. of species	Median no. of hectare records	Maximum no. of hectare records	Median no. of records 100 m buffer in SSSI	Median % records 100 m buffer in SSSI
England	274	6	103	3	62
Wales	201	4	132	3	75
Scotland	324	8	79	6	79
GB	433	14	175	9	73
BAP	77	6	94	3	62
Red List	112	7	103	4	61
NS	174	4	57	3	58
Hornwort	1	3	3	0	0
Liverwort	73	6	94	3	66
Moss	200	5	103	3	60

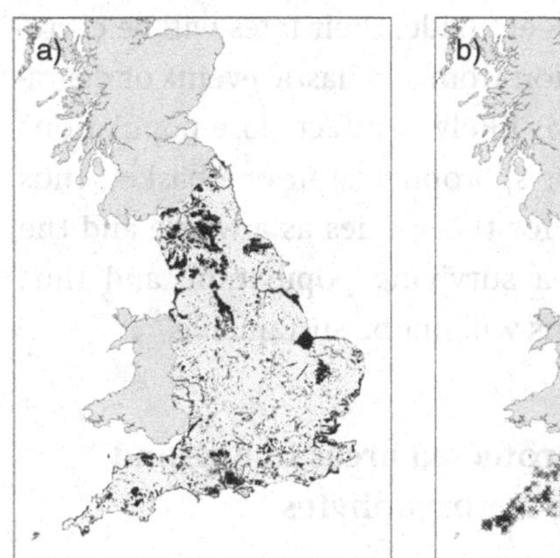

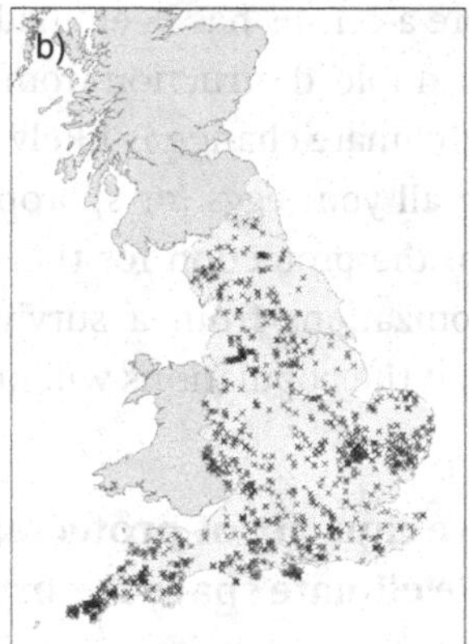

Fig. 20.1. Distributions of (a) SSSIs (Sites of Special Scientific Interest) in England; (b) hectare bryophyte records with 100 m buffer that do NOT overlap with a SSSI boundary; (c) hectare bryophyte records with a 100 m buffer that overlap with a SSSI boundary. (b) and (c) together show the sum of all hectare records within England. Note 1. Although there are concentrations of both SSSIs (a) and of hectare bryophyte records (b and c combined) these do not always overlap (b). Comparison of (b) and (c) indicates regions where there are a lot of records but not many of them are in SSSIs (e.g., southern Lancashire).

The number of records per species varies widely from the minimum of one (species with no hectare records were excluded from this analysis) to a maximum of 175 records (for Great Britain). However, the number of records was very left skewed, with most species having very few records; the median for England was three records. Although the species investigated here are genuinely rare, there is likely to be a considerable amount of under-recording at this fine resolution so these results should be interpreted with caution. The percentage of bryophyte records within 100 m of a SSSI was lower in England than for either Wales or Scotland or Great Britain as a whole (Table 20.1). However, the land area that is SSSI within England is also lower (7% compared with 12% and 13% for Wales and Scotland). In addition there may be: (i) differences in the nature of SSSIs within the three countries (with those in England tending to be designated for species other than bryophytes and/or smaller and therefore less likely to catch bryophytes by accident); (ii) differences in the recording nature of volunteers in the three countries (with those in Wales and Scotland tending to visit SSSIs more often) or (iii) a combination of both.

Care should be taken in interpreting these results. At this fine resolution under-recording combined with clustering of hectare records (spatial autocorrelation) is likely to bias results. Not all records for the same species will be independent populations (e.g., adjacent hectare records) and where many records for the same species are recorded in close proximity within a single SSSI or even neighboring SSSIs (e.g., where a census has been made) their fates will be correlated. This means that, for example, destruction from stochastic events or degradation due to habitat loss or climate change is likely to affect close populations equally – effectively putting all your eggs (or sporophytes) in one basket, thus reducing the effectiveness of the protection for the species as a whole and the likelihood of chance recolonizations from a surviving population, and thus increasing the probability that the populations will not be sustainable.

Current and future capacity of protected areas in England to provide suitable climate space for bryophytes

As expected, different statistical methods were able to model the current distribution of the species with varying success (Table 20.2). The distribution of some species (e.g., *Campyliadelphus elodes, Philonotis arnellii*) was consistently poorly predicted by almost all methods, whereas the distributions of other species (e.g., *Ephemerum recurvifolium, Herzogiella seligeri*) were consistently well predicted by most methods. The average proportion of protected area in grid cells modeled as climatically suitable by the bioclimate models ranged from 2.7% for *Leptobarbula berica* to 19.2% for *Didymodon acutus* (Table 20.2). Across all 43 species, an average of 8.9% of their climatically suitable grid cells is designated as protected area. Species with a high proportion of protected area in grid cells that are climatically suitable for them are generally found at higher elevations with two notable exceptions: the two species with the highest proportion of protected area (*Didymodon acutus* and *Campyliadelphus elodes*) do have climatically suitable areas at rather low elevations (mean elevation 80 m and 121 m, respectively; Table 20.2).

In the 1991–2020 period the average proportion of protected area in climatically suitable grid cells ranged from 0.7% for *Leptobarbula berica* to 19.5% for *Pylaisia polyantha* (Table 20.3). The overall average proportion of protected area for the suitable climate space of all species is projected to increase from 8.9% in 1991–2020, to 9.2% in 2021–2050, and 10.2% in 2051–2080. This indicates that on average, the climate space currently occupied by these 43 British bryophyte species will overlap marginally more with protected areas with increasing climate change over the next 80 years. Suitable climate space of these species will on average move upwards and further north and 100 km^2 cells at both higher elevations and higher latitudes within Great Britain tend to have a higher proportion of protected area. This suggests that many species would need to migrate north and upslope in order to find analogous climate space (Table 20.3). This can already be observed in a number of southern European species moving into central Europe (Frahm & Klaus 1997). Although this might seem a positive indication it should be borne in mind that these species, e.g., from lowland

Table 20.2. *Summary of bryophyte characteristics, and recent climate space models*

Phyla (M, moss; L, liverwort), Order (*abbreviated*), and biogeographic element (BGE) are taken from BRYOATT (Hill *et al.* 1991). Results for best fitting of the nine BioMod methods used (model). Goodness of fit measured by test AUC value (AUC): values below 0.50 are considered worse than random; values below 0.70 are considered poor; between 0.70 and 0.79, fair; between 0.80 and 0.89, good; between 0.90 and 1.0, excellent. "Cut" is the threshold used to convert the continuous "probability of occurrence" values into binary presence/absence values. Mean percentage of protected area (PA), mean elevation (Elev.), mean latitude (Lat.), and mean longitude (Long.) of grid cells modeled as being climatically suitable in the reference period (1961–1990).

Species	Phyla	Order	BGE	Model	AUC	Cut	PA (%)	Elev. (m)	Lat. [°N]	Long. (°E)
Aloina ambigua	M	Pott	83	RF	0.815	0.619	6.0	66	52.32	− 1.41
Aloina rigida	M	Pott	43	GBM	0.795	0.048	3.8	79	53.01	− 1.16
Amblystegium humile	M	Hypn	76	GAM	0.844	0.050	4.4	61	52.12	− 0.75
Brachydontium trichodes	M	Grim	72	GAM	0.827	0.051	15.1	310	54.78	− 3.21
Brachythecium salebrosum	M	Hypn	36	GLM	0.796	0.082	4.0	65	52.30	− 0.63
Bryum intermedium	M	Brya	75	RF	0.715	0.692	11.1	98	53.22	− 2.01
Bryum torquescens	M	Brya	91	GAM	0.828	0.026	7.8	59	51.60	− 2.03
Calypogeia integristipula	L	Jung	56	GBM	0.724	0.045	13.7	267	54.48	− 2.42
Campyliadelphus elodes	M	Hypn	73	RF	0.706	0.694	17.5	121	53.11	− 2.34
Campylophyllum calcareum	M	Hypn	73	GAM	0.829	0.065	9.6	133	51.97	− 3.44
Cephalozia macrostachya	L	Jung	42	RF	0.712	0.671	12.3	195	52.91	− 2.79
Cephaloziella stellulifera	L	Jung	92	GLM	0.831	0.029	12.3	176	53.04	− 2.89
Cladopodiella francisci	L	Jung	43	GLM	0.711	0.037	10.0	162	54.71	− 3.35
Cololejeunea rossettiana	L	Pore	92	GBM	0.859	0.040	8.7	156	53.40	− 1.70
Dicranella crispa	M	Dicr	26	MDA	0.673	0.017	6.4	76	51.99	− 1.61
Dicranum spurium	M	Dicr	43	MDA	0.819	0.033	9.3	180	53.96	− 2.64
Didymodon acutus	M	Pott	86	GLM	0.804	0.051	19.2	80	53.72	− 2.89
Discelium nudum	M	Funa	46	GLM	0.919	0.026	4.8	76	51.85	− 0.91
Drepanocladus sendtneri	M	Hypn	26	RF	0.623	0.737	8.1	92	52.01	− 2.72
Ephemerum recurvifolium	M	Pott	83	RF	0.874	0.638	9.8	191	54.33	− 3.47
Grimmia orbicularis	M	Grim	92	GBM	0.764	0.045	5.1	71	51.79	− 0.30
Hamatocaulis vernicosus	M	Hypn	46	GAM	0.812	0.051	9.9	176	53.47	− 1.98
Herzogiella seligeri	M	Hypn	53	GLM	0.884	0.045	5.9	71	51.64	− 0.97

Table 20.2. (*cont.*)

Species	Phyla	Order	BGE	Model	AUC	Cut	PA (%)	Elev. (m)	Lat. [°N]	Long. (°E)
Hypnum imponens	M	Hypn	73	GLM	0.769	0.027	17.5	253	56.06	−4.23
Leptobarbula berica	M	Pott	91	GLM	0.864	0.048	2.7	83	52.28	−1.74
Nardia geoscyphus	L	Jung	26	ANN	0.768	0.068	10.0	176	53.20	−3.31
Octodiceras fontanum	M	Dicr	73	GBM	0.887	0.057	15.5	230	55.42	−4.39
Orthotrichum striatum	M	Orth	53	GBM	0.765	0.153	10.1	167	53.61	−3.45
Philonotis arnellii	M	Brya	73	CTA	0.604	0.099	4.2	90	52.12	−1.10
Philonotis caespitosa	M	Brya	56	MDA	0.726	0.035	8.3	82	51.55	−2.59
Platygyrium repens	M	Hypn	76	RF	0.901	0.680	7.7	127	52.41	−2.34
Pleurochaete squarrosa	M	Pott	92	GLM	0.893	0.049	4.7	60	52.83	−1.03
Pohlia lescuriana	M	Brya	74	GAM	0.706	0.042	6.2	134	53.28	−2.21
Pterygoneurum ovatum	M	Pott	86	GBM	0.831	0.085	13.1	157	55.05	−4.23
Pylaisia polyantha	M	Hypn	76	GAM	0.694	0.047	10.6	135	54.35	−2.87
Rhynchostegiella curviseta	M	Hypn	91	GAM	0.920	0.061	5.4	71	51.73	−0.99
Riccia cavernosa	L	Ricc	76	RF	0.754	0.676	10.6	80	52.83	−1.95
Ricciocarpos natans	L	Ricc	86	GLM	0.903	0.058	3.7	52	52.48	−0.52
Seligeria pusilla	M	Grim	53	GAM	0.783	0.053	15.0	256	53.83	−2.85
Sphaerocarpos michelii	L	Sphae	91	GBM	0.918	0.030	5.4	42	52.08	−0.24
Tortella inflexa	M	Pott	91	GLM	0.968	0.045	7.0	84	51.46	−0.87
Tortula atrovirens	M	Pott	86	ANN	0.929	0.043	9.2	54	51.82	−3.38
Weissia squarrosa	M	Pott	72	GAM	0.846	0.036	3.2	77	52.19	−1.57

England, would need to migrate considerable distances both north and upslope to find suitable habitats, substrates etc. in these places.

Species that occupy a smaller proportion of the available analogous climate space are likely to have distributions that are more affected by other factors (e.g., land management, habitat, substrate, microclimate, or an interaction between any of these factors and climate). This does not preclude climate as a significant limiting factor in the distribution of these species but suggests that other factors are also important. Investigating the exact nature of these interactions and which factors play what role is an important area for future work.

The overlap between current distribution and future projected analogous climate decreases gradually over the three time slices for most species (Fig. 20.2a). For instance, the median overlap of all species is projected to be *c.* 21% for 1991–2020, *c.* 15% for 2021–2050, and *c.*10% for 2051–2080 (Fig. 20.2a). The minimum and first quartile (lower whisker and lower edge of box, respectively) for all three time slices is zero, indicating that at least a quarter of all species have

Table 20.3. *Mean percentage of protected area (PA), mean elevation (Elev), mean latitude (Lat) and mean longitude (Long) of grid cells modeled as being climatically suitable in each of the three future periods (1991–2020, 2021–2050, 2051–2080) by bioclimate modeling of 43 species for which a sufficient number of records was available*

Results are presented for best fitting method from Table 20.2. Life form, Order, and biogeographic element of the species (not shown) are the same as for Table 20.2. NAn indicates that no grid cells were climatically analogous to those from the recent modeled distribution (see text for caveats).

	1991–2020				2021–2050				2051–2080			
Species	**PA**	**Elev**	**Lat**	**Long**	**PA**	**Elev**	**Lat**	**Long**	**PA**	**Elev**	**Lat**	**Long**
Aloina ambigua	NAn	NAn	NAn	NAn	0.3	100.5	52.1	0.4	NAn	NAn	NAn	NAn
Aloina rigida	5.9	85.0	51.7	− 1.4	8.1	91.8	51.7	− 2.4	9.2	108.7	52.1	− 2.5
Amblystegium humile	4.3	43.4	52.3	0.2	NAn	NAn	NAn	NAn	NAn	NAn	NAn	NAn
Brachydontium trichodes	11.0	170.1	54.1	− 3.3	1.2	101.8	54.6	− 3.5	8.3	123.1	54.8	− 3.6
Brachythecium salebrosum	5.3	114.5	53.0	− 1.7	10.8	214.6	54.0	− 2.3	15.2	269.3	54.3	− 2.7
Bryum intermedium	NAn	NAn	NAn	NAn	NAn	NAn	NAn	NAn	NAn	NAn	NAn	NAn
Bryum torquescens	6.5	84.7	52.2	− 1.7	6.9	99.2	52.7	− 2.0	8.5	116.2	53.5	− 2.4
Calypogeia integristipula	8.9	228.8	56.2	− 3.7	NAn	NAn	NAn	NAn	NAn	NAn	NAn	NAn
Campyliadelphus elodes	NAn	NAn	NAn	NAn	NAn	NAn	NAn	NAn	NAn	NAn	NAn	NAn
Campylophyllum calcareum	NAn	NAn	NAn	NAn	NAn	NAn	NAn	NAn	14.6	144.8	51.5	− 1.8
Cephalozia macrostachya	NAn	NAn	NAn	NAn	NAn	NAn	NAn	NAn	NAn	NAn	NAn	NAn
Cephaloziella stellulifera	11.6	83.7	53.4	− 3.7	12.6	116.0	54.3	− 3.9	13.4	124.9	55.1	− 3.9
Cladopodiella francisci	15.0	236.6	54.6	− 3.5	7.2	123.0	52.5	− 2.0	4.9	97.3	51.9	− 1.7
Cololejeunea rossettiana	6.6	89.5	52.0	− 2.3	9.9	120.0	52.7	− 2.9	9.0	121.7	53.0	− 2.5
Dicranella crispa	8.3	127.9	53.9	− 2.5	11.1	141.8	54.6	− 3.8	11.6	171.5	54.5	− 4.1
Dicranum spurium	9.4	112.8	54.1	− 2.8	13.6	168.9	55.5	− 3.1	16.8	176.2	55.2	− 3.1
Didymodon acutus	6.5	76.4	52.0	− 2.4	8.5	56.4	52.4	− 2.5	8.3	131.9	54.8	− 3.6
Discelium nudum	6.7	106.9	52.8	− 2.3	8.2	92.0	53.1	− 2.1	10.5	152.6	54.2	− 3.2
Drepanocladus sendtneri	NAn	NAn	NAn	NAn	NAn	NAn	NAn	NAn	NAn	NAn	NAn	NAn

Table 20.3. (*cont.*)

	1991–2020				2021–2050				2051–2080			
Species	**PA**	**Elev**	**Lat**	**Long**	**PA**	**Elev**	**Lat**	**Long**	**PA**	**Elev**	**Lat**	**Long**
Ephemerum recurvifolium	NAn	NAn	NAn	NAn	NAn	NAn	NAn	NAn	NAn	NAn	NAn	NAn
Grimmia orbicularis	8.4	115.1	52.7	− 2.6	9.3	130.1	53.7	− 2.8	10.5	136.2	54.0	− 2.9
Hamatocaulis vernicosus	13.6	181.6	55.0	− 4.1	14.7	199.7	55.9	− 4.3	17.1	228.0	56.3	− 4.5
Herzogiella seligeri	NAn	NAn	NAn	NAn	NAn	NAn	NAn	NAn	NAn	NAn	NAn	NAn
Hypnum imponens	9.1	263.1	56.3	− 4.9	NAn	NAn	NAn	NAn	6.1	184.7	55.5	− 4.6
Leptobarbula berica	0.7	206.1	51.6	− 3.3	8.6	268.2	52.8	− 2.8	6.0	280.4	52.9	− 2.9
Nardia geoscyphus	15.3	183.3	56.3	− 4.4	15.9	193.9	56.3	− 4.7	15.6	190.9	56.4	− 4.9
Octodiceras fontanum	4.3	78.8	52.3	− 1.1	5.7	95.6	52.4	− 2.0	6.9	110.9	52.7	− 2.5
Orthotrichum striatum	10.5	142.2	53.5	− 3.4	11.3	134.1	54.0	− 3.8	12.4	146.4	54.3	− 3.9
Philonotis arnellii	13.7	169.9	55.1	− 4.1	14.2	168.3	55.4	− 4.3	14.0	165.6	55.7	− 4.2
Philonotis caespitosa	8.7	126.0	53.4	− 2.7	10.6	167.9	54.2	− 3.2	12.5	165.5	54.8	− 3.6
Platygyrium repens	NAn	NAn	NAn	NAn	NAn	NAn	NAn	NAn	NAn	NAn	NAn	NAn
Pleurochaete squarrosa	10.7	56.9	51.5	− 3.2	11.8	41.8	50.8	− 3.6	8.6	71.4	52.9	− 3.5
Pohlia lescuriana	10.0	121.6	51.8	− 2.5	7.7	100.8	52.2	− 2.2	7.4	102.6	52.5	− 2.1
Pterygoneurum ovatum	3.6	87.7	52.8	− 1.3	4.5	112.6	53.8	− 2.0	6.8	131.4	55.0	− 2.3
Pylaisia polyantha	19.5	336.5	55.2	− 2.6	12.5	316.2	55.4	− 3.3	NAn	NAn	NAn	NAn
Rhynchostegiella curviseta	NAn	NAn	NAn	NAn	12.4	396.6	51.8	− 3.5	NAn	NAn	NAn	NAn
Riccia cavernosa	NAn	NAn	NAn	NAn	NAn	NAn	NAn	NAn	NAn	NAn	NAn	NAn
Ricciocarpos natans	5.6	90.0	52.6	− 1.7	6.2	113.2	53.7	− 1.8	8.5	162.8	55.1	− 2.7
Seligeria pusilla	9.0	165.3	53.9	− 2.9	NAn	NAn	NAn	NAn	7.8	94.1	52.0	− 2.9
Sphaerocarpos michelii	9.4	53.0	51.2	− 1.7	7.1	66.5	51.7	− 1.6	6.1	73.5	52.1	− 1.7
Tortella inflexa	NAn	NAn	NAn	NAn	NAn	NAn	NAn	NAn	NAn	NAn	NAn	NAn
Tortula atrovirens	10.5	82.2	52.9	− 3.0	8.4	86.1	53.4	− 2.4	9.4	105.9	54.1	− 2.6
Weissia squarrosa	6.5	92.9	52.4	− 1.8	7.2	114.5	53.0	− 2.3	8.6	136.2	53.4	− 2.8

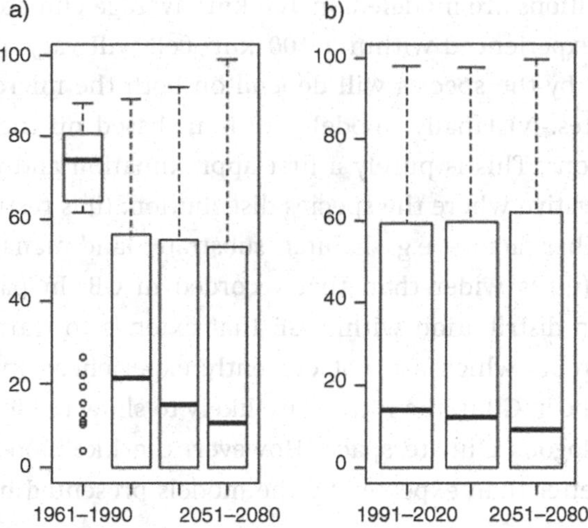

Fig. 20.2. Mean overlap between: (a) occupied 100 km^2 cells and climatically suitable squares for 1961–1990, 1991–2020, 2021–2050, and 2051–2080; (b) climatically suitable squares in the reference period (1961–1990) and climatically suitable squares for 1991–2020, 2021–2050, and 2051–2080. Solid horizontal lines within the box represent the median over all species considered; upper and lower edges of boxes represent the interquartile range with 50% of species falling within the box; dashed lines and caps represent the range, and unfilled circles outliers.

NO overlapping distribution with future analogous climate space. Comparing the overlap between the climatically suitable cells in the reference period with that of the future projected analogous areas follows the same trend (Fig. 20.2b). As the reduction in overlapping climate space increases, more individuals and populations of a species will experience climate conditions outside the conditions in which they currently occur within Great Britain. This poses a potential threat to these species if they are unable to (i) adapt to the new conditions or (ii) migrate or disperse to new areas where conditions are still suitable.

Projections of future climatically suitable space always carry a large amount of uncertainty, and those that form the basis of this analysis are no exception. For instance: (i) Climate projections vary and in any case are only projections with all the uncertainties inherent in both the climate model and SRES scenario. (ii) Species do not experience average conditions and year to year variation in climate may cause species to remain longer at a site, colonize new sites sooner, or become locally extinct sooner than expected. (iii) Species distributions should not be expected to instantaneously reflect the changing climatic conditions; lags should be expected where populations gradually decline, are recolonized through a rescue effect or remain for the lifetime of individuals. (iv) Distributions are modeled on 100 km^2 average climate values. In reality the climate experienced within a 100 km^2 cell will vary and local conditions experienced by the species will depend on both the microclimate and habitat at local sites. (v) Finally, models are built based on the known Great Britain distributions. This is purely a first approximation and the relationship will be conservative where the species distribution: (i) is restricted or has been reduced by other factors, e.g., habitat, substrate, land management; (ii) is under-recorded; (iii) is wider than that recorded in GB. In particular, species with a southern distribution within GB that extends to warmer and drier conditions in Europe, which are not currently experienced in GB but which are projected to be in GB in the future, are likely to show reduced, or in extreme cases no, analogous climate space. However, the likelihood is that these species will do better than expected by the models presented here.

Conclusions

The proportion of hectare records within 100 m of an SSSI is generally low for all the regions, phyla, and conservation lists we investigated. This may be due to myriad different factors. Recording absences as well as presences would help distinguish between these factors and provide valuable information. An inventory of the SSSIs for bryophyte presence and absence would help establish which SSSIs were important for the current distribution of the species and which species, even with a complete inventory of the SSSI network, would still be largely outside the protected areas.

Most of the 43 species for which bioclimatic models were built had a relatively small overlap between their current distribution (100 km^2 cells) and future projected suitable climate space. Species with an already restricted range (most species considered here), and a low overlap with future climatically suitable space should be evaluated as to their likely ability to adapt to new conditions or migrate to new areas. Note that species with a more southern distribution whose northern range margin currently lies within Great Britain may have little or no analogous climate space under future scenarios based on their current GB distribution. However, where non-analogous climates match the climate elsewhere in the species' distribution the species may be unaffected; where the new non-analogous climate is more similar to conditions at the core of the species distribution they may benefit. Note, however, that even these species are likely to experience changes in competitive interactions and disturbances due to climate change. A species by species evaluation considering the European distribution of species, the current range, habitat specificity, and probability of dispersal is required. Where species currently occupy the only suitable habitat within Great Britain the diminishing area with suitable climate space coinciding with this suitable habitat will pose a serious threat to their continued existence in Great Britain (Travis 2003).

Acknowledgements

The overlay of hectare records with the SSSI network was done by Corrado Topi, with thanks. The protected areas layer from which we took the proportion of protected area in each 100 km^2 grid cell was collated originally by Sarah Jackson (UKPopNet, University of Sheffield) and the proportion within each 100 km^2 grid cell was calculated originally by Mark Parnell and Felix Eigenbrod (UKPopNet, University of Sheffield) under the supervision of Kevin Gaston as part of UKPopNet project: *Linking biodiversity and ecosystem services: processes, priorities and prospects*. The BRC bryophyte data were collated and provided by Mark O. Hill and Chris D. Preston. We would like to thank Mark O. Hill, Ron Porley, Chris D. Preston, Nancy Slack, and Chris D. Thomas for much valuable discussion. This chapter is based on a Natural England report (SAE03–02–106) and is a partial product of UKPopNet project: *Linking biodiversity and ecosystem services: processes, priorities and prospects* (NERC R8-H12–01 and English Nature). Finally we would like to thank all the many dedicated recorders without whom none of these analyses would be possible.

References

Araújo, M. B., Cabeza, M., Thuiller, W., Hannah, L. & Williams, P. H. (2004). Would climate change drive species out of reserves? An assessment of existing reserve-selection methods. *Global Change Biology* **10**: 1618–26.

Bates, J. W., Thompson, K. & Grime, J. P. (2005). Effects of simulated long-term climatic change on the bryophytes of a limestone grassland community. *Global Change Biology* **11**: 757–69.

Dockerty, T., Lovett, A. & Watkinson, A. (2003). Climate change and nature reserves: examining the potential impacts, with examples from Great Britain. *Global Environmental Change* **13**: 125–35.

Early, R. & Thomas, C. D. (2007). Multispecies conservation planning: identifying landscapes for the conservation of viable populations using local and continental species priorities. *Journal of Applied Ecology* **44**: 253–62.

Eigenbrod, F., Anderson, B. J., Armsworth, P. R. *et al.* (2009). Ecosystem service benefits of contrasting conservation strategies in a human-dominated region. *Proceedings of the Royal Society of London B* **276**: 2903–11.

Elith, J., Graham, C. H., Anderson, R. P. *et al.* (2006). Novel methods improve prediction of species' distributions from occurrence data. *Ecography* **29**: 129–51.

Frahm, J. P. & Klaus, D. (1997). Moose als Indikatoren von Klimafluktuationen in Mitteleuropa. *Erdkunde* **51**: 181–90.

Franco, A. M. A., Anderson, B. J., Roy, D. B. *et al.* (2009). Surrogacy and persistence in reserve selection: landscape prioritization for multiple taxa in Britain. *Journal of Applied Ecology* **46**: 82–91.

Gaston, K. J., Charman, K., Jackson, S. F. *et al.* (2006). The ecological effectiveness of protected areas: the United Kingdom. *Biological Conservation* **132**: 76–87.

Gignac, L. D. (2001). Bryophytes as indicators of climate change. *Bryologist* **104**: 410–20.

Guisan, A. & Zimmermann, N. E. (2000). Predictive habitat distribution models in ecology. *Ecological Modelling* **135**: 147–86.

Hannah, L., Midgley, G., Andelman, S. *et al.* (2007). Protected area needs in a changing climate. *Frontiers in Ecology and the Environment* **5**: 131–8.

Hickling, R., Roy, D., Hill, J. K. & Thomas, C. D. (2005). A northward shift of range margins in British Odonata. *Global Change Biology* **11**: 502–6.

Hill, M. O., Preston, C. D. & Smith, A. J. E. (1991). *Atlas of the Bryophytes of Britain and Ireland: Liverworts (Hepaticae and Anthocerotae)*, vol. 1. Colchester: Harley Books.

Hoegh-Guldberg, O., Hughes, L., McIntyre, S. *et al.* (2008). Assisted colonization and rapid climate change. *Science* **321**: 345–6.

IPCC (2007). *Climate Change 2007: The Physical Science Basis. Contribution of Working Group I to the Fourth Assessment Report of the Intergovernmental Panel on Climate Change.* Cambridge and New York: Cambridge University Press.

IUCN (2001). *IUCN Red List Categories and Criteria: Version 3.1.* Gland, Switzerland and Cambridge, UK: IUCN.

Kremen, C., Cameron, A., Moilanen, A. *et al.* (2008). Aligning conservation priorities across taxa in Madagascar with high-resolution planning tools. *Science* **320**: 222–6.

Mitchell, R. J., Morecroft, M. D., Acreman, M. *et al.* (2007). *England Biodiversity Strategy – towards adaptation to climate change.* Final report to Defra for contract CRO327. 194 pp.

Parmesan, C. (2006). Ecological and evolutionary responses to recent climate change. *Annual Review of Ecology, Evolution and Systematics* **37**: 637–69.

Pearson, R. G., Raxworthy, C. J., Nakamura, M. & Peterson, A. T. (2007). Predicting species' distributions from small numbers of occurrence records: a test case using cryptic geckos in Madagascar. *Journal of Biogeography* **34**: 102–17.

Pyke, C. R., Andelman, S. J. & Midgley, G. (2005). Identifying priority areas for bioclimatic representation under climate change: a case study for Proteaceae in the Cape Floristic Region, South Africa. *Biological Conservation* **125**: 1–9.

Root, T. L., MacMynowski, D. P., Mastrandrea, M. D. & Schneider, S. H. (2005). Human-modified temperatures induce species changes: joint attribution. *Proceedings of the National Academy of Sciences of the United States of America* **102**: 7465–9.

Rosenzweig, C., Karoly, D., Vicarelli, M. *et al.* (2008). Attributing physical and biological impacts to anthropogenic climate change. *Nature* **453**: 353–7.

Thuiller, W., Araujo, M. B. & Lavorel, S. (2004). Do we need land-cover data to model species distributions in Europe? *Journal of Biogeography* **31**: 353–61.

Thuiller, W., Lafourcade, B., Engler, R. & Araujo, M. B. (2008). BIOMOD – A platform for ensemble forecasting of species distributions. *Ecography* **32**: 369–73. doi: 10.1111/j.1600-0587.2008.05742.x

Travis, J. M. J. (2003). Climate change and habitat destruction: a deadly anthropogenic cocktail. *Proceedings of the Royal Society of London B* **270**: 467–73.

Warren, M. S., Hill, J. K., Thomas, J. A. *et al.* (2001). Rapid responses of British butterflies to opposing forces of climate and habitat change. *Nature* **414**: 65–9.

21

Modeling the Distribution of *Sematophyllum substrumulosum* (Hampe) E. Britton as a Signal of Climatic Changes in Europe

CECÍLIA SÉRGIO, RUI FIGUEIRA AND RUI MENEZES

Introduction

Attention to climate change has significantly increased in the past 20 years, both on global and on regional scales. A great deal of research has been carried out relative to global warming based on alteration of species distributions. Examples are a study supported by a large number of African vascular plant species (McClean *et al.* 2005), another using amphibian and reptile distribution (Araújo *et al.* 2006), and also, on a European scale, diadromous fish distribution (Lassalle *et al.* 2008). In a more narrow range, we can cite research using alpine plants in the Swiss Alps (Guisan & Theurillat 2000), on the effects on rare lichens in the UK (Binder & Ellis 2008), or identifying the dynamic in snowbed bryophytes related to the duration of snow-lie in Scotland (Woolgrove & Woodin 1994).

Bryophytes are important ecologically; they constitute an important component of biodiversity and are recognized as keystone species of ecosystem monitoring. Many bryophyte species have been adversely affected by human activities, principally because of deterioration in essential habitats (Bates & Farmer 1992) or water quality (Vanderpoorten & Klein 1999), as well as increased nitrate (Lee *et al.* 1998) and air pollution (Zechmeister *et al.* 2007). These adverse effects have been widely cited as to why some bryophytes are now considered endangered.

Bryophyte species have, in general, tolerance to wide ranges of temperature. This attribute is largely due to their water relations, since they can survive better

Bryophyte Ecology and Climate Change, eds. Zoltán Tuba, Nancy G. Slack and Lloyd R. Stark. Published by Cambridge University Press.

at higher temperature extremes when dry than when wet. However, as in other organisms, the optimum temperature range of bryophytes is reflected in their geographic distribution, their altitudinal preferences, or climatic characteristics. Various additional factors are also important, but regions of similar latitude that have comparable annual average temperatures often have similar bryophyte flora components. In fact, all major vegetation zones are essentially correlated with latitude, mean annual temperature, and precipitation. Furthermore, in oceanic areas with high atmospheric humidity over the year, this is an essential factor determining the distribution of Atlantic bryophytes in the UK (Porley & Hodgetts 2005).

There is ample evidence that geographical ranges of particular habitats sensitive to temperature and water availability tend to shift to higher elevations or different latitudes owing to climate warming, for example *Sphagnum*-dominated bogs (Gignac 2001). Nevertheless, little is known about the impact of global warming on single bryophyte species.

For lichen epiphytes, van Herk *et al.* (2002) produced the first evidence for redistribution of some species of the Netherlands due to climate change over a period of about 25 years. According to this last study some recent changes in diversity are correlated significantly with a measured increase in air temperature. Other research aims to monitor species diversity in epiphyte communities in relation to changing climate (Ellis & Coppins 2007). The future diversity composition of this important woodland community, mainly composed of lichens and bryophytes, may depend upon the interaction of contemporary and historic habitat structure, but is also allied to more recent climatic warming. Therefore the continued monitoring of bryophyte distribution trends serves as a valuable indicator of regional-scale climate change. These indicators can help us to understand how the vegetation might be affected and how to deal with the new areas of occurrence of particular species. However, there is still no published evidence validating global warming in European territory based on shifts to northern latitudes of epiphytic bryophyte species.

Some authors have discussed the expansion of more than 40 new bryophyte species with subtropical tendencies in central Europe in the past 20 years (Frahm & Klaus 2001; Frahm 2005), which may indicate effects of global climate changes. One of the cited species is the epiphytic moss *Sematophyllum substrumulosum* (Hampe) E. Britton. Nevertheless, this has not been confirmed by experimental evidence. *Sematophyllum substrumulosum* is an epiphytic moss that lives in aero-hygrophytic communities and is reasonably common in Portugal and Spain (Casas *et al.* 1985; Sim-Sim & Sérgio 1998). From what is known about its ecology, it seems to prefer acidic substrates and develops generally on conifer trunks or at the base of different species of *Pinus*, *Cryptomeria*,

Cupressus, and *Juniperus*. In terms of its phytogeography, it is considered an oceanic–Mediterranean species, with preferences for coastal and temperate areas. It can be sensitive to air pollution (Sim-Sim & Sérgio 1998) but, as it is clearly acidophilic, it can expand its distribution into the frequent *Pinus pinaster* plantation areas. Actually, new locations have been reported for Portugal (Garcia 2006).

Sematophyllum substrumulosum appears to be much more frequent in the Macaronesia region, with citations for different islands of the Azores, Madeira, and Canary Island archipelagos. On the other hand, it has a large distribution in Europe, from the north of France to the Mediterranean area (Portugal, Spain, Italy, Corsica, Croatia, and Montenegro), expanding to Tunisia, Algeria, and Morocco. It always exists in areas of Atlantic influence, and the cartography of the Iberian Peninsula and Micronesia confirms this trend. However, it also occurs in the Algeciras and Balearic areas (Guerra & Gallego 2005) in enclaves with Macaronesian vegetation. In analyses made in the 1980s (Sérgio *et al.* 1989) about the selection of bryophytes as indicators of bioclimatic domains in Portugal, this species was connected to communities that develop in areas where the Emberger Index (Rivas-Martinez 2005) displays high values. The Emberger Index, which is a pluviothermic quotient, is a ratio between the average of precipitation and the temperature range between squared maximum and minimum temperatures.

The discovery of this moss in new localities, each time in more northerly latitudes from its original distribution, drove some authors (Frahm 2005) to consider that these new areas should be related to the increase of temperature due to global warming.

Since 1995, and after being discovered for the first time in the UK (in the Scilly Islands; Holyoak 1996), *Sematophyllum substrumulosum* was repeatedly found in Sussex (Een 2004; Matcham *et al.* 2005). It was also found in Belgium (De Beer & Arts 2000) and in The Netherlands (van Melick 2003; van Zanten 2003) in two different regions. During the first part of 2008, the species was reported from six new localities in France, all in Brittany (De Zuttere & Wattez 2008).

The recent expansion of the geographical distribution of this thermophilic moss species, gradually towards northern latitudes, has provided the opportunity to model its distribution in order to validate the use of bryophytes as indicators of future climate changes.

Objectives

The goals of this study are: (1) to fit a series of models for predicting the spatial occurrence in Portugal of *Sematophyllum substrumulosum* based on the total number of plant records (herbarium and bibliographic); (2) to determine which are the most important environmental factors in defining its distribution in Portuguese territory; (3) to derive climate change impact scenarios by changing the climatic parameters in the scenarios produced for Europe by IPCC (2001); and (4) to discuss different distribution models of this moss species on the basis of two future scenarios of temperature warming, validated by means of new localities where the plant was recently found.

In order to produce the most reliable results, a combined approach with two methods was used in the study, using presence-only models that include improvements in several respects over previous work using ordinal abundance models (Sérgio *et al.* 2007). The first step includes the identification of the main environmental factors that determine the species' distribution, using the ENFA approach (Hirzel *et al.* 2002). The second stage, performed after the validation by the ecological and taxonomic experts concerning the relevant environmental variables, produces the best models possible by the Maxent method (Phillips *et al.* 2006).

Data and methods

Species distribution data

The species to be tested is *Sematophyllum substrumulosum* and the known distribution of the species in Portugal is based mainly on records from specimens from the LISU herbarium database BROTERO (Casa *et al.* 2005) and bibliography references, with a total of 101 occurrences (Fig. 21.1). All occurrences were geo-referenced according to the specimen information of the locality description, using geographic gazetteers and supporting cartography on a GIS. The remaining European and Mediterranean data were gathered from bibliographic references and from herbarium revision data on the first Cartography European project (C. Sérgio 1988, unpublished data).

For Europe we can show the new localities (Fig. 21.2) reported by Holyoak (1996), De Beer & Arts (2000), van Zanten (2003), van Melick (2003), Een (2004), Matcham *et al.* (2005), and De Zuttere and Wattez (2008). Other references were also obtained from the GBIF data portal (MA-Musci 20986, Pando *et al.* 2003–2010). Samples were divided into two data sets, one for distribution modeling, and the other for model validation, based on the date of sampling. The samples selected for validation are all records sampled after 1995.

Climate data

Future climate data are based on the A1F1 Special Report on Emission Scenarios by the Intergovernmental Panel on Climate Change (IPCC 2000, 2001).

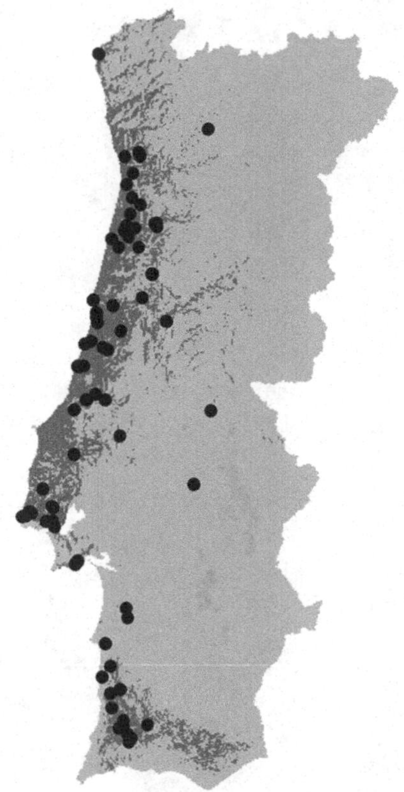

Fig. 21.1. Actual distribution of *Sematophyllum substrumulosum* in Portugal based on herbarium and bibliographic records (101 localities). Dark grey areas indicate potential distribution predicted by Maxent.

Present climate data were also obtained from the Worldclim database, including averages for the period 1965–78 (Hijmans *et al.* 2005). The distribution models built are based on the determination of the ecological niche of *Sematophyllum substrumulosum*, taking as reference the species' distribution. In this work, two approaches were used, based on the methods ENFA (Hirzel *et al* 2002) and Maxent (Phillips *et al.* 2006). The first method is applied to Portugal's continental area, which is part of the species' natural occurrence region, in order to determine which environmental factors are important in defining its distribution.

The Maxent model was used afterwards to determine the present and future potential distribution of the species in the Euro-Mediterranean area. For this exercise, the IPCC scenarios A1F1 for years 2020 and 2050 were used.

For the determination of the best model for the present potential distribution in Europe, the references and herbarium records with collection data after 1995

Fig. 21.2. Actual distribution of *Sematophyllum substrumulosum* (Hampe) E. Britton in Europe, based on herbarium and bibliographic records. Closed dots, records of collections before 1995; open dots, records of collections from 1996 to 2007.

were excluded, in order to make the distribution data compatible with climate data, which corresponds to averages for the period 1965–78. Based on the best distribution models for Portugal, calculated with Maxent, the cut-off that allows a 10% omission error was determined.

Results

The ENFA results indicate that the main factors controlling the distribution of *Sematophyllum substrumulosum* are the average minimum temperature of the coldest month (TMMF), the Emberger Index, with a positive score (Table 21.1), and elevation, with a negative score. Only the first two factors are subject to change in time, and reflect the species' ecological relationship to extreme temperature and water availability. The marginality value is 0.62, which is a relatively low value, indicating that the species is not restricted to

Table 21.1. *Marginality and specificity coefficients for the ENFA (Ecological Niche Factor Analysis) for the first three factors, with percentage of explained variance, on the distribution modeling of* Sematophyllum substrumulosum *in Portugal*

The environmental factors are elevation, slope, aspect North–South (Exp_NS), aspect East–West (Exp_EO), Emberger Index (IEmb), Termicity Index (ITerm, Rivas-Martinez 2005), total annual precipitation (PrecTotal), and minimum average temperature of the coldest month (TMMF). Variables are sorted by decreasing values of the marginality factor.

	Marginality (14%)	Specificity 1 (6%)	Specificity 2 (2%)
Elevation	− 0.475	0.247	0.266
Slope	0.253	− 0.037	− 0.221
Exp_EO	− 0.224	0.069	− 0.002
Exp_NS	0.292	− 0.034	0.071
IEmb	0.459	− 0.51	− 0.292
ITerm	− 0.106	− 0.44	0.274
PrecTotal	0.165	0.532	0.772
TMMF	0.571	0.442	0.344

specific classes of values of the environmental variables. The map of potential distribution for Portugal (Fig. 21.1), determined by Maxent (Area Under Curve, AUC, 0.945), indicates that the species prefers areas nearer to the coast, where extreme negative temperatures are less likely to occur.

The model correctly predicts the species absence in the inland high-altitude part of the country. The altitudinal range where it does occur is rarely higher than 500 m. This distributional restriction is also consistent with the European mapping of the present data (Fig. 21.2).

The significance of the models obtained for any of the tested dates is high, since the AUC indicator for the training data was above 0.987 for all of the models. The potential distribution models for Europe, for the present climate (Fig. 21.3a), which excludes occurrence records from 1995 on, reveal the potential distribution of the species from northern areas to the most northern known location (France, Ile d'Oléron). This result already indicates the potential presence in the Scilly Isles, the islands located off southwest Great Britain. The models can predict the occurrence for the majority of sites where species records after 1995 were observed (Fig. 21.2). This prediction was integral in the 2050 scenarios as well.

The 2020 and 2050 potential occurrence maps (Fig. 21.3b and 21.3c) show a northern drift in the distribution, although it is restricted to lowlands and to the western limit of the continent. There are new potential areas for the species in

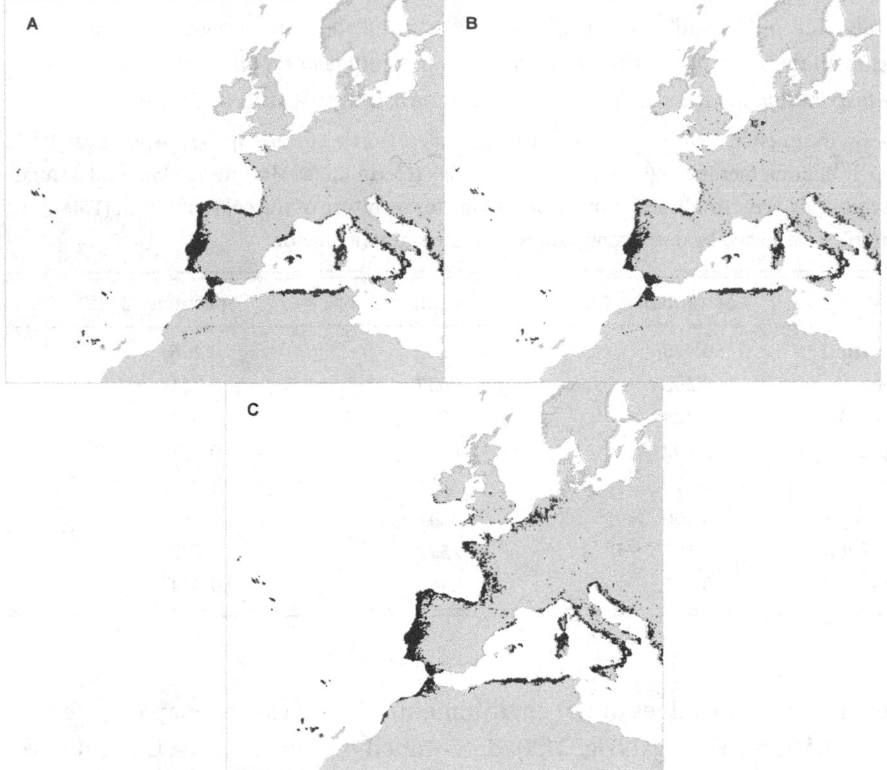

Fig. 21.3. Distribution of potentially suitable habitats for *Sematophyllum substrumulosum* predicted by Maxent. (A) Present climate; (B) scenario for 2020; (C) scenario for 2050. The first map in (A) shows the probabilistic prediction by the model based on the species data in Europe up to 1995. Using a species-specific calibrated threshold, these probabilities are cut back into a binary yes/no (0/1) map showing suitable versus unsuitable habitats.

Great Britain, Ireland, Belgium, and The Netherlands that are already predicted for the 2020 scenario. These areas are greatly enlarged in the 2050 map, with expansion of the coastal areas of occurrence in northwest Continental Europe and in the British Isles. On the other hand, there is a contraction of the distribution in North Africa, where its predicted presence remains restricted to the coastal areas in the future scenarios.

Conclusions

For species interactions some authors (such as Baarda 2006) have discussed thoughts for the future of UK Atlantic oak woods associated with climate change but have not included any particular bryophyte species. Therefore, this study reports the first attempt to use a single epiphytic bryophyte species as an indicator of current climate change.

The environmental variables most relevant for the distribution modelling of *Sematophyllum substrumulosum* are the minimum temperature of the coldest month and the Index of Emberger, indicating that the species is likely to occur in places with higher precipitation and lower temperature ranges. This reveals that this species might present good features as an indicator of climate change. The species is dependent on factors that are likely to be better surrogates of changes in climate, as is the case of minimum temperature of the coldest month, or the Emberger Index, which includes precipitation and temperature range in the calculation.

The identification of the determinant environmental factors of the species' distribution is a critical step in modeling its distribution (Sérgio *et al.* 2007). We adopt a stratified approach, using different geographic areas for the identification of such factors and the future scenarios distribution modeling. Based on the original phytogeographic area of the species, the ENFA analysis allows an intuitive and verifiable identification of environmental variables whose contribution to the distribution model is assessed by the factor loadings. With this approach, only the relevant variables are retained from a battery of factors to support the model of future distribution, thus ensuring that it remains of high ecological significance.

The study was performed above without accounting for the effects of direct anthropogenic intervention in the forests, and thus, the models cannot perfectly predict the potential habitat of species unless no anthropogenic changes occur in forests where they are presently found. The future evolution of landscape and European forest policy can be important constraining factors for *Sematophyllum substrumulosum* expansion.

Furthermore, this approach will yield valid reasons for the expansion or regression of bryophyte species if we include the species' interactions as well as its reproductive strategy. *Sematophyllum substrumulosum* is a monoicous (autoicous) plant that frequently develops sporophytes. It has very small spores (about 12 μm in diameter) and wind is probably its main vehicle of dispersion. Recently, Muñoz *et al.* (2004) tested the correlation between near-surface wind direction and speed with the cryptogamic vegetation similarity of certain areas. They found a stronger correlation of floristic similarity with maximum wind connectivity than with geographic proximity. They concluded that wind is the main force determining current cryptogamic plant distribution. Thus we propose that wind is the long-distance dispersal mechanism for *S. substrumulosum* and that the new habitat conditions promoted by climatic changes will provide new areas of occurrence in Europe.

As can be expected, the interaction between landscape structure and climatic change is even more critical for organisms that are less mobile than animals. However, *S. substrumulosum* presents a high dispersal ability to colonize new suitable habitats such as new forest fragments. On the other hand, spore production, dispersal capacity, and availability of habitat correspond to important parameters for metapopulation dynamics in bryophytes (Söderström & Herben 1997).

This study includes a visual prognostic of suitable climatic expansion or reduction of this epiphytic species. In addition these results should help identify the most probable new areas of occurrence that can be expected in Europe and recognize the remarkable trends of predicted disappearance of all suitable areas for this species in Northern Africa. In the latter case, we consider that this moss might well be able to persist in certain locations, less affected by desertification, but its survival in the long term will become seriously affected, especially in the case of small, fragmented populations. With this study, we suggest that the decrease of moss populations exposed to the global climate change factor of progressive drought might be an important feature in the consideration of the Mediterranean region for particular bryophyte habitats and for epiphytic species.

It is interesting to compare our models with new records for *Sematophyllum substrumulosum* in France, Belgium, The Netherlands, and Great Britain (Holyoak 1996; De Beer & Arts 2000; van Melick 2003; van Zanten 2003; De Zuttere & Wattez 2008), which were not included in the working dataset. We consider that the environmental circumstances that made these new occurrences possible are already a result of changes in climate, and thus the data can be used for model validation as defined by Araújo and Guisan (2006).

In the future, we suggest including more epiphytic environmental predictors in stationary distribution models in order to make them more comprehensive. With this in mind, an effort should be made to observe new potential areas of occurrence, especially to validate the new outputs indicated by the model. This is the case for the new data in France, Belgium, and The Netherlands, where we can confirm the expansion of the species to northern latitudes, but it is impossible at present to certify that the species is absent from the Atlas Mountains in Morocco without new fieldwork. North African species are likely to be more sensitive to decreasing precipitation and humidity than to climate warming. This is one reason why suitable climate spaces were projected to increase for this bryophyte species with climate warming in Atlantic areas. Thus in these contexts we suggest the use of the moss *Sematophyllum substrumulosum* as a suitable indicator to anticipate and to validate global change in climate according to IPCC scenarios.

References

Araújo, M. B. & Guisan, A. (2006). Five (or so) challenges for species distribution modeling. *Journal of Biogeography* **33**: 1677–88.

Araújo, M. B., Thuiller, W. & Pearson, R. G. (2006). Climate warming and the decline of amphibians and reptiles in Europe. *Journal of Biogeography* **33**: 1712–28.

Baarda, P. (2006). Atlantic Oakwoods in Great Britain: factors influencing their definition, distribution and occurrence. *Botanical Journal of Scotland* **57**(1+2): 1–20.

Bates, J. W. & Farmer, A. M. (eds.) (1992). *Bryophytes and Lichens in a Changing Environment*. Oxford: Clarendon Press.

Binder, M. D. & Ellis, C. J. (2008). Conservation of the rare British lichen *Vulpicida pinastri*: changing climate, habitat loss and strategies for mitigation. *Lichenologist* **40**: 63–79.

Casa, J., Carvalho, P., Figueira, R., Sérgio, C. & Calaim, J. (2005). Brotero, Herbaria Information System, Version 1.0.1.2., .NET C#. Lisbon: J. Casa and FFCUL. http://brotero.politecnica.ul.pt/broteroonline, March 2009.

Casas, C., Brugués, M., Cros, R. M. & Sérgio, C. (1985). *Cartografia de Briòfits. Península Ibèrica i les Illes Balears, Canàries, Açores i Madeira*. Barcelona: Institut d'Estudis Catalans.

De Beer, D. & Arts, T. (2000). *Sematophyllum substrumulosum* (Musci, Sematophyllaceae), nieuw voor de Belgische Flora. *Belgian Journal of Botany* **133** (1–2): 15–20.

De Zuttere, P. & Wattez, J. R. (2008). La présence méconnue de *Sematophyllum substrumulosum* (Hampe) E. Britton dans la région Carnacoise (Département du Morbihan, Bretagne méridionale, France). Sa répartition actuelle en Europe. *Nowellia Bryologica* **35**: 2–13.

Een, G. (2004). *Sematophyllum substrumulosum* new to mainland Britain. *Bulletin of the British Bryological Society* **84**: 6–7.

Ellis, C. J. & Coppins, B. J. (2007). Changing climate and historic-woodland structure interact to control species diversity of the *Lobarion* epiphyte community in Scotland. *Journal of Vegetation Science* **18**: 725–34.

Frahm, J.-P. (2005). Bryophytes and global change. *Bryological Times* **115**: 8–10.

Frahm, J.-P. & Klaus, D. (2001). Bryophytes as indicators of recent climatic fluctuations in Central Europe. *Lindbergia* **26**: 97–104.

Garcia, C. (2006). *Briófitos epífitos de ecossistemas florestais em Portugal. Biodiversidade e conservação*. Unpublished Ph.D. thesis, Faculdade de Ciências da Universidade de Lisboa.

Gignac, L. D. (2001). Bryophytes as indicators of climate change. *Bryologist* **104**: 410–20.

Guerra, J. & Gallego, M. T. (2005). An overview of *Sematophyllum* (Bryopsida, Sematophyllaceae) in the Iberian Peninsula. *Cryptogamie, Bryologie* **26**: 173–82.

Guisan, A. & Theurillat, J.-P. (2000). Assessing alpine plant vulnerability to climate change: a modelling perspective. *Integrated Assessment* **1**: 307–20.

Hijmans, R. J., Cameron, S. E., Parra, J. L., Jones, P. G. & Jarvis, A. (2005). Very high resolution interpolated climate surfaces for global land areas. *International Journal of Climatology* **25**: 1965–78.

Hirzel, A. H., Hausser, J., Chessel, D. & Perrin, N. (2002). Ecological-niche factor analysis: How to compute habitat-suitability maps without absence data? *Ecology* **83**: 2027–36.

Holyoak, D. T. (1996). *Sematophyllum substrumulosum* (Hampe) Broth. in the Isles of Scilly: a moss new to Britain. *Journal of Bryology* **19**: 341–5.

IPCC (2000). *Special Report on Emissions Scenarios: A Special Report of Working Group III of the Intergovernmental Panel on Climate Change*, ed. N. Nakicenovic & R. Swart. Cambridge, UK and New York: Cambridge University Press.

IPCC (2001). *Climate Change 2001: The Scientific Basis. Contribution of Working Group I to the Third Assessment Report of the Intergovernmental Panel on Climate Change*, ed. J. T. Houghton, Y. Ding, D. J. Griggs *et al.* Cambridge, UK, and New York: Cambridge University Press.

Lassalle, G., Béguer, M., Beaulaton, L. & Rochard, E. (2008). Diadromous fish conservation plans need to consider global warming issues: an approach using biogeographical models. *Biological Conservation* **141**: 1105–18.

Lee, J. A., Caporn, S. J. M., Carroll, J. *et al.* (1998). Effects of ozone and atmospheric nitrogen deposition on bryophytes. In *Bryology for the Twenty-First Century*, ed. J. W. Bates, N. W. Ashton & J. G. Duckett, pp. 331–41. Leeds: Maney Publishing and British Bryological Society.

Matcham, H. W., Porley, R. D. & O'Shea, B. J. (2005). *Sematophyllum substrumulosum* – an overlooked native? *Field Bryology. Bulletin of the British Bryological Society* **87**: 5–8.

McClean, C. J., Lovett, J. C., Küper, W. *et al.* (2005). African plant diversity and climate change. *Annals of the Missouri Botanical Garden* **92**: 139–52.

Muñoz, J., Felicísimo, A. M., Cabezas, F., Burgaz, A. R. & Martínez, I. (2004). Wind as a long-distance dispersal vehicle in the southern hemisphere. *Science* **304**: 1144–7.

Pando, F., Lopez-Galán, J. & Dueñas, M. (2003–10). MA Cryptogamic collections online databases. www.rjb.csic.es/herbario/crypto/crypdb.htm (accessed via the GBIF portal, http://data.gbif.org/datasets/resource/235, 23 Sept 2010).

Phillips, S. J., Anderson, R. P. & Schapire, R. E. (2006). Maximum entropy modelling of species geographic distributions. *Ecological Modelling* **190** (3–4): 231–59.

Porley, R. & Hodgetts, N. (2005). *Mosses and Liverworts*. London: Collins.

Rivas-Martinez, S. (2005). *Clasificación Bioclimática de la Tierra*. http://www.ucm.es/info/cif/book/bioc/global_bioclimatics_2.htm. March 2009.

Sérgio, C., Figueira, R., Draper, D., Menezes, R. & Sousa, J. (2007). The use of herbarium data for the assessment of red list categories: modelling bryophyte distribution based on ecological information. *Biological Conservation* **135**: 341–51.

Sérgio, C., Sim-Sim, M. & Santos-Silva, C. (1989). Briófitos epifíticos como indicadores dos domínios bioclimáticos en Portugal. Tratamento estatístico de áreas seleccionadas. *Anales del Jardín Botánico de Madrid* **46**: 457–67.

Sim-Sim, M. & Sérgio, C. (1998). Distribution of some epiphytic bryophytes in Portugal. Evaluation and present status. *Lindbergia* **23**: 50–4.

Söderström, L. & Herben, T. (1997). Dynamics of bryophyte metapopulations. *Advances in Bryology* **6**: 205–40.

van Herk, C. M., Aptroot, A. & van Dobben, H. F. (2002). Long-term monitoring in the Netherlands suggests that lichens respond to global warming. *Lichenologist* **34**: 141–54.

van Melick, H. M. H. (2003). *Sematophyllum substrumulosum* ook in Zuidoost-Brabant. *Buxbaumiella* **63**: 14–15.

van Zanten, B. O. (2003). *Sematophyllum substrumulosum* (Hampe) Britt. new to The Netherlands and first record of *Lophocolea semiteres* for the Prov. of Drenthe. *Buxbaumiella* **63**: 7–14.

Vanderpoorten, A. & Klein, J. P. (1999). A comparative study of the hydrophyte flora from the Alpine Rhine to the Middle Rhine. Application to the conservation of the Upper Rhine aquatic ecosystems. *Biological Conservation* **87**: 163–72.

Woolgrove, C. E. & Woodin, S. J. (1994). Relationships between the duration of snowlie and the distribution of bryophyte communities within snowbeds in Scotland. *Journal of Bryology* **18**: 253–60.

Zechmeister, H. G., Dirnböck, T., Hülber, K. & Mirtl, M. (2007). Assessing airborne pollution effects on bryophytes – lessons learned through long-term integrated monitoring in Austria. *Environmental Pollution* **147**: 696–705.

22

Modeling Bryophyte Productivity Across Gradients of Water Availability Using Canopy Form–Function Relationships

STEVEN K. RICE, NATHALI NEAL, JESSE MANGO, AND KELLY BLACK

Introduction

Bryophytes can dominate plant–atmosphere exchange surfaces in mesic to hydric Arctic, boreal, and temperate ecosystems and can contribute up to 50% of gross primary production (Goulden & Crill 1997; Bisbee *et al.* 2001; O'Connell *et al.* 2003a), although estimates in more dense forests are lower (Skre & Oechel 1979; Kolari *et al.* 2006). Soils in these systems store approximately one third of the world's reactive pool of soil carbon (McGuire *et al.* 1995) with a major contribution coming from bryophytes (Gorham 1991; O'Neill 2000; Turetsky 2003). Within these systems, the bryophyte layer also influences hydrology, nutrient uptake and cycling, and soil temperature.

In the boreal zone, significant research has been undertaken to determine how forest bryophytes affect carbon exchange and sequestration. These studies have focused on the influence of environmental forcing variables (e.g., temperature, light, and water availability) on the productivity and carbon dynamics of feathermoss (*Pleurozium*) and *Sphagnum* moss species, the two most dominant groups ecologically. This work has led to a better understanding of temporal variation in bryophyte function and has provided insights into how the performance of individual species varies across gradients of temperature, light intensity, and water availability (Skre & Oechel 1979; Trumbore & Harden 1997; Bisbee *et al.* 2001; O'Connell *et al.* 2003a, b; Heijmans *et al.* 2004; Kolari *et al.* 2006; Kulmala *et al.* 2008).

Bryophyte Ecology and Climate Change, eds. Zoltán Tuba, Nancy G. Slack and Lloyd R. Stark. Published by Cambridge University Press.

However, these studies have neglected the causes and consequences of intraspecific variation, which can be similar in magnitude to differences among species (O'Connell *et al.* 2003a; Kolari *et al.* 2006; Benscoter & Vitt 2007; Bond-Lamberty & Gower 2007), contribute to spatial variation in community function (Heijmans *et al.* 2004), and may serve as sensitive and rapid indicators of environmental change. In addition, species-based investigations generally employ species as categorical independent variables and the functional traits that underlie and cause continuous physiological variation remain poorly known. More general knowledge of form–function relationships in bryophytes will not only lead to a better understanding of the underlying physiological mechanisms that cause functional variation, but these relationships may form the basis of quantitative and predictive models of function, as they have in vascular plants (Grime *et al.* 1997; Smith *et al.* 1997; Wright *et al.* 2004; Shipley *et al.* 2006). The boreal zone, where bryophytes often dominate the forest floor, is expected to experience changes in precipitation and temperature (Christensen *et al.* 2007) that will alter forest floor hydrological conditions, and mechanistic models may help predict species and ecosystem responses to those changes. Although models may be developed and validated for particular species, general patterns and trade-offs that emerge from the modeling effort may guide the development of bryophyte functional models that apply more broadly.

In this chapter, we propose a general model of bryophyte canopy productivity, based on continuously varying characteristics of the canopy, that combines water and carbon components. The consequences of water–carbon dynamics on plant productivity across a range of water availabilities are explored. Parameterizing the model for the common boreal forest feathermoss *Pleurozium schreberi* and using a simulation approach, we examine functional trade-offs associated with variation in canopy-level traits.

Carbon dynamics in bryophytes

Our understanding of the carbon exchange process and its control in bryophytes lags behind that of their vascular plant counterparts. Bryophyte shoot systems differ from vascular plant leaves or canopies in the scale and organization of their photosynthetic tissue, mechanisms they use to govern gas exchange, demands on shoots for nutrient uptake and retention, their reliance on external water storage and transport, and their ability to withstand tissue desiccation. Consequently, bryophyte shoot systems do not function as vascular plant leaf or canopy analogs (Proctor 2000; Cornelissen *et al.* 2007; Rice *et al.* 2008).

Unlike vascular plants, which exert control over carbon and water exchange via stomata, bryophytes lack the ability to regulate these processes over the short term. Instead, bryophyte shoot systems function as three-dimensional exchange surfaces and variation in the organization of the canopy (i.e., the size, arrangement, and orientation of leaves, branches and branch systems) influences the exchange of matter and energy. Shoot systems in bryophytes, however, also serve to store water in extracellular capillary spaces or in specialized hyaline or alar cells. Either through allocation to water storage cells or via the retention of water extracellularly, enhanced water storage ability may impede the acquisition of light or carbon by the canopy, leading to functional trade-offs associated with variation in canopy structure. Such trade-offs between water storage and carbon assimilation are differentially expressed across gradients of water availability and have been demonstrated in *Sphagnum* (Titus & Wagner 1984) and in cushion moss species (Alpert & Oechel 1987; Zotz *et al.* 2000; Hamerlynck *et al.* 2002; Rice & Schneider 2004). Although those studies sought to understand the physiological mechanisms that are responsible for the relationship between water and carbon balance, there remains no investigation into how canopy structural characteristics influence such observed trade-offs. Relating variation in canopy structural or functional characteristics to these trade-offs will not only inform our understanding of the physiological implications of life form variation in bryophytes, but could also serve as the basis of predictive models of bryophyte function.

Productivity model structure

In bryophytes, the canopy (i.e., green shoots and their organization) has served as the principal functional unit for evaluating water and carbon dynamics. The canopy-level model presented here is composed of two interacting component models that control the pools of plant carbon and water (Fig. 22.1). Net photosynthetic assimilation of the canopy expressed on an area basis (A, mol CO_2 m^{-2} h^{-1}) forms the basis of the carbon model. Bryophyte canopies exhibit typical saturating photosynthetic light response curves (Proctor 2000), and many species saturate at quite low light levels, often less than 100 μmol photons m^{-2} s^{-1}. Consequently, incident light (I_o) is an environmental input to the model. When the canopy is light-saturated, photosynthetic tissues throughout the canopy are similarly saturated. Consequently, maximal A should vary directly as a function of shoot area expressed per unit ground area, or the shoot area index (SAI, m^2 shoot/m^2 ground area), a

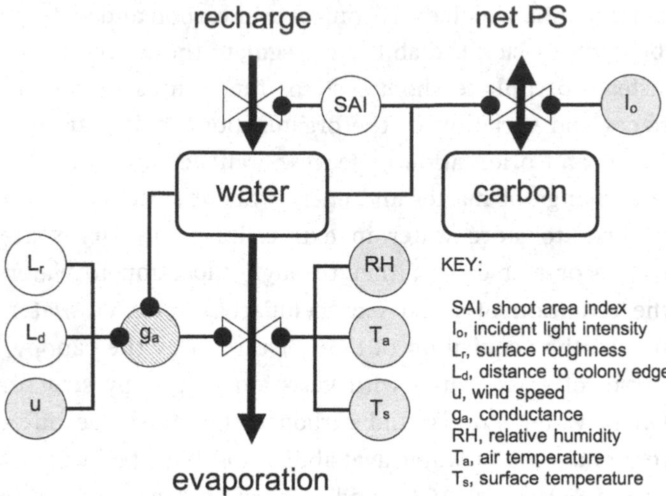

Fig. 22.1. Bryophyte productivity model structure. State variables include pools of water and carbon in the living bryophyte canopy and are shown in boxes. Arrows show water and carbon fluxes associated with recharge (e.g., rainfall), evaporation and net photosynthesis (net PS). Hourglass symbols denote control points for each flux. Intrinsic and extrinsic controls are shown as open and filled circles, respectively. Abbreviations for these are shown in the figure key.

characteristic that varies by over a factor of three within the feathermoss *Pleurozium schreberi* (Bond-Lamberty & Gower 2007; S. Rice *et al.*, unpublished data).

In addition, net photosynthesis in bryophytes exhibits a unimodal response to water content and this provides the basis of the direct interaction between the carbon and water component models. At high water contents, surface water films enclose shoots and impede diffusion of CO_2 to cellular sites of carboxylation. This can reduce photosynthesis by up to 50% of its maximum (Dilks & Proctor 1979). As plants dry, they reach an optimum photosynthetic rate at a particular water content that varies among species. With further drying, plant tissues desiccate and respiratory demands and cellular damage cause a decline in photosynthesis. Thus, the net carbon uptake model component requires inputs of I_o, *SAI*, and plant water content. Other environmental factors and plant traits will also affect canopy carbon assimilation, especially temperature, pigment concentrations, patterns of cell and tissue allocation, and respiratory demands of desiccation tolerance (Skre & Oechel 1979; Proctor 2000; Hamerlynck *et al.* 2002; Marschall & Proctor 2004; Rice *et al.* 2008). Such complexities could be added to the base model and improve its predictive ability.

The water content component model determines canopy water content based on the water holding capacity together with the difference between water recharge and water loss through evaporation. Recharge of canopy water models rainfall or other water additions, which can add water up to, but not exceeding, the maximal water holding capacity of the canopy. Given that canopy water is stored within shoots, it is modeled as a function of canopy *SAI*, a parameter that affects both the carbon and water component models directly.

Water lost through evaporation from plant shoots is modeled on the relationship between canopy structure and conductance to water loss, a process that has been particularly well described. Proctor (1981) applied principles of fluid dynamics to model water loss as a diffusional process occurring through boundary layers developed in air adjacent to the bryophyte canopy surface. Under conditions of flow commonly experienced by terrestrial bryophytes, water loss from bryophytes exhibiting compact surfaces with dense leaves and branching (i.e., low surface roughness, L_r, see Rice *et al.* (2005) for more on this parameter and its measurement) follows patterns expected for flat surfaces of similar dimension in laminar flows. Under these conditions, boundary layer conductance (g_a) is a function of only one intrinsic property of the canopy, the distance to the edge of the canopy (here referred to as L_d) expressed to the –0.5 power, a relationship shared by vascular plant leaves (Schuepp 1993).

Shoot systems with more irregular, rough surfaces (high L_r) interact more with the flow, reduce boundary layer thickness, and increase the power function associated with water loss (Rice *et al.* 2001). Under particular gradients of water vapor pressure between the plant surface and the atmosphere, which are based on surface and air temperatures and the relative humidity, the relationship between conductance and the product of wind speed and L_r is a log-linear function. Rice *et al.* (2001) established that the $g_a - L_r$ relationship meets the general form of: $g_a = aL_r^{\,b}$, where a incorporates environmental conditions and a constant and b is evaluated empirically. In bryophytes, the scaling function (b) is positive and greater than −0.5, the scaling function associated with changes in the horizontal extent of flat plates and cylinders. This difference is caused by the interaction of air flow with vertical shoots that project into the flow and generate turbulence, thereby enhancing water loss. Consequently, variation in both the vertical dimension (L_r) and the horizontal dimension (L_d) affect g_a and water loss.

Lastly, g_a is influenced by canopy water content. In *Sphagnum* and *P. schreberi* samples, g_a is constant above a saturated water content, but decreases linearly below water contents of around 8 g H_2O g dw^{-1}, where dw is dry weight

(Williams & Flanagan 1996). This decrease in g_a is caused by at least two phenomena. First, as surface water films evaporate, water becomes held within cells and cell walls, where it is bound at a lower water potential (Proctor *et al.* 1998), thereby reducing the gradient that leads to water loss. Second, because there is resistance to water movement along stems and shoots, apices will dry first, leaving saturated tissues within the canopy with longer diffusive paths, hence reducing overall canopy conductance (Dilks & Proctor 1979). Thus, the overall form of the equation for g_a combines these effects: $g_a = aL_d^{\,b}L_r^{\,c}(dWC)$, where *a*, *b*, *c*, and *d* are constants in the equation and *WC* is the water content, which has a linear effect with slope *d* below a critical water content WC_c, and a constant effect, dWC_c, above that critical value. Water loss through evaporation will also be affected by environmental factors that influence the concentration gradient: surface temperature, air temperature, and atmospheric relative humidity.

Parameterizing the productivity model for *P. schreberi*

The productivity model was parameterized for the common boreal forest feathermoss *Pleurozium schreberi*, an important forest floor species with a circumboreal distribution. Using known quantitative relationships together with environmental data, the productivity model could be parameterized to predict net carbon exchange in the field. However, that is not the present goal. Instead, we explore the relationship between canopy structure and productivity, its interaction with environmental conditions, especially water availability, and evaluate functional trade-offs associated with variation in canopy structure that emerge from those interactions.

Carbon component. The relationship between *SAI* and light saturated levels (500 μmol photons m^{-2} s^{-1}) of net photosynthesis at optimal plant water contents has been investigated by S. Rice *et al.* (unpublished data) using 26 field collected samples of *P. schreberi* from northern New York State, USA. A linear relationship explains 31% of variation in net photosynthesis (Fig. 22.2a). Although light response curves for this species have been published (Skre & Oechel 1981; Williams & Flanagan 1996; Whitehead & Gower 2001), not enough is known about the canopy organization of those samples to incorporate variation in photosynthetic light response curves into the present model. Consequently, we use only light saturated values of net photosynthesis in our model, realizing that differences in *SAI* or L_r may also influence photosynthesis under conditions of low or variable light intensity. Water content also influences carbon uptake. Williams and Flanagan (1996) present a drying curve for *P. schreberi*, which is well characterized by a third-order polynomial (Fig. 22.2b).

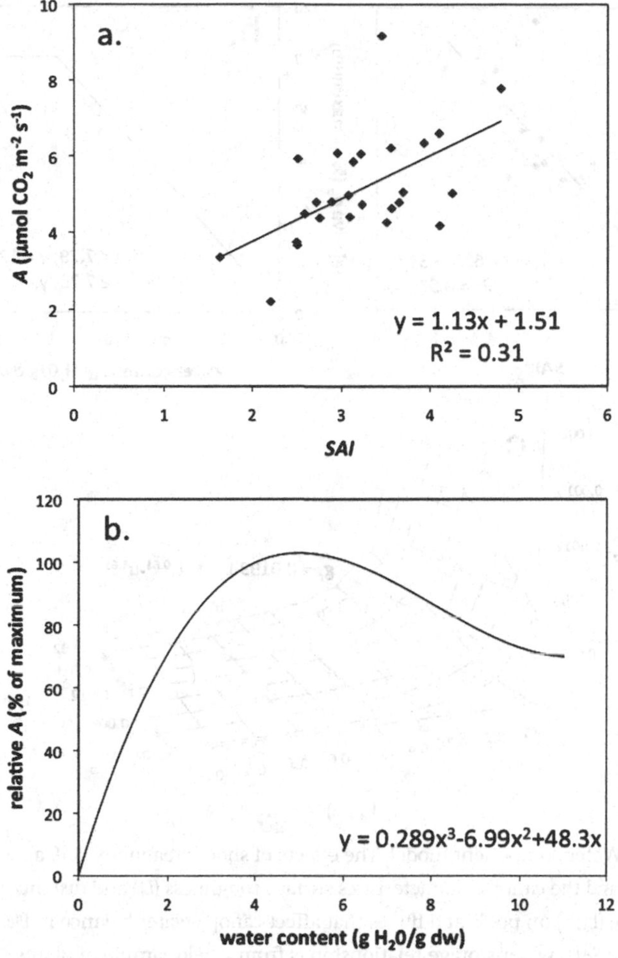

Fig. 22.2. Carbon component model. The effects of variation in shoot area index (*SAI*, a) and water content (b) on net assimilation of CO_2 (*A*) in *Pleurozium schreberi*. The *SAI* – *A* relationship is from a sample of field collected plants. The water content – *A* relationship is derived from Williams and Flanagan (1996) and fit with a third-order polynomial through the origin.

Water content component. Since water within the canopy is held primarily in extracellular capillary spaces between leaves, maximal water content (g H_2O m^{-2} g dw^{-1}) is a function of the *SAI*. S. Rice *et al*. (unpublished data) evaluated this for a set of field-collected samples and found a linear relationship between the two (Fig. 22.3a). Shoots were assumed to achieve their maximum water content during each recharge event.

Water loss through evaporation from the canopy surface is the product of g_a and the concentration gradient. The boundary layer conductance combines the

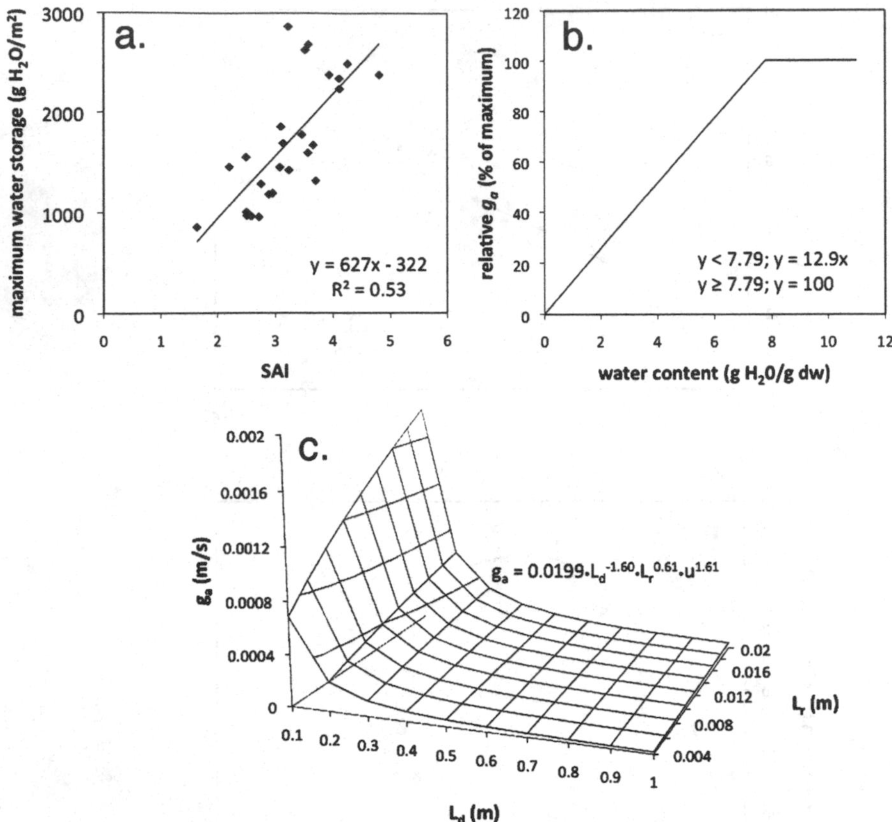

Fig. 22.3. Water component model. The effects of shoot area index (*SAI*, a), water content (b) and the canopy characteristics surface roughness (L_r) and distance to the canopy edge (L_d, c) on pools and fluxes that affect canopy water balance in *Pleurozium schreberi*. The *SAI* – water storage relationship is from a field sample of plants. The dependence of conductance to water loss (g_a) on water content is based on data presented in Williams and Flanagan (1996). Canopy structure also influences g_a. The relationship shown combines effects of L_r and L_d described in Rice *et al.* (2001) and Rice (2006).

effects of L_r, L_d, and water content. The L_r parameterization is based on a forced convection model developed with a set of 11 bryophyte species where the relationships among flow conditions, canopy characteristics, and conductance to water were evaluated in a laminar flow wind tunnel (Rice *et al.* 2001). Within a given set of environmental conditions, the model $g_a = 0.794\ L_r^{0.61}\ u^{1.61}$, where u is wind speed and all variables are expressed using m and s, explains 85% of variation in g_a (Rice *et al.* 2001). The value 0.794 is based on the kinematic viscosity of air and diffusivity of water at 20 °C, and a constant.

In the present formulation, the conductance derived from the equation above is modified by L_d and water content. The former effect was derived from the measurement of g_a across a range of L_d in *P. schreberi* (Rice 2006). At windspeeds between 0.9 and 3.8 $m\,s^{-1}$, g_a decreased proportionally to the −1.47 to −1.70 power of L_d. The value −1.60 was used in the present study, which assumes minimal effects on g_a at the mat edge, but decreasing conductance towards the mat interior. The effect of water content is based on a $g_a - WC$ curve in Williams and Flanagan (1996) for *P. schreberi*. In those data, *WC* has no effect on g_a until below *WC* of 7.8 g H_2O g dw^{-1}. Below that *WC*, there is a linear decrease with a slope of 12.9 % *A* g H_2O^{-1} g dw^{-1}. This piecewise linear function was set to vary from 0 to 100% and used in the model (Fig. 22.3b). The form of the full g_a model combines the effects of L_r, L_d, and *WC* (Fig. 22.3c).

Dynamics of the productivity model

Productivity was modeled across *SAI* (1.5 to 5.0) and L_r (0.004 to 0.02 m) values that are within the range of variation found in *P. schreberi*. Environmental conditions were held constant at 20 °C surface and air temperature, 75% relative humidity, 1 $m\,s^{-1}$ windspeed, and 500 μmol photons $m^{-2}\,s^{-1}$, and the canopy was 1 m from the edge (L_d = 1 m). Simulations began with canopies saturated with water. Rates of net photosynthesis and water loss were calculated at the start and assumed to be constant for 0.25 h. Following that interval, a new water content was calculated, thereby altering rates of photosynthesis and g_a. Using this difference equation approach, productivity was simulated over a 1600 h time period by summing net photosynthesis over the elapsed time, and calculating an average productivity (mmol CO_2 $m^{-2}\,h^{-1}$). The model was run at different frequencies of water recharge that simulated rainfall of sufficient duration to saturate canopy water contents. Recharge frequencies over the duration of the model runs were varied from 1 to 64 times in multiples of two over the time period and also set to maintain saturated water content permanently. Results were explored by graphing productivity as a function of canopy structure as characterized by *SAI* and L_r.

Under all recharge frequencies, productivity varied as a function of *SAI* (Fig. 22.4a–h). However, productivity was only associated with L_r at low to intermediate recharge frequencies, but not at permanent saturation. When plants were water limited at a low recharge frequency, productivity exponentially increased at low L_r and high *SAI*. This pattern arose from the interaction between factors that control water balance and carbon gain. Canopies with lower L_r had reduced rates of water loss and maintained photosynthetic assimilation for longer duration. When combined with elevated *SAI*, which increases both water storage and net photosynthesis, productivity was optimized at high *SAI*.

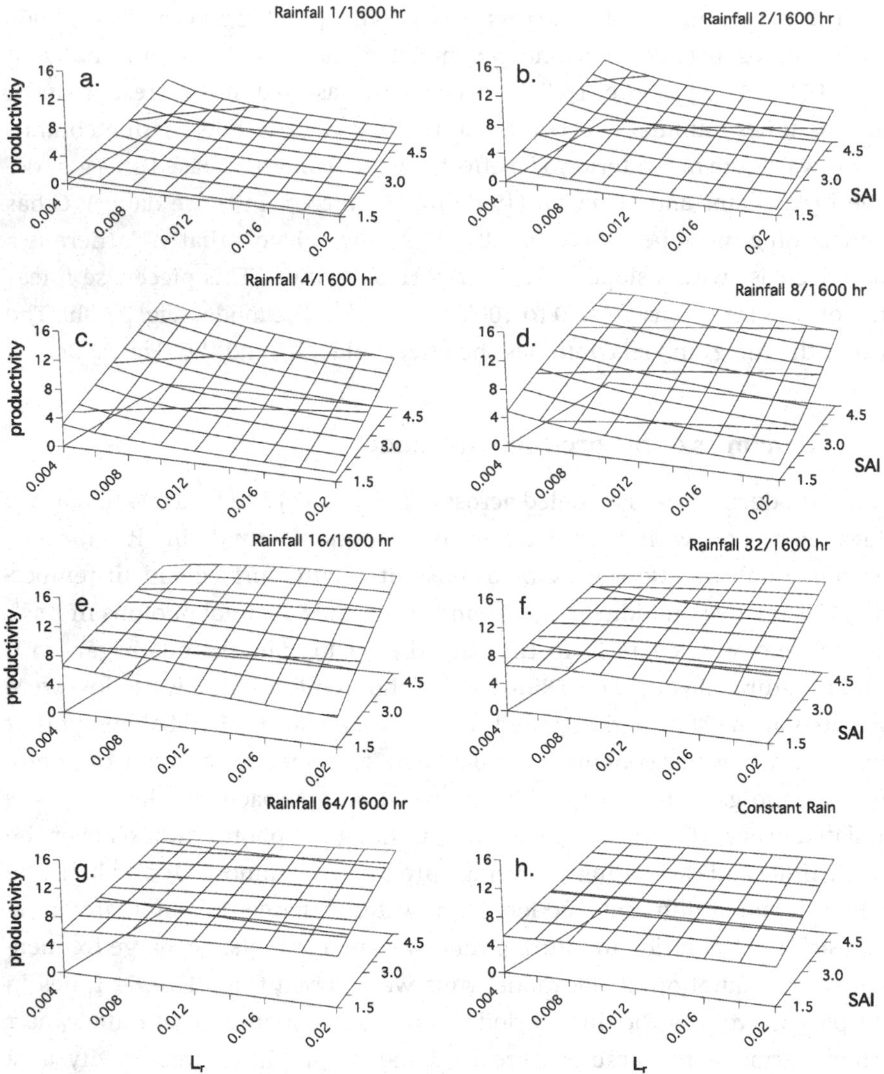

Fig. 22.4. Productivity in different rainfall regimes, with output of the model shown in Fig. 22.1 parameterized for *Pleurozium schreberi*. Runs (a–g) simulated a 1600 h time period with rainfall occurring once during that interval (1/1600 h) through 64 times per interval (64/1600 h). Constant recharge in each 0.25 h time step is also shown (h). Plant canopy and environmental conditions modeled were: L_d = 1 m, surface and air temperature 20 °C, relative humidity 75%, and wind speed 1 m s^{-1}. Units for productivity, L_r, and *SAI* are mmol CO_2 m^{-2} h^{-1}, m, and m^2 shoot area m^{-2} ground area, respectively.

When canopies were subject to more frequent water recharge, the model predicted a different response. At intermediate levels of recharge, productivity remained a positive function of *SAI*, although the relationship changed from exponential with a scaling exponent above 1 to a function with the

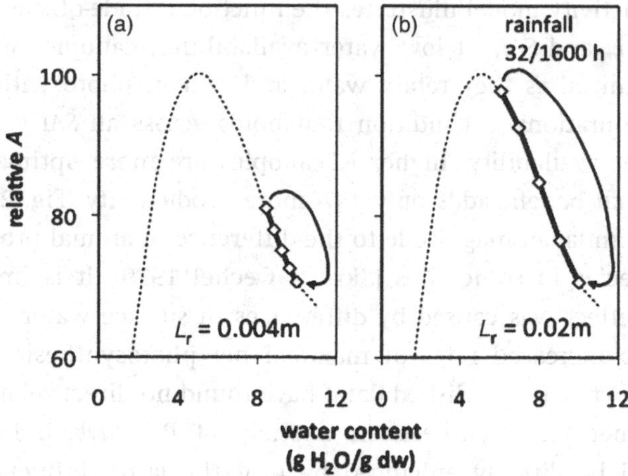

Fig. 22.5. Water content and carbon assimilation cycles. The relative net photosynthesis (A) during drying and rewetting cycles is shown by thick lines for plants with low (a) and high (b) L_r. At intermediate frequencies of water recharge, plants with a lower L_r remain saturated and do not achieve maximal rates of net photosynthesis as do plants with higher L_r (b). Symbols represent the water content and net photosynthesis at 10 h intervals. Thin lines indicate the return to saturated conditions at a water content of 10 g H_2O g^{-1} dry weight at each interval. The dashed lines represent a portion of the drying curve shown in Fig. 22.2(b).

exponent less than one. The relationship of productivity with L_r, however, changed more dramatically. At moderate levels of water availability, productivity increased at higher L_r, opposite to the pattern observed at low water availability. This transition occurs first at high *SAI*, then at higher frequencies of water recharge; it happens at lower *SAI* as well.

The model predicts that low L_r should optimize productivity at low water availability, but that at intermediate water availability, high L_r would be favored. This pattern emerges from the effects of two aspects of the model that control the magnitude of net photosynthesis and its duration. First, the dependence of photosynthesis on water content (Fig. 22.2b) exhibits a strong depression at high water contents. Consequently, net photosynthesis will increase if plants dry more quickly, as long as they do not desiccate. This is achieved at higher L_r under intermediate levels of water recharge. This is further enhanced by the negative feedback that reduces conductance at low water contents (Fig. 22.3b). This results in longer duration of maximal carbon gain in plants with high L_r that dry and attain high net photosynthesis more quickly (Fig. 22.5). As water recharge frequencies increase, plant canopies remain fully saturated, resulting in a depression of carbon uptake. When permanently saturated, higher net photosynthesis associated linearly with *SAI*, but there is no interaction with L_r.

Thus, the productivity model illustrates the functional trade-offs associated with variation in L_r and *SAI*. At low water availabilities, canopies with low L_r values are optimized as they retain water and remain photosynthetically active for longer durations, a condition that holds across all *SAI* values. At intermediate water availability, higher L_r canopies are more optimal. Even though the maximal benefit adds only 14% more productivity (Fig. 22.4f, g), this difference is similar in magnitude to the difference in annual production between moss species in some sites (Skre & Oechel 1979). It is important to note that this effect was caused by differences in surface water films on shoots and not to increased rates of maximal net photosynthesis in high L_r canopies. S. Rice *et al.* (unpublished data) have found no direct relationship between L_r and net photosynthesis in a study of *P. schreberi*. However, productivity could be directly enhanced by L_r if the latter influenced the light response of bryophyte canopies. With less self-shading, high L_r canopies may increase the photochemical efficiency and reduce the light compensation point, which may benefit bryophytes in shaded habitats. Bryophyte canopies that achieve higher L_r through greater vertical growth may also have an ecological advantage as they compete more effectively with other bryophytes and lichens (During 1992; Sulyma & Coxson 2001).

Discussion

In *P. schreberi*, the canopy structural characteristics *SAI* and L_r control plant water status, which directly affects rates of maximal carbon uptake, and characterize independent qualities of the canopy (correlation coefficient $r=0.13$, for $n=26$ samples) (S. Rice *et al.*, unpublished data). Their combined effect on carbon dynamics depends on water availability. Canopies with lower L_r optimize productivity at low water availability, but this is reversed at intermediate rates of water recharge (Fig. 22.5). In contrast, plants with higher *SAI* always show higher productivity across all water availabilities. Why, then, does a three-fold variation in *SAI* exist? Why don't all plants have a high *SAI*? Young plants may possess low *SAI* as they have not developed their shoot system fully. However, in a companion study (S. Rice *et al.*, unpublished data), all canopies were supported on brown shoot tissue, indicating that they were not newly established. A better explanation is that *SAI* may influence plant light relations. Plants with low *SAI* may have lower light compensation points and/or photochemical efficiencies. This would allow them to take advantage of variable or low light, as is common in boreal forests. With more information on the variation in light response curves and how this relates to canopy structure, light availability would be a useful environmental dimension to add to the present model.

The productivity modeling suggests that both *SAI* and L_r could serve as important structural parameters for field studies. Although neither is simple to quantify, contemporary research tools make their evaluation more straightforward. Bond-Lamberty and Gower (2007) present a protocol for measuring *SAI* for bryophytes that employs a flatbed scanner to calculate shoot area – dry mass relationships that are used to convert dry mass to *SAI*. High-resolution leaf area meters can also be used on samples with sparse branching. Given the dependence of photosynthetic assimilation on shoot area, *SAI* might be the best measure of photosynthetic tissue, especially when comparisons among species or functional groups are of interest. However, within a species, it may be useful to use shoot mass per area (*SMA*, $g\,m^{-2}$) instead, owing to its ease of measurement. In *P. schreberi*, this relates linearly to *SAI* with a high R^2 value (*SAI* = 0.014 *SMA*+0.38; S. Rice *et al.*, unpublished data).

Advances in digital optics and computational ability have also made measurement of L_r more available. Earlier studies based the calculation of L_r on measurements of the depth of the canopy surface obtained using a contact probe (Hayward & Clymo 1983; Rice *et al.* 2001). It is easy to construct benchtop contact probes that perform this task, but the measurements are time consuming and are difficult to employ in the field. In response to these problems, Rice *et al.* (2005) developed an optical approach using laser scanning to generate canopy depth measurements. In this technique, a plane of laser light is trained orthogonally to the bryophyte surface and a camera takes an image of the canopy–light intersection from a non-orthogonal angle. After transformation, these points are converted to Cartesian coordinates. Such instruments are available commercially, but are not difficult to construct (see Rice *et al.* 2005). This technique provides a density of canopy depth points at a grid scale less than 0.4 mm and L_r measurements correspond well with those obtained using a contact probe. Recently, Krumnikl *et al.* (2008) have developed a stereo photogrammetric technique that provides even more detail, although the algorithms are not yet available. Both of these methods can be employed in the field and provide robust canopy depth measures that can be used to calculate L_r (see Rice *et al.* 2005 for details of the calculations) for use in functional models.

Conclusions

In the boreal forest, climate change is expected to lead to changes in forest floor hydrologic regimes through its effects on annual precipitation, the frequency of rainfall events, and rates of evapotranspiration (Plummer *et al.* 2006). For boreal forests in central to eastern Canada, most climate change scenarios predict increased, yet more variable rainfall (Christensen *et al.* 2007). However, this change will be insufficient to compensate for elevated temperatures, leading to increased drought frequency and severity (Girardin & Mudelsee 2008). Broad-scale studies of vegetation–environment relationships may be useful to explore how such changes will alter community composition and subsequent ecosystem function. However, that approach assumes that vegetation is at equilibrium with local climate, a condition that will not be met in the near term. Changes in local water availability will have immediate consequences on carbon exchange and the productivity model presented in this chapter provides a context for evaluating that effect. Results suggest that the short-term response to increased precipitation frequency depends on the interaction of rainfall frequency and the canopy characteristics of shoot area index (*SAI*) and surface roughness (L_r). Increased rainfall that leads to frequent saturation of the moss canopy will decrease productivity as carbon uptake is reduced by external water films. However, below a critical frequency that depends on the canopy *SAI* and L_r, increases in rainfall lead to greater average productivity. Field surveys across existing precipitation gradients would determine the distribution of *SAI* and L_r with respect to present climate and implementation of the model may help predict how the contribution of forest floor bryophytes may respond to altered climate.

References

Alpert, P. & Oechel, W. C. (1987). Comparative patterns of net photosynthesis in an assemblage of mosses with contrasting microdistributions. *American Journal of Botany* **74**: 1787–96.

Amthor, J. S., Chen, J. M., Clein, J. S. *et al.* (2001). Boreal forest CO_2 exchange and evapotranspiration predicted by nine ecosystem process models: intermodel comparisons and relationships to field measurements. *Journal of Geophysical Research* **106**: 33623–48.

Benscoter, B. W. & Vitt, D. H. (2007). Evaluating feathermoss growth: a challenge to traditional methods and implications for the boreal carbon budget. *Journal of Ecology* **95**: 151–8.

Bisbee, K. E., Gower, S. T., Norman, J. M. & Nordheim, E. V. (2001). Environmental controls on ground cover species composition and productivity in a boreal black spruce forest. *Oecologia* **129**: 261–70.

Bond-Lamberty, B. & Gower, S. T. (2007). Estimation of stand-level leaf area for boreal bryophytes. *Oecologia* **151**: 584–92.

Christensen, J. H., Hewitson, B., Busuioc, A. *et al.* (2007). Regional climate projections. In *Climate Change 2007: The Physical Science Basis. Contribution of the Working Group I to the Fourth Assessment Report of the Intergovernmental Panel on Climate Change*, ed. S. Solomon, D. Quin, M. Manning, *et al.*, pp. 847–940. Cambridge: Cambridge University Press.

Cornelissen, J. H. C., Lang, S. I., Soudzilovskaia, N. A. & During, H. J. (2007). Comparative cryptogam ecology: a review of bryophyte and lichen traits that drive biogeochemistry. *Annals of Botany* **99**: 987–1001.

Dilks, T. J. K. & Proctor, M. C. F. (1979). Photosynthesis, respiration and water content in bryophytes. *New Phytologist* **82**: 97–114.

During, H. J. (1992). Ecological classification of bryophytes and lichens. In *Bryophytes and Lichens in a Changing Environment*, ed. J. W. Bates & A. M. Farmer, pp. 1–31. Oxford: Clarendon Press.

Gimingham, C. H. & Birse, E. M. (1957). Ecological studies on growth-form in bryophytes: I. Correlation between growth-form and habitat. *Journal of Ecology* **45**: 533–45.

Girardin, M. P. & Mudelsee, M. (2008). Past and future changes in Canadian boreal wildfire activity. *Ecological Applications* **18**: 391–406.

Gorham, E. (1991). Northern peatlands: role in the carbon cycle and probable responses to climatic warming. *Ecological Applications* **1**: 182–95.

Goulden, M. L. & Crill, P. M. (1997). Automated measurements of CO_2 exchange at the moss surface of a black spruce forest. *Tree Physiology* **17**: 537–42.

Grime, J. P., Thompson, K., Hunt, R. *et al.* (1997). Integrated screening validates primary axes of specialization in plants. *Oikos* **79**: 259–81.

Hamerlynck, E. P., Csintalan, Z., Nagy, Z. *et al.* (2002). Ecophysiological consequences of contrasting microenvironments on the desiccation tolerant moss *Tortula ruralis*. *Oecologia* **131**: 498–505.

Hayward, P. M. & Clymo, R. S. (1983). The growth of *Sphagnum*: experiments on, and simulation of, some effects of light flux and water-table depth. *Journal of Ecology* **71**: 845–63.

Heijmans, M. M. P. D., Arp, W. J. & Chapin, F. S. III (2004). Carbon dioxide and water vapour exchange from understory species in boreal forest. *Agricultural and Forest Meteorology* **123**: 135–47.

Kolari, P., Pumpanen, J., Kulmala, L. *et al.* (2006). Forest floor vegetation plays an important role in photosynthetic production of boreal forests. *Forest Ecology and Management* **221**: 241–8.

Krumnikl, M., Sojka, E., Gaura, J. & Motyka, O. (2008). A new method for bryophyte canopy analysis based on 3D surface reconstruction. *Seventh International Conference on Computer Information Systems and Industrial Management Applications Proceedings*, ed. V. Snášel, A. Abraham, K. Saeed & J. Pokorný, pp. 210–11. Los Alamitos, CA: IEEE Computer Society.

Kulmala, L., Launiainen, S., Pumpanen, J., *et al.* (2008). H_2O and CO_2 fluxes at the floor of a boreal pine forest. *Tellus* **60B**:167–78.

Marschall, M. & Proctor, M. C. F. (2004). Are bryophytes shade plants? Photosynthetic light responses and proportions of chlorophyll *a*, chlorophyll *b* and total carotenoids. *Annals of Botany* **94**: 593–603.

McGuire, A. D., Melillo, J. M., Kicklighter, D. W. & Joyce, L. A. (1995). Equilibrium responses of soil carbon to climate change: empirical and process-based estimates. *Journal of Biogeography* **22**: 785–96.

O'Connell, K. E. B., Gower, S. T. & Norman, J. M. (2003a). Comparison of net primary production and light-use dynamics of two boreal black spruce forest communities. *Ecosystems* **6**: 236–47.

O'Connell, K. E. B., Gower, S. T. & Norman, J. M. (2003b). Net ecosystem production of two contrasting boreal black spruce forest communities. *Ecosystems* **6**: 248–60.

O'Neill, K. P. (2000). Role of bryophyte-dominated ecosystems in the global carbon budget. In *Bryophyte Biology*, ed. A. J. Shaw & B. Goffinet. pp. 344–68. Cambridge: Cambridge University Press.

Plummer, D. A., Caya, D., Frigon, A. *et al.* (2006). Climate and climate change over North America as simulated by the Canadian RCM. *Journal of Climate* **19**: 3112–32.

Proctor, M. C. F. (1981). Diffusion resistance in bryophytes. In *Plants and their Atmospheric Environment*, ed. J. Grace, E. D. Ford & P. G. Jarvis, pp. 219–29. Oxford: Blackwell Scientific.

Proctor, M. C. F. (2000). Physiological ecology. In *Bryophyte Biology*, ed. A. J. Shaw & B. Goffinet, pp. 225–47. Cambridge: Cambridge University Press.

Proctor, M. C. F., Nagy, Z., Csintalan, Zs. & Takács, Z. (1998). Water-content components in bryophytes: analysis of pressure-volume curves. *Journal of Experimental Botany* **49**: 1845–54.

Rice, S. K. (2006). Towards an integrated understanding of bryophyte performance: the dimensions of space and time. *Lindbergia* **31**: 42–53.

Rice, S. K. & Schneider, N. (2004). Cushion size, surface roughness, and the control of water balance and carbon flux in the cushion moss *Leucobryum glaucum* (Leucobryaceae). *American Journal of Botany* **91**: 1164–72.

Rice, S. K., Gutman, C. & Krouglicof, N. (2005). Laser scanning reveals bryophyte canopy structure. *New Phytologist* **166**: 695–704.

Rice, S. K., Collins, D. & Anderson, A. M. (2001). Functional significance of variation in bryophyte canopy structure. *American Journal of Botany* **88**: 1568–76.

Rice, S. K., Aclander, L. & Hanson, D. T. (2008). Do bryophyte shoot systems function like vascular plant leaves or canopies? Functional trait relationships in *Sphagnum* mosses (Sphagnaceae). *American Journal of Botany* **95**: 1366–74.

Schuepp, P. H. (1993). Leaf boundary layers. *New Phytologist* **125**: 477–507.

Shipley, B., Lechowicz, M. J., Wright, I. & Reich, P. B. (2006). Fundamental trade-offs generating the worldwide leaf economics spectrum. *Ecology* **87**: 535–41.

Skre, O. & Oechel, W. C. (1979). Moss production in a black spruce *Picea mariana* forest with permafrost near Fairbanks, Alaska, as compared with two permafrost-free stands. *Holarctic Ecology* **2**: 249–54.

Skre, O. & Oechel, W. C. (1981). Moss functioning in different taiga ecosystems in interior Alaska. I. Seasonal, phenotypic, and drought effects on photosynthesis and response patterns. *Oecologia* **48**: 50–9.

Smith, T. M., Shugart, H. H. & Woodward, F. I. (eds.) (1997). *Plant Functional Types*. Cambridge: Cambridge University Press.

Sulyma, R. & Coxson, D. S. (2001). Microsite displacement of terrestrial lichens by feather moss mats in late seral pine-lichen woodlands of North-central British Columbia. *Bryologist* **104**: 505–16.

Titus, J. E. & Wagner, D. J. (1984). Carbon balance for two *Sphagnum* mosses: water balance resolves a physiological paradox. *Ecology* **65**: 1765–74.

Trumbore, S. E. & Harden, J. W. (1997). Accumulation and turnover of carbon in organic and mineral soils of the BOREAS northern study area. *Journal of Geophysical Research* **102**: 28817–30.

Turetsky, M. (2003). The role of bryophytes in carbon and nitrogen cycling. *Bryologist* **106**: 395–409.

Whitehead, D. & Gower, S. T. (2001). Photosynthesis and light-use efficiency by plants in a Canadian boreal forest ecosystem. *Tree Physiology* **21**: 925–9.

Williams, T. G. & Flanagan, L. B. (1996). Effect of changes in water content on photosynthesis, transpiration and discrimination against $^{13}CO_2$ and $C^{18}O^{16}O$ in *Pleurozium* and *Sphagnum*. *Oecologia* **108**: 38–46.

Wright, I. J., Reich, P. B., Westoby, M., *et al.* (2004). The worldwide leaf economics spectrum. *Nature* **428**: 821–7.

Zotz, G., Schweikert, A., Jetz, W. & Westerman, H. (2000). Water relations and carbon gain are closely related to cushion size in the moss *Grimmia pulvinata*. *New Phytologist* **148**: 59–67.

VIII CONCLUSIONS

23

Bryophytes as Predictors of Climate Change

L. DENNIS GIGNAC

Introduction

Increases in atmospheric greenhouse gases that produce positive radiative forcings are having large-scale impacts on the global climate system, such as increases in temperatures, changes in precipitation patterns, and changes in the frequency of severe weather events. The greatest impact is occurring at high latitudes in the northern hemisphere, at high altitudes throughout the world, and in Antarctica and the sub-Antarctic Islands. Because of their efficient long-distance dispersal mechanisms and their high fidelity to climatically sensitive habitats, bryophytes should react quickly to changes to their environment and thus have the potential of being effective predictors of current climate change.

Since the beginning of the industrial era approximately 200 years ago, carbon dioxide (CO_2), methane (CH_4), and nitrous oxide (N_2O) concentrations have been increasing exponentially in the atmosphere (IPCC 2001). Concentrations of other greenhouse gases such as halocarbons and perfluorocarbons that are produced by anthropogenic activities are also increasing in the atmosphere. As a result of greenhouse gas enrichment, global mean surface air temperatures have risen by about 0.3–0.6 °C since the nineteenth century and by 0.2–0.3 °C over the past 40 years (IPCC 2001). The largest temperature increases have occurred over the northern hemisphere land mass north of 40° latitude (Serreze *et al.* 2000), on widespread high-altitude mountains throughout the world (Böhm *et al.* 2001), and in the Antarctic (Nyakatya & McGeoch 2008). Temperature increases have also been observed at low latitudes, although to a lesser extent than at high latitudes (Dash *et al.* 2007; Malhi *et al.* 2008). Warming, however, is spatially heterogeneous with

Bryophyte Ecology and Climate Change, eds. Zoltán Tuba, Nancy G. Slack and Lloyd R. Stark. Published by Cambridge University Press.

some areas such as northern Europe having higher increases than the global average while others such as southern coastal Greenland have undergone a cooling trend (Chylek *et al.* 2004).

Changes in precipitation patterns that are likely linked to enhanced concentrations of atmospheric greenhouse gases have also been observed on a global scale. In regions close to the equator, some of the most important climatic changes are related to changes in precipitation. Changes to the Indian and African monsoons and the El Niño – Southern Oscillation (ENSO) appear to be related to enhanced greenhouse conditions and are affecting the quantity and timing of precipitation (Giannini *et al.* 2008; Malhi *et al.* 2008). Changes in precipitation are also spatially and temporally heterogeneous, and predictions of future patterns are generally uncertain and often difficult to coherently relate to climate change (Paeth *et al.* 2008).

The temporal variability of other smaller-scale climatic events such as wind direction and speed, number of frost events, number and timing of rain events, and dry season mist frequency is increasing as the climate changes. The timing and frequency of extreme weather events such as droughts and high maximum temperatures are also changing as a result of global warming. All of those factors have impacts on biological processes (Le Roux & McGeoch 2008).

Climatic changes that have already been observed on a global scale are expected to become greater throughout the twenty-first century as atmospheric concentrations of CO_2 continue to increase. Recent models by the Intergovernmental Panel on Climate Change (IPCC) based on results from several General Circulation Models and different CO_2 increase scenarios predict that mean annual global temperatures are expected to rise between 1.1 and 6.4 °C by the end of this century (IPCC 2007). Warming is expected to be greatest at most northern latitudes and least over the Southern Ocean near Antarctica and in the northern North Atlantic.

Bryophytes should respond relatively quickly to changes in climate because of their considerable variety of dispersal mechanisms (Frahm 2008). They can propagate both sexually by means of spores and asexually by using several types of propagules such as gemmae, brood bodies, and leaf buds. Furthermore, fragments from parts of the plants can detach and produce new plants. Most propagules can disperse over long distances, particularly spores, which are generally very small and can be produced in large quantities. Spores remain viable for many years in the soil spore bank and can readily take advantage of changing growing conditions or new habitats. Many bryophytes have strong affinities to their habitat and most rare species are restricted by narrow habitat requirements and are habitat-limited

(Söderström & During 2005). If habitats should disappear or change as a result of global warming, bryophyte distribution and perhaps their ranges should contract. Thus, bryophytes should be excellent predictors of climate change.

Effects of increased greenhouse gases can produce direct and indirect effects on bryophytes (Fig. 23.1). Direct effects are those produced by the gases themselves; indirect effects are those produced by changes to the climate, principally in temperature, but also in precipitation and extreme climatic events. In this chapter, direct effects of atmospheric CO_2 and N_2O enrichment as well as effects of current climate change on bryophytes are linked to: (1) growth, growth form, and abundance; (2) shifts in species' ranges; (3) changes in species' interactions with other bryophytes and vascular plants; and (4) changes in species' abundance in bryophyte-dominated ecosystems. Bryophyte predictors of each of those linkages are selected and their effectiveness is discussed based on the documented evidence. Areas for future research that would advance our use of bryophyte predictors of climate change will also be explored. Of note, most of the documented evidence of effects on bryophytes comes from high latitudes (>40°) and high altitudes as well as Antarctica and the sub-Antarctic Islands, where effects of climate change are having and will have the greatest impact.

Bryophytes as predictors of atmospheric greenhouse gas enrichment

The concentration of atmospheric CO_2 is one of the principal factors influencing photosynthesis and productivity in C_3 plants such as bryophytes. Therefore, one could expect that increased CO_2 concentrations would have a fertilization effect on growth and productivity. Results of several studies have demonstrated that the response is species-specific (van der Heijden *et al.* 2000; Heijmans *et al.* 2001, 2002; Toet *et al.* 2006). For example, Jauhiainen *et al.* (1994, 1998b) found CO_2-induced increases in production for *Sphagnum angustifolium* and *S. warnstorfii*, but not for *S. fuscum*. On occasion, when CO_2 enrichment produced growth in length in *Sphagnum*, it did not translate into increased production (Berendse *et al.* 2001; Hoosbeek *et al.* 2001). In some species, the stem morphology changed and the distance between branches increased, whereas in other species the number of capitula increased per unit area but the size of the capitula diminished (Aerts *et al.* 1992; Jauhiainen *et al.* 1994, 1998a; Berendse *et al.* 2001; Heijmans *et al.* 2001).

In Arctic, alpine, and many peatland ecosystems, plant production is nutrient-limited, usually by N and P (Shaver *et al.* 2001). The availability of N has recently increased in many of those ecosystems because of higher

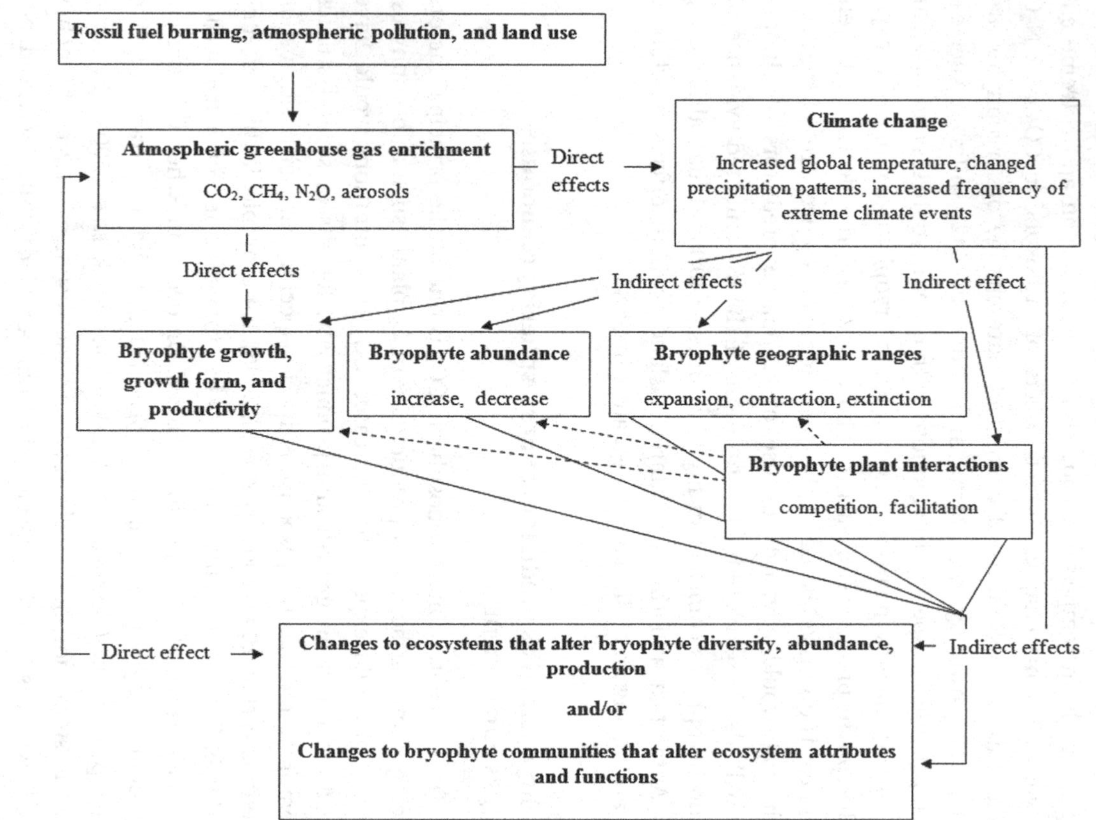

Fig. 23.1. Conceptual diagram of some direct and indirect effects of increased atmospheric concentrations of greenhouse gases on bryophytes.

atmospheric deposition, particularly in northwest Europe, and because of increased mineralization produced by warmer soils and higher evapotranspiration. As a result, many experiments that examined potential impacts of climate change on those ecosystems had fertilizer treatments incorporated into the design.

Experiments have shown that effects of warming and additional N loadings on *Sphagnum* growth and abundance varied between locations (Aerts *et al.* 1992). In areas where N loadings were low, addition of N increased *Sphagnum* growth at least in the short term (Aerts *et al.* 1992), but not in the long term (Aerts *et al.* 2001), whereas in areas where N deposition was moderate, *Sphagnum* growth was not affected by increasing N because other elements such as P (Aerts *et al.* 1992; Gerdol *et al.* 2007; Phuyal *et al.* 2008) and/or K (Hoosbeek *et al.* 2002; Bragazza *et al.* 2004) became limiting. Beyond a maximum loading, there is overwhelming evidence that growth and/or abundance of most bog and poor fen *Sphagnum* species was adversely affected (Jauhiainen *et al.* 1994, 1998b; Gunnarsson & Rydin 2000; Berendse *et al.* 2001; Heijmans *et al.* 2002; Limpens *et al.* 2003, 2004; Wiedermann *et al.* 2007). High N deposition and loadings produce a toxic effect on most *Sphagnum* and as a result, N becomes available for vascular plant uptake. That uptake increases vascular plant growth and the introduction of N-demanding vascular species, resulting in shading and litter production that also reduces *Sphagnum* growth (Lamers *et al.* 2000; Berendse *et al.* 2001).

Bryophytes should not be lumped together as a single functional group in relation to N loadings but rather should be considered as two groups, nutrient-philic species and all other bryophytes including *Sphagnum* (Gordon *et al.* 2001). Positive and negative indicators of high N deposition in conjunction with climate warming are listed in Table 23.1. High N loadings may also affect bryophyte growth forms. For example, Sandvik and Heegard (2003) observed that high fertilizer loadings produced a more lax growth form in *Pohlia wahlenbergii*. Generally, on high Arctic sites, bryophytes responded positively to nutrient additions (Robinson *et al.* 1998; Van Wijk *et al.* 2003; Wasley *et al.* 2006).

Bryophytes as predictors of increased temperature and evapotranspiration

A positive relationship between growth and productivity of some bryophytes and climatic variables was demonstrated by analyzing growth in specimens obtained from a broad range of climatic areas and relating them to meteorological data obtained from weather stations. Growth of *Hylocomium*

Table 23.1. *Bryophyte indicators of high atmospheric N deposition*

Effects include growth, production, and abundance.

Negatively affected	Positively affected	Reference
Most *Sphagnum* species	*Sphagnum fallax*	Limpens *et al.* (2003)
	S. flexuosum	
Hylocomium splendens	*Brachythecium oedipodium*	Dirkse & Martakis (1992)
Ceratodon purpureus	*B. reflexum*	
Dicranum polysetum	*B. starkei*	
D. fucescens	*Lophocolea denticulatum*	
Ptilidium ciliare	*Plagiothecium denticulatum*	
	P. laetum	
	Polytrichum formosum	
	P. longisetum	
H. splendens	*P. commune*	Potter *et al.* (1995)
Dicranum scoparium	*P. juniperinum*	Gordon *et al.* (2001)
Schistidium sp.	*Pohlia wahlenbergii*	
	Bryum caespetitium	Robinson *et al.* (1998)
	Tetraplodon mnioides	
Racomitrium lanuginosum		Jones *et al.* (2002)

splendens in the Arctic (Callaghan *et al.* 1997), in boreal and oceanic regions north of 49° latitude (Vitt 1990), and along an altitude gradient in the Alps (Zechmeister 1995) had highly significant relationships with Mean Annual Temperature (MAT) or evapotranspiration (Table 23.2). Growth of *Pleurozium schreberi* and *Hypnum cupressiforme* was also related to MAT (Zechmeister 1995). The growth form of *H. splendens* can vary between temperate and boreal regions and between mid- and high-arctic regions (Callaghan *et al.* 1997). In temperate and boreal regions, it has sympodial growth whereas in the mid- and high-Arctic it has a monopodial or pseudomonopodial growth. *Tomenthypnum nitens* also has an Arctic–alpine expression that differs from the more common type by its pattern of branching (Miller 1980). Stapper and Kricke (2004) found that the proportion of epiphytic mosses to lichens decreased with increasing nocturnal temperatures in urban environments in Germany. Although they associated this change with heat pollution caused by urban environments, the relationship could also be used to detect increases in night-time temperatures that result from global warming.

Several experimental studies examined effects on growth by artificially raising temperature, usually 1–4 °C above ambient, in order to simulate global warming. Results from these experiments are quite varied and may depend on

Table 23.2. *Bryophyte predictors of temperature-related variables*

Growth responses of bryophytes (L, length; M, mass; P, production; GF, growth form; A, abundance) from various habitats to direct and indirect effects of increases in temperature and evapotranspiration. MAT, Mean Annual Temperature; RNT, Relative Nocturnal Temperature; >, increase, <, decrease. See text for references.

Species	Response	Habitat	Variable
Hylocomium splendens	>L, GF	Arctic	MAT
H. splendens	> L, M	alpine	MAT
H. splendens	-lessthan- M	boreal oceanic	evapotranspiration
Pleurozium schreberi	>L, M	alpine	MAT
Hypnum cupressiforme	>L	alpine	MAT
Ratio mosses/lichens	<ratio	epiphytes	RNT
Sphagnum fuscum	> L, GF	sub-Arctic bog	summer warming
Pohlia wahlenbergii	> L, GF	alpine	growing season warming
Polytrichum strictum	> A, P	boreal bog	growing season warming
P. strictum	> A	boreal bog	evapotranspiration, drawdown
Tomenthypnum nitens	GF	alpine, Arctic	MAT
S. magellanicum	> A	boreal bog	growing season warming
Hollow *Sphagnum* species	< L, M	peatlands	evapotranspiration, drawdown
Rhytidiadelphus squarrosus	< A	limestone grassland	winter warming
Lophocolea bidentata	< A	limestone grassland	winter warming
Campyliadelphus chrysophyllus	> A	limestone grassland	winter warming
Fissidens dubius	> A	limestone grassland	winter warming

factors other than a direct relationship with temperature. Positive effects of temperature on growth in length were observed in *Polytrichum strictum* and *Sphagnum fuscum* in peatlands as well as *Pohlia wahlenbergii* in an alpine environment but only when moisture conditions were not sub-optimal (Weltzin *et al.* 2001; Dorrepaal *et al.* 2003; Sandvik & Heegard 2003). Even when moisture conditions were favorable for growth in length of *S. fuscum*, increased temperatures produced a distinct reduction in capitulum bulk density and greater distance between the branches (Dorrepaal *et al.* 2003). Bates *et al.* (2005) observed a decrease in abundance of *Rhytidiadelphus squarrosus* and *Lophocolea bidentata* and an increase in abundance of *Campyliadelphus chrysophyllus* and *Fissidens dubius* in a winter warming treatment on a limestone grassland community.

Several studies have demonstrated an indirect negative effect of temperature increase on growth and productivity of peatland bryophytes, mostly related to increased evapotranspiration that produced water table drawdown and thus negative water balance in the mosses. Gunnarsson *et al.* (2004) found that increases in temperature produced higher evapotranspiration that reduced the production of *Sphagnum balticum*, a species that grows in peatland hollows. Hummock *Sphagnum* species seemed to be better able to grow and survive increased water stress than hollow species because they avoid complete drying out of their capitula (Schipperges & Rydin 1998). Reduced *Sphagnum* growth as a result of warming and increased evapotranspiration that produced water table drawdown also favored the growth and increased abundance of *Polytrichum strictum* in bog habitats (Weltzin *et al.* 2001).

Expansions and contractions of species ranges as predictors of climate change

Increased temperatures and climatic extremes are expected to drive species and species assemblages towards the poles and higher altitudes. Initial signs of changes in vegetation are expected to be found in high altitudes before they occur in other regions since diversity exists within short distances and species migration can therefore happen relatively quickly. There is abundant evidence that vegetation zones have already shifted to higher altitudes in mountains (Soja *et al.* 2007). However, almost all of the evidence for a shift of vegetation zones towards higher altitudes comes from the distribution of vascular plants and there is very little documented evidence as yet of altitudinal shifts in bryophyte distribution, although such studies are in progress. Among the few published examples of bryophyte species shifts in mountain environments are *Dicranoweisia cirrata*, which increased its altitudinal range in the Vosges Mountains, France, and in the Austrian Alps, and *Orthotrichum pallens*, also in the Vosges Mountains (Frahm & Klaus 2001). Hohenwallner *et al.* (2002) proposed a method for detecting the timing of the disappearance of snowbeds in the high alpine regions of Austria using the abundance of three bryophytes, *Andreaea rupestris*, *Pohlia filum*, and *Dicranoweisia crispula* in permanent quadrats. *Andreaea rupestris* is an indicator of early snowmelt, *P. filum* is an indicator of late snowmelt while *D. crispula* is intermediate between the other species. *Polytrichum norvegicum*, an obligate indicator of late-lying snowbeds in Europe, and *P. alpinum* have been used to reconstruct past climate change (Birks 1982).

Lewis Smith (2005) reported that several colonies of *Polytrichastrum longisetum* and *Polytrichum strictum*, which had never previously been recorded on Signy Island in the Antarctic, had recently been found there. Recent range contractions have been reported for *Sphagnum falcatulum*, a species of boggy

Table 23.3. *Range expansion of Atlantic, Mediterranean–Atlantic, and Mediterranean bryophyte species into central Europe*

Atlantic	Mediterranean–Atlantic	Mediterranean
Cololejeunea minutissima	*Cephaloziella baumgartneri*	*Bryum torquescens*
Fissidens monguillonii	*Cryphaea heteromalla*	*Campylopus oerstedianus*
Fissidens rivularis	*Leptobarbula berica*	*Dicranella howei*
Lejeuna lamacerina	*Tortella inflexa*	*Didymodon australasiae*
Lepidozia cupressina		*Entosthodon hungaricus*
Leptodontium flexifolium		*Phascum leptophyllum*
Leptodontium gemmascens		*Pottia recta*
Lophocolea fragrans		*Riccia gougetania*
Orthotrichum pulchellum		*Scorpiurium circinatum*
Orthotrichum sprucei		*Tortula brevissima*
Oxystegus hibernicus		*Tortula cuneifolia*
Platyhypnidium lusitanicum		*Tortula pagorum*
Pottia wilsonii		*Tortula princeps*
Scleropodium cespitans		
Sematophyllum micans		
Ulota phyllantha		
Zygodon conoideus		

(Data from Frahm & Klaus, 2001).

mire habitats, on Macquarie Island in the sub-Antarctic, its most southerly occurrence (Whinam & Copson 2006). *Sphagnum falcatulum* had disappeared or declined in coverage in many small patches where it was previously located on the island. The contractions and disappearances of the species in many localities between 1992 and 2000 were linked to higher evapotranspiration produced by higher MAT and lower than average precipitation.

Frahm and Klaus (2001) gathered an impressive body of evidence that showed that several bryophyte species with Atlantic, Atlantic–Mediterranean, or Mediterranean distributions had expanded their ranges up to 400 km to the east into southeast Germany (Table 23.3). They related those range extensions to increases in winter temperatures of 1.5 °C. Since 2001, 11 species have been added to the list (Frahm 2007). Frahm (2007) also observed that many of the species that were present in 2001 had become more abundant in central Europe over the course of recent years and a few had produced sporophytes for the first time. As a result of the addition of several species and further range expansion of species already present in 2001, Mediterranean and Mediterranean–Atlantic species can now be found in northwest regions of Germany.

Hedenäs *et al.* (2002) used a method based on the frequency of collection of species in a herbarium to identify temporal increases or decreases in abundance of some of the more common species. Their results suggested that the temporal decrease in the sub-oceanic species *Rhytidiadelphus loreus* and *Loeskeobryum brevirostre* in southwestern Sweden may be attributable to climate change.

Changes in species interactions as predictors of climate change and N deposition

Species interactions often play a key role in determining effects of climate change on bryophytes (Fig. 23.1). Competition in particular has been shown to affect plant growth, species exclusion, the spatial distribution of species, and community composition (Brooker 2006). Simulated climate change and increased N and S deposition experiments have demonstrated that competition among bryophytes has resulted in a decrease in growth and abundance of species such as *Dicranum elongatum*, *Sphagnum balticum*, *S. magellanicum*, and *S. papillosum* on mires (Sonesson *et al.* 2002; Limpens *et al.* 2003; Gunnarsson *et al.* 2004). In those experiments, reduced growth and abundance resulted from overtopping by more minerotrophic competitors such as *S. lindbergii*, *S. flexuosum*, and *S. fallax* that were favored by additional N loadings or species such as *S. fuscum* and *S. papillosum* that were better adapted to increases in temperature and resulting drawdown on peatlands.

The effects of one of the first short-term (5–10 years) climate change experiments produced a significant decrease in moss cover on Arctic tundra (Chapin *et al.* 1995). Increases in nutrient availability had produced a shift in the competitive balance where mosses were eliminated by the increased dominance of deciduous shrubs. Subsequently, many warming, N, and CO_2 enrichment experiments have demonstrated that bryophytes decline in abundance in bog, montane, Arctic, or alpine ecosystems as a result of increased shading and litterfall by deciduous shrubs and/or grasses and sedges (Molau & Alatalo 1998; Press *et al.* 1998; Hobbie *et al.* 1999; Berendse *et al.* 2001; Cornelissen *et al.* 2001; Graglia *et al.* 2001; Jägerbrand *et al.* 2003, 2006; van der Wal *et al.* 2003; van Wijk *et al.* 2003; Hollister *et al.* 2005; Jónsdóttir *et al.* 2005; Wahren *et al.* 2005; Schuur *et al.* 2007). Bryophyte growth, abundance, and diversity generally suffer when vascular plant cover reaches between 50% and 60% or more (Aerts *et al.* 1992; Heijmans *et al.* 2002). Recent evidence has shown that shrubs have already increased in abundance in many arctic ecosystems (Cornelissen *et al.* 2001; Sturm *et al.* 2001). Cornelissen *et al.* (2001) demonstrated that lichen abundance had declined naturally in many sub-Arctic and mid-Arctic ecosystems as a result of increased vascular plant biomass; the same effect has not as yet been demonstrated for mosses. Although there is a potential decline in bryophyte abundance in high arctic and alpine ecosystems as a result of increased shrub abundance, not all species are negatively impacted. *Polytrichum* and *Brachythecium* species have been shown to increase in abundance on some experimental plots as a result of warming and CO_2 enrichment experiments (Berendse *et al.* 2001; Wahren *et al.* 2005).

Changes in bryophyte-dominated ecosystems as predictors of climate change

Bryophyte-dominated ecosystems and the composition and production of species therein can provide signals of current climate change. Several studies have examined the relatively short-term changes (10–50 years) to peatlands based on vegetation and environmental surveys from two separate time frames. There is strong evidence that some of the observed changes are allogenic and driven by anthropogenic effects on the systems. Two such effects are N inputs from the atmosphere and water table drawdown, which may be produced by increased evapotranspiration that results from increased temperatures.

Climate warming can change the microtopography on peatlands, favoring hummock *Sphagnum* species at the expense of hollow species (Schipperges & Rydin 1998; Gunnarson *et al.* 2004; Robroek *et al.* 2007). Less precipitation and higher evapotranspiration favors the growth of trees on open bogs (Gunnarsson & Flodin 2007). Such changes were observed in a 40-year study of a *Sphagnum*-dominated mire in southern Sweden; drawdown in this case resulted from a drop in the regional water table produced by increased consumption of water (Gunnarsson *et al.* 2002). Eutrophication of bogs produced by N deposition increased the occurrence of *S. fallax* as well as the abundance of vascular species indicative of fens (Gunnarsson & Flodin 2007). Hogg *et al.* (1995) observed poor growth and extinction of *Sphagnum* in a peatland in England and attributed those losses in part to extremely dry summers and increased asymmetric competition with higher plants. Evidence also indicates that *Sphagnum* abundance is decreasing on bogs in southeastern Sweden and Denmark and that the balance between mosses and vascular plants is shifting towards higher plants (Gunnarson *et al.* 2000; Malmer *et al.* 2003).

Permafrost degradation and changes to bryophyte communities in peatlands of the boreal forest and the tundra are also indicative of a changing climate. In the zone of discontinuous permafrost in the boreal plains of western Canada, the boreal forest of central Alaska, and the mountains of southern Norway, isolated peat plateaus and palsas underlain by permafrost are thawing, producing relatively rapid changes to bryophyte communities (Camill 1999; Sollid & Sørbell 1998; Osterkamp *et al.* 2000; Vitt *et al.* 2000a). Permafrost produces volume expansion of the peat thus raising the surface above the water table. On occasion the permafrost may be several meters thick, forming small peat mounds or palsas; in other situations the ice lenses are only about 1 m thick and give rise to peat plateaus that may be localized or cover several kilometers (Vitt *et al.* 2000a). The thaw rate and stability of these permafrost features are to a large extent controlled by MAT, thus making them particularly sensitive to climate change (Vitt & Halsey 1994; Camill & Clark 1998; Vitt *et al.* 2000b; Camill 2005). At the southern boundary of the discontinuous zone of permafrost, these features exist out of equilibrium with the climate and are disappearing. They are being replaced by thermokarst features such as bogs and fens with internal lawns, peat plateaus with collapse scars and edge collapse scars, and various other thaw features on the landscape (Vitt & Halsey 1994; Camill 1999; Jorgenson & Osterkamp 2005). The vegetation on the surface of peatland permafrost features at the southern boundary of the discontinuous permafrost zone is indicative of drier conditions as the surface is disconnected from the water table by the ice. Peat plateaus and palsas are covered by a dense canopy of conifers with a moss layer dominated by boreal forest species with occasional to abundant *S. fuscum* hummocks and *Tomenthypnum nitens* (Camill & Clark 1998; Vitt *et al.* 2000a; Beilman 2001; Jorgenson *et al.* 2001; Turetsky *et al.* 2007; Table 23.4). When the permafrost degrades, it subsides to produce a thermokarst feature that is below the surrounding non-permafrost peatland surface because less peat had accumulated at the top of the permafrost than in the surrounding peatland. As a result, the thermokarst features are wetter environments that are characterized by bryophyte communities different from those of the surrounding peatland.

If the peatland completely surrounding the thermokarst feature is a bog, poor fen or peat plateau, the lower, wetter areas will support species from wet environments as well as occasional small hummocks of *S. magellanicum* and *S. fuscum* that are found closer to the edges of the thermokarst feature (Vitt & Halsey 1994; Camill 1999; Beilman 2001). If the permafrost collapse should occur at the edge of the bog or peat plateau, it may become connected to water from the surrounding fen that may have a high pH, in which case the moss community is dominated by rich fen species (Camill 1999). Similar thawing effects producing wetter environments and subsequent shifts of

Table 23.4. *Bryophyte predictors of the peat surface above permafrost and of recent thermokarst features such as bogs and fens with internal lawns, peat plateaus with collapse scars and edge collapse scars in boreal and sub-Arctic peatlands*

Surface of permafrost feature	Internal lawns and edge collapse scars bogs and poor fens	Edge collapse scars rich fens
Ceratodon purpureus	*Cladopodiella fluitans*	*Calliergon giganteum*
Dicranum polysetum	*Sphagnum angustifolium*	*Campylium stellatum*
Dicranum undulatum	*Sphagnum fallax*	*Cladopodiella fluitans*
Hylocomium splendens	*Sphagnum fuscum*	*Drepanocladus aduncus*
Pleurozium schreberi	*Sphagnum jensenii*	*Drepanocladus exannulatus*
Polytrichum strictum	*Sphagnum lindbergii*	*Hamatocaulis vernicosus*
Ptilium crista-castrensis	*Sphagnum magellanicum*	*Scorpidium revolvens*
Sphagnum fuscum	*Sphagnum majus*	*Scorpidium scorpioides*
Tomenthypnum nitens	*Sphagnum obtusum*	*Straminergon stramineum*
	Sphagnum riparium	
	Sphagnum russowii	

(Data from Camill 1999; Vitt *et al.* 2000a; Beilman 2001; Jorgenson *et al.* 2001; Turetsky 2007).

species abundances have also been observed on sub-Arctic peatlands (Christensen *et al.* 2004; Payette *et al.* 2004; Malmer *et al.* 2005).

Racomitrium lanuginosum moss–sedge montane heaths are communities that are affected by high N deposition. High N loadings have a negative impact on the growth of *R. lanuginosum* (Pearce & van der Wal 2002) that, along with competitive interactions with vascular plants and herbivory, are responsible for the loss of many moss-dominated montane communities in the UK in the past 50 years (Van der Wal *et al.* 2003). Such losses of moss-dominated communities could also be occurring on a wide range of mountain habitats elsewhere in the world.

Effectiveness of using bryophytes as predictors of greenhouse gas enrichment and climate change

Inferences about increased atmospheric greenhouse gas enrichment are often difficult to make from responses of bryophyte species and indicator species in general because of the high noise to signal ratio. The chances of identifying effects of greenhouse gases on growth and growth forms in particular can be quite small since effects are often masked by wide fluctuations in between-year temperatures and precipitation. For example, Gerdol *et al.* (2007)

and Aldous (2002) found that climatic conditions such as drought and subsequent water table drawdown overrode any effects of nutrient addition on the production of *Sphagnum*. Changes to other factors such as degree of shading and water chemistry may also confuse the signal since they can affect the uptake of CO_2 and N (Limpens *et al.* 2003; Vitt *et al.* 2003). Therefore, particular care should be taken when interpreting bryophyte growth and growth form signals before attributing them to effects of greenhouse gas enrichment, particularly in peatlands.

Changes to the growth, growth form, range, and distribution of a bryophyte species at a few locations cannot unequivocally be attributed to climate change. A species may be responding to microclimatic or other conditions within its environment that are not related to those of the regional climate. It is only when signals are occurring at a regional scale that the probability that the signal is responding to climate change becomes compelling. In those cases, the growth variables of bryophytes listed in Table 23.2 become useful and even powerful predictors of changes to climatic variables such as temperature and evapotranspiration. Also, the changes in species ranges, growth form, and distribution as observed by Frahm (2007) and Frahm and Klaus (2001) can clearly be attributable to changes in the regional climate since many species at many different locations are used as signals (Table 23.3).

Other clear signals at a regional scale are: decrease in bryophyte abundance in low and mid Arctic tundra as a result of asymmetric competition with shrubs; increase in bryophyte abundance in the high Arctic where they expand into areas that were bare of vegetation; replacement of dry boreal species by species of wet peatland habitats in the discontinuous permafrost zone as a result of thaw (Table 23.4); and loss of hollow species in bogs and increased abundance of poor fen *Sphagnum* species as a result of regional water table drawdown and increased N deposition. Although many species can be used as predictors of atmospheric greenhouse gas enrichment and climate change, *Hylocomium splendens*, *Pohlia wahlenbergii*, *Sphagnum fallax*, and species of the genus *Polytrichum* and *Brachythecium* in particular appear to provide some of the clearest signals in many regions (Tables 23.1 and 23.2).

Conclusions

Monitoring impacts of climate change utilizing bryophyte signals is an effective and yet under-utilized method. Although large-scale studies of vegetation changes in alpine habitats in Europe that include bryophytes are now underway, there is a general lack of information pertaining to effects of global warming on bryophytes in alpine habitats in other regions particularly in North America, in Asia, and in tropical mountain ranges. In fact, a meta-analysis of range expansions of a wide variety of organisms in alpine habitats does not even list bryophytes (Root *et al.* 2003), although bryophyte abundance usually increases and plays an important role in species richness in high-altitude environments. Convey *et al.* (2006) suggested that the lack of more confirmed examples of non-indigenous bryophytes in Antarctica is partly a function of past scientific research focus. The same rationale may also explain the lack of documented examples of bryophyte expansion or contraction in high altitude environments. Permanent quadrats that are used to monitor changes in high-altitude vegetation for regions other than the Alps and in Sweden (where studies are presently occurring) should incorporate bryophytes since their responses may be indicative of climate change effects that are difficult to determine based on the distribution of vascular plants (Hohenwallner *et al.* 2002).

Bryophyte range expansions and contractions are effective predictors of climate change, as demonstrated by Frahm (2007) and Frahm and Klaus (2001), which are also under-utilized. Bryophytes are small organisms that cannot be easily identified to the species level in the field and the number of professionals and amateurs interested in bryophytes is relatively small when compared with those who study vascular plants (Hedenäs *et al.* 2002). Even though the number of specialists is limited, there is an abundance of published reports of new records of bryophytes at the regional or national scales. Perhaps it would be more efficient to network those limited resources into a centralized system that collates reports of new discoveries of bryophyte range expansions with the express purpose of relating them to climate change.

Acknowledgements

I thank Jan-Peter Frahm for his help and suggestions, Donald Ipperciel for translating articles, Madeleine Wallace for organizing and correcting the references, and Aria Hahn for a preliminary literature search.

References

Aerts, R., Wallén, B. & Malmer, N. (1992). Growth-limiting nutrients in *Sphagnum*-dominated bogs subject to low and high atmospheric nitrogen supply. *Journal of Ecology* **80**: 131–40.

Aerts, R., Wallén, B., Malmer, N. & Caluwe, H. D. (2001). Nutritional constraints on *Sphagnum* growth and potential decay in northern peatlands. *Journal of Ecology* **80**: 292–9.

Aldous, A. R. (2002). Nitrogen retention by *Sphagnum* mosses: responses to atmospheric nitrogen deposition and drought. *Canadian Journal of Botany* **80**: 721–31.

Bates, J. W., Thompson, K. & Grime, J. P. (2005). Effects of simulated long-term climatic change on the bryophytes of a limestone grassland community. *Global Change Biology* **11**: 757–69.

Beilman, D. W. (2001). Plant community and diversity change due to localized permafrost dynamics in bogs of western Canada. *Canadian Journal of Botany* **79**: 983–93.

Berendse, F., Van Breeman, N., Rydin, H. *et al.* (2001). Raised atmospheric CO_2 and increased N deposition cause shifts in plant species composition and production in *Sphagnum* bogs. *Global Change Biology* **7**: 591–8.

Birks, H. J. B. (1982). Quaternary bryophyte paleo-ecology. In *Bryophyte Ecology*, ed. A. J. E. Smith, pp. 273–490. New York: Chapman & Hall.

Böhm, R., Auer, I., Brunetti, M. *et al.* (2001). Regional temperature variability in the European Alps: 1760–1998 from homogenized instrumental time series. *International Journal of Climatology* **21**: 1779–801.

Bragazza, L., Tahvanainen, T., Kutnar, L. *et al.* (2004). Nutritional constraints in ombrotrophic *Sphagnum* plants under increasing atmospheric nitrogen deposition in Europe. *New Phytologist* **163**: 609–16.

Brooker, R. W. (2006). Plant-plant interactions and environmental change. *New Phytologist* **171**: 271–89.

Callaghan, T. V., Carlsson, B. Å., Sonnesson, M. & Temesváry, A. (1997). Between-year variation in climate-related growth of circumarctic populations of the moss *Hylocomium splendens*. *Functional Ecology* **11**: 157–65.

Camill, P. (1999). Patterns of boreal permafrost peatland vegetation across environmental gradients sensitive to climate warming. *Canadian Journal of Botany* **77**: 721–33.

Camill, P. (2005). Permafrost thaw accelerates in boreal peatlands during late-20th century climate warming. *Climatic Change* **68**: 135–52.

Camill, P. & Clark, J. S. (1998). Climate change disequilibrium of boreal permafrost peatlands caused by local processes. *American Naturalist* **151**: 207–22.

Chapin III, F. S., Shaver, G. R., Giblin, A. E., Nadelhoffer, K. J. & Laundre, J. A. (1995). Responses of arctic tundra to experimental and observed changes in climate. *Ecology* **76**: 694–711.

Christensen, T. R., Johansson, T., Åkerman, H. J. *et al.* (2004). Thawing sub-arctic permafrost: effects on vegetation and methane emissions. *Geophysical Research Letters* **31**: 1–4.

Chylek, P., Box, J. E. & Lesins, G. (2004). Global warming and the Greenland Ice Sheet. *Climatic Change* **63**: 201–21.

Convey, P., Frenot, Y., Gremmen, N. & Bergstrom, D. M. (2006). Biological invasions. In *Trends in Antarctic Terrestrial and Limnetic Ecosystems*, ed. D. M. Bergstrom, P. Convey & A. L. Huiskes, pp. 193–220. Dordrecht, The Netherlands: Springer.

Cornelissen, J. H. C., Callaghan, T. V., Alatalo, J. M. *et al.* (2001). Global change and Arctic ecosystems: is lichen decline a function of increases in vascular plant biomass? *Journal of Ecology* **89**: 984–94.

Dash, S. K., Jenamani, R. K., Kalsi, S. R. & Panda, S. K. (2007). Some evidence of climate change in twentieth-century India. *Climatic Change* **85**: 299–321.

Dirkse, G. M. & Martakis, G. F. P. (1992). Effects of fertilizer on bryophytes in Swedish experiments on forest fertilization. *Biological Conservation* **59**: 155–61.

Dorrepaal, E., Aerts, R., Cornelissen, J. H. C., Callaghan, T. V. & Van Logtestijn, R. S. P. (2003). Summer warming and increased winter snow cover affect *Sphagnum fuscum* growth, structure and production in a sub-arctic bog. *Global Change Biology* **10**: 93–104.

Frahm, J.-P. (2007). Moose als Indikatoren des Klimawandels. *Gefahrstofee-Reinhaltung der Luft-Ausgabe* **6**: 269–73.

Frahm, J.-P. (2008). Diversity, dispersal and biogeography of bryophytes (mosses). *Biodiversity Conservation* **17**: 277–84.

Frahm, J. -P. & Klaus, D. (2001). Bryophytes as indicators of recent climate fluctuations in Central Europe. *Lindbergia* **26**: 97–104.

Gerdol, R., Petraglia, A., Bragazza, L., Iacumin, P. & Brancaleoni, L. (2007). Nitrogen deposition interacts with climate in affecting production and decomposition rates in *Sphagnum* mosses. *Global Change Biology* **13**: 1810–21.

Giannini, A., Biasutti, M., Held, I. M. & Sobel, A. H. (2008). A global perspective on African climate. *Climatic Change* **90**: 359–83.

Gordon, C., Wynn, J. M. & Woodin, S. J. (2001). Impacts of increased nitrogen supply on high Arctic heath: the importance of bryophytes and phosphorus availability. *New Phytologist* **149**: 461–71.

Graglia, E., Jonasson, S., Michelsen, A. *et al.* (2001). Effects of environmental perturbations on abundance of subarctic plants after three, seven and ten years of treatments. *Ecography* **24**: 5–12.

Gunnarsson, U. & Flodin, L-Å. (2007). Vegetation shifts towards wetter site conditions on oceanic ombrotrophic bogs in southwestern Sweden. *Journal of Vegetation Science* **18**: 595–604.

Gunnarsson, U. & Rydin, H. (2000). Nitrogen fertilization reduces *Sphagnum* production in bog communities. *New Phytologist* **147**: 527–37.

Gunnarsson, U., Granberg, G. & Nilsson, M. (2004). Growth, production and interspecific competition in *Sphagnum*: effects of temperature, nitrogen and sulphur treatments on a boreal mire. *New Phytologist* **163**: 349–59.

Gunnarsson, U., Malmer, N. & Rydin, H. (2002). Dynamics or constancy in *Sphagnum* dominated mire ecosystems? A 40 year old study. *Ecography* **25**: 685–704.

Gunnarsson, U., Rydin, H. & Sjörs, H. (2000). Diversity and pH changes after 50 years on the boreal mire Skattlösbergs Stormosse, Central Sweden. *Journal of Vegetation Science* **11**: 277–86.

Hedenäs, L., Bisang, I., Tehler, A. *et al.* (2002). A herbarium-based method for estimates of temporal frequency changes: mosses in Sweden. *Biological Conservation* **105**: 321–31.

Heijmans, M. M. P. D., Berendse, F., Arp, W. J. *et al.* (2001). Effects of elevated carbon dioxide and increased nitrogen deposition on bog vegetation in the Netherlands. *Journal of Ecology* **89**: 268–79.

Heijmans, M. M. P. D., Klees, H. & Berendse, F. (2002). Competition between *Sphagnum magellanicum* and *Eriophorum angustifolium* as affected by raised CO_2 and increased N deposition. *Oikos* **97**: 415–25.

Hobbie, S. E., Shevtsova, A. & Chapin, F. S. (1999). Plant response to species removal and experimental warming in Alaskan tussock tundra. *Oikos* **84**: 417–34.

Hohenwallner, D., Zechmeister, H. G. & Grabherr, G. (2002). Bryophyten und ihre Eignung als Indikatoren für den Klimawandel im Hochgebirge – erste Ergebnisse. *Österreichisches Botanikertreffen* **10**: 19–21.

Hogg, P., Squires, P. & Fitter, A. H. (1995). Acidification, nitrogen deposition and rapid vegetational change in a small valley mire in Yorkshire. *Biological Conservation* **71**: 143–53.

Hollister, R. D., Webber, P. J. & Tweedie, C. E. (2005). The response of Alaskan arctic tundra to experimental warming: differences between short- and long-term responses. *Global Change Biology* **11**: 525–36.

Hoosbeek, M. R., van Breemen, N., Berendse, F. *et al.* (2001). Limited effect of increased atmospheric CO_2 concentration on ombrotrophic bog vegetation. *New Phytologist* **150**: 459–63.

Hoosbeek, M. R., van Breemen, N., Vasander, H., Buttler, A. & Berendse, F. (2002). Potassium limits potential growth of bog vegetation under elevated atmospheric CO_2 and N deposition. *Global Change Biology* **8**: 1130–8.

IPCC (Intergovernmental Panel on Climate Change). (2001). *Climate Change 2001: The Scientific Basis. Contribution of Working Group I to the Third Assessment Report of the Intergovernmental Panel on Climate Change*. New York: Cambridge University Press.

IPCC (Intergovernmental Panel on Climate Change). 2007. *Climate Change (2007): The Physical Science Basis. Contribution of Working Group I to the Fourth Assessment Report of the Intergovernmental Panel on Climate Change*. New York: Cambridge University Press.

Jägerbrand, A. K., Lindblad, K. E. M., Björk, R. G., Alatalo, J. M. & Molau, U. (2006). Bryophyte and lichen diversity under simulated environmental change compared with observed variation in unmanipulated alpine tundra. *Biodiversity and Conservation* **15**: 4453–75.

Jägerbrand, A. K., Molau, U. & Alatalo, J. M. (2003). Responses of bryophytes to simulated environmental change at Latnjajaure, northern Sweden. *Journal of Bryology* **25**: 163–8.

Jauhiainen, J., Vasander, H. & Silvola, J. (1994). Response of *Sphagnum fuscum* to N deposition and increased CO_2. *Journal of Bryology* **18**: 183–95.

Jauhiainen, J., Vasander, H. & Silvola, J. (1998a). Nutrient concentration in *Sphagna* at increased N-deposition rates and raised atmospheric CO_2 concentrations. *Plant Ecology* **138**: 149–60.

Jauhiainen, J., Wallén, B. & Malmer, N. (1998b). Potential NH_4^+ and NO_3^- uptake in seven *Sphagnum* species. *New Phytologist* **138**: 287–93.

Jones, M. L. M., Oxley, E. R. B. & Ashenden, T. W. (2002). The influence of nitrogen deposition, competition and desiccation on growth and regeneration of *Racomitrium lanuginosum* (Hedw.) Brid. *Environmental Pollution* **120**: 371–8.

Jónsdóttir, I. S., Magnússon, B., Gudmundsson, J., Elmarsdóttir, Á. & Hjartarson, H. (2005). Variable sensitivity of plant communities in Iceland to experimental warming. *Global Change Biology* **11**: 553–63.

Jorgenson, M. T., Racine, C. H., Walters, J. C. & Osterkamp, T. E. (2001). Permafrost degradation and ecological changes associated with a warming climate in central Alaska. *Climatic Change* **48**: 551–79.

Lamers L. P. M., Bobbink, R. & Roelofs, J. G. M. (2000). Natural nitrogen filter fails in raised bog. *Global Change Biology* **6**: 583–6.

Le Roux, P. & McGeoch, M. A. (2008). Changes in climate extremes, variability and signature on sub-Antarctic Marion Island. *Climatic Change* **86**: 309–29.

Lewis Smith, R. I. (2005). Bryophyte diversity and ecology of two geologically contrasting Antarctic islands. *British Bryophyte Society* **25**: 195–206.

Limpens, J., F. Berendse & H. Klees. (2004). How phosphorus availability affects the impact of nitrogen deposition on *Sphagnum* and vascular plant bogs. *Ecosystems* **7**: 793–804.

Limpens, J., Tomassen, H. B. M. & Berendse, F. (2003). Expansion of *Sphagnum fallax* in bogs: striking the balance between N and P availability. *Journal of Bryology* **25**: 83–90.

Malhi, Y., Roberts, T., Betts, R. A. *et al.* (2008). Climate change, deforestation, and the fate of the Amazon. *Science* **319**: 169–72.

Malmer, N., Albinsson, C., Svensson, B. M. & Wallén, B. (2003). Interferences between *Sphagnum* and vascular plants: effects on plant community structure and peat formation. *Oikos* **100**: 469–82.

Malmer, N., Johansson, T., Olsrud, M. & Christensen, T. R. (2005). Vegetation, climatic changes and net carbon sequestration in a North-Scandinavian subarctic mire over 30 years. *Global Change Biology* **11**: 1895–909.

Miller, N. G. (1980). Fossil mosses of North America and their significance. In *The Mosses of North America*, ed. R. J. Taylor & A. E. Leviton, pp. 9–36. San Francisco, CA: Pacific Division of the American Association for the Advancement of Science.

Molau, U. & Alatalo, J. M. (1998). Responses of subarctic-alpine plant communities to simulated environmental change: biodiversity of bryophytes, lichens and vascular plants. *Ambio* **27**: 322–9.

Nyakatya, M. J. & McGeoch, M. A. (2008). Temperature variation across Marion Island associated with a keystone plant species (*Azorella selago* Hook. (Apiaceae)). *Polar Biology* **31**: 139–51.

Osterkamp, T. E. & Jorgenson, M. T. (2005). Response of boreal ecosystems to varying modes of permafrost degradation. *Canadian Journal of Forest Research* **35**: 2100–11.

Osterkamp, T. E., Viereck, L., Shur, Y. *et al.* (2000). Observations of thermokarst and its impact on boreal forests in Alaska, U.S.A. *Arctic, Antarctic, and Alpine Research* **32**: 303–15.

Paeth, H, Scholten, A., Friederichs, P. & Hense, A. (2008). Uncertainties in climate change predictions: El Niño-Southern Oscillation and monsoons. *Global and Planetary Change* **60**: 265–88.

Payette, S., Delwaide, A., Caccianiga, M. & Beauchemin, M. (2004). Accelerated thawing of sub-arctic peatland permafrost over the last 50 years. *Geophysical Research Letters* **31**: 1–4.

Pearce, I. S. K. & van der Wal, R. (2002). Effects of nitrogen deposition on growth and survival of montane *Racomitrium lanuginosum* heath. *Biological Conservation* **104**: 83–9.

Phuyal, M., Artz, R. R. E., Sheppard, L., Leith, I. D. & Johnson, D. (2008). Long-term nitrogen deposition increases phosphorus limitation of bryophytes in an ombrotrophic bog. *Plant Ecology* **196**: 111–21.

Potter, J. A., Press, M. C., Callaghan, T. V. & Lee, J. A. (1995). Growth responses of *Polytrichum commune* and *Hylocomium splendens* to simulated environmental change in the sub-arctic. *New Phytologist* **131**: 533–41.

Press, M. C., Potter, J. A., Burke, M. J. W., Callaghan, T. V. & Lee, J. A. (1998). Responses of a subarctic dwarf shrub heath community to simulated environmental change. *Journal of Ecology* **86**: 315–27.

Robinson, C. H., Wookey, P. A., Lee, J. A., Callaghan, T. V. & Press, M. C. (1998). Plant community responses to simulated environmental change at a high arctic site. *Ecology* **79**: 856–66.

Robroek, B. J. M., Limpens, J., Breeuwer, A., Crushell, P. H. & Schouten, M. G. C. (2007). Interspecific competition between *Sphagnum* mosses at different water tables. *Functional Ecology* **21**: 805–12.

Root, L. T., Price, J. T., Hall, K. H. *et al.* (2003). Fingerprints of global warming on wild animals and plants. *Nature* **421**: 57–60.

Sandvik, M. & Heegaard, E. (2003). Effects of simulated environmental changes on growth and growth form in a late snowbed population of *Pohlia wahlenbergii* (Web. et Mohr) Andr. *Arctic, Antarctic, and Alpine Research* **35**: 341–8.

Schipperges, B. & Rydin, H. (1998). Response of photosynthesis of *Sphagnum* species from contrasting microhabitats to tissue water content and repeated desiccation. *New Phytologist* **140**: 677–84.

Schuur, E. A. G., Crummer, K. G., Vogel, J. G. & Mack, M. C. (2007). Plant species composition and productivity following permafrost thaw and thermokarst in Alaskan tundra. *Ecosystems* **10**: 280–92.

Serreze, M. C., Walsh, J. E., Chapin, F. S. III, *et al.* (2000). Observational evidence of recent change in the northern high-latitude environment. *Climatic Change* **46**: 159–207.

Shaver, G. R., Bret-Harte, M. S., Jones, M. H. *et al.* (2001). Species competition interacts with fertilizer to control long-term change in tundra productivity. *Ecology* **82**: 3163–81.

Söderström, L. & During, H. J. (2005). Bryophyte rarity viewed from the perspectives of life history strategy and metapopulation dynamics. *Journal of Bryology* **27**: 261–8.

Soja, A. J., Tchebakova, N. M., French, N. H. F., *et al.* (2007). Climate-induced boreal forest change: predictions versus current observations. *Global and Planetary Change* **56**: 274–96.

Sollid, J. L. & Sørbell, L. (1998). Palsa bogs as climate indicators – examples from Dovrefjell, southern Norway. *Ambio* **27**: 287–91.

Sonesson, M., Carlsson, B. Å., Callaghan, T. V., *et al.* (2002). Growth of two peat-forming mosses in subarctic mires: species interactions and effects of simulated climate change. *Oikos* **99**: 151–60.

Stapper, N. J. & Kricke, R. (2004). Epiphytische Moose und Flechten als Bioindikatoren von städtischer Überwärmung, Standorteutrophierung und verkehrsbedingten Immissionen. *Limprichtia* **24**: 187–208.

Sturm, M., Racine, C. & Tape, K. (2001). Increasing shrub abundance in the Arctic. *Nature* **411**: 546–7.

Toet, S., Cornelissen, J. H. C., Aerts, R., *et al.* (2006). Moss responses to elevated CO_2 and variations in hydrology in a temperature lowland peatland. *Plant Ecology* **182**: 27–40.

Turetsky, M. R., Wieder, R. K., Vitt, D. H., Evans, R. J. & Scott, K. D. (2007). The disappearance of relict permafrost in boreal north America: effects on peatland carbon storage and fluxes. *Global Change Biology* **13**: 1922–34.

Van der Heijden, E., Verbeek, S. K. & Kuiper, P. J. C. (2000). Elevated atmospheric CO_2 and increased nitrogen deposition: effects on C and N metabolism and growth of the peat moss *Sphagnum recurvum* P. Beauv. var. *mucronatum* (Russ.) Warnst. *Global Change Biology* **6**: 201–12.

Van der Wal, R., Pearce, I., Brooker, R., *et al.* (2003). Interplay between nitrogen deposition and grazing causes habitat degradation. *Ecology Letters* **6**: 141–6.

Van Wijk, M. T., Clemmensen, K. E., Shaver, G. R., *et al.* (2003). Long-term ecosystem level experiments at Toolik Lake, Alaska, and at Abisko, northern Sweden: generalizations and differences in ecosystem and plant type responses to global change. *Global Change Biology* **10**: 105–23.

Vitt, D. H. (1990). Growth and production dynamics of boreal mosses over climatic, chemical and topographic gradients. *Botanical Journal of the Linnean Society* **104**: 35–59.

Vitt, D. H. & Halsey, L. A. (1994). The bog landforms of continental Western Canada in relation to climate and permafrost patterns. *Arctic and Alpine Research* **26**: 1–13.

Vitt, D. H., Halsey, L. A. & Zoltai, S. C. (2000a). The bog landforms of continental western Canada in relation to climate and permafrost patterns. *Arctic and Alpine Research* **26**: 1–13.

Vitt, D. H., Halsey, L. A. & Zoltai, S. C. (2000b). The changing landscape of Canada's western boreal forest: the current dynamics of permafrost. *Canadian Journal of Forest Research* **30**: 283–7.

Vitt, D. H., Wieder, K., Halsey, L. A. & Turetsky, M. (2003). Response of *Sphagnum fuscum* to nitrogen deposition: a case study of ombrogenous peatlands in Alberta, Canada. *Bryologist* **106**: 235–45.

Wahren, C.-H. A., Walker, M. D. & Bret-Harte, M. S. (2005). Vegetation responses in Alaskan arctic tundra after 8 years of a summer warming and winter snow manipulation experiment. *Global Change Biology* **11**: 537–52.

Wasley, J., Robinson, S. A., Lovelock, C. E. & Popp, M. (2006). Climate change manipulations show Antarctic flora is more strongly affected by elevated nutrients than water. *Global Change Biology* **12**: 1800–12.

Weltzin, J. F., Harth, C., Bridgham, S. D., Pastor, J. & Vonderharr, M. (2001). Production and microtopograhy of bog bryophytes: response to warming and water-table manipulations. *Oecologia* **128**: 557–65.

Whinam, J. & Copson, G. (2006). *Sphagnum* moss: an indicator of climate change in the sub-Antarctic. *Polar Record* **42**: 43–9.

Wiedermann, M. M., Nordin, A., Gunnarsson, U., Nilsson, M. B. & Ericson, L. (2007). Global change shifts vegetation and plant-parasite interactions in a boreal mire. *Ecology* **88**: 454–64.

Zechmeister, H. G. (1995). Growth rates of five pleurocarpous moss species under various climatic conditions. *Journal of Bryology* **18**: 455–68.

24

Conclusions and Future Research

NANCY G. SLACK AND LLOYD R. STARK

Although it is difficult to draw general conclusions from such a variety of studies as those presented in this book, it is very obvious that there are already valuable data on bryophyte ecology in relation to many aspects of predicted climate change. There are baseline data from ongoing monitoring studies, as well as experimental research comparing bryophyte responses to ambient environmental factors, with responses to projected changes to those factors under various climate change models. In addition there are some reports of changes in bryophyte distribution (in relation to climatic factors) that have already occurred. In a concluding chapter, Gignac (Chapter 23, this volume) reviews much of the evidence of changing climate, indicates the advantage of utilizing bryophytes as indicators of such change, and provides an overview of bryological research relating to climate change. As Proctor notes in Chapter 3 of this volume, "the only certainty [with climate change] is change itself", and "normal conditions" are an illusion. As he points out, determining the causes of bryophyte distributional changes in terms of concurrent climate data is fraught with cause/effect and correlational problems, leading scientists to be cautious in their evaluation of data in this field.

A great deal of recent research on bryophytes includes not only ecology and physiology but also the molecular aspects of bryophyte biology. The complete sequencing of the genome of the moss *Physcomitrella patens* is making newly possible the understanding of how special physiological traits of bryophytes are important to their ecology (Cuming 2009).

With increased funding for climate change research, bryophyte studies, in addition to those of other plant and animal organisms, have multiplied. The sheer volume of studies reported by Jägerbrand and colleagues for the Arctic and alpine regions in this book (Chapter 11) surprised even the editors. Much of this research includes long-term studies that are ongoing. For example, research on

Bryophyte Ecology and Climate Change, eds. Zoltán Tuba, Nancy G. Slack and Lloyd R. Stark. Published by Cambridge University Press.

the bryophyte ecology of remarkably desiccation-tolerant desert mosses, including field experimental studies, continues. In Chapter 8 of this volume, Stark *et al.* suggest that the delicate symbiosis between the dominant desert moss *Syntrichia caninervis* and the cyanobacterium *Microcoleus* may be adversely affected by changes in precipitation and N deposition patterns in the Mojave Desert. Similarly, ongoing experimental research is continuing with UV-B radiation. Lappalainen *et al.* (Chapter 5, this volume) present the seasonal variation of the bryophyte response to UV acclimation *in situ*, and find the responses to be species-specific. The better the models for these aspects of climate change, the more valuable this research will become. As climate changes, precipitation regimes are expected to change, and as outlined by Rice *et al.* (Chapter 22, this volume), the productivity model at the level of the moss canopy will have value in predicting moss productivity.

The use of bryophytes (as well as vascular plants) in this research has been emphasized by many authors in this volume. It has been made clear that bryophytes, which have evolved separately from vascular plants for hundreds of millions of years, show great differences not only in morphology but in physiology, as well as in reproduction and dispersal. Many of these characteristics, and in particular their very close contact with environmental factors have made them ideal plants for study of air pollution in the past, and currently of the many aspects of climate change. Their small size and relatively fast growth rates also make them ideal for experimental studies. These so-called "canaries in the coal mine" make excellent environmental monitors. Many studies have drawn attention to the myriad ecological functions of bryophytes. Jácome *et al.* in Chapter 10 of this volume, using transplant experiments in the tropics, find that epiphytic bryophyte communities will not collapse due to projected climate change, but the structure and ecological function of such communities will no doubt change markedly. The authors experimentally found that bryophyte species react more rapidly to parameters of climate change than vascular plant species.

In alpine systems, the date of snowmelt is an excellent abiotic response variable for use in climate change progressions, and Hohenwallner *et al.* (Chapter 12, this volume) note that snowbed bryophyte species can serve as excellent climate change indicators because coverage of such bryophyte communities is determined largely by the date of snowmelt! Seppelt (Chapter 13, this volume) provides evidence of significant ice and snow regression in many sub-Antarctic islands, and makes a case for bryophytes to serve as ecological markers, since in this region the vitality of mosses is directly linked to soil moisture. Recent drought on subantarctic islands has restricted the most healthy moss communities to rarer moist microhabitats, whereas mosses in continental Antarctica are physiologically buffered from such perturbations in climate.

As discussed by Tuba (Chapter 2), as the concentration of CO_2 increases experimentally, photosynthetic rates in bryophytes respond quickly over the short term (increasing), but acclimate downward over the long term. This issue is addressed experimentally in the context of stress tolerance by Brinda *et al.* (Chapter 9) for a desert moss. Several authors in this volume remind us that bryophytes survive in nutrient-poor habitats, which may limit their abilities to response to elevated atmospheric CO_2. However, their rapid response rates to climate change variables such as CO_2, temperature, and precipitation changes are due to their poikilohydric nature, their limited development of source–sink differentiation, and their perennial habit. These characteristics allow mosses to react immediately to favorable periods. Tuba *et al.* (Chapter 4) remind us also that we should keep in mind that the elements of climate change, namely temperature, CO_2, N-deposition, and precipitation, are not necessarily straightforward, but represent a dynamic of interactions among these parameters that may challenge experimental designs. Several authors in this volume comment on the species-dependent nature of response to climate change. Interestingly, Tuba *et al.* (Chapter 4, this volume) note that, due to fundamental differences in desiccation tolerance between bryophytes and vascular plants, and the profound effects of desiccation tolerance on the photosynthetic apparatus of both groups, the responses to climate change variables need not be assumed parallel between bryophytes and vascular plants.

It is also important to be able to monitor changes in distribution and biodiversity with altered climate for many kinds of plants and animals. Until recently there had been relatively little published research on bryophytes and climate change, with fewer such studies on bryophytes than on other groups (Lovejoy & Hannah 2005). Global warming has already adversely affected some animals, with the polar bear as the poster child, but also corals and thus coral reef ecosystems. Corals survive in only a limited temperature range; increased temperatures harm them in a variety of ways.

Bryophytes as a group inhabit a great variety of temperature regimes, and need to be studied individually or in particular communities. Forman (1964) did classic experiments on the moss *Tetraphis pellucida*'s environmental niche factors in relation to its geographic distribution; for other such studies since that time, see Slack (1997). Recently such studies have included climatic factors, some by the authors of this book, (e.g., Gignac *et al.* 1991, 1998; Jägerbrand *et al.* 2006). Many others cited in this book involve Arctic, alpine, and peatland mosses in ecosystems dominated by bryophytes. Aquatic bryophytes present different challenges (Glime, Chapter 6, this volume). Should climate change lead to elevated water temperatures, this would increase respiration rates and decrease photosynthetic rates. Further, the frequency of desiccating events is expected to increase for aquatic bryophytes, and these elements in combination will intensify competition from algae and tracheophytes, ultimately forcing bryophytes to higher latitudes and elevations. Assessing the effects of UV radiation (UVR) on aquatic bryophytes is beset with methodological concerns (Martínez-Abaigar & Núñez-Olivera, Chapter 7, this volume), although chlorophyll fluorescence appears to be a viable response variable. Interestingly, biomass production in peatlands does not appear to be at risk from higher levels of UVR. The latter authors note that bryophytes have value in aquatic settings as UV biomonitors, and propose *Jungermannia exsertifolia* as the appropriate species to use.

Peatlands, dominated by bryophytes (especially *Sphagnum*) and representing large stores of the world's carbon, are an obvious focus for several chapters in this volume. In western Canada, drought is a manifestation of climate change, and Vile *et al.* (Chapter 14) demonstrate that changes in water table depth caused by drought have greater effects on carbon fluxes than periodic wetting of the peat surface, thus implicating water availability as the key factor in peatland bryophyte survival and perennation. In NW Siberian peatlands, warming has caused a thawing of permafrost, a tendency of *Sphagnum* bogs to migrate northward, and thus leaving southern peatlands vulnerable to the effects of warming (Naumov & Kosykh, Chapter 15, this volume). The southern peatlands of Europe are highlighted for monitoring (Nagy, Chapter 16, this volume), with the author contending that the most important practical aspect of conservation management of these southern bogs is two-pronged: (a) management of the surrounding vascular vegetation, and (b) optimizing water supply to these peatlands. Given that climate change projections call for a 60% increase in drought years, and 600 mm rainfall/year is required to sustain a peatland in this region, Jakab and Sümegi (Chapter 17, this volume) predict that Hungarian peatlands will vanish in the near future.

Conservation biology has recently become an established interdisciplinary field. Although not only bryophytes but all cryptogams were the subject of only about 4% of publications in leading conservation journals between 2000 and 2005 (Hylander & Johansson 2007), the inclusion of bryophytes in both conservation programs and in research is increasing. Rare species have been the major focus (Vanderpoorten & Hallingbäck 2009). Major threats of previous concern have been land use, forestry, drainage, peat extraction, and other aspects of habitat destruction rather than climate change. Experimental work on the biology of rarity is valuable for all of these concerns (Cleavitt 2005).

For assessing present and future distributional changes of rare and other bryophytes with global warming and other aspects of climate change (Frahm & Klaus 2001), it is essential to have reliable baseline data. That is the case for the excellent study by Sergio *et al.* of the present and projected future distribution of one moss species, *Sematophyllum substrumulosum* (Chapter 21, this volume). It is also true of the excellent distributional data presented here by Anderson and Ohlemüller (Chapter 20) for British bryophytes. In Europe there are good data for much of northern Europe, including Sweden, Switzerland, Germany, and the Netherlands, but less complete data for southern Europe, from which bryophytes are moving north with increasing temperatures. In addition, as Bates and Preston point out in Chapter 19 of this volume, care must be taken in attributing changes in distribution only to climate change; the recent reduction in air pollution in Britain has also resulted in such changes. The latter authors point out that species expansions are easier to detect than contractions of a species range. That these difficulties arise in the UK, which represents the most reliable regional database for bryophyte distributions in the world, is notable. For Hungarian peatlands, Pócs (Chapter 18, this volume) makes a strong case for nine recent bryophyte species arrivals and six species disappearances that correlate with climate change, with the arrivals based on a plant's occurrence in an "unsaturated" environment.

Although single species whose ecology and current distribution are well known can be used to monitor future changes, it is important to also monitor populations and communities. New methods are available for such studies (Rydin 2009). The problems for Arctic and alpine communities, often bryophyte-dominated and with increasing global temperatures, are of widespread concern and the subject of much European monitoring (Jägerbrand *et al.*, Chapter 11, this volume). In the northeastern United States, "Mountain Watch" programs have been monitoring alpine vascular plants with nowhere to go if temperature increases continue and more competitive lower elevation plants move upwards. This is equally true for alpine bryophytes; future monitoring is under discussion. In these alpine and Arctic communities of both Europe and North America both population sizes of bryophyte species and biodiversity of bryophyte-dominated communities should also be monitored.

In terms of the preservation of rare species in particular, the biodiversity of bryophytes in general, and the use of current distribution patterns as baselines to determine changes in distribution as a result of climate change, more data are urgently needed. As noted, good bryophyte data are generally available in most of Europe. In the United States there are good distributional data for towns and counties only in Maine and Vermont, with good species lists for other parts of the Northeast, e.g., New Hampshire, Massachusetts, and New York. But even in New York, which has had its share of good bryologists, species new to the state are still being discovered, and in studies of specialized bryophyte habitats such as rich fens and cliffs of particular rock types, very rare species and species thought to have been extirpated have recently been found (Wesley & Slack 1991; Slack 1992; Cleavitt *et al.* 2006). Great progress on the bryophyte flora of California, Colorado, Nevada, and Ohio has been made in the last ten years, but most other western and midwestern states lack species lists or distributional maps. Much bryological exploration has been undertaken in China, where both Chinese and foreign bryologists are providing baseline data for various regions. But there, as elsewhere in the world, more needs to be done. Since not only global temperature change, but also changes in precipitation worldwide are predicted, progress in obtaining baseline data for the both the Old World and the New World tropics is needed.

We feel that this book has made a real contribution to our present knowledge of bryophyte ecology and its use in climate change studies. It will be valuable to all workers in this field and to others interested in the many aspects and consequences of climate change. Much future research is surely forthcoming.

References

Cleavitt, N. L. (2005). Patterns, hypotheses, and processes in the biology of rare bryophytes. *Bryologist* **108**: 554–6.

Cleavitt, N. L., Williams, S. A. & Slack, N. G. (2006). *Updating the Rare Moss List for New York State: Ecological Community and Species-centered Approaches*. Final Report for the Biodiversity Research Institute. Albany, NY: NY State Museum.

Cuming, A. C. (2009). Mosses as model organisms for developmental, cellular, and molecular biology. In *Bryophtye Biology*, ed. B. Goffinet & A. J. Shaw, pp. 199–236. New York: Cambridge University Press.

Forman, R. T. (1964). Growth under controlled conditions to explain the hierarchical distributions of a moss, *Tetraphis pellucida*. *Ecological Monographs* **34**: 1–25.

Frahm, J.-P. & Klaus, D. (2001). Bryophytes as indicators of recent climate fluctuations in Central Europe. *Lindbergia* **26**: 97–104.

Gignac, L. D., Vitt, D. H. & Bayley, S. E. (1991). Bryophyte response surfaces along ecological and climatic gradients. *Vegetatio* **93**: 29–45.

Gignac, L. D., Nicholson, B. J. & Bayley, S. E. (1998). The utilization of bryophytes in bioclimatic modeling: predicted northward migration of peatlands in the Mackenzie River Basin, Canada, as a result of global warming. *Bryologist* **101**: 572–87.

Hylander, K. & Jonsson, B. G. (2007). The conservation ecology of cryptogams. *Biological Conservation* **135**: 311–14.

Jägerbrand, A. K., Lindblad, K. E. M., Bjork, R. B., Alatallo, J. M. & Molau, U. (2006). Bryophyte and lichen diversity under simulated environmental change compared with observed variation in unmanipulated Alpine tundra. *Biodiversity and Conservation* **15**: 4453–75.

Lovejoy, T. E. & Hannah, L. (2005). *Climate Change and Biodiversity*. New Haven, CT: Yale University Press.

Rydin, H. (2009). Population and community ecology of bryophytes. In *Bryophyte Biology*, ed. B. Goffinet & A. J. Shaw, pp. 393–444. New York: Cambridge University Press.

Slack, N. G. (1992). Rare and endangered bryophytes in New York State and eastern United States: current status and preservation strategies. *Biological Conservation* **59**: 233–41.

Slack, N. G. (1997). Niche theory and practice: bryophyte studies. *Advances in Bryology* **6**: 169–204.

Vanderpoorten, A. & Hallingbäck, T. (2009). Conservation biology of bryophytes. In *Bryophyte Biology*, ed. B. Goffinet & A. J. Shaw, pp. 487–533. New York: Cambridge University Press.

Wesley, F. R. & Slack, N. G. (1991). *Paludella squarrosa* rediscovered in New York. *Evansia* **8**: 52.

Index

Printed in the United States
by Baker & Taylor Publisher Services

Printed in the United States
by Baker & Taylor Publisher Services